Springer Climate

Springer Climate is an interdisciplinary book series dedicated to climate research. This includes climatology, climate change impacts, climate change management, climate change policy, regional climate studies, climate monitoring and modeling, palaeoclimatology etc. The series publishes high quality research for scientists, researchers, students and policy makers. An author/editor questionnaire, instructions for authors and a book proposal form can be obtained from the Publishing Editor.

Now indexed in Scopus® !

More information about this series at http://www.springer.com/series/11741

Md. Nazrul Islam • André van Amstel

Bangladesh II: Climate Change Impacts, Mitigation and Adaptation in Developing Countries

Springer

Md. Nazrul Islam
Department of Geography and Environment
Jahangirnagar University
Savar, Dhaka, Bangladesh

André van Amstel
Wageningen University, Environmental
Systems Analysis SENSE Research School
Wageningen, The Netherlands

ISSN 2352-0698 ISSN 2352-0701 (electronic)
Springer Climate
ISBN 978-3-030-71948-7 ISBN 978-3-030-71950-0 (eBook)
https://doi.org/10.1007/978-3-030-71950-0

This Springer imprint is published by the registered company Springer Nature Switzerland AG
The registered company address is: Gewerbestrasse 11, 6330 Cham, Switzerland

Dedicated
To my only Daughter
SABABA MOBASHIRA ISLAM
(Only Daughter of Professor
Dr. Md. Nazrul Islam)

Preface

Bangladesh is one of the world's most vulnerable nations, and will become even more vulnerable because of climate change. According to the Global Climate Risk Index 2020, Bangladesh is the seventh most climate-affected nation in the world. Moreover, the people of Bangladesh are also worried about how a changing climate could dramatically alter current weather patterns in such a way that diseases spread from regions where they are now epidemic to areas where they are currently of little concern. However, all these will badly harm social status, economic growth, and public health. Governmental and nongovernmental organizations (NGOs) can also play a vital role in preventing and managing the impacts of climate change. However, extreme climatic events such as floods, cyclones, riverbank, erosion, drought, salinity, intrusion, and communicable diseases have adverse socio-economic effects on the people affected and are therefore of great concern for Bangladesh's national sustainable development. Additionally, not only climate change issues but also local matters are responsible for these changes. Several structural and bioengineering approaches are suitable for reducing land erosion and damage to the road networks, where awareness building is required to combat human health risks. Finally, this study focuses on real field conditions with the perceptions of communities of people who are field survivors for all climate change issues, rather than computer-aided simulation models, and the outcomes will be helpful for academicians, researchers, and national and international stakeholders in developing climate-resilient, sustainable coastal cities in the near future. These issues are critically analyzed and discussed in this book.

Measuring resilience to climate change impacts is essential to find ways to help communities to adapt to climate change and improve their livelihoods. Several chapters of this book have identified specific attributes and types of capital for policy intervention to reduce vulnerability and increase the resilience of coastal fishing communities. In order to scale up, more such studies need to be conducted in different contexts. Bangladesh is experiencing an increased number of hydro-meteorological disasters caused by climate change, losing lives and property, and adopting survival mechanisms to tackle the worst situations caused by natural disasters regarding the climate change issue. Some empirical study findings illustrate the effects of flooding in Bangladesh and the struggling mechanisms against this havoc associated with the climate change phenomenon. At the beginning of the twenty-first century Bangladesh was already facing numerous disastrous floods and other natural disasters as well. Also, the natural setting of Bangladesh in the South Asian subcontinent and the characteristics of the tropical monsoon climate are greatly responsible for the riverbank erosion and flood hazards. The livelihood security of the households of the southwestern coastal region of Bangladesh has been immensely affected by various climatic conditions and disasters. The present research is intended to conduct a very in-depth study on the impact of cyclonic disasters on all types of livelihood capital (i.e., natural, physical, human, social, and financial capital) and the relevance of such impacts to the food security conditions of the study area. The results of the present study show that the percentages of livelihood related to food production are decreasing because of the cyclonic disaster-induced unfavorable environment in the study area. Nonetheless, the findings establish a strong reason to consider lightning as an emerging disaster, and will help to draw attention to and extend the views of researchers related to the climate change field to find out the relationship with climatic variability that will contribute to formulating appropriate adaptation strategies to minimize the changing requirements.

The coastal area of Bangladesh is frequently affected by natural hazards. For that reason, the people of this zone regularly needed to migrate and adapt to an unfavorable environment. Owing to the rapid rise in the carbon emissions, all around the weather conditions are shifting, sea levels are rising, weather disasters are getting more severe. Although every country around the world is facing the same problem, the issue has become a grave concern to Bangladesh as it is one of the most vulnerable countries in terms of impacts of climate change because of its geographical location, dense population, economic and socioeconomic conditions, and lack of technological capability to mitigate the problem. Throughout the world climate change has become an issue that is given little importance given the severity of the problem and compared with other crises that exist all around the world. Some of chapters of this book illustrate and analyze the Landsat Satellite-derived land use maps, which indicate the large-scale destruction of mangrove, the mixed aquaculture–vegetation development of the wetlands, and plantations on the coast from 1996 to 2019 due to climate and anthropogenic change. However, to live with the climatic impacts in the longer term and maintaining sustainable livelihoods, long-term, proactive, and effective adaptation strategies need to be implemented, e.g., enhancing cooperation among the community of people, improving infrastructure, incorporating modern fishing technology, and producing new as well as diversified livelihood activities, amongst others.

Bangladesh is struggling with an increasing trend of environmental degradation due to the adverse shocks of climate change. Local environmental governance plays a significant role in attaining local sustainable development. In Bangladesh, the local government authorities perform a wide range of development activities in the cities. To understand local sustainable development, the environmental impact of these development activities must be explored. The main purpose of this chapter is to investigate the extent to which environmental governance promotes sustainable development considering environmental issues in development programs. Adaptation to climate change is a critical issue in Southeast Asia. Owing to geophysical conditions and socio-economic–demographic backwardness, South Asia is projected to be one of the worst affected regions from global warming and climate change. Much like the sub-continent, Bangladesh has fallen victim to climate change and its impact is going to be disastrous if not taken under control. The surrounding affected regions of Bangladesh has teamed up to form an alliance to study, adapt, and mitigate such calamities. Academic and policy attention are increasing sharply as a result of the growing evidence on unavoidable impacts in the region. Climate change will affect every sector across South Asian countries severely. The overwhelming dependence on agriculture and natural resources for living makes the people of this region very vulnerable to climate change. South Asia is under serious threat from sea-level rising and increasing incidence of extreme events such as floods, droughts, cyclones, storms, and the irregularity of monsoons. Rapid urbanization has only fueled the situation, which may become desperate because of the large-scale influx of "climate refugees" to the ill-equipped cities in the near future. There is a need for global, regional, and local-level mitigation and adaptive strategies to face the reality of climate change in Bangladesh and in other South Asian countries as well.

Dhaka, Bangladesh — Md. Nazrul Islam

Wageningen, The Netherlands — André van Amstel

Contents

About the Authors

Md. Nazrul Islam is a permanent Professor at the Department of Geography and Environment in Jahangirnagar University, Savar, Dhaka-1342, Bangladesh. Prof. Nazrul has completed his PhD from the University of Tokyo, Japan. In addition, he has completed a 2-year Standard JSPS Postdoctoral Research Fellowship from the University of Tokyo, Japan. Prof. Nazrul's fields of interest are environmental and ecological modeling, the impact of climate change on the aquatic ecosystem, the modeling of phytoplankton transition, harmful algae, and marine ecosystems regarding dealing with the coupled hydrodynamic ecosystems model on coastal seas, bays, and estuaries, the application of computer-based programming for numerical simulation modeling, etc. Prof. Nazrul is an expert on scientific research techniques and methods of developing the models for environmental systems analysis research. Prof. Nazrul visited as an invited speaker several foreign Universities in Japan, USA, Australia, UK, Canada, China, South Korea, Germany, France, the Netherlands, Taiwan, Malaysia, Singapore, and Vietnam, etc. Prof. Nazrul was awarded "Best Young Researcher Award" by the International Society of Ecological Modeling (ISEM) for his outstanding contribution to the Ecological Modeling fields, 2013, Toulouse, France, and he was also awarded "Best Paper Presenter Award 2010" by SautaiN in Kyoto, Japan. In addition, he was awarded a "Best Poster Presenter Award" at the Techno Ocean Conference, Kobe City Exhibition Hall, Kobe, in Japan.

Prof. Nazrul has made more than 40 scholarly presentations in more than 20 countries around the world, authored more than 100 peer-reviewed articles and 15 books and research volumes. Recently, Prof. Nazrul published an excellent textbook entitled "Environmental Management of Marine Ecosystems" jointly with Prof. Sven Erik Jorgensen by the CRC Press (Taylor & Francis). He also recently published an excellent book entitled: "Bangladesh I: Climate Change Impacts, Mitigation and Adaptation in Developing Countries" (Springer Publications, the Netherlands & Germany). Prof. Nazrul is currently serving as Executive Editor-in-Chief of the journal "Modeling Earth Systems and Environment," (Springer International Publications, journal no. 40808). E-mail: nazrul_geo@juniv.edu

André van Amstel is an Assistant Professor at Wageningen University and Research, the Netherlands. Andre van Amstel studied Physical Geography and Planning in Amsterdam and has long-standing expertise in Integrated Environmental Assessment. He contributed to the development of the Integrated Model to Assess the Global Environment at the Institute of Public Health and the Environment in the Netherlands. He is particularly interested in the global environment and the risk of a runaway greenhouse effect by methane from melting of the permafrost in the Arctic. Van Amstel contributed to the IPCC 2006 Guidelines on Agriculture, Forestry, and Land Use, and as such contributed to the Nobel Peace Prize 2007 for Al Gore and the IPCC. Since 2014, he has been a member of the Nordforsk Scientific Advisory Board for the Arctic Council of Ministers on Arctic Integrated Environmental Research. E-mail: andre.vanamstel@wur.nl

Chapter 1
Climate Change Impact and Comprehensive Disaster Management Approach in Bangladesh: A Review

Md. Nazrul Islam, Sahanaj Tamanna, André van Amstel, Md. Noman, Md. Sajid Saadat Ali, Ditiya Malek Aparajita, Prantanu Roy, Sadia Rahman Tanha, Nayem Sarkar, Md. Ashiquzzaman, Sobuj Kumar Ghosh, Syeda Raisa Hasnat, Md. Farhanuzzaman Bhuiyan, Md. Sayef, Protyaee Saha, Gazi Rokibul Hasan, and Adipta Ghosh

Abstract Due to its geographical condition and geophysical location Bangladesh is one of the world's most vulnerable country, which will become more vulnerable to the impact of climate change. According to the Global Climate Risk Index 2020 and Intergovernmental Panel on Climate Change (IPCC) 2011, Bangladesh is the seventh most climate change-affected nation in the world. This chapter elaborates on the possible impacts of climate change in Bangladesh through various natural disasters, i.e., increasing temperature, sea level rise, salinity intrusion, cyclone, storm surges, drought, etc. and also discusses the comprehensive disaster management approach in Bangladesh. It is now a worry in the scientific community that climate change could dramatically change weather patterns like the disease spread

M. N. Islam (✉)
Department of Geography and Environment, Jahangirnagar University, Savar, Dhaka, Bangladesh
e-mail: nazrul_geo@juniv.edu

S. Tamanna
Bangladesh Environmental Modeling Alliance (BEMA), Non-Profit Research Organization, Dhaka, Bangladesh

A. van Amstel
Wageningen University, Environmental Systems Analysis SENSE Research School, Wageningen, The Netherlands
e-mail: andre.vanamstel@wur.nl

M. Noman
Department of Geography and Environment, Jahangirnagar University, Dhaka, Bangladesh

M. S. S. Ali · D. M. Aparajita · P. Roy · S. R. Tanha · N. Sarkar · M. Ashiquzzaman · S. K. Ghosh · S. R. Hasnat · M. F. Bhuiyan · M. Sayef · P. Saha · G. R. Hasan · A. Ghosh
North South University, Dhaka, Bangladesh

M. N. Islam, A. van Amstel, *Bangladesh II: Climate Change Impacts, Mitigation and Adaptation in Developing Countries*, Springer Climate,
https://doi.org/10.1007/978-3-030-71950-0_1

of epidemics (such as COVID-19) from vulnerable regions to invulnerable regions. All sectors will be affected by the impact of climate change, not only Bangladesh but also other South Asian countries. In Bangladesh, both the government and non-governmental organizations (NGOs) are trying to prevent and alter the impacts of climate change by enhancing several adaptation and mitigation approaches. But still, coastal districts and northern areas in Bangladesh are facing many climatic issues, such as flash floods, super cyclones, salinity intrusion, storm surges, drought and riverbank erosion etc. Moreover, the government is taking the immediate response of shifting people in a cyclone center at the moment of extreme natural events but most of the peoples of the coastal districts in Bangladesh are illiterate so that they very careless about the awareness. On the basis of current information, it is suggested that the government should make some policy in disaster management for a sustainable solution for coastal areas in Bangladesh.

Keywords Climate change · Disaster vulnerability · Global scenarios · Adaptation · Bangladesh

1.1 Introduction

Climate change is one of the major challenges in the world today and makes more pressure on the environment (Ali et al. 2011; Sarker et al. 2020; Hoque et al. 2020). The Sea level rise, changing weather patterns, and frequent natural events threaten agricultural food production and sustainable development on an unprecedented scale in Bangladesh (Islam and Neelim 2010; Mahmood 2012a, b; Jakariya et al. 2016; Christopher and Siverd 2020). Without a proper action plan, adaptation and management strategies, coping with these impacts in the future will become more complicated and will create a food crisis at the coastal districts of Bangladesh (Mahmood 2012a, b; Habiba et al. 2014). This overview deals with the concept of Global Climate Change, the associated terms, causes, consequences, solutions and its potential health impact, such as the current COVID-19 global pandemic (Mehedi et al. 2010; Asgary and Halim 2011; Jakariya et al. 2016; Abu Abdullah 2016; Ahmad 2019; Christopher and Siverd 2020). It expresses the urgency of acting to mitigate permanent greenhouse gases (GHGs) and global warming to protect human health, global economy, and society as well. Thus, addressing climate change impact and management approaches requires cooperation between not only different level of governments, private sectors, but also among the regional and global cooperation of other countries (Islam and Neelim 2010; Jakariya et al. 2016; Abu Abdullah 2016; Ahmad 2019). Extreme natural disasters that take place every year in Bangladesh create more vulnerability in the deltas owing to the flat and low-lying landscape, population density, poverty, geographical location, etc. (Islam and Neelim 2010; Denissen 2012; Habiba et al. 2013; Sarker et al. 2020; Hoque et al. 2020). Besides, 80% of the total land area consists of flood plains and most of the country is prone to flooding during rainy season (Hoque et al. 2020). In addition to

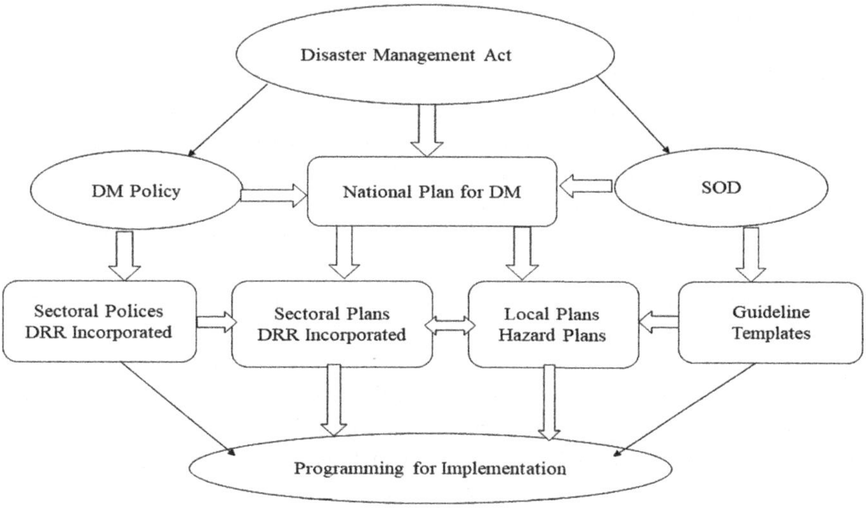

Fig. 1.1 Disaster management regulative framework

flooding, Bangladesh experiences various types of natural disasters almost every year because of global warming issues such as cyclones and storm surges, salinity intrusion, extreme temperatures, and drought. In a changing climate, the pattern of impacts is eroding our assets, investment, and future projects and plans also (Habiba et al. 2014; Abu Abdullah 2016; Ahmad 2019). Every year, riverbank erosion, permanent inundation, and sea level rise are increasing, which destroys the settlements and make numerous people homeless in the world as well as the coastal zones in Bangladesh. Considering this situation, Bangladesh has also followed some disaster management regulative frameworks (Fig. 1.1).

The major climatic conditions in Bangladesh, summers are becoming hotter, monsoons irregular, with untimely or heavy rainfall, crop damage, increasing mortality, extreme cold, outbreaks of dengue, malaria, diarrhea, etc., are endangering our generation every year (Mehedi et al. 2010; Asgary and Halim 2011; Abedin and Shaw 2013; Habiba and Shaw 2013). Climate change and climate variability worsen existing poverty, exacerbate inequalities, and trigger both new vulnerabilities and some opportunities for individuals and communities. Poor people are poor for different reasons and thus are not all equally affected, and not all vulnerable people are poor. Climate change interacts with nonclimatic stressors and entrenched structural inequalities to shape vulnerabilities and uncertainties regarding food, water, life, property settlement, and livelihood assets in Bangladesh (Mehedi et al. 2010; Asgary and Halim 2011; Jakariya et al. 2016; Abu Abdullah 2016; Ahmad 2019; Christopher and Siverd 2020). Extreme events occurring in climate change directly affect coastal people, making them refugees by causing them to lose their homes, land, and property. It is said that by 2050, one in every seven persons in Bangladesh will have been displaced by climate change according to Intergovernmental Panel of Climate Change (IPCC) 2019. Up to 18 million people may have to move because of sea level rise alone (Abedin and Shaw 2013; Habiba and Shaw 2013; Environmental Justice Foundation 2020).

The urban poor are also directly affected by these changes in climate and disasters (Ali et al. 2011). The risk of natural disasters is being enhanced by the impacts of climate change, especially in the absence or shortage of the necessary infrastructure as well as employment opportunities for them in the major cities of Bangladesh. Gender inequality and socio-economic status make women more vulnerable than men (Jakariya et al. 2016; Abu Abdullah 2016; Ahmad 2019). Similarly in coastal areas in Bangladesh, since 1991 cyclone and storm surge in Bangladesh, the death rate in women was almost five times higher than that in men. Although vulnerability depends on religion, ethnicity, gender, and the size of their farm operations, social power and local influences often prevent them from being affected long term. The government is not always fully capable of dealing with their problematic criticisms, and relief is rarely being delivered to the victims (Mehta 2007). Also sometimes local government overlooks local dynamics such as bringing people into a shelter early, ensuring dry food and drinking water. Some people of society who are already under stress could become vulnerable due to lack of proper planning, addressing to them, and overlooking the peoples. As Bangladesh seeks ways to adapt to climate change, it could set an example of inclusive planning for other nations to follow several guidelines which are provided by IPCC and Global Comprehensive Disaster Management Approaches (Mehta 2007; Chowdhury 2019).

1.2 Statement of the Problem

Anthropogenic activities and human interference are the main factors responsible for climate change. GHGs are increasing owing to human activities and affects ecosystems and human health on a global scale (IUCN 2010). Third-world countries experience the greatest impacts of climate change and it depends inequalities with other countries (Abedin and Shaw 2013; Habiba and Shaw 2013), besides its impact on the future of health of the people. During this century, the earth's average surface temperature rises are likely to surpass the safe threshold of 2 °C above the preindustrial average temperature. Higher latitude countries will become more vulnerable, predicting 2° to 3 °C rises by 2090, and 4 to 5 °C rises in northern Canada, Greenland, and Siberia (Costello et al. 2009; Abedin and Shaw 2013; Habiba and Shaw 2013). The average temperature of today's world has already increased by 0.6 °C since the middle of the 1800s. In the last century, the average temperature of the earth increased by 1.5–4.5 °C leading to melting of polar and mountain ice and thus to a rise in sea level (Fig. 1.2). It has also been shown that if climate change continues unabated, by the year 2050, the production of rice will decrease by 8% and that of wheat by 32% (Alam et al. 2010; Abedin et al. 2019; Ahmed et al. 2019; Sarker et al. 2020; Hoque et al. 2020). An update on the Intergovernmental Panel on Climate Change (IPCC)'s fourth assessment, identified that if there is no action to cut emissions, there is the potential for a temperature rise of as much as 7 °C by

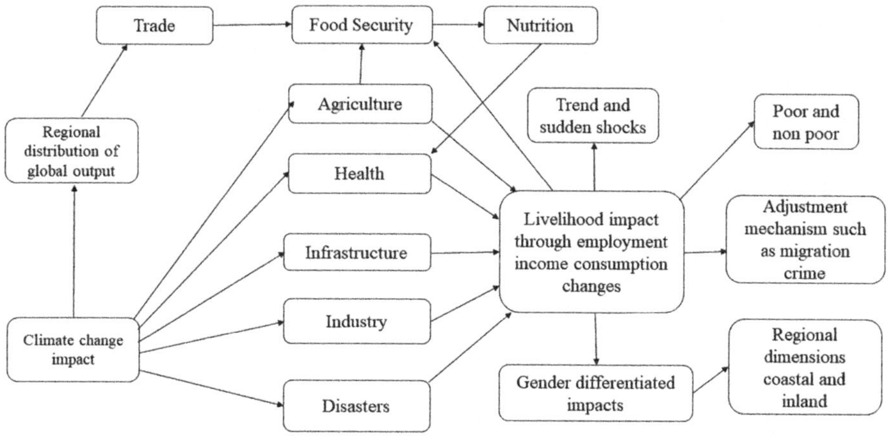

Fig. 1.2 Intensity of impacts of different sectors due to the climate (NAPA 2009)

2100. The IPCC's fourth assessment report in 2007 stated that it is clear that humans play an active role in increasing global warming. As a result in last 50 years, the cold days and nights as well as frost have seemed less frequent. In contrast, hot days and nights as well as heat waves have seemed more frequent. The linear warming trend over the last 50 years of on average 0.13 °C per decade is nearly twice that for the last 100 years. From the period 1850 to 1899 to the period 2001 to 2005 the temperature has been rising 0.76 °C (Hasnain 2000; WWF 2005) mentioned in their studies that since the mid-1970s the average air temperature measured at 49 stations of the Himalayan region rose by 10 °C, with high elevation sites warming the most.

According to the Global Climate Risk Index 2020 analyses to what extent countries and regions have been affected by impacts of weather-related loss events and climate change, i.e., Puerto Rico, Sri Lanka, and Dominica (Eckstein et al. 2018; Alam et al. 2010; Abedin et al. 2019; Ahmed et al. 2019). Bangladesh is in seventh position. According to Cable News Network (CNN), Puerto Rico and Dominica were severely hit by hurricane Maria, which is the strongest hurricane on record to make landfall in 2017. Massive rainfall has caused flooding across Bangladesh, India, and Nepal, which caused 40 million people to become displaced and 1200 people to lose their lives in these regions (Dewan 2014). Another impact for climate change is mudslides (Mehedi et al. 2010; Asgary and Halim 2011; Jakariya et al. 2016; Abu Abdullah 2016; Ahmad 2019; Christopher and Siverd 2020). A devastating mudslide took 500 lives in Sierra Leone. Hence, the people of that country faced health issues such as cholera and malaria. The climate risk report also shows that 80 people died and 270,000 people lost their homes after a major flood and storm in Madagascar in the same year. Climate change is taking a toll, not only on the ecology of nations around the world but also on their political, economic, and social stability, with the poorest nations and the poorest of the rich nations being the worst sufferers (Sarker et al. 2020; Hoque et al. 2020). The World Bank stated that a 1-m sea rise would affect 84 developing countries, and 17% of

Bangladesh's land area could be flooded like Cape Town and elsewhere, as well as small island developing states from the Maldives to Tuvalu being overwhelmed, with storm surges as well. Recent studies have found that up to 12% of the world's GDP is already at risk from existing climate patterns (Leontief 1976; Choudhury et al. 2004; Lloyd-Jones 2006; Khan 2015).

For example, in a tropical cyclone alone, the value of GDP exposed more than tripled from US$525.7 billion in the 1970s to US$1.6 trillion in the first decade of the 2000s. However, it has been felt that sufficient experience is lacking on how to connect dots, how to bring together concepts such as climate change and poverty eradication, climate change and food security, and climate change and access to water (Alam et al. 2010; Abedin et al. 2019; Ahmed et al. 2019). Ultimately, if climate change is not being solved, then poverty eradication, food security, and access to water cannot be solved either (Leontief 1976; Choudhury et al. 2004; Lloyd-Jones 2006; Khan 2015; Islam and Islam 2017). Geographically, Bangladesh is located in the tropical region. Every year, Bangladesh faces many natural disasters as a common phenomenon because of climate change. It is important that Bangladesh lies on a deltaic plain with five major river systems, i.e., the Jamuna-Brahmaputra, the Padma-Ganges, the Surma-Meghna, the Padma-Meghna, and the Karnaphuli. Although altitudes up to 105 m above sea level have been observed in the northern part of the plain, most elevations are less than 10 m above sea level, but the elevations show a decreasing trend in the coastal south, where the topography is generally at sea level. These geographical features make Bangladesh vulnerable to natural disasters, such as floods and cyclones, and the high levels of poverty increase the enormity of the challenges that the country is likely to face from climate change (Mahmood 2012a, b; Abedin and Shaw 2013; Habiba and Shaw 2013).

In an Action Aid research report Bangladesh was ranked the fifth most vulnerable country to climate change and hunger. According to the World Risk Index 2017, Bangladesh ranked sixth. There is a very high chance (19.17%) of disasters. Bangladesh will be warmer than today by 0.5 to 0.2°C by the year 2030, which means that the "temperature is relatively higher during the monsoon than during summer" (Mahmood 2012a, b; Abedin and Shaw 2013; Habiba and Shaw 2013). If we consider the 30-year mean winter rainfall volumes of about 64 mm with a variability of around 53%, by 2030, the best estimated projection is for monsoon rainfall to increase by 10–15% and winter rainfall by 5–10% (Rahman et al. 2013; Kafiluddin 2005). Besides, Bangladesh is a very low energy-consuming country, and it is pursuing a low carbon growth path, while building its resilience to climate change and reducing the risk of climate change, which shows national development. The comprehensive disaster management system of Bangladesh is aimed at reducing the risk of disaster. The government and the United Nations Development Programme (UNDP) issued funds to avoid and fight disaster in both rural and urban areas through structural and nonstructural settlements and awareness. In the absence of stable and transparent institutions, those steps remain on paper (Sarker et al. 2020; Hoque et al. 2020). Research published on the Intech Open website by Haque and Uddin (2013) finds the real scenario of the disaster management system of Bangladesh. It says that only a very limited culture of partnership in disaster management has been established so far (Leontief 1976; Choudhury et al. 2004; Lloyd-Jones 2006; Khan 2015). The core reason is politicians and poor governance. The policymakers, government agencies, and nongovernmental organizations (NGOs) know less about the impacts of climate, disasters, vulnerabilities, and how to support in extreme situations.

1.3 Literature Review

The adverse ecological, social, and economic impacts of climate change are increasing and have wide-ranging effects on the future sustainability of the earth (Stern 2006; Catherine and McMullen 2009; Alam et al. 2010; Abedin et al. 2019; Ahmed et al. 2019). According to the IPCC, GHG emissions by industries into the atmosphere, and anthropogenic emissions of gases such as carbon dioxide and methane are considered to be the major factors behind global warming (IPCC 2007; Abedin and Shaw 2013; Habiba and Shaw 2013). The physical extent of climate-related impacts is different from region to region (Burtis 2006). Coastal areas and communities will be amongst the highest at risk because of their proximity to the sea (Habiba et al. 2013; Siverd et al. 2020). Recent studies estimated that the rapid impact of global warming is irreversible and that it affects not only humans but also species and ecosystems (Adedeji et al. 2014). Per decade, the average temperature is increasing by 0.2 °C reaching the threshold of 2 °C above pre-industrial levels by 2050 (Adedeji et al. 2014). Global average precipitations have increased by 2% in the last 100 years and are likely to increase in the future (IPCC 2013). Also, the loss of biodiversity due to climate change has directly or indirectly changed the pattern and dynamics of energy flow and material circulation (Zhong and Wang 2017). Many species do not cope with the unstable changes in the environment, probably creating unsuitable conditions, and therefore will be faced with the threat of extinction (Correa 2015). It is predicted that in the next century CO_2 will be the main driving factor for global biodiversity loss (Sala et al. 2000). Many species ranges have moved poleward and upward in elevation in the last century (Parmesan and Yohe 2003) and this is likely not to cease. Biological resources are changing into useful goods and services in the same way as grasslands and forests are converted into cropland as a consequence of climate change (Lambin and Meyfroidt 2011). Several GHGs are generated by the production of biological resources for fuels or fibers and the reduction of grasslands (Hector and Bagchi 2007; Burnham and Ma 2015). Changes in stocks of biomass can increase the carbon emission rate (Hector and Bagchi 2007). It has been well-known that disasters damage the entire economy when they predominantly take place in developing countries (Kulatunga et al. 2012). Although disaster damages the whole world on an unprecedented scale, the poor developing countries are the most severely affected victims (UN/ESCAP 2006; Lloyd-Jones 2006). In spite of making strides in achieving sustainable development, Bangladesh has to face significant challenges posed by climate change (Ahmed and Haque 2002). According to the Third Assessment Report of the IPCC, its geographical location makes Bangladesh the most vulnerable country to climate change in the world (McCarthy et al. 2001). The scientific community also recognizes Bangladesh as one of the highest ranking vulnerable countries on earth (Islam and Neelim 2010; Ahmed et al. 2016). Because Bangladesh remains a developing nation that faces diverse environmental problems ranging from low income; lack of resources such as land and permanent housing to accommodate the people; shortages of clean water and adequate food; inability to participate in commercial activity; high population density (120 million people living in an area of 144,000 km^2), human health, and illiteracy (Mclean and Moore 2005; Ali 1999). There are almost 855 people per km^2 living in Bangladesh.

Climate change is a long-lasting problem and will be exaggerated owing to natural calamities. It has also turned natural hazards into disastrous situations (Ali 1999). Bangladesh is one of the most disaster-prone countries in the world because of its geographical setting and environment, with 80% of the total floodplain area facing a high risk of several hazards (Choudhury 2002; Shimi et al. 2010). In spite of the zero contribution to carbon emissions in Bangladesh, the country has to suffer from the massive effects of global climate change. It is estimated that frequent natural disaster will be extremely severe in Bangladesh, especially in the coastal zones (Khan et al. 2015; Ali 1999). However, Bangladesh already faces tropical cyclones and storm surges, floods and riverbank erosion every year (Ali 1999). Owing to the extreme disaster magnitude, the vulnerability of the people in general, and of the economy to natural disasters, various regimes of the government of Bangladesh (pre- and post-independence) have developed an institutional infrastructure to deal with natural hazards and their potential losses (PRSP 2005). Identifying urgent and immediate research needs for evidence-based action is also an important step in reducing the health risks of climate change. A sustainable management policy requires research gaps to be filled and objectives to be updated (Zimm et al. 2018).

1.4 Concept and Approaches to the Climate System and Natural Disaster

Climate change is sometimes mentioned directly as a hazard or disaster, but it could correspondingly be viewed as the background against which hazard and disasters arise. By shifting context, the risk is shifted as well (Leontief 1976; Choudhury et al. 2004; Lloyd-Jones 2006; Khan 2015). Climate (a concept that is essentially statistical in nature) and weather are related concepts, but climate is more than "average weather," being composed of seasonal and diurnal variations in weather elements that contribute to the ideas of norm, anomalies, and extreme events. Shifts in weather result in variations in climate. As a simple analogy, the climate system can be considered as a heat engine, distributing excess radiation energy received by the earth from the sun in tropical latitudes toward polar latitudes where a radiation energy deficit is experienced (Mehedi et al. 2010; Asgary and Halim 2011; Jakariya

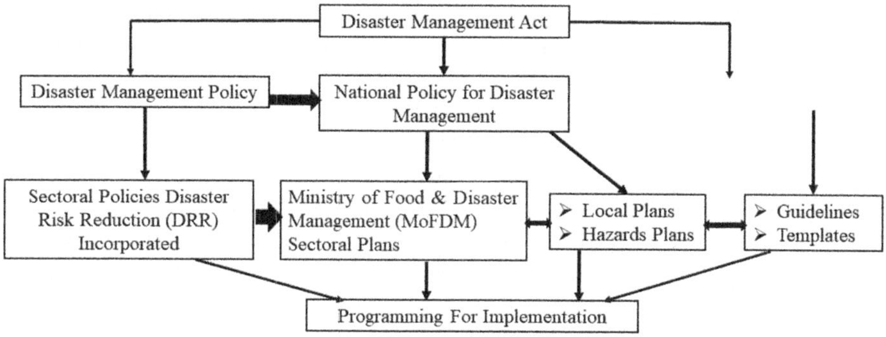

Fig. 1.3 Regulatory framework in implementing disaster risk management in Bangladesh. (Source: modified and adapted from Ministry of Environment and Forests, Government of the People's Republic of Bangladesh, Financial Support: DFID and DANIDA)

et al. 2016; Abu Abdullah 2016; Ahmad 2019; Christopher and Siverd 2020). The energy difference results in a latitudinal temperature gradient that drives the general circulation (winds in the atmosphere and currents in the oceans) in the climate system. Fluctuations in these energy transport processes on longer-than-synoptic time scales and at various spatial locations constitute the cause of climate anomalies and extremes.

The most prominent and visible characteristic of the climate system is that it is variable. It fluctuates in response to external forcing (such as solar radiation from the sun) and to internal nonlinear dynamical processes on many different time scales, ranging from microscale processes in both space and time to increasingly larger scales that are global in extent (Leontief 1976; Choudhury et al. 2004; Lloyd-Jones 2006; Khan 2015). The impact of climate variability on various components that comprise the climate system, as well as on human socio-economic structures, depends significantly on time scale and spatial extent. In more recent times, extreme weather events (Fig. 1.3) that have caused large disasters include:

(1) Cyclone Sidr in November 2007, which devastated the coastal zone of Bangladesh;
(2) The flood in 1998, when 75% of the total area of Bangladesh was flooded;
(3) Cyclone Aila in May 2009, which destroyed the south-west zone of Bangladesh (Sarker et al. 2020; Hoque et al. 2020).

Of course, other parts of the world have also been influenced by extreme weather conditions. For example, parts of Africa have experienced prolonged droughts and in 2006 the UN warned of a potential humanitarian catastrophe in eastern Africa. Parts of Australia have also been hit by a prolonged drought in recent years, with a devastating impact on its agricultural production (Karoly et al. 2003; Potgieter et al. 2005).

Natural disasters occur because of interactions among the natural, built, and social systems. If climate change causes an increase in the frequency and/or intensity of some hazards (or a shift in their geographical location) then the number of natural disasters can be expected to increase (Abedin and Shaw 2013; Habiba and Shaw 2013; Sarker et al. 2020; Hoque et al. 2020). What does climate change science say about this issue? There is compelling evidence in our observed climate data and from the outputs of climate models that points toward a warming climate (Leontief 1976; Choudhury et al. 2004; Lloyd-Jones 2006; Khan 2015). Basic physics indicates that as the atmosphere warms, evaporation of water from the ocean surface will increase, adding more moisture to the atmosphere. This should result in an intensification of the hydrological cycle as a warmer atmosphere can hold more moisture, and is precisely what global climate models show as a consequence of a warming atmosphere. Expectations are, therefore, that along with a gradually warming climate there will be accompanying increases in many hazards (Intergovernmental Panel on Climate Change-WG1 2007). Some hazards are confidently expected to become worse, such as heat waves, flood, and drought. Confidence in predictions about others, such as tornadoes or hurricanes, is lower.

Bangladesh is the most vulnerable country and well known for its disaster management and framework. Bangladesh has had its own disaster management BUREAU since 1993. To achieve technical monitoring and capacity building it has a good number of institutional structures. The ministry of food and disaster management takes control of all the disaster agencies (Abedin and Shaw 2013; Habiba and Shaw 2013). Bangladesh disaster management policy has been formulated to state the national perspective on disaster risk mitigation and describe management framework and principles. The disaster management council has planned a national plan regarding climate changes and disaster management (Leontief 1976; Choudhury et al. 2004; Lloyd-Jones 2006; Khan 2015). The main aim is to reduce the vulnerability of the poor during disasters. This plan has been developed with the government mission. Over the last three decades GOB has donated US$10 billion to make the climatic condition more resilient and less vulnerable to natural disaster. Climate change affects human health in various ways. The floods in Bangladesh in 2004 are a good example of the direct effect as it caused 800 deaths. The recent cyclone Sidr affected 8.5 million people and killed 3500 people (Leontief 1976; Choudhury et al. 2004; Lloyd-Jones 2006; Khan 2015). The impacts of climate change on Bangladesh have significant implications for its economic development. There are many human dimensions to climate change. Climate change has a severe effect on our agriculture. Bangladesh is an agricultural country and is very much dependent on its agricultural sector. Climate change also affects our forests and biodiversity. Bangladesh has a diverse range of forest ecosystems, including bamboo, freshwater swamp forest, and mangrove. Climate change has a detrimental effect on all of these.

After the Industrial Revolution, carbon gas emissions from fossil fuel, deforestation, and agricultural practices have made the world change with regard to global warming and the climate. Several anticipated impacts of climate change such as temperature rises, changes in rainfall patterns, changes in the distribution of weather events such as droughts, storms, floods, and heat waves, sea level rise, and consequent impacts on human and natural systems are increasing day by day (Habiba et al. 2014; Sarker et al. 2020; Hoque et al. 2020). Many scientists estimated that climate change results will be devastating for both human and natural systems. It will ultimately threaten human civilization by changing the earth system. Natural disasters will cause heat increases every year with a massive destructive power in coastal countries such as Bangladesh (Leontief 1976; Choudhury et al. 2004; Lloyd-Jones 2006; Khan 2015). However, an action plan for management strategies has been slow to evolve. Climate change draws attention to making a relationship between countries, governments, science and society, and triggers engagement of all stakeholders in the country. Engagement with climate change by social scientists is prompting conceptual renewal in areas such as disaster management studies.

1.5 Organization of This Chapter

This chapter is based on the basic terms and ideas of the climate change impacts and natural hazards and disasters in Bangladesh. In this chapter could help the academician and researchers to get to know the causes and impact of climate change in Bangladesh. There are two important forms of geography, physical geography and human geography (Abedin and Shaw 2013; Habiba and Shaw 2013; Sarker et al. 2020; Hoque et al. 2020). The scientific study is of the natural features of the Earth's surface, especially in its current aspects, including land formation, climate, currents, and distribution of flora and fauna. It is also called physiographic. Alternatively, human geography is the branch of geography dealing with how human activity affects or is influenced by the earth's surface. Thus, in this chapter we get to know climate changes that have taken place in both the physical and human geography (Leontief 1976; Choudhury et al. 2004; Lloyd-Jones 2006; Khan 2015). We also learn about the vulnerability of Bangladesh. There is a climate calendar that shows that flooding occurs from May to October, with severe flooding during the period July to August. Therefore, flood is already listed as one of the effects of climate change. Flash floods, draught, riverbank erosion, cyclone, tornadoes, and cold waves have become very common in Bangladesh. In physical geography phenomena studies include rocks and minerals, landforms, soils, animal, plant, water atmosphere, environment, climate, and weather. In humans we also see population, settlements, economic activities, transportations, urban systems, human migrations, and social traditions (Leontief 1976; Choudhury et al. 2004; Lloyd-Jones 2006; Khan 2015). Holistic synthesis is the common major sub-disciplines of human and physical geography. Climate change is the biggest problem at the moment, not just in Bangladesh but also throughout the entire world. But it has become one of the biggest problems for Bangladesh in particular as we are suffering so much more than other countries. Numerous floods, cyclones, and typhoons are occurring almost every year and millions of people are losing houses and farming land because of them (Abedin et al. 2014, 2019; Alam et al. 2010; Ahmed et al. 2019). Therefore, we should be very careful about climate change and do everything necessary to stop it from getting worse or else we will be in deep danger in the near future.

1.6 Data Sources and Methodology

Bangladesh is a land of rivers, more than 700 rivers that engage with the coastal area covered all the delataic lands of the country. The Padma, Meghna, and Jamuna are the main rivers that lead into the Bay of Bengal. Bangladesh has a long coastline of about 700 km and an exclusive economic zone of 200 nautical miles in the Bay of Bengal. Marine fisheries contribute 19% of the total fish production of the country. The Bay of Bengal has the most resources concerned with an ocean-based business, or blue economy. Bangladesh is blessed with the bounty of natural resources like the Bay of Bengal, it is also highly vulnerable to natural hazards. It has a long history of resilience and capacity of coping with major natural disasters. The government and the people have a wealth of experience in preparing for, and responding to, disaster events. (Leontief 1976; Choudhury et al. 2004; Lloyd-Jones 2006; Khan 2015). However, nowadays from the past experience of the coastal area and the Mangrove forest, Sundarbans have been exploited by human nature and this has a direct effect on the mainland of southern and western parts of Bangladesh where there is severe damage. Bangladesh suffers from different types of disasters such as floods, cyclones, storm surges, and riverbank erosion, earthquakes, drought, salinity intrusion, fire, tsunamis, and many more hazards (Mehedi et al. 2010; Asgary and Halim 2011; Jakariya et al. 2016; Abu Abdullah 2016; Ahmad 2019; Christopher and Siverd 2020). The coastal zone of Bangladesh consists of 19 coastal districts: Jessore, Norail, Gopalganj, Shariatpur, Chandpur, Satkhira, Khulna, Bagerhat, Pirojpur, Jhalakati, Barguna, Barishal, Patuakhali, Bhola, Lakshmipur, Noakhali, Feni, Chittagong, and Cox's Bazar and these are the most affected sea-coast low line areas (Sarker et al. 2020; Hoque et al. 2020). These are the areas that are mostly affected during times of natural hazards and calamities. Major issues in riverine districts are lower soil erosion, crop damage due to excessive rain, earthquakes due to lack of adherence to proper building codes (RAJUK). Most of the people of Bangladesh are not aware of this major environmental change and natural hazards and disasters impacts. Although the government of Bangladesh has tried to address these problems, many related organizations, agencies, and ministries have the major lack of integration and co-ordination to solve these issues and problems.

As a coastal country, Bangladesh has always suffered from many extreme natural events. In historically deprivation as a result of colonization, imposed war, economic vulnerability, and fragile governance, Bangladesh is considered to be more vulnerable (Abedin and Shaw 2013; Habiba and Shaw 2013). The global level of CO_2 has surpassed 400 parts per million (PPM) owing to massive emissions of carbon where the safe level is 350 PPM mentioned by The National Aeronautics and Space Administration (NASA). This change will make more climate refugees from generation to generation (Sarker et al. 2020; Hoque et al. 2020). Therefore, every country should come forward to address the problem and make a framework for taking responsibility for all countries on a global platform. Despite a number of mitigation steps being undertaken, it is clear that the detrimental impacts of climate change are mounting alarmingly. It has been mentioned that "climate change can make some hazards worse, for example, tropical cyclones seem to be decreasing in frequency but increasing in intensity due to climate change." This kind of negative frequent cyclone with higher intensity has the greatest potential to initiate catastrophic disasters (Leontief 1976; Choudhury et al. 2004; Lloyd-Jones 2006; Khan 2015). It is also obvious that coastal Bangladesh will be seriously affected by climate change over the next decades, with elevated air and sea temperatures, sea-level rises, and more intense cyclones, all threatening livelihoods and lives.

1.6.1 Primary Data Sources and Collection Methods

Data were collected from both primary and secondary sources. Primary data were collected by direct interview and secondary data through research reports, journals, brochures, and websites. In order to review the content policy, the content analysis method was used. For an overview of the existing condition of the institutions, observation and in-depth interview methods were used. Some disaster experts were also interviewed.

1.6.2 Secondary Data Sources and Collection Methods

A comprehensive literature review was used as the research methodology for this study. A key work search for natural disasters, hazards, vulnerability, disaster risk reduction, and Bangladesh was used to find literature from various sources such as electronic library databases, table of contents of journals, online journals, and e-books. Literature review was structured and presented within different categories of natural hazards and disaster risk reduction strategies used for natural hazards. In order to make an overall assessment of disaster severity, a set of questionnaire surveys has been conducted in different parts of the country. Three districts from each of the eight divisions of Bangladesh have been selected for the data collection using the cluster sampling method. In this survey, a questionnaire was prepared on the topic of different forms of natural disasters such as: flood, cyclone, drought, riverbank erosion, landslide and salinity (Abedin et al. 2014, 2019; Alam et al. 2010; Ahmed et al. 2019). Cumulative data of about 100 questionnaires were collected from each division. From the questionnaire survey conducted, it was found that different regions of the countries mostly face different disaster types. Analyzing the severity of different forms of disasters, the regions were crossed matched with the severity indexes.

1.7 Scenarios of Natural Hazards and Disaster in Bangladesh

Before 1970, the climatic scenario of Bangladesh was difficult to explain owing to the forefront of political attention. The backwardness of technology, research, and policy also make it complicated to provide a proper explanation of the previous scenario. Early work was done, such as mathematical simulation or qualitative exercises (Meadows et al. 1972; Herrera et al. 1976; Mesarovic et al. 1977), input–output analysis (Leontief 1976), and eclectic approaches (Barney 1980). Reviews and critiques of global assessments introduced fresh insight (Pirani 1974; Meadows et al. 1982). More recently, a second wave of global scenario studies have included narrative scans of alternative futures (Burrows et al. 1991; Milbrath 1989), a possible analysis by the Dutch Central Planning Bureau (1992), the pessimistic analysis by Kaplan (1994), a consideration of surprising futures, and the United Nations Global Outlook (United Nation 1990). The climate change issues have made several models including world energy scenarios, most importantly, those of the Intergovernmental Panel on Climate Change (IPCC 1992), which remain a traditional concept of long-range development. In addition, authors have revisited first-wave studies and affirmed their essential findings, despite intense and sometimes rancorous criticism in the interim (Barney 1993; Meadows et al. 1992a, b).

1.7.1 Tropical Cyclones and Storm Surges

From the record of casualties, tropical cyclones could be the number one extreme natural event, while floods ranked the second most devastating natural event in Bangladesh (Asgary and Halim 2011; Shimi et al. 2010). The exceptional geographical location of Bangladesh where the Himalayas are to be found on the north side and the Bay of Bengal touches on the southern side toward a funnel-shaped coastline (Choudhury 2002; Sarker et al. 2020; Hoque et al. 2020). Bangladesh has suffered approximately 178 severe cyclones with wind speeds of more than 87 km/h formed in the Bay of Bengal from 1891 to 1998 and 38 severe cyclones from 1970 to 1998 (Alam and Colin 2010). The cyclones of 1970, 1985, 1991 and 1997 are some notable events in the recent past (Khan 2008; Sarker et al. 2020; Hoque et al. 2020). Cyclones and tidal surges have caused major devastations in human lives and property in Bangladesh for generations (Alam and Colin 2010). Severe cyclones in Bangladesh have claimed thousands of human lives with millions of people being affected. In 1970 a huge cyclone killed 500,000 people and in April 1991 an enormous cyclone was responsible for about 140,000 lives (Choudhury 2002). Another dreadful cyclone, Sidr, which occurred in November 2007, claimed more than 3000 lives, approximately 53,000 people were reported missing, and 8.7 million people were directly affected in 30 districts of Bangladesh (IFRC & RCS 2010). Besides, the storm surge in the Bay of Bengal, with an abnormal rise in seawater associated with a tropical cyclone, also caused severe devastation in coastal regions (Alam and Colin 2010).

1.7.2 Flood Scenarios in Bangladesh

Although floods are a devastating, worldwide natural disaster, Asian countries such as India, China, Philippines, Iran, Bangladesh, and Nepal are recognized as highly vulnerable to floods (WWAP 2006). As a riverine country, Bangladesh has to face devastating floods every year with different magnitudes (Choudhury et al. 2004; Hossain 2003). Although the intensity and magnitude of floods vary from year to year, a study shows that Bangladesh will face an increasing level of floods (Mclean and Moore 2005). In Bangladesh, 80% of land is considered to be floodplains, where 34% of land is flooded for about 5–7 months each year (Islam 2004 cited in Shimi et al. 2010). During the monsoon season 20–25% of the land is inundated in Bangladesh (Choudhury et al. 2004; Hossain 2003). Almost 20% of the total landmass is affected by annual flooding, whereas over 50% of the landmass was under water in the 1992 floods (ISDWC 2002 cited in Mclean and Moore 2005). From 1954 to 2004, Bangladesh has faced 29 major floods, 11 of which were "devastating" and six were "most devastating" floods (Choudhury et al. 2004). Some remarkable floods such as 1987, 1988, 2004, and 2010 resulted in massive destruction in Bangladesh (Khan 2008; Habiba et al. 2012) (Fig. 1.4).

Regular floods make barriers to economic and social development. Normal functions of life such as daily activities, water supply and sanitation, washing away crops, polluting groundwater stocks etc., are hampered because of frequent floods (Shimi et al. 2010; BSHF 2001 cited Mclean and Moore 2005). Extreme events result in severe floods that bring damage to affected areas where the damage is higher if the event is prolonged (Fig. 1.3) (Rahman 2015; Khan 2008). In Bangladesh, floods can be classified into four types: river floods, rainwater floods, flash floods, and cyclonic/storm-surge floods whereas in general floods are classified into three types: "normal" and "abnormal" or "extreme" events (Khan 2008; Choudhury et al. 2004). Although heavy monsoon rainfalls and melting snow in the upper catchment areas of the major rivers of Bangladesh result in river floods, rainwater floods occur owing to heavy rainfalls that affect floodplains and other low-lying regions (Choudhury et al. 2004). Flash floods originate from heavy rainfalls that mainly take place in the eastern and northern hill streams in Bangladesh; cyclonic/storm-surge flooding affects the coastal regions of Bangladesh and is formed because of tropical cyclones in the Bay of Bengal and (Choudhury et al. 2004). River floods and flash floods are of utmost concern to Bangladesh (Khan 2008). Although the normal river floods, together with seasonal variability in flow and water levels, are somewhat beneficial to the ecosystem because of carrying alluvium, extreme floods have so far resulted in severe losses.

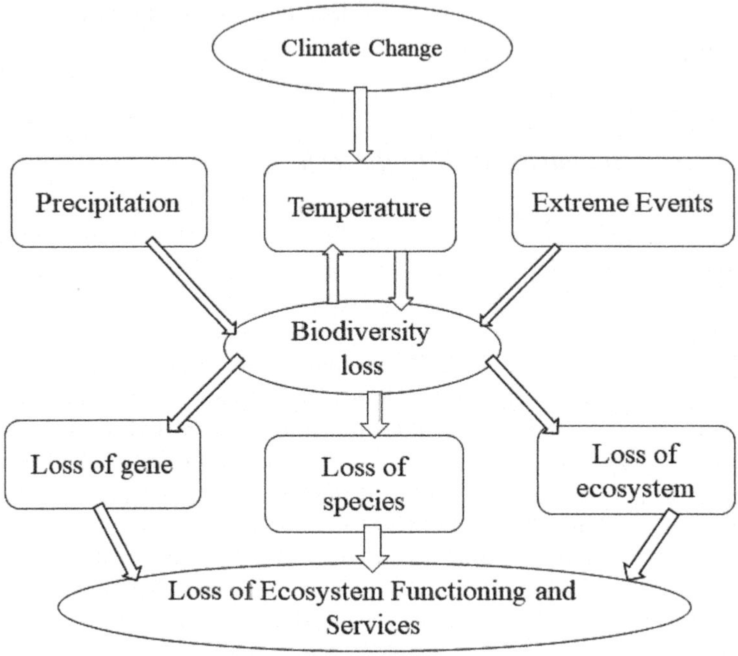

Fig. 1.4 Link between climate change and its impacts on biodiversity and ecosystem services, and the impact of biodiversity loss on climate change (Sintayehu 2018)

1.7.3 Scenarios of Droughts in Bangladesh

Drought is another result of the impact of climate change. When low precipitation occurs over an extended period of a season or more, it creates a shortage of water for activities, groups, or environmental sectors (Habiba and Shah 2010). Although it is not a frequent hazard in Bangladesh, it forms a part of the natural disaster list in Bangladesh because droughts occur occasionally, causing extensive damage to crops (Choudhury 2002; Sarker et al. 2020; Hoque et al. 2020). It is not an old phenomenon like floods and it occurs for 7 months, from November to May, as the rate of rainfall is low in these months, but it afflicts the country with low agricultural production (Paul 1998; Habiba and Shah 2010; Karim 2004). In the last 50 years, Bangladesh has experienced the effects of droughts at least 20 times (Choudhury 2002; Habiba and Shaw 2010). There was a severe drought in the 1998–1999 dry season in some areas of the northwestern, southwestern, and central parts of the country (Karim 2004; Sarker et al. 2020; Hoque et al. 2020). After that, in 1979, 1989, and 1994–1995 the severest drought was experienced in Bangladesh as well (Choudhury 2002; Karim 2004).

As an agriculture-based country, the drought caused massive destruction for agricultural production, as well as economic, social, and environmental imbalance. Frequent drought in the northwestern zone in Bangladesh in recent times reduced rice production by 3.5 million tons in the 1990s (Karim 2004; Islam and Neelim 2010). Droughts being closely related to climate change, the effects of global climate change predictions indicate that the dry seasons will become drier and hotter (Karim 2004; Sarker et al. 2020; Hoque et al. 2020). Mirza and Pal (1992) and Das (1997) classify droughts in Bangladesh into three groups (Meteorological drought, Agricultural drought, Hydrological drought) depending on their impact (Habiba and Shaw 2010):

1.7.4 *Soil Erosion and Deposition in Low-Lying Land in Bangladesh*

Soil erosion is a common phenomenon of the mighty rivers in Bangladesh, such as the Padma and the Meghna. The river takes away chunks of riverbanks causing displacement of people and the reduction of property (Leontief 1976; Choudhury et al. 2004; Lloyd-Jones 2006; Khan 2015). Owing to recurrence of such erosion, displaced people are forced to come to cities for their earnings. In last 34 years, it is estimated that the Jamuna, Ganges, and Padma river sides were submerging about 219,286 acres, 69,135 acres and 95,119 acres respectively. In 2007, the massive erosion of the Jamuna river submerged 3,408 acres of land, 543 localities, and 3,360 m of embankment, 5,160 m of roads, and two marketplaces into the body of water (Abedin and Shaw 2013; Habiba and Shaw 2013). In the same way, the Ganges' erosion took place and submerged 1,778 acres of land, 136 acres of localities, and 570 m of roads while the Padma had caused 1600 acres of lands, 370 acres of localities, 3930 meters of roads, nine educational institutions, five marketplaces, and one Union Council office (CEGIS 2000). The unstable characteristics of Bangladesh Rivers have led to great erosion in the coastal areas. The braided pattern of rivers in Bangladesh are separated by many small islands along their courses (Sarker et al. 2020; Hoque et al. 2020). In monsoon seasons large overbank spills, riverbank erosion, and bank-line shifts are typical (Mehedi et al. 2010; Asgary and Halim 2011; Jakariya et al. 2016; Abu Abdullah 2016; Ahmad 2019). The major river channels in Bangladesh have gradual migration and shift anywhere between 60 m and 1.6 km annually. Normally, major erosion occurs in a typical year along about 2400 km of the bank line. The unstable nature of the shifting of the riverbanks affects the coastal population, agriculture production, and destroys properties and infrastructure.

1.8 Climate Change and Natural Disaster Impacts in Bangladesh

Bangladesh is one of the most vulnerable countries (MVCs) owing to climate change and a weak natural resource-based economy (Jakariya et al. 2016). From 1999 to 2009, Bangladesh was the worst suffering countries of climate change-induced disasters (Mehedi et al. 2010). Hydro-meteorological disasters are becoming more frequent in Bangladesh, e.g., cyclones, tidal surges, flooding, sea level rise, coastal and riverbank erosion, ground and surface water scarcity, salinity intrusion, drought, heat stress and rainfall-induced landslides and flash flooding owing to changes in the temperature and rainfall pattern (Leontief 1976; Choudhury et al. 2004; Lloyd-Jones 2006; Khan 2015). These natural events are causing problems such as land soil degradation, loss of crops and agricultural productivity, food insecurity, waterborne diseases, threatening livelihoods, gender inequality, stress on human health, human displacement and migration, social instability, unemployment, poverty, and ultimately triggering conflicts.

The casualty of tropical cyclones and storm surges are associated not only with its geographical location but also with its demographic and socio-economic conditions (Islam and Neelim 2010; Haque and Jahan 2016). It has been estimated that Bangladesh will lose 17% of its land and 15 million of people will displaced by 2050. From 1877 to 1995 Bangladesh experienced only 1% of the world's total tropical cyclone, but total deaths constituted approximately 53% of the total world's deaths by tropical cyclone (Ali 1999; Islam and Neelim 2010). A severe cyclone hits Bangladesh every 3 years on average (Abdullah et al. 2016). Moreover, the coastal areas and the Bay of Bengal are located at the northern tip of the Indian Ocean and are frequently hit by cyclonic storms, generating high tidal surges, floods, and storm surges, which lead to permanent or temporary human displacement (Abedin and Shaw 2013; Habiba and Shaw 2013; Subhani and Ahmad 2019). As a consequence of frequent cyclone and storm surges at least 20 million Bangladeshis may be forced to displace, with ramifications for food and water supplies as well as disease epidemics (Mehedi et al. 2010).

Bangladesh, with the Bay of Bengal on its southern border, has been ranked as the most vulnerable country to tropical cyclones (Islam and Neelim 2010; Haque and Jahan 2016). The coastal zone of Bangladesh, an area covering 47,211 km^2 facing the Bay of Bengal or having proximity to the Bay, and the exclusive economic zone in the Bay (PDO-ICZMP 2004:13), is generally perceived to be a zone of multiple vulnerabilities (Mallick et al. 2009). A record of the last 200 years shows that almost 70 strong cyclones hit the coastal belt and a record of the last 35 years estimated that 900,000 people lost lives owing to tropical cyclones (PDO-ICZMP 2004:14). It is forecast that, by 2050, about 17% of coastal areas will be inundated (Islam and Mehedi 2016). Many sources of vulnerability to disaster still exist. Poverty, settlement in low-lying coastal areas, inadequate availability of cyclone shelters, and overdependence on traditional livelihoods are still considered major sources of vulnerability (Alam and Colin 2010; Mallick et al. 2017).

In July 2016, hundreds of villages and several districts were severely affected by flooding in Bangladesh. The disastrous situation lasted for a few months, and approximately four million people were stranded by floodwater, losing their houses and crops. On the other hand, the tropical cyclones and associated tidal surges are usual in the coastal region (with approximately 585 km of coastline) of Bangladesh, which has nearly 27 million people, with an average density of 854 people per km^2 (Leontief 1976; Choudhury et al. 2004; Lloyd-Jones 2006; Khan 2015). Frequent cyclones and inundation of coastal area are increasing by the magnitude of the increase of global warming. Coastal communities are washed out by cyclone and water surges every year (Islam and Neelim 2010). Most recently, monsoon rains have triggered flooding across 21 districts (out of 64) in north-eastern Bangladesh and killed at least 122 people as of 21 August 2017. Every year, almost 50 million of people are seriously affected by floods. Besides, deadly landslides triggered by torrential monsoons in southeastern Bangladesh took 160 lives in June 2017. This landslide took place in the coastal area of Bangladesh just 2 weeks after cyclone Mora, killed 9 people and caused significant damage. Thus, river flooding, flash flooding, landslides, storm surges, and cyclones devastate at least one-third of Bangladesh each year (Leontief 1976; Choudhury et al. 2004; Lloyd-Jones 2006; Khan 2015). Climate change impacts have the potential to exacerbate this catastrophic hydro-meteorological disaster scenario in Bangladesh (Table 1.1).

It is estimated that around five million people were either seriously affected or displaced by the climate-induced disasters in Bangladesh in recent years. Frequent disasters hit coastal areas and forced climate refugees to migrate to urban areas. There they were highly exposed to other hazards such as flooding and water pollution (Choudhury et al. 2004; Lloyd-Jones 2006; Khan 2015). The population pressure is increasing in the major cities of Bangladesh owing to the excessive rate of climate refugees and cities are rapidly getting larger. This extra burden of cities is creating unplanned urbanization and degrading water bodies and natural systems. The pressure of excess population in cities now make them highly vulnerable to problems of urban life such as water logging, flash flooding, encroachment of floodplain areas, development of informal settlements, urban heat islands, urban landslides, traffic jams, air and water pollution, and scarcity of drinking water, and the city dwellers even lack other city life facilities (Leontief 1976; Choudhury et al.

Table 1.1 Recent major disasters and casualty in Bangladesh

Year	Disaster	Deaths
1991	Cyclone	≥138,868
1996	Tornado	545
1998	Flood	1,050
2004	Flood	747
2007	Cyclone	3,363
2009	Cyclone	339

Source: Alam and Colin 2010; Haque and Jahan 2016; Mallick et al. 2017

2004; Lloyd-Jones 2006; Khan 2015). All these problems are evident in the capital of Bangladesh, which was ranked as one of the worst capitals in the world.

1.9 Combating Climate Change and Disaster Vulnerabilities in Bangladesh

The 21st session of the Conference of the Parties (COP21) was held in Paris, France, from 30 November to 12 December 2015. At COP21, parties to the United Nations Framework Convention on Climate Change (UNFCCC) reached a historic agreement to address the climate change issues. The agreement highlights keeping the rise in global temperature below 2 °C (even though this rise in temperature could potentially inundate the coastal region of Bangladesh), making action plans to reduce GHGs, raising US$100 billion (in loans and donations) each year from 2020 to finance projects that enable vulnerable countries to adapt to the impacts of climate change (such as a rise in sea level or droughts). This establishes an obligation for the industrialized countries to fund the victim countries through climate finance. This agreement will finally enter into force once it is ratified by the 55 countries, representing at least 55% of emissions (UNFCCC 2016). Most importantly, on 3 September 2016, the USA and China, together responsible for 40% of the world's carbon emissions, both formally joined the agreement. The Paris Agreement came into force on 4 November 2016, and as of 21 August 2017, as many as 160 parties out of 197 have ratified the convention. Unfortunately, on 1 June 2017, the USA announced that it would cease all participation in the 2015 Paris Agreement. It proves that the progress toward a global consensus to combat climate change is under serious threat (Abedin et al. 2014; Alam et al. 2010; Abedin et al. 2019; Ahmed et al. 2019).

This squeezing trend of the global economy is horrifying, and it indicates that the rich are getting richer and the poor are becoming poorer, and that the world's net wealth is being concentrated. Professor Muhammad Yunus pointed out this problem and asked for a review of the world's current economic system (Abedin and Shaw 2013; Habiba and Shaw 2013). This is an example of how the current global economic system can make other countries economically vulnerable. In addition to this, over-exploitation of resources and, thus, CO_2 emissions by the industrialized countries trigger climate change. The IPCC, as part of its future pathways for adaptation, mitigation, and sustainable development, has clearly signposted that "adaptation can reduce the risks of climate change impacts, but there are limits to its effectiveness, especially with greater magnitudes and rates of climate change (Abedin et al. 2014, 2019; Alam et al. 2010; Ahmed et al. 2019. This scenario applies to Bangladesh, and the country is now seeing an overwhelming increase in the number of climate refugees than previously recorded. It is becoming impossible for the climate victims to return to their normal livelihoods in the disaster affected area (Habiba et al. 2014). The biggest challenges in the upcoming years to

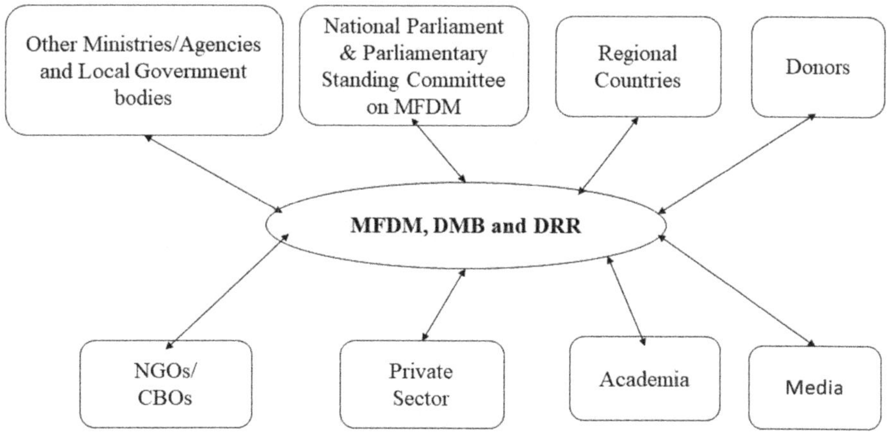

Fig. 1.5 Proposed partnership framework for disaster management in Bangladesh

combating climate change and implementing effective initiatives such as the COP21 agreement include the global monopoly of the economic system, and the negligence by a major portion of politicians and corporations depicts how Bangladesh (an example of a victim country) became socio-economically vulnerable after centuries of inflicted wars and tyranny by the developed world, and how Bangladesh is now facing discriminating problems because of the climate change-induced extreme events that are being triggered externally by the same developed world (Fig. 1.4). This vicious cycle of oppression, disaster, and poverty is hindering the overall sustainable development and Directorate of Relief and Rehabilitation progress in Bangladesh (Fig. 1.5).

1.10 Addressing Climate Refugees and Climate Migration

It is estimated that from 2011 to 2050, around 15 to 25 million people will migrate from affected areas to urban areas because of flooding, storm surges, and riverbank erosion in Bangladesh (Habiba et al. 2013). Among them, a significant portion would migrate solely because of climate change, and the remaining because of the current context of climatic disasters, economic needs, and population growth (Abedin et al. 2014, 2019; Alam et al. 2010; Ahmed et al. 2019). Traditionally, migration has been treated with fear, but the Refugee and Migratory Movements Research Unit (RMMRU) at the University of Dhaka has suggested that migration should be considered one of the climate change adaptation techniques.

The scientific community predicts that by 2050, one in every seven people in Bangladesh and one in every 45 people in the world will be displaced by climate change. Extreme climatic events create climate refugees who have lost their homestead, arable land, or livelihoods in the rural settings. In 2015, the government of

Bangladesh developed a "National Strategy on the Management of Disaster and Climate-Induced Internal Displacement (NSMDCIID)" to address the multiple human rights challenges faced by migrating people in the aftermath of climatic disasters (Islam and Neelim 2010; Abedin and Shaw 2013; Habiba and Shaw 2013). This is a great initiative in terms of identifying a vulnerable community, although it only focuses on national level displacement or migrated populations in Bangladesh due to natural hazards. This should be considered the first milestone, but there is still a need to develop such formal strategies through international consensus to share the burden of and responsibilities for climate refugees by the top climate-polluting countries.

1.11 Comparison with Other Countries and Global Perspective

Global climate change has already had massive effects on the environment, including icebergs melting, sea levels rising, plant and animal ranges shifting, natural disasters increasing, etc. The previous predictions on climate change are now truly happening to the earth: loss of sea ice, sea levels rising so fast, the seas and ocean waves becoming longer, and more heat waves in the open seas and coastal zones also. Global temperatures will continuously rise owing to GHGs being produced caused by increased industrialization. The IPCC, which includes more than 1300 scientists from the United States and other countries, forecasts a temperature rise of 2.5 to 10°F over the next century (Pirani 1974; Meadows et al. 1982; Islam and Neelim 2010).

Mainstreaming disaster risk management into development planning can overturn the current trend of rising disaster impact. Furthermore, when countries reconstruct stronger, faster, and more inclusively after disasters, they can diminish the influence on people's livelihoods and well-being by as much as 31%, potentially cutting global average losses. If countries work in cooperation, they can mitigate the impact of climate change and save lives. In contrast, many developing countries have a lack of ability to provide tools, expertise, and instruments to factor the potential impacts of disasters into their investment decisions (Disaster Risk Management 2019).

Bangladesh is a disaster-amenable country of an area of about 147,570 km^2 with a population nearing 180 million. The country is well within the ecliptic and is the largest delta in the world formed by the mighty rivers the Ganges, the Brahmaputra, and the Meghna. Bangladesh has a special geographical landmark, i.e., the Himalayan range to the north. The combined effects of the role played by these special geographical features have a momentous bearing on the weather system of Bangladesh. The weather system is not always fortunate (Islam and Neelim 2010; Kabir et al. 2016). Bangladesh has become the worst victim for this kind of weather system.

The Center for Research on the Epidemiology of Disasters publishes an Annual Disaster Review to provide golden information on the grandeur of natural disasters and their influence on society. In 2014, they named China, the United States, the Philippines, Indonesia, and India as the five countries most repeatedly hit by natural disasters. From recent years these same countries have regularly featured at the top of this annual list (Park 2015).

1.11.1 *Natural Disaster Risk Management in India*

According to a recent study by the United Nations Office for Disaster Risk Reduction (UNISDR), India was ranked third amongst the top five most disaster-hit countries of the world in 2015. India faced around 19 natural disasters of varying intensity in 2015 alone, resulting in severe damage costing over US$3.30 billion. In the Indian context, there are challenges aplenty in effective disaster management (Kabir et al. 2016; Sarker et al. 2020; Hoque et al. 2020). The primary responsibility for disaster management is entrusted to the government authorities, with representatives at national, state, and district levels. In addition to the government and its agencies, the military, NGOs, and voluntary organizations play a major role in disaster relief and rescue activities. A key challenge is to ensure and achieve effective coordination and collaboration amongst these stakeholders (Islam and Neelim 2010). Given the significant risk, as well as the diversity of disasters, capacity building of various stakeholder groups is yet another formidable challenge. Providing the right information at the right time to the concerned stakeholders, especially in the initial hours after a disaster, is the cornerstone of an effective and efficient disaster management system. From relief and rehabilitation, the scope of disaster management has expanded over the years at a global level to include prevention, preparedness, response, and recovery. The Hyogo Framework for Action (HFA) 2005–2015 focused on disaster resilience by attempting to reduce disaster losses.

For 2 years, the south Indian state Kerala has experienced heavy floods, which have affected one sixth of its population and caused massive numbers of casualties. The disaster also has a significant financial impact as it washed away nearly 2.6% of the state's gross domestic product. Between 1 June and 19 August 2018, Kerala received unusually high rainfall, about 42% above normal, resulting in the worst-ever floods since 1924, which affected almost 5.4 million people. After the 2018 floods, the Kerala state government induced the Rebuild Kerala Development Program to rebuild areas impinged by the flood and it also worked to build a more elastic, green, extensive and vibrant Kerala state. In India, its focus is on increasing capacity to take up ecosystem reinstatement as part of the Mahatma Gandhi Rural Employment Guaranteed Scheme, which was a nation-wide program that employed 2.6 million women in Kerala. This will be accomplished by partnership with the Kerala State Disaster Management Authority and the Kerala Institute of Local Administration (Promoting livelihoods through nature-based disaster risk reduction in India, 2019).

1.11.2 Natural Disaster Risk Management in Indonesia

Indonesia is one of the most disastrous countries in Asia. Its risk is driven by the country's high exposure to a range of geophysical and hydro-meteorological hazards, combined with grave vulnerabilities resulting from population growth, unequal economic development, urbanization, a lack of social and environmental considerations within development processes, and other drivers (Kabir et al. 2016). Disasters caused by environmental hazards are becoming increasingly costly and severe in Indonesia. Every year natural disasters come in turn including earthquakes, tsunamis, floods, volcanoes, landslides, forest fires, tornadoes, and social conflicts, and covering the entire state of Indonesia. Although efforts to manage disaster impacts and reduce disaster risk have long been considered, the 2004 Indian Ocean tsunami transformed the way disasters are viewed and how the risks are managed and reduced (Sarker et al. 2020; Hoque et al. 2020). To deal with these natural calamities, the Government of Indonesia has developed community-based disaster management (CBDM). Between 1815 and 2014 (200 years), 291,427 people were killed by disasters, which occurred 13.172 times.

They adopted the Hyogo Framework for Action 2005–2015: Building the Resilience of Nations and Communities to Disasters and then the Sendai Framework for Disaster Risk Reduction 2015–2030 (Djalante et al. 2012). Budget becomes a problem when the outcome of natural disasters cost over 20% of the municipal budget and the additional costs cannot be covered by the provincial government (Fig. 1.6) (Deserter Risk Reduction and Management in Indonesia 2019).

1.11.3 Natural Disaster Risk Management in the Philippines

The Philippines is highly prone to natural disasters because of the virtue of its geographic circumstances, such as earthquakes, tropical cyclones and floods, and volcanic eruptions, making it one of the most disaster-amenable countries in the world (Mehta 2007). The Philippine institutional arrangements and disaster management systems rely on a response, or a respondent approach, in contrast to a more efficient proactive approach, in which disasters are staved off by proper land-use planning, construction, and other pre-event measures that distort the procreation of disaster-prone conditions (Mehta 2007; Islam and Neelim 2010; Kabir et al. 2016). It is important to evolve a more proactive role, so that a national framework for extensive disaster risk management is prepared and implemented. The framework should incorporate the necessary steps of compact risk management, which comprise risk identification, risk reduction, and risk sharing/financing the Government and individual households who pay the majority of costs accrued by natural disasters. More efficient options for financing disaster risk, and taking the burden of disasters away from the public sector should be pursued, including the idea of an insurance scheme

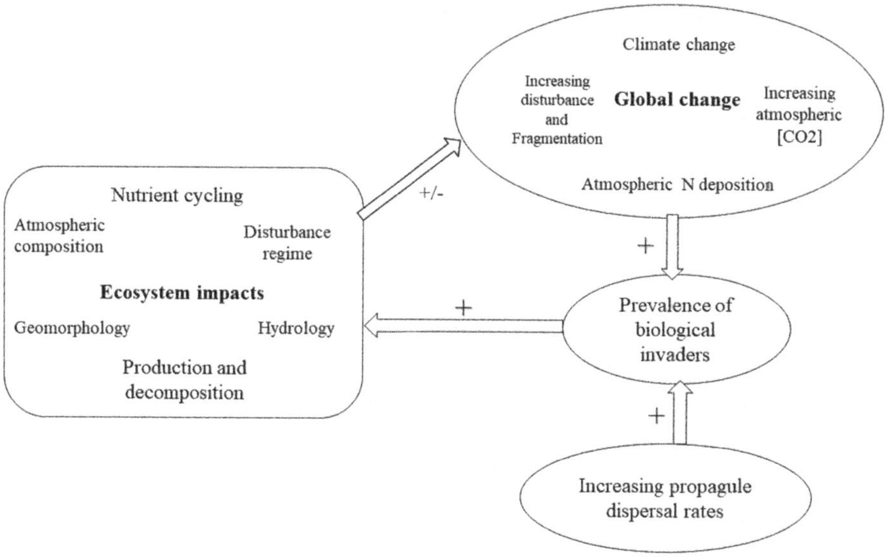

Fig. 1.6 Schematic diagram showing the pathways by which climate change and other factors influence the impact of invasive species

for those affected, or subsidiary credit facilities. Despite the high risk of danger in the Philippines, the insurance coverage for catastrophes affecting residential dwellings is almost non-existent.

1.11.4 Natural Disaster Risk Management in the United States

The United States is highly vulnerable to natural disasters. Every state experiences to one or more hazards – earthquakes, floods, hurricanes, droughts, landslides, tsunamis, volcanoes, tornadoes, and wildfires – every year. As Hurricane Hugo disclosed in 1989, natural disasters can destroy years of development and destroy natural resources in minutes or hours. The Committee has taken many propounded, multidisciplinary initiatives for the nation to mitigate the impacts of natural hazards (Mehta 2007). Key elements of the Decade program include hazard and risk appraisals; awareness and education; subsidence; preparedness for emergency response, healing and reconstruction; prediction and dissemination of warnings; strategies for learning from disasters; and international cooperation (Mehta 2007; Kabir et al. 2016). The Decade's progress should judge in the short term by using delegate measures, for example, the number of states and local authorities that improve their hazard and risk assessments, train replication teams, develop and exercise emergency response and rally plans, or take steps to emphasize building codes or their enactment; documented changes in the acquaintance and actions of such groups as the media, health workers, architects, engineers, policy makers, and the public; qualitative and quantitative changes in efforts to transfer technology and enhance professional skills through conferences and workshops; and the number of new bilateral and multilateral projects (The US Decade for Natural Disaster Reduction 1991).

1.11.5 Natural Disaster Risk Management in China

Floods, earthquakes, droughts, and typhoons are affecting life and property across China. The 2008 earthquake that struck the Sichuan region, for example, killed at least 69,000 people, injured hundreds of thousands, and left 15 million homeless. From 2006 to 2011, The Asia Foundation worked with the Ministry of Civil Affairs (MOCA), local Departments of Civil Affairs, Chinese and American business associations, and Chinese charity organizations to verify disaster management through increased public–private partnerships. By the end of 2012, there were over 1200 such communities around China, and the central government aims to increase this to a total of 5000 by the end of 2015 (Islam and Neelim 2010). Under this project, the Ministry of Civil Affairs also issued Regulations on the Relief of Natural Disasters, which motivated village committees, community centers, businesses, NGOs, and the public to help in natural disaster relief efforts. As China expands the capacity of its disaster management system at home, it is also helping other countries to do the same (Disaster Management in China).

1.11.6 Natural Disaster Risk Management in Bangladesh

A large number of poor people are living in the vulnerable areas of the southern part of Bangladesh. The vulnerability is so pathetic that they have to go and settle in the newly anointed land in the Bay of Bengal and its surrounding areas, which is rarely hit by tidal floods or demolishing cyclones (Abedin and Shaw 2013; Habiba and Shaw 2013; Sarker et al. 2020; Hoque et al. 2020). The untoward impacts of all the natural hazards affecting socio-economic conditions need to be attenuated for sustainable development. To overcome this situation, the government of Bangladesh has taken up many plans and programs for disaster reduction through disaster management (Disaster Management Bureau 2008; Habiba et al. 2012). The government introduced a project "Support to Comprehensive Disaster Management" in 1993, with the overall goal of minimizing the human, economic, and environmental costs of disaster in Bangladesh. One of the main components for the development objective of the project was to increase the capacities of the households and local communities in the highly risky areas through establishment of Local Disaster Action Plans (LDAPs) to cope with cyclones, floods, and other potentially disastrous situations (Rahman 2017; Ahmad 2019; Christopher and Siverd 2020). The project was finished on 30 June 2001, creating various opportunities for the Comprehensive

Disaster Management Program of more direction in risk management with cooperation from development partners and international agencies. The high-powered National Disaster Management Council (NDMC) and In-Ministerial Disaster Management Co-ordination Committee (IMDMCC) developed as operative bodies to promote and co-ordinate risk reduction, preparedness activities, and subsidence measures, meet twice and eventually four times a year (Abedin et al. 2014, 2019; Alam et al. 2010; Ahmed et al. 2019). When the NDMC was formulated and showed disaster management policy to all stakeholders, the IMDMCC took the initiative to implement the policy consisting of inter-ministerial coordination, supervising the services of the Armed Forces as well as NGOs working in the field of disaster management in the country (Sarker et al. 2020; Hoque et al. 2020). A well-established organization named Directorate of Relief and Rehabilitation under the mechanism of the NDMC within the administrative control of the Ministry of Disaster Management and Relief (MDMR) wherein the Emergency Operation Center is located (Islam and Neelim 2010). The MDMR has a small dynamic professional unit known as the Disaster Management Bureau to perform specialist functions and assure coordination with line agencies and NGOs by convening meetings of the Disaster Management Training and Public Awareness Building Task Force, the Focal Point Operational Co-ordination Group on Disaster Management, the NGO Co-ordination Committee on Disaster Management, and the Committee for Speedy Dissemination of Disaster-Related Warning Signals regularly every 3 months (Disaster Management in Bangladesh 2003).

According to other countries in the world, Bangladesh is one of the disaster-amenable countries of the world, with exceptionally limited resources, its real development is not possible without the integration of disaster subsidence programs as in other countries (Abedin and Shaw 2013; Habiba and Shaw 2013). Bangladesh is fighting hard to establish a well-performed and experienced disaster management system from national down to community level to relieve the effects of disasters. Being aware of the limitations and the vulnerability of the country to natural disaster, the government has been making continuous efforts to make Bangladesh a safer part of the world in the twenty-first century and seek the help of development partners (Sarker et al. 2020; Hoque et al. 2020). We know that our country has limited resources, but the members of our government are trying their best to protect our country.

1.12 Climate Change and Natural Hazards: Future Scenarios and Concerns

Day by day Bangladesh will become an even more vulnerable country owing to the impact of climate change. According to the Global Climate Risk Index 2020, Bangladesh is the seventh most climate-affected nation in the world (Rahman 2017; Ahmad 2019; Christopher and Siverd 2020). Nearly every year Bangladesh encounters different forms of natural disasters due to climate change. According to the New York Times, 0.3% of the pollution in Bangladesh is related to climate change, but the country is facing some of the biggest consequences of rising sea levels. By 2050, oceans could flood 17% of the land of Bangladesh and displace approximately 18 million of its residents (Islam and Neelim 2010).

Because of global warming the ice in the Himalayas is melting. Thus, by the end of this century, sea levels are likely to rise up to 1.5 m along the coastline of Bangladesh, which will be accompanied by more seasonal fluctuations in sea levels. Today, catastrophic hurricanes and exceptionally high tides occur once a decade, but by 2100 it could become as normal as three to fifteen times a year (Park 2019). Almost 80% of the area of Bangladesh consists primarily of floodplains, which is prone to flooding during the rainy season. In addition, the adverse effects of climate change will be rising temperatures, sea level rise, salinity intrusion, cyclones, storm surges, drought, heavy monsoon downpours, etc. (Islam and Neelim 2010; Denissen 2012).

1.12.1 Impact of Climate Change on Forest and Vegetation

The Sundarbans forest, the home of many endemic species, including the Royal Bengal Tiger (*Panthera tigris*), will be severely affected under climate change. The Sundarbans largely depends on freshwater supply along with the Ganges system. But climate change creates low-flow conditions and does not supply adequate freshwater (Abedin et al. 2014, 2019; Alam et al. 2010; Ahmed et al. 2019). The problem is likely to be compounded by the withdrawal of surface flows in the upstream areas to offset increasing moisture stress. Like this scenario, the salinity intrusion rate is increasing, which massively hampers the succession process of forest vegetation and food production. It is inferred that poor-quality shrubs will dominate, with increasing salinity, and high-value timber species will gradually disappear (Ahmed et al. 1998b).

1.12.2 Impact of Climate Change on Human Health

There exists limited information toward understanding the impacts of climate change on human health. Climate variability is strongly linked with the activity of pathogens (World Bank 1999; Koelle et al. 2005). Rodó et al. (2002) found increases in cholera cases in Bangladesh, with increases in the intensity of El Niño events. Incidences of pathogen-induced diseases would likely be increased under climate change. Malaria and dengue fever are common in Bangladesh. Increases in surface temperature would encourage parasites such as mosquitoes. One may therefore infer that deadly diseases such as malaria, dengue, etc., would put human health at higher risk under climate change (Agrawala et al. 2003). The Bangladesh National Adaptation Program of Action (NAPA), however, considered that the implications of climate change on human health was rather uncertain, although it echoed Agrawala et al. (2003), commenting that the health risk to the poor would be disproportionate given the government's spending on the health sector.

1.12.3 Impact of Climate Change on Small and Megacities in Bangladesh

There were four cyclones in the Bay of Bengal in 2016 – Roanu, Kyant, Nada, and Vardah, whereas normally there is only one per year. Riverbank erosion is the number one cause of inland climate displacement. Up to 50% of those who are now living in urban slums in Bangladesh may be there because they were pressured to escape their rural homes owing to riverbank erosion (Rahman 2017; Ahmad 2019; Christopher and Siverd 2020). Particular effects can occur due to many events increasing flooding, water logging, drainage congestions, etc. These natural events may affect the urban infrastructure, industry, trade, commerce, and utility services, etc. As a result, normal productivity may be disrupted during a major flood (Denissen 2012).

1.12.4 Impact of Climate Change on Coastal Areas in Bangladesh

Nearly one-fourth of the country's total population lives in Bangladesh's coastal areas, whereas the majority of the population is somehow directly or indirectly affected by coastal floods or tidal surges, riverbank flooding, salinity, cyclones, etc. In fact, with the rise of the sea level to a mere 1 m, Bangladesh could lose up to 15% of its land area under the sea and about 30 million people living in Bangladesh's coastal area will become refugees because of global climate change (Abedin et al. 2014, 2019; Alam et al. 2010; Ahmed et al. 2019). However, most of the country is less than 10 m above sea level and approximately 10% of the country's population lives below 1-m elevation; thus, the entire coastal area is extremely vulnerable to high tides and storm surges (Sarker et al. 2020; Hoque et al. 2020. In addition, the Bay of Bengal is located at the tip of the northern Indian Ocean, where extreme cyclonic storms and long tidal waves are frequently produced and have extreme impacts due to the shallow and conical shape of the Bay near Bangladesh.

1.12.5 Impact of Climate Change on Population Migration and Displacement

As rainfall patterns change, Bangladesh's drier northwestern regions are at risk of drought, which drives people away through destruction of crops and disruption of livelihood. Currently, there is no major factor in this displacement, but this risk is expected to increase as climate change progresses (Abedin et al. 2014, 2019; Alam et al. 2010; Ahmed et al. 2019). Landslides, which are also caused by increasingly erratic rainfall, affect Bangladesh's hilly northeastern and southeastern regions and can cause displacement by damaging homes and property and disrupting farming.

1.12.6 Impact of Climate Change on Water Resources and Hydrology

Generally, the country is highly prone to natural hazards, a few of which take on disastrous proportions. In most cases, however, disasters manifest as hydrological events caused by climatic extremes (Khan 2000). The country's geographical location and its high dependence on the overall GBM regional hydrology, spatial and temporal distribution of water resources contribute to the high degree of susceptibility of Bangladesh to water-related extreme events (Ahmed et al. 1998a). In the same way, it is necessary to create a relationship between the climatic regime and the associated risks in the form of water-related disasters (Islam and Neelim 2010). The effect of climate change on the surface and ground water sources may be very extreme and alarming, as Bangladesh is an overpopulated country. Changes in water resources and hydrology will have a significant impact on the economy of the country, where people are mostly dependent on the surface water for irrigation, fishing, industrial production, navigation, and other similar activities.

1.12.7 Impact of Climate Change on Agriculture, Aquaculture, and Fisheries

Bangladesh's economy is primarily based on agricultural activities, as two-thirds of the population are directly or indirectly engaged in agricultural activities. Besides extreme temperatures and drought, salinity intrusion in the coastal area may create serious implications for the coastal land that has historically been used to grow rice (Abedin et al. 2014, 2019; Alam et al. 2010; Ahmed et al. 2019). The fisheries sector contributes around 3.5% of Bangladesh's GDP and people here also rely on fish products to meet most of their daily protein needs. If the climate changes drastically then it may create an adverse impact on the fisheries sector. Floods affect agriculture production considerably. Karim et al. (1996) reported that the 1988 flood caused a reduction of agricultural production by some 45%. In the case of the most devastating flood of recorded history, in 1998, the production potential of Aman of some 2 to 2.3 Mha could not be realized. The prolonged flood of 1998 did not allow the farmers to transplant seedlings at the appropriate time, rendering a loss of about 3.5 MMT (Million Metric Tons). With the possibility of an increasing frequency of such high-intensity floods, it may be argued that Aman production is likely to suffer heavy damage under climate change.

In recent years, aquaculture has shown great prospects, especially in areas that are now virtually flood free owing to the presence of embankments. A large proportion of pond/water tanks inside embankments are now utilized for fish culture (pisciculture). Fish culture, in most areas, is a year-round activity. If there is any breach in the embankment, the culture ponds may get washed off and fish released into open water (Sarker et al. 2020; Hoque et al. 2020). This may also happen during a high-intensity flood, when water from outside easily overflows the embankments. Although the fish thrive in open water, it inflicts losses to the fish-farmers. It is already known that climate change would increase the extent of monsoon flooding. Therefore, the potential threat to culture fisheries would also increase under climate change (GOB 2005).

As climate change will have a significant influence on water-related hazards and disasters, peoples' livelihoods will also be severely affected (RVCC 2003a, b; Ahmed and Schaerer 2004; Asaduzzaman et al. 2005). Not only production of food will be adversely affected but waterborne disease will also affect human health. Their settlements and property will also be destroyed and the overall quality of life will fall. Drawing from Asaduzzaman et al. (2005), the Bangladesh NAPA also commented that the anticipated adverse impacts of climate change on peoples' livelihoods will fall disproportionately on the poor (GOB 2005). Frequent natural events trigger displacement of rural people and create climate refugees in Bangladesh.

1.13 Discussion

Countries that are most industrialized and developed consume the world's resources. Biocapacity is limited to 1.7 global hectares per person, which should be equal to the world's biological footprints. Bangladeshi citizens consume 50% fewer resources, whereas Australian citizens consume approximately 5.5% more than the average. The scenario is a rather hazardous in the UK (Islam and Neelim 2010). If the world followed their lifestyle, three additional spheres would be required. GHGs, carbon dioxide emission, concentrations of methane and nitrous oxide, burning of fossil fuels, cement, and flaring are mostly responsible for climate change. This results in elevated air and sea temperatures, a rise in the sea level, and more intense cyclones threatening all livelihoods in Bangladesh. Likewise, floods, landslides, storm surges, and cyclones are common phenomena there Rahman 2017; Ahmad 2019; Christopher and Siverd 2020). In particular, major cities such as

Dhaka, Chittagong, and Sylhet suffer from floods and landslides that cause traffic jams and several diseases. Even unplanned urbanization, degrading the natural vegetation and water bodies, is mostly responsible for this problem, which is caused by the excessive immigration and population pressure in Bangladesh. The world has ranked Dhaka city, which is the capital of Bangladesh, as one of the worst livable cities because of all these problems (Sarker et al. 2020; Hoque et al. 2020). The climate change in Bangladesh will make the hazards coming to us even worse. The British and Pakistani governments have left us with nothing but poverty for 315 years. Over the last few decades, Bangladesh achieved a huge change in public health, female empowerment, and education. That is how our country has progressed over the last few decades.

Climate change is something the whole world, as well as Bangladesh, needs to fight against. Bangladesh, together with the French government came up with the "Paris Climate Change Agreement." Bangladesh has also come into negotiations with Bangkok to combat climate change (Rahman 2017; Ahmad 2019; Christopher and Siverd 2020). It is stated that Bangladesh needs almost US$3 billion to achieve climate adaptation and US$2 billion dollars to mitigate against the effects of climate change. Over the last few years, the Bangladesh government has taken necessary steps against climate change by creating some expertise workshops against climate change filled with finance mechanisms and instruments (Abedin and Shaw 2013; Habiba and Shaw 2013). The Bangladesh government has also already taken some necessary steps for all refugees. It is said that before 2050, a total of 16 to 26 million refugees who will be suffered because of floods and all hazards and the affected people will be semi-migrated and some of them will be completely migrated (Rahman 2017).

Disasters vary in size, from smaller events such as a storm affecting a single suburban area, to large-scale events that can affect whole rural areas or large regions that cross state boundaries. As a result of various disasters, people may be killed or injured, or may lose their homes, property, and valuable possessions. They suffer various problems because of this. It is necessary to protect your family, home, business, and assets from such events. Insurers provide society with the means to do this. Taking steps to prepare for a disaster can help to get you back to normal faster (Mehedi et al. 2010; Asgary and Halim 2011; Jakariya et al. 2016; Abu Abdullah 2016; Ahmad 2019; Christopher and Siverd 2020). If you are unprepared, the devastation and financial loss caused by natural disasters can be magnified, hampering development of a country and creating an economic crisis.

The great step is to identify the different types of disasters we are exposed to, the likelihood of these occurring, and their potential impact (Abedin et al. 2014, 2019; Alam et al. 2010; Ahmed et al. 2019). Learning the most common disasters in our area, particularly if we are new to the region, can help our focus on preparation plans for each disaster. There are knowledge gaps, right from downscaling of the climate model to the considerations of glacial melting or blending of meteorological science. Capacity building in these areas will be of prime importance (Ahmad 2019; Christopher and Siverd 2020). Local councils and emergency management authorities will identify risks in our area and outline the local plans and recommendations for each, and for those people of that particular area to help them. To free our assets from risk, we should review our insurance policies and confirm our cover against various events, right after we have identified the disasters we are exposed to. If someone does not have insurance, they should consider the types of policies they may require to cover the risk.

If someone lives in a flood or bushfire zone, it would be judicious to insure their assets accordingly. For example, if someone's home is next to a river or creek and their property is at risk of flooding, they should check their policy to see if those are covered. If you are unsure, contact your insurer to find out (Ahmad 2019; Christopher and Siverd 2020). Many insurers will prohibit insurance when natural disasters are considered forthcoming. This means if a cyclone is heading your way you may not be able to purchase insurance cover if the insurer takes up restrictions on new policies. The insurance companies did that because some property owners buy insurance just before a disaster and then cancel it once the risk has passed. Therefore, if someone lives in a coastal area and their property could be harmed because of a natural disaster they should buy insurance ahead of time, and not just before the disaster (Abedin et al. 2014, 2019; Alam et al. 2010; Ahmed et al. 2019).

There is an evident need for more effective and accurate weather forecasting processes and mechanisms in Bangladesh. The Bangladeshi authorities have already been recognized this. The research community and the other international institutes approve the need for more timely forecasting. With regard to forecasting, there are a number of issues that still need to be addressed (Sarker et al. 2020; Hoque et al. 2020). For agricultural sectors, accurate seasonal weather forecasting is not available. This should be available to the researcher and the farmers to help them to prepare for natural disasters such as droughts and floods. For floods, the lead time for forecasting is only 72 h which is not sufficient for planners to make a move or take any decisions (Leontief 1976; Choudhury et al. 2004; Lloyd-Jones 2006; Khan 2015). There are also no reports available on the Internet regarding this to improve the research. Long-term forecasting may help to mitigate loss and damages.

1.14 Conclusion

Climate change is now affecting in every sector in Bangladesh. It causes obstruction to national economic progress and distress in people's lives, poor people, rural communities, and countries today and even more in the future. The result of climate change is found everywhere, such as changing weather patterns, rising sea levels, weather events are becoming more extreme, and GHG emissions are now at their highest levels in history. Without a proper action plan, the world's average surface temperature is likely to surpass 2°C this century. The poor countries and their people will be the most vulnerable. Affordable, scalable solutions are now available to enable countries to leapfrog to cleaner, more resilient economies. The pace of change is quickening as more people are turning to renewable energy and a range of other measures that will reduce emissions and increase adaptation efforts. Global change is a world-wide problem that is not limited within a country's borders. It is an issue that requires solutions that need to be coordinated at the international level to help developing countries to move toward a low-carbon economy. As decision-makers respond to these risks, the nation's scientific enterprise can contribute through research that improves understanding of the causes and consequences of climate change and that is also useful to decision-makers at the local, regional, national, and international levels. Climate change exists in every country because of human activities, poses risks, and has already had an effect on human and natural systems on a large scale. The compelling case for these conclusions is provided in many studies, part of a congressionally requested suite of studies. Although noting that there is always more to learn and that the scientific process never ends, much research shows that hypotheses on climate change are supported by multiple lines of evidence and have stood firm in the face of serious debate and careful evaluation of alternative explanations.

References

Abdullah ANM, Zander KK, Myers B, Stacy N, Garnett ST (2016) A short-term decrease in household income inequality in the Sundarbans, Bangladesh, following cyclone Aila. Nat Hazards 83(2):1103–1123

Abedin MA, Shaw R (2013) Safe water adaptability for salinity, arsenic and drought risks in southwest of Bangladesh. Risk Hazards Crisis Public Policy 4(2)):62–82

Abedin MA, Habiba, U., Shaw, R (2014) Community perception and adaptation to safe drinking water scarcity: salinity, arsenic, and drought risks in coastal Bangladesh. Int J Disaster Risk Sci 5(2):110–124

Abedin MA, Collins, A. E, Habiba U, Shaw R (2019) Climate change, water scarcity, and health adaptation in Southwestern coastal Bangladesh. Int J Disaster Risk Sci 10(1)):28–42

Abu Abdullah KZ (2016) A short-term decrease in household income inequality in the Sundarbans, Bangladesh, following Cyclone Aila. Nat Haz 83(2):1103–1123

Adedeji O et al (2014) Global climate change. J Geosci Environ Protect 2(02):114–122

Agrawala S, Ota T, Ahmed AU, Smith J, Aalst MV (2003) Development and climate change in Bangladesh: focus on coastal flooding and the Sundarbans 2003. Organisation for Economic Co-operation and Development (OECD), Paris

Ahmed I (2019) The National Plan for disaster Management of Bangladesh: gap between production and promulgation. Int J Disaster Risk Reduction 37

Ahmed AU, Haque N (2002) Climate change and sustainable development. Paper presented at the Dhaka meet on sustainable development, March 14–18, 2002. Incorporated in Q.K. Ahmad and A.U. Ahmed (eds.), Citizens' Perspectives on Sustainable Development, Bangladesh Unnayan Parishad (BUP)

Ahmed AU, Schaerer C (2004) Sustaining livelihoods opportunities for the coastal poor under climate change: a case study from Bangladesh. In: Anonymous (ed) Proceedings of coastal zone Asia Pacific, Brisbane, 7-9 September 2004

Ahmed AU, Alam M, Rahman AA (1998a) Adaptation to climate change in Bangladesh: future outlook. In: Huq S, Karim Z, Asaduzzaman M, Mahtab F (eds) Vulnerability and adaptation to climate change for Bangladesh. Kluwer Academic Publishers, Dordrecht, pp 125–143

Ahmed AU, Siddiqi NA, Choudhuri RA (1998b) Vulnerability of Forest ecosystems of Bangladesh to climate change. In: Huq S, Karim Z, Asaduzzaman M, Mahtab FU (eds) Vulnerability and adaptation to climate change for Bangladesh. Kluwer Academic Publishers, Dordrecht, pp 93–111

Ahmed M, Kabir A. N. et al (2016) A comprehensive assessment of arsenic in commonly consumed foodstuffs to evaluate the potential health risk in Bangladesh. Sci Total Environ 544:125–133

Alam, E Collins, A. E. (2010). Cyclone disaster vulnerability and response experiences in coastal Bangladesh. Disasters 34(4), 931–9954

Alam E et al (2010) Cyclone disaster vulnerability and response experiences in coastal Bangladesh. Disasters 34(4):931–954

Ali A (1999) Climate change impacts and adaptation assessment in Bangladesh. Clim Res 12:109–116

Ali KMB, Sarwar MI, Molla MS (2011) Quality of urban environment: a case of citizens perception in Satkhira Municipality, Bangladesh. Chittagong University J Biol Sci 6(1 & 2):65–76

Asaduzzaman M, Ahmed AU, Haq E, Chowdhury SMZI (2005) Climate change and Bangladesh: livelihoods issues for adaptation. Bangladesh Institute for Development Studies (BIDS), Dhaka

Asgary A, Halim M (2011) Measuring people's preferences for cyclone vulnerability reduction measures in Bangladesh. Disaster Prev Manag 20:186–198. https://doi.org/10.1108/09653561111126111

Bagchi AH (2007) Biodiversity and ecosystem multifunctionality. Nature 448:188–190

Barney GO (1980) The global 2000 report to the president of the U.S. entering the 21st century: the technical report. US Government Printing Office, Washington, DC

Barney GO (1993) Global 2000 revisited: what shall we do? A report on the critical issues of the 21st century prepared for the 1993 Parliament of world's religions. Millennium Institute, Arlington

BER (2019) Disaster risk Reduction and Management in Indonesia. BER staff

Braganza K, Karoly D, Hirst AC, Mann M, Stott P, Ronald S, Tett S (2003) Simple indices of global climate variability and change: part I - variability and correlation structure. Clim Dyn 20:491–502. https://doi.org/10.1007/s00382-002-0286-0

Burnham M, Ma Z (2015) Linking smallholder farmer climate change adaptation decisions to development. Clim Dev:1756–5537

Burrows BC et al (1991) Into the 21st century: a handbook for a sustainable future. Adamantine, Twickenham

Burtis B (2006). Cross border indicators of climate change over the past century. Northeastern United States and Canadian Maritime Region

Catherine P, McMullen JR (2009) Climate change science compendium 2009. United Nations Environment Programme, Nairobi, Kenya/London, UK

CEGIS (2000) Centre for environment and geographic information services. Retrieved from cegisbd.com: https://www.cegisbd.com/

Choudhury DA (2002) Managing natural disasters in Bangladesh. In: The Dhaka meet on sustainable development. Bangladesh Unnayan Parishad, pp 16–18

Choudhury NY, Paul A, Paul BK (2004) Impact of costal embankment on the flash flood in Bangladesh: a case study. Appl Geogr 24(3):241–258

Chowdhury ZH (2019) The conversation. Retrieved 5 Apr 2020, from theconversation.com: https://theconversation.com/climate-change-impacts-in-bangladesh-show-how-geography-wealth-and-culture-affect-vulnerability-128207

Correa M (2015) Correa et al., 2012 SFN abstract. UK

Costello A et al (2009) Managing the health effects of climate change. Lancet 373:1693–1733

Das KM (1997) Disaster management with particular reference to Bangladesh, DMB. Dhaka, Bangladesh

Denissen A-K (2012) Climate change & its impacts on Bangladesh, 3 April. Retrieved from NCDO.: http://www.ncdo.nl/artikel/climate-change-its-impacts-bangladesh

Dewan H (2014) Societal impacts and vulnerability to floods in Bangladesh and Nepal. Weather Clim Extrem 7:36–42

Disaster Management Bureau (2008) National disaster management policy. Ministry of Food and Disaster Management, The Government of the People's Republic of Bangladesh

Djalante R et al (2012) Building resilience to natural hazards in Indonesia: progress and challenges in implementing the Hyogo framework for action. Nat Hazards 62:779–803

Eckstein D, Hutfils M-L, Winges M (2018) Global climate risk index 2019, December. Retrieved from Germanwatch: https://germanwatch.org/sites/germanwatch.org/files/Global%20Climate%20Risk%20Index%202019_2.pdf

Environmental Justice Foundation (2020) Retrieved 2020, from Ejfoundation.org. https://ejfoundation.org/reports/climate-displacement-in-bangladesh

Habiba U, Shaw R (2010) Drought scenario in Bangladesh. Environ Disaster Risk Manage 13:213–245

Habiba U, Takeuchi Y, Shaw R (2010) Chapter 3 overview of drought risk reduction approaches in Bangladesh. Comm Environ Disaster Risk Manag 5:37–58. https://doi.org/10.1108/S2040-7262(2010)0000005009

Habiba U, Shaw R (2013) Farmers' response to drought in Northwestern Bangladesh. Risks and conflicts: local responses to natural disasters, pp 131–161

Habiba U, Shaw R, Takeuchi Y (2012) Farmer's perception and adaptation practices to cope with drought: perspectives from Northwestern Bangladesh. Int J Disast Risk Red 1:72–84

Habiba U, Shaw R, Hassan AWR (2013) Drought risk and reduction approaches in Bangladesh. Disaster Risk Reduction Approaches in Bangladesh, pp:131–164

Habiba, U., Shaw, R., Takeuchi, Y. (2014). Farmers' adaptive practices for drought risk reduction in the northwest region of Bangladesh. Nat Hazards 72, 2, 337–359

Haque E, Uddin MS (2013) Disaster management discourse in Bangladesh: a shift from post-event response to the preparedness and mitigation approach through institutional partnerships. IntechOpen Limited, Princes Gate Court, London, UK. https://doi.org/10.5772/54973

Haque A, Jahan S (2016) Regional impact of cyclone Sidr in Bangladesh: a multi-sector analysis. Int J Disast Risk Sci 7:312–327

Hasnain SI (2000) Status of the glacier research in the HKH region. ICIMOD, Katmandu. 140 pp

Herrera, A. O. et al (1976) Catastrophe or new society? A Latin American world model. IDRC, Ottawa

Hoque MA-A, Pradhan, B., Ahmed N (2020) Assessing drought vulnerability using geospatial techniques in northwestern part of Bangladesh. Sci Total Environ 705:135957

Hossain AA (2003) Integrated flood management case study Bangladesh: flood management. World Meteorological Organization, Geneva

Hurricane Maria – CNN (2017) Retrieved from CNN. https://edition.cnn.com/specials/weather/hurricane-maria

Intergovernmental Panel on Climate Change-WG1 (2007) Detection and attribution of climate change: from global to regional. Intergovernmental Panel on Climate Change, Geneva

IPCC (1992) Climate change 1992: the supplementary report to the IPCC scientific assessment. Intergovernmental Panel on Climate Change, Rio de Janeiro

IPCC (2007) Intergovernmental Panel on Climate Change, Working Group II: impacts, adaptation and vulnerability, fourth assessment report. Intergovernmental Panel on Climate Change, Geneva

IPCC (2013) Impacts of 1.5°C global warming on natural and human systems. Intergovernmental Panel on Climate Change, Geneva

IPCC (2019) In: Shukla PR, Skea J, Buendia EC, Masson-Delmotte V, Pörtner H-O, Roberts DC, Zhai P, Slade R, Connors S, van Diemen R, Ferrat M, Haughey E, Luz S, Neogi S, Pathak M, Petzold J, Pereira JP, Vyas P, Huntley E, Kissick K, Belkacemi M, Malley J (eds) Climate change and land: an IPCC special report on climate change, desertification, land degradation, sustainable land management, food security, and greenhouse gas fluxes in terrestrial ecosystems. In press

Islam A, Islam M (2017) Climate change and its impact on sustainable development in Bangladesh. Int J Econ Ener Environ 2:90–95. https://doi.org/10.11648/j.ijeee.20170206.11

Islam M, Mehedi H (2016) Climate-induced human displacement: a case study of cyclone Aila in the south-west coastal region of Bangladesh. Nat Hazards:81. https://doi.org/10.1007/s11069-015-2119-6
Islam T, Neelim A (2010) Climate change in Bangladesh: a closer look into temperature and rainfall data. The University Press, Dhaka
IUCN (2010) The nature of progress: annual report 2010. Gland, Switzerland: International Union for Conservation of Nature (IUCN), Dhaka, Bangladesh
Jakariya, M. et al (2016) Vulnerability analysis of the cyclone Aila affected community in the South-West Belt of Bangladesh. J Health Environ Res 2(2):5–12
Kabir R, Khan HTA, Ball E, Caldwell K (2016) Climate change impact: the experience of the coastal areas of Bangladesh affected by cyclones Sidr and Aila. J Environ Pub Health 2016:9654753. https://doi.org/10.1155/2016/9654753
Kafiluddin AKM (2005) Human health, disease, and the environment. 57/1, Lake Circus, Kalabagan, Dhaka, 1205. Bangladesh. Accessed on July 23, 2011
Kaplan, R. D. (1994). The coming anarchy. The Atlantic Monthly
Karim L (2004) Democratizing Bangladesh: state, NGOs, and militant Islam. Cult Dyn 16(2–3):291–318. https://doi.org/10.1177/0921374004047752
Karim Z et al (1996) Agricultural vulnerability and poverty alleviation in Bangladesh. In: Climate change and world food security, NATO ASI series, vol 137. Springer, Berlin, pp 307–346
Ahmed, B., Kelman, I., Kamruzzaman, M Mohiuddin, H., Rahman M. M., Das, A., Fordham, M., Shamsudduha, M. (2019) Indigenous people's responses to drought in Northwest Bangladesh. Environ Develop 29, 55–66
Khan MNU (2000) Disaster Management System in Bangladesh. Bangladesh Army Journal, Dhaka
Khan MS (2008) Disaster preparedness for sustainable development in Bangladesh. Disaster Prev Manag 17(5):662–671
Khan MH, Bryceson I, Faruque F et al (2015) Natural disasters and land-use/land-cover change in the southwest coastal areas of Bangladesh. Reg Environ Chang 15(2):241–250
Koelle K et al (2005) Pathogen adaptation to seasonal forcing and climate change. Proc R Soc B Biol Sci 272(1566):971–977
Kulatunga U, Wedawatta G, Amaratunga D et al (2012) Disaster risk reduction measures in Bangladesh. Global challenges in construction industry: world construction symposium. University of Huddersfield, Colombo
Lambin EF, Meyfroidt, P. (2011) Global land use change, economic globalization, and the looming land scarcity. Proc Natl Acad Sci 108:3465–3472
Leontief WW (1976) The future of the world economy: a study on the impact of prospective economic issues and policies on the international development strategy. United Nations, New York
Lloyd-Jones, T (2006) Mind the gap! Post-disaster reconstruction and the transition from humanitarian relief. RICS, London
Mahmood S (2012a) Impact of climate change in Bangladesh: the role of public administration and government's integrity. J Ecol Nat Environ:223–240
Mahmood SAI (2012b) Impact of climate change in Bangladesh: the role of public administration and government's integrity. J Ecol Nat Environ 4(8):223–240
Mallick BJ et al (2009) Local adaptation strategies of a coastal community during cyclone Sidr and their vulnerability analysis for sustainable disaster mitigation planning in Bangladesh. J Bangladesh Instit Planners 2:158–168
Mallick BJ et al (2017) Living with the risks of cyclone disasters in the south-western coastal region of Bangladesh. Environments 4(1):1–17
McCarthy JJ et al (2001) Climate change 2001: impacts, adaptation, and vulnerability. Contribution of Working Group II to the Third Assessment Report of the Intergovernmental Panel on Climate Change (IPCC). Glob Ecol Biogeogr 12:87–88
Mclean SN, Moore DR (2005) A mitigation strategy for the natural disaster of poverty in Bangladesh. Disaster Prevent Manage 14(2):223–232
Meadows DH, Meadows DL, Randers J, Behrens WW (1972) The limits to growth. Potomac Associates, New York

Meadows D et al (1982) Groping in the dark: the first decade of global modelling. Wiley, New York

Meadows D et al (1992a) Limits to growth. Universe Books, New York

Meadows DH et al (1992b) Beyond the limits: global collapse or a sustainable future. Earthscan Publications, London

Mehedi, H., et al. (2010) Climate induced displacement: case study of cyclone Aila in the Southwest coastal region of Bangladesh. Khulna

Mehedi H, Mukta Z, Roy K, Nag A, Farhana S (2010) Climate induced displacement: case study of cyclone Aila in the southwest coastal region of Bangladesh. https://doi.org/10.13140/RG.2.1.3490.9921

Mehta M (2007) Gender matters: lessons for disaster risk reduction in South Asia. ICIMOD, Kathmandu. http://www.preventionweb.net/files/2406_GenderandDisasters.pdf

Mesarovic M, Pestel E, Mihram G (1977) Mankind at the turning point. In: Systems, Man and Cybernetics, IEEE Transactions on, vol 7, pp 73–74. https://doi.org/10.1109/TSMC.1977.4309596

Milbrath LW (1989) SUNY series in environmental public policy. In: Envisioning a sustainable society: learning our way out. Press, State University of New York

Mirza MMQ, Paul S (1992) Pakritik Durjog O Bangladesher Paribesh (in Bengali). Centre for Environmental Studies and Research (CESR), Dhaka

NAPA (2009) Bangladesh: national adaptation programme of action. Ministry of Environment and Forests, Dhaka, Bangladesh

Park J (2015) Climate change and capitalism. Consilience 14:189–206. Retrieved May 12, 2021, from http://www.jstor.org/stable/26188749

Park W (2019) Rising sea levels are making traditional ways of life impossible. Rural Bangladeshis are having to adapt to survive. The country disappearing under rising tides, 2 September. Retrieved from https://www.bbc.com/future/article/20190829-bangladesh-the-country-disappearing-under-rising-tides

Parmesan C, Yohe G (2003) A globally coherent fingerprint of climate change impacts across natural systems. Nature 421:37–42

PDO-ICZMP (2004) Living in the coast: problems, opportunities and challenges. Working Paper WP011, Dhaka, pp 13–15

Pirani FA (1974) Thinking about the future: a critique of the limits to growth. Sussex University Press, London

Potgieter AB, Hammer GL, Meinke H, Stone RC, Goddard L (2005) Three putative types of el niño revealed by spatial variability in impact on Australian wheat yield. J Clim 18:1566–1574

PRSP (2005) Bangladesh: poverty reduction strategy paper. International Monetary Fund, Washington, DC

Rahman A-U (2015) Introduction and disaster risk reduction approaches in Pakistan. Springer, Japan, Tokyo. https://doi.org/10.1007/978-4-431-55369-4

Rahman Khan M (2017) Climate change governance: Bangladesh perspective. Int J Horticult Agricult 2(1):1–8. https://doi.org/10.15226/2572-3154/2/1/00110

Rodó, X. et al. (2002) ENSO and cholera: a nonstationary link related to climate change? National Academy of Sciences of the United States of America, pp 12901–12906

RVCC (2003a) The Reducing Vulnerability to Climate Change (RVCC) project. The RVCC Project is managed by CARE Canada and implemented by CARE, Dhaka

RVCC (2003b) Report of a community level vulnerability assessment conducted in Southwest Bangladesh. Reducing Vulnerability to Climate Change, Dhaka

Sala OE et al (2000) Biodiversity: global biodiversity scenarios for the year 2100. Science 287:1770–1774

Sarker M, Nichol J, Siti M, Baharin A, Shamsuddin S, Chung E-S, Reid J, Siswanto E (2020) An integrated method for identifying present status and risk of drought in Bangladesh. Remote Sens 12:17. 2686

Shamim A, Hasan ABM, Rahman M et al (2013) Change in temperature over Bangladesh associated with degrees of global warming. Asian J Appl Sci Eng 2:62–75

Shimi AC et al (2010) Impact and adaptation to flood: a focus on water supply, sanitation and health problems of rural community in Bangladesh. Disaster Prev Manag 19(3):298–313
Sintayehu DW (2018) Impact of climate change on biodiversity and associated key ecosystem services in Africa: a systematic review. Ecosyst Health Sustain 4:225–239
Siverd C, Hagen S, Bilskie M, Braud D, Twilley R (2020) Quantifying storm surge and risk reduction costs: a case study for Lafitte, Louisiana. Clim Chang 161:1–23. https://doi.org/10.1007/s10584-019-02636-x
Stern N (2006) The stern review: economics of climate change. Cambridge University Press, Cambridge
Subhani R, Ahmad MM (2019) Socio-economic impacts of cyclone Aila on migrant and non-migrant households in the Southwestern coastal areas of Bangladesh. Geosciences 9(11):482
UNFCCC (2016) Aggregate effect of the intended nationally determined. United Nations Framework Convention on Climate Change, Rio de Janeiro, Brazil
UN/ESCAP (2006) Enhancing regional cooperation in infrastructure development including that related to disaster management. United Nations Economic and Social Commission for Asia and the Pacific, Bangkok
World Bank (1999) World development report 1999/2000: entering the 21st century. Oxford University Press, New York
WWAP (2006) UN world water development report. World Water Assessment Programme, Paris
WWF (2005) WWF/World Bank global Forest Alliance - annual report 2005. World Wildlife Fund (WWF)
Zhong, L., Wang, J (2017) Evaluation on effect of land consolidation on habitat quality based on InVESTModel. Trans Chin Soc Agric Eng 33:250–255
Zimm C et al (2018) Identifying sustainability and knowledge gaps in socio-economic pathways vis-à-vis the sustainable development goals. Economies 6(2):20

Chapter 2
Climate Change Impacts and Mitigation Approach: Coastal Landscape, Transport, and Health Aspects

Md. Mozahidul Islam, Md. Shahin, Md. Miraj, Subarna Ghosh, Md. Nazrul Islam, and Israt Islam

Abstract Planet earth is warming over time and a warmer planet favorable for its winners for sure rather than suffering countries. Bangladesh, a lesser climate change contributor, is the worst victim country of the world, facing significant natural calamities almost every year, which has consequences from the country's coastal margin to parliament. Insights into how climate change impacts coastal areas landscape, transport network, and human health, which is crucial for sustainable development, are identified in this study. The coastal landscape's southwestern 170 km length and 20 km width, the mid-central zone's existing road network, and the overall impact on human health due to climate change are identified by a semi-structured questionnaire survey with professional clusters, satellite image processing, and secondary data. Study finds, islands, and exposed sub-districts such as "Patharghata' are facing high currents, a cause of rapid erosion–accretion, road networks are severally affected by cyclones, storm surge, erratic rainfall, extreme heat, and flood where salinity and extremely hot weather have adverse consequences for human health. Additionally, not only climate change issues but also local implications are responsible for these changes. Several structural and bioengineering approaches are suitable for reducing land erosion and road network damage where awareness is

M. M. Islam (✉) · I. Islam
Department of Environmental Science and Management (DESM), SELS, Independent University, Bangladesh, Dhaka, Bangladesh
e-mail: mozahid@iub.edu.bd

M. Shahin · M. Miraj
Department of Disaster Resilience and Engineering, Patuakhali Science and Technology University, Patuakhali, Bangladesh

S. Ghosh
Department of Public Health Nutrition (PHN), Primeasia University, Dhaka, Bangladesh

M. N. Islam
Department of Geography and Environment, Jahangirnagar University, Savar, Dhaka, Bangladesh

M. N. Islam, A. van Amstel, *Bangladesh II: Climate Change Impacts, Mitigation and Adaptation in Developing Countries*, Springer Climate,
https://doi.org/10.1007/978-3-030-71950-0_2

building suitable for appeasing human health risks. Finally, this study focuses on real field conditions with community perceptions who are field survivors for all climate change issues, rather than computer-aided simulation models, and the outcome would be helpful for academicians, researchers, national and international stakeholders to develop climate-resilient, sustainable coastal cities in the near future.

Keywords Climate change · Coastal landscape · Public health · Sustainable development · Bangladesh

Abbreviations

ADB	Asian Development Bank
ANN	Artificial neural network
BBS	Bangladesh Bureau of Statistics
BPCZ	Barguna Patuakhali Coastal Zone
BST	Bangladesh Standard Time
CA	Cellular automata
DFID	Department for International Development, UK
GBM	Ganga-Brahmaputra Meghna
GDP	Gross Domestic Product
ICZM	Integrated Coastal Zone Management
ICZMP	Integrated Coastal Zone Management Plan
IFAD	International Fund for Agricultural Development
JICA	Japan International Cooperation Agency
LCA	Land Cover Analysis
LGED	Local Government Engineering Department
MLP	Multi-Layer Perception
NDMA	National Disaster Management Authority, India
OBIA	Object-based image analysis
RCC	Reinforced cement concrete
RHD	Roads & Highway Department, Bangladesh
UNISDR	United Nations International Strategy for Disaster Reduction
USA	United States of America
USGS	United States Geological Survey

Definitions

Term	Description
Coast	The coast, also known as the coastline or seashore, is the area where land meets the sea or ocean or a line that forms the boundary between the land and the ocean or a lake

	(Merriam-Webster, 2015). A precise line that can be called a coastline cannot be determined owing to the coastline paradox. The term coastal zone is a region where interaction of the sea and land processes occurs.
Coastal area	Coastal area is a notion that is geographically broader than the coastal zone, the borders of which require a less strict definition. This notion indicates that there is a national or sub-national recognition that a distinct transitional environment exists between the ocean and terrestrial domains.
Coastal zone	Coastal zone is most frequently defined as "land affected by its proximity to the sea and that part of the sea affected by its proximity to the land" or, in other words, the area where the processes that depend on the sea–land interaction are the most intensive. This interface is taking place along two axes: the axis running along the coast and the axis perpendicular to the coastline (ICZM).
Frontiers	A frontier is the political and geographical area near or beyond a boundary. The term came from French in the fifteenth century, with the meaning "borderland" – the region of a country that fronts onto another country. Frontier can also be referred to as a "front." A difference has also been established in academic scholarship between frontier and border, the latter denoting a fixed, rigid, and clear-cut form of state boundary.
Coastal Margin	Coastal margins are composed of watersheds, estuaries, and the ocean continental shelf. Some of the fascinating processes that affect them include plume dynamics, nutrient fluxes, and tides. In addition to land-dwelling humans, fish, waterfowl, and mammals, coastal margins are home to diverse and microbial communities. From the level of these microbes, through biological systems, to global biogeochemical processes and seeking to revolutionize our understanding of coastal margins. They are currently developing models that will predict everything from algal blooms to salmon runs and improve the health of our rivers and oceans.
Coastline	Coastline is the contact line dividing the land from the water bodies. It usually coincides with the line marking the landward extent of tidal influence.
Coastal uplands	Coastal uplands are defined as an area of the interior between the shorelands and most frequently, the highest peak of the closest mountain range. Sometimes, the depth of the belt is limited (for example, in the USA the limit is 5 miles).
Inland	Inland may be any area outside the coastal belts. However, it should not be considered as an altogether unimportant

	zone, as many processes affecting the state of the coastal zone originate in that area. That coastal waters, intertidal area, coastline, shore lands area, and coastal uplands are the elements of the coastal zone.
Pavement	Pavement refers to the flexible pavement, and rigid pavement.
Flexible Pavement	Flexible pavement is made up of aggregate, sand, and asphalt. Asphalt – a sticky black substance – is used as the binding material in asphalt pavements.
Transport	Transportation infrastructure such as roads, highways, and road embankments.
Preterm birth	Preterm is defined as babies born alive before 37 completed weeks of gestation, or fewer than 259 days since the first day of the woman's last menstrual period (World Health Organization 2012).
Food security	Food security exists when all people, at all times, have physical and economic access to sufficient, safe, and nutritious food that meets their dietary needs and food preferences for an active and healthy life" (Summit 1996).
Heat stress	Heat stress is defined as the perceived discomfort and physiological strain associated with exposure to a hot environment, especially during physical work (Byard and Payne-James 2015).
Vector-borne disease	Vector-borne diseases are diseases caused by viruses (e.g., yellow fever), bacteria (e.g., plague), protozoa (e.g., malaria), or nematodes (e.g., onchocerciasis) that are transmitted from one vertebrate host to the next by insects, such as mosquitoes, or arachnids, such as ticks (Kermack and McKendrick 1927).

2.1 Introduction

Climate change is an issue of concern, awareness, development, and debate in the twenty-first century. It comprises many unclear, biased, and bargaining issues related to contributing materials and equipment, the country's percentage of gas emission, outcome of computer simulation models, whether the measures taken are significant or not, and many other matters, but the crystal-clear part is "Impact", the suffering of mankind, specifically the low-lying sea-facing country's coastal community loss of property and endangered of flora and faunal diversity in the coastal areas. Beyond these, the planet earth is notoriously warming, which is alternatively indicating that this "peaceful' home is going to be "harmful' and ultimately destroying its habitats. Climate change is also an unavoidable natural phenomenon that is continuing for decades and human intervention is making these situations complex in a certain percentage. As nature maintains its most important part and mankind does not have control of it, so that only these certain portions are kept to a minimum is a workable measure that was realized only a couple of years ago.

Bangladesh, a sea-facing, low-lying, mainly riverine South Asian country of this world located across the region of the Tropic of Cancer. Physiographic features and geographic locations bring some advantage and curses as well over time, as the tropical monsoon climate brings a heavy seasonal rainfall, high temperatures, and high humidity, making this land fertile for cultivation and suitable for growing almost all crops in this area. Still, an excessive flux of these climatic parameters cause flood, waterlogging, drought, heat stress, pest attack, and related other hazards and disasters. Climate change impact differs per region, whereas the cyclone- and flood-prone southern part, the drought-prone northern part, the low-lying Haor depression in the northeast, several districts among the major rivers, particularly the Jamuna, and several southeastern districts including Chittagong Hill are the high risk zones, making the situation complex for taking measures. In this regard, zoning-based specific mitigation measures are required to address the issues fruitfully.

The coastal area of Bangladesh is very dynamic for its physical geography and practices of the local community, which are changing over time (Brammer 2014a). Abundant coastal resources such as shrimp, fisheries, and forest resources attracting communities who are directly or indirectly dependent on these resources for short- or long-term income sources and sometimes over-exploitation of these resources, threatening the coastal environment (Hoq 2003). Indirect stakeholders such as city dwellers are dependent on these coastal resources for protein sources, medicinal plants, and for other possessions (Mozumder et al. 2018); to continue this flow, a resilient coastal community together with sustainable building along the coast is very important. Moreover, most of the coastal districts connecting with the capital by road have waterways whereas a few districts have air and railway. So far, speedy, safe transportation network development and maintenance is not only a matter of GDP and communication, but is also required to reduce the loss of life, specifically during emergency periods for evacuation and distributing relief. Recently, there have been reports of climate change also impacting socio-economic patterns, including human health in Bangladesh, whereas 20% of miscarriages occurred because of an increase in the salinity level in the coastal areas of Bangladesh (BBC 2018).

By the way, as the coastal community is most vulnerable to climate change, a different sectoral collaborative approach is required for its sustainable development. In order to take proper initiative, the study combines three major segments: identifying and predicting the coastal landscape changes from 1988 to 2018 toward 2050 by using the Remote Sensing Geographic Information System (GIS) approach, identifying the climate change-induced impact on road network and mitigation measures, and identifying the impacts on human health of climate change over time.

2.1.1 Coastal Boundary, Study Extent, and Resources

The coastline of Bangladesh extends from the southwestern (local name Nildumur) to the south-eastern part toward Chattogram, a 710–km-long area (Ahsan 2013) that is divided into three major zones named the western, central, and eastern coastal zones (Ahmed et al. 2018). Among all 65 districts of Bangladesh, 19 are considered as coastal districts; these are severely affected by natural calamities almost every year. Among all the sub-districts of these coastal districts, 48 sub-districts are sea-facing and called the "first tier of coastal upazilas" (Table 2.1), whereas another 81 are considered interior sub-districts (Uddin and Kaudstaal 2003), secondarily affected by the high/low tide, waterlogging, cyclone hazards and disasters. To identify the real scenario, the southwestern area measuring 170 km long and 20 km wide (shaded area in Fig. 2.1) is considered for coastal landscape change analysis to fill the gap in the historical analysis, the central coast's Barguna and Patuakhali districts (road network in Fig. 2.1) are considered for road network analysis, and the remaining coastal part, along with major areas of urban development, the human health impacts due to climate change are taken into consideration for this study.

The western and central coasts are part of the Ganges–Brahmaputra–Meghna (GBM) delta system, which is rich in marine resources, whereas the east coast is nondeltaic (Shamsuzzaman et al. 2017). In general, each of the three zones has one dominant, or controlling, factor, which in the west is "mangrove forests' dominating the coastal fringes. The fate of 33.87 million (BBS and SID 2011) among a population of 164 million (i.e., 27% of the total population) who are directly or indirectly dependent on this area's coastal resources, which also contributes significantly to the national GDP (Admiral and Alam 2016). Proper protection and utilization of these resources can boost the fortune of its dependents, political, strategic, and social advantages over other states in reaping benefits from those resources, including shipping of goods, fishing, hydrocarbon, and mineral extraction, naval missions, and scientific research. This income-generating and lifesaving coastline of Bangladesh contains distinct coastal landforms and specific geological, physical, and biological characteristics (Mascarenhas 2011).

2.1.2 Aspects of the Coastal Area of Bangladesh

The coastal environment is a very complex and diverse ecosystem compared with the mainland (Kirui et al. 2013). Climate change and sea level rises are common phenomena in most coastal environments around the world and pose threats to the inhabitant (IPCC 2007). The relatively warm Indian ocean, tropical climate, and physical geography combined with climate change are conducive to disasters occurring almost every year along the coast of Bangladesh (Kim et al. 2012). Although some countries on this planet the characteristics almost same like Bangladesh, but the country's funnel shaped bay originating in the warm Indian ocean is more

Table 2.1 Sea-facing coastal districts, exposed, and interior sub-districts to the coast; 48 exposed sub-districts

Districts	Sub-districts (Upazilas)	
	Exposed	Interior
Bagerhat	Mongla, Saran Khola, Morrelganj	Bagerhat Sadar, Chitalmari, Fakirhat, Kachua, Mollahat, Rampal
Barguna	Amtali, Barguna Sadar, Patharghata, Bamna	Betagi
Barisal	Agailjhara, Babuganj, Bakerganj, Gaurnadi, Hizla, Mehendiganj, Muladi, Wazirpur, Banari Para, Barisal Sadar	
Bhola	Bhola Sadar, Burhanuddin, Char Fasson, Daulatkhan, Lalmohan, Manpura, Tazumuddin	
Chandpur	Chandpur Sadar, Faridganj, Haimchar, Hajiganj, Kachua, Matlab, Shahrasti	
Chittagong	Anowara, Banshkhali, Chittagong port, Double Mooring, Mirsharai, Pahartali, Panchlaish, Sandwip, Sitakunda, Patenga, Halisahar, Kotwali, Boijid Bostami,	Boalkhali, Chandanaish, Lohagara, Rangunia, Chandgaon, Fatikchhari, Hathazari, Patiya, Raozan, Satkania, Bakalia, Karnaphuli, Kulshi
Cox's Bazar	Chakaria, Cox's Bazar Sadar, Kutubdia, Ukhia, Maheshkhali, Ramu, Teknaf	
Feni	Sonagazi	Chhagalnaiya, Feni Sadar, Parshuram,Daganbhuiyan
Gopalganj	Gopalganj Sadar, Kashiani, Kotali Para, Muksudpur, Tungipara	
Jessore	Bagher Para, Chaugachha, Jhikargacha, Manirampur, Abhaynagar, Keshabpur, Jessore Sadar, Sharsha	
Jhalokati	Jhalokati Sadar, Kanthalia, Nalchity, Rajapur	
Khulna	Dacope, Koyra	Batiaghata, Daulatpur, Dumuria, Dighalia, Khalishpur, Khan Jahan Ali, Khulna Sadar, Paikgachha, Phultala, Rupsha, Sonadanga, Terokhada
Lakshmipur	Ramgati	Lakshmipur Sadar, Raipur, Ramganj
Narail	Lohagara, Narail Sadar, Kalia, Narigati	
Noakhali	Companiganj, Hatiya, Noakhali Sadar	Chatkhil, Senbagh, Begumganj
Patuakhali	Dashmina, Rangabali, Galachipa, Kala Para	Bauphal, Mirzaganj, Patuakhali Sadar

(continued)

Table 2.1 (continued)

Districts	Sub-districts (Upazilas)	
	Exposed	Interior
Pirojpur	Mathbaria	Bhandaria, Kawkhali, Nazirpur, Pirojpur Sadar, Nesarabad (Swraupkati)
Satkhira	Assasuni, Shyamnagar	Debhata, Kalaroa, Kaliganj, Satkhira Sadar, Tala
Shariatpur	Bhederganj, Damudya, Goshairhat, Naria, Palong, Zanjira	

Source: ICZMP

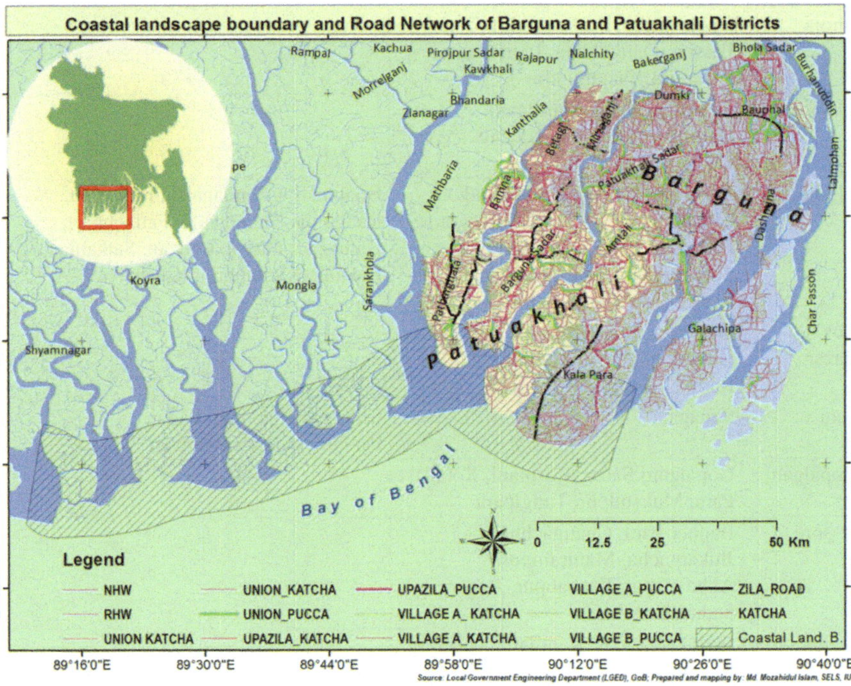

Fig. 2.1 Study area: 170 km long with 7 km lower and 13 km upper from the coastal margin of 2018 (shaded area) along with the road network of the Barguna and Patuakhali sub-districts

unique, which is very favorable for cyclone formation[1] and in recent years this region has been affected by high-intensity and -magnitude cyclones, floods, storm surges, along with other natural calamities (Ralph and Gough 2009). Another

[1] The Indian Ocean, which is relatively warmer is playing a vital role in cyclone formation in this region as temperatures of 26 °C or above required for mass evaporation from sea water created an abrupt flow of air as a preliminary stage/cause of cyclone eye generation. Several eye accumulations make a successful cyclone for hitting land. Global temperature rise is intensifying more cyclone eye generation at one time on a sea surface.

significant problem is that the coastal areas of Bangladesh facing extreme salinity intrusion and strong back water effects from the Bay of Bengal, as the water flow significantly reduces (Habiba et al. 2013) owing to illegal upstream water being withdrawn by India (Mirza 2002; Kothari and Poongavanam 2019). Combining all aspects, this region makes a unique hotspot for climate change and sea level rise. Therefore, the inhabitants in this region are continuously coping with different disasters and altering their livelihood patterns; some of these are using indigenous tools and techniques to survive with life-threatening nature (Islam et al. 2016).

2.1.3 Is the Coastal Area/Margin Changing?

Humans have traditionally altered the natural character of the Bangladeshi coastal zone through their exploitation of its natural resources (Ahamed et al. 2012). The type and scale of human modifications have changed dramatically during the past 40 years, as traditional Bengali practices have given way to modern technologies. These traditional and modern human modifications may have a profound effect on the vulnerability and resilience of the coastal zone to climate change and sea-level rise (Rubio-Cisneros et al. 2019). Shoreline changes also constitute a considerable phenomenon in Bangladesh. Barguna Patuakhali Coastal Zone (BPCZ) is situated at the lower confluence point of the GBM basin. Annual accretion of the Barguna Patuakhali coastal zone (BPCZ) shoreline is 77.087 m and erosion is 98.26 m (Sarwar and Woodroffe 2013). From these circumstances it is clear that the coastal landscape is changing and its analysis demands special attention owing to conservation of the naturally grown mangrove fringe "Sundarban" and "Tengragiri wildlife sanctuary." Apart from this, a prediction model of the area change needs to be developed to take appropriate conservation measures in the future to save the coastal margin, not only for these resources but also its function as a lifesaving first barrier for coastal habitats from any hydro-wind-based hazard and disasters (Islam et al. 2020).

2.2 Current Scenario of the Coastal Landscapes of Bangladesh

Climate change in Bangladesh differs from region to region, which makes the situation more complex and it is difficult to take a single initiative for all regions. These diverse climate change impacts are not limited to landscape changes, which is also lead to deteriorating health conditions, poverty levels, professional behavior (Torikul 2015), migration, and overall livelihood and socio-economic status that deserve special attention (Islam et al. 2019). In this context, this study considers historical change analysis of the most impactful coastal region measuring 170 km wide and 20 km from the southwest to southeastern region (local name Nildumur, Burigoalini Forest Range) (Fig. 2.1).

2.2.1 Historical Landscape: A Changing Scenario from 1988 to 2018

In the historical analysis of these regions of 1988, 1998, 2007, 2018, 4 years of Landsat satellite imagery was used for tiles-cluster-based (Several logic blocks are grouped together into a cluster. Clusters are connected to other clusters via Switch Matrices) further analysis and modeling in the southwestern exposed subdistricts (Table 2.1) using remote sensing and GIS technology. Field samples selected in 2017 and 2018 along with randomly generated samples by ArcMap 10.5 were used for accuracy assessment, which further cross matched with Google Earth Pro 7.0's very high-resolution (2.5 m) imagery. Object-based image analysis (OBIA) technology was incorporated where segmentation, parameter selection, and the Nearing neighborhood classification scheme were used as per standard OBIA studies in eCognition (Liu et al. 2018) for vegetation and coastal landscape change analysis purposes. Further analysis and multi-layer perception (MLP) done by Idrisi TerrSet and estimated by the cellular automata (CA) Markov model including the five distinct land cover classes named agri-grassland, forest, sandbar, settlement, and waterbody, as per the Anderson Classification scheme.[2]

After extensive analysis of the 4 years of imagery, the study findings, the increasing settlement rate is very high (an average 88 ha/year and a change rate per annum of 1.18; sandbar is a more fluctuating class owing to deposited land-based island creation and submerged portions, but agri-grass is a converting class and the annual conversion rate is 93 ha/year, where the forest area reduction rate is the highest (93 ha/year). Waterbody is the interchanging class with forest as the mangroves face it and almost every disaster with a high/low tidal force eroding the forest area eroding an average of 60 ha/year from 1988 to 2018. It is worth mentioning that soil is also deposited over time, but the erosion rate is higher than accretion. Conversion potentials are presented in Fig. 2.3, the predicted conversion cross matrix in Table 2.3, and the change rate per annum with overall change rate in Table 2.2.

Coastal erosion–accretion is a phenomenon that has a short- and long-term impact on the dependent community's livelihood, cultural pattern, homestead activity (Laurance et al. 2012), and the faunal diversity of these protected areas (Islam et al. 2020).

Change mapping from 1988 to 2018 shows that the lower part near Patharghata sub-district has faced constant erosion (Fig. 2.2) from 1988 to 1997, 1997 to 2007, and 2007 to 2018. Accretions prominent in the islands locally named Bangabandhu

[2]Anderson classification scheme first published in 1972 by USGS, USA for unifying land cover classes into different regions. As remote sensing was becoming populous but different regions were mixing distinct classes, this scheme was used with flexibility worldwide. However, as this paper is not the core work of the LCA, only mentioning perspective classes and its inclusion-like settlement, including all urban lands, houses, infrastructure, market; agri-grass refers to cultivated lands for different crops; forest combines top canopy vegetation patches; sandbar usually indicates the sea–land confluence, sandy mass, and waterbody includes all bays, channels, rivers, and natural ditches.

Table 2.2 Study area landmass change rate, change rate per annum from 1988 to 1998, 1998 to 2007, 2007 to 2018 and overall changes from 1988 to 2018

	1988–1998			1998–2007			2007–2018			1988–2018		
Class/interval	ha	ha/year	CRPA	ha	ha/year	CRPA	ha	ha/year	CRPA	ha	ha/year	CRPA
Agriculture	−1799	−200	−0.95	−555	−62	−0.32	−95	−9	−0.05	−2449	−82	−0.39
Forest	−490	−54	−0.08	582	65	0.10	−2873	−261	−0.38	−2780	−93	−0.14
Sandbar	934	104	2.35	2244	249	4.66	−2383	−217	−2.85	795	27	0.60
Settlement	2399	267	3.58	64	7	0.07	180	16	0.16	2643	88	1.18
Waterbody	−1045	−116	−0.08	−2336	−260	−0.18	5172	470	0.32	1790	60	0.04

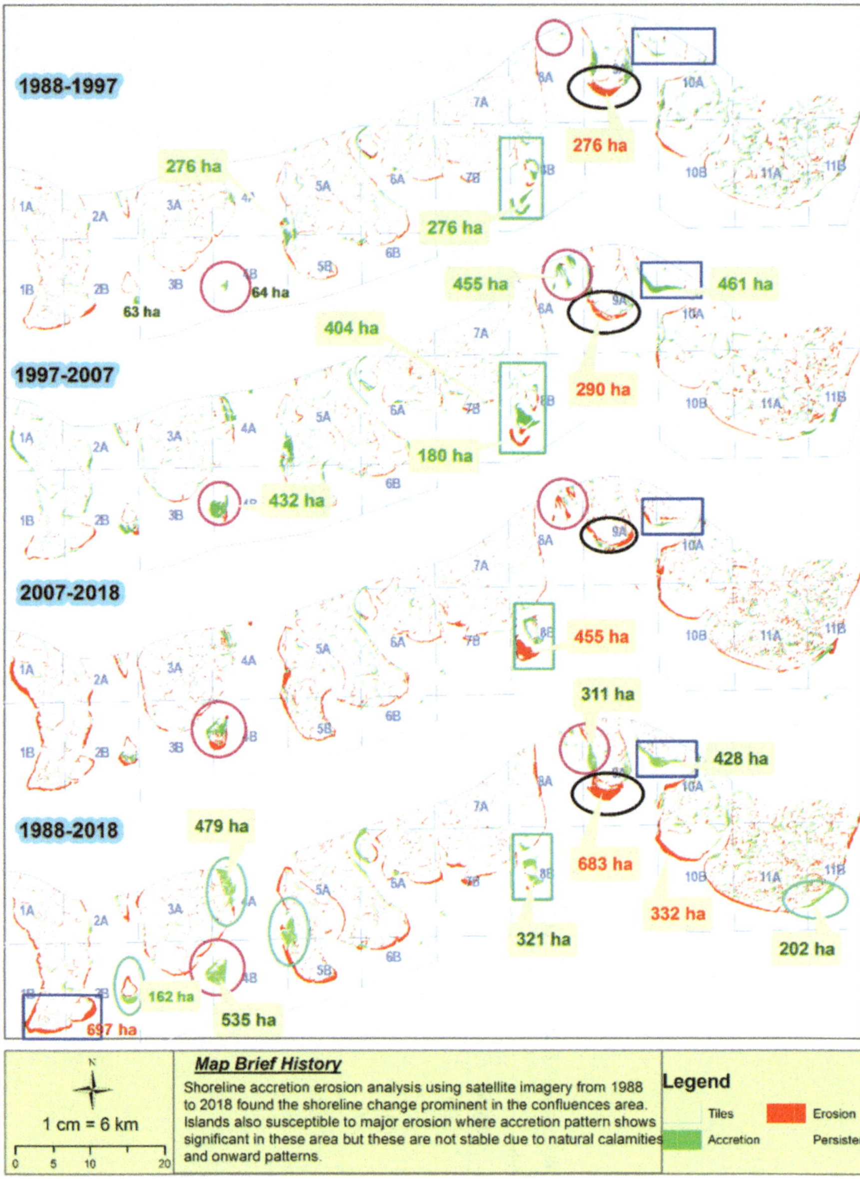

Fig. 2.2 Shoreline change mapping from 1988 to 1998, 1998 to 2007, 2007 to 2018, and the overall period 1988 to 2018

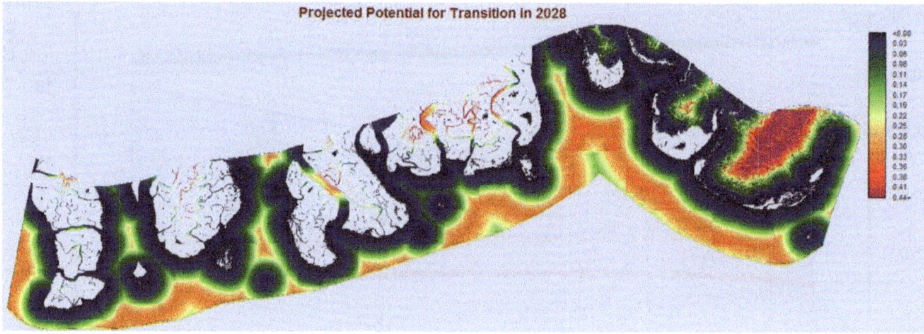

Fig. 2.3 Soft prediction (potential for transition) for the year 2028

Island, Dimer-char's northwestern part. The confluence point of Nishanbaria and Khuttar char faced accretion from 1997 to 2007, which eroded in 2007 and overall it faced an accretion trend of about 300 ha (Fig. 2.2).

2.2.2 Prediction Estimation of the Future Changes

Spatio-temporal resolution helps to figure our probable changes of an area based on different dynamic and static variables. Thus, variable selection, specifically the dynamic variable selection, is an important step for prediction. The study uses the multi-layer perception (MLP)-based artificial neural network (ANN)[3] module with population, road network, elevation as a layer for dynamic variables, then the pre-set algorithms used as artificial intelligence for prediction estimations. The transition potentiality[4] map provides the information about the susceptibility of an area to changes and the high trend of probability of change. The red color indicates a high trend of change, yellow presents a medium level, and green and blue present the low level of transitions. Several channels and narrow rivers are in a high trend of change, especially in the forest area. Figure 2.3 also shows the mask out area (mask out area means the compositional shapes are nonporous and obstruct the ink through the mesh/grid) where land cover will be almost the same. Northeastern parts are prone to a high trend of change by settlement (i.e., lower part of Taltoli) and all watery area will face a medium level of change, especially the island areas.

[3] Artificial neural network is compared to the human nervous system owing to its connectivity with nodes such as the human neuron for performing a specific work. Like its name, it easily incorporates results of different layers and combines using its artificial thinking, weighted machine learning algorithms, overlay and spatial distribution, and patterns based on static and dynamic variables.

[4] Transition probability map data presents based on the predefined sub-models' structure; here, all class to erosion and waterbody to accretion sub-model are used, as it is too technical and this work only deals with the outcomes rather than focusing more on methods it has omitted. However, those interested are suggested to follow the field guide of Idrisi TerrSet, published by Clark Lab, USA for more details.

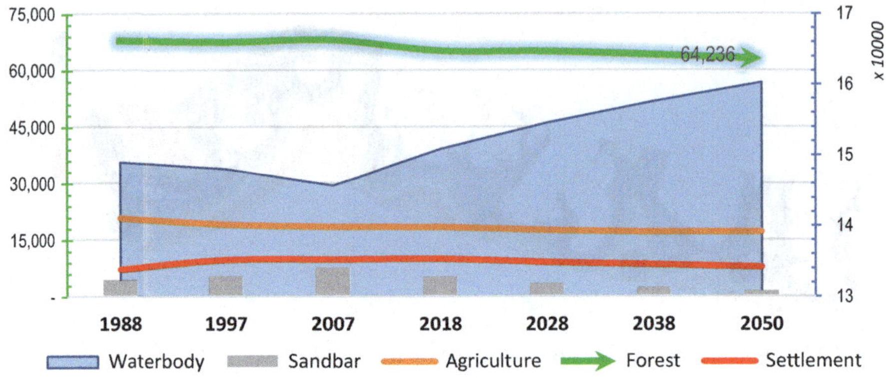

Fig. 2.4 Probable coastal frontiers area changes between 1988 and 2050

In order to present historical outcomes with predicted changes two axis stacked area bar charts with line charts are presented in Fig. 2.4. As per the study outcome, the waterbody area is a constantly increasing class that is three times greater than the second highest class forest cover and it would be five times greater than the agriculture area. Furthermore, the forest area will be reduced over time; however, the model failed to predict 2038–2050 results in a more sophisticated way.

2.2.3 Predicted Area and Probability Conversion Statistics

Prediction estimation by the CA-Markov chain procedure presents the highest amount of area intact in forest (0.95) and waterbody (0.96) and the most interchanging class was sandbar (0.32) between 1988 and 1998, which was used to predict the area in 2007 (Table 2.3). The interval 1998–2007, which was used for predicting the study area of 2018, presents that the unchanged area transition was in the forest and waterbody classes (0.95), but like the previous transition, the sandbar class was the most interchanging class. The prediction for 2028 was prepared from the imagery of the 2007 and 2018 transitions where the waterbody class presented the highest rate of remaining intact, even though it increased from the previous 2 years.

Settlement areas also present the highest transition; the probability number remains constant, which alternatively indicates that it might have failed to identify the conversions for further years at a very sophisticated level (Table 2.3). The Markov chain procedure presents the highest probability of the waterbody remaining intact in 2038 and 2050; after that, forest and settlement remain at 0.98, 0.91, and 0.92 respectively.

Table 2.3 Transition probability distribution and probable land area change statistics using cellular automata-Markov chain (Ag.=Agriculture; Frst.=Forest; Sandb.=Sandbar; Sett.=Settlement; Wat.=Waterbodies)

Transition probability distributions							Probable land area changes				
	Ag.	Frst.	Sandb.	Sett.	Wat.		Ag.	Frst.	Sandb.	Sett.	Wat.
1988–1998 used to predict the land cover of 2007											
Ag.	0.77	0.00	0.02	0.19	0.02	Ag.	212,487	–	–	–	355
Frst.	0.00	0.95	0.02	0.00	0.04	Frst	–	747,848	–	–	2679
Sndb.	0.05	0.23	0.32	0.08	0.33	Sndb.	–	–	57,444	–	1948
Set.	0.21	0.05	0.04	0.64	0.06	Set.	–	–	–	108,635	701
Wat.	0.01	0.01	0.02	0.01	0.96	Wat.	954	1,825	2582	1135	1,638,218
1998–2007 used to predict the land cover of 2018											
Ag.	0.76	0.00	0.02	0.20	0.02	Ag.	206,091		–	–	483
Frst.	0.00	0.95	0.02	0.00	0.02	Frst.	–	755,192	–	–	1839
Sndb.	0.06	0.26	0.32	0.03	0.33	Sndb.	–	–	81,585	–	2767
Set.	0.34	0.02	0.04	0.53	0.08	Set.	–	–	–	109,330	876
Wat.	0.00	0.01	0.03	0.00	0.95	Wat.	259	2250	4872	582	1,610,816
2007–2018 used to predict the land cover of 2028											
Ag.	0.79	0.00	0.00	0.16	0.04	Ag.	204,465	63	–	–	848
Frst.	0.00	0.93	0.02	0.00	0.04	Frst.	–	721,847	–	–	3161
Sndb	0.04	0.07	0.32	0.04	0.52	Sndb.	–	404	54,453	–	3024
Set.	0.30	0.02	0.01	0.60	0.07	Set.	–	230	–	111,182	777
Wat.	0.00	0.01	0.01	0.01	0.98	Wat.	–	1240	–	–	1,675,256
2018–2028 used to predict the land cover of 2038											
Ag.	0.96	0.02	0.00	0.00	0.03	Ag.	195,766	–	–	–	500
Frst.	0.00	0.98	0.00	0.00	0.02	Frst.	–	722,327	–	–	1469
Sndb.	0.00	0.04	0.64	0.00	0.31	Sndb.	–	–	36,148	–	1169
Set.	0.00	0.02	0.00	0.91	0.07	Set.	–	–	–	101,442	677
Wat.	0.00	0.00	0.00	0.00	1.00	Wat.	–	463	–	–	1,716,993
2028–2038 used to predict the land cover of 2050											
Ag.	0.97	0.00	0.00	0.00	0.03	Ag.	191,266	–	–	–	–
Frst.	0.00	0.98	0.00	0.00	0.02	Frst.	–	701,317	–	–	12,419
Sndb.	0.00	0.00	0.62	0.00	0.38	Sndb.	–	–	15,966	–	9658
Set.	0.00	0.00	0.00	0.92	0.08	Set.	–	–	–	87,797	7552
Wat.	0.00	0.00	0.00	0.00	1.00	Wat.	–	–	–	–	1,750,976

2.2.4 *Causal Analysis of These Changing Factors and Climate Change Implications*

The average condition of the atmosphere near the earth's surface over a long period of time considers temperature, precipitation, humidity, wind, cloud, and barometric pressure, mainly where geo-physical settings also govern the climate of any country.

– Seasonal reversals between summer and winter in Bangladesh have been fast changing over the last decades; during winter, a stream of cold air flows eastward from this high pressure and enters the country through its northeastern corner by changing its course clockwise, almost at a right angle. Generally, winds are stronger in summer (8–16 km/h) than in winter (3–6 km/h). The mean pressure is 1020 millibars in January and 1005 millibars during March through September, which is changing over time and these freeze-thaws are making land more prone to erosion.

- Extreme temperatures during the summer season combine with a center of low-pressure flow over the west-central part as a result, a stream of warm and moist air from the Bay of Bengal flows toward the above-mentioned low pressure through Bangladesh, frequently causing cyclones, storm surges, and other natural calamities.
- Rainfall and relative humidity are important factors for coastal erosion and land cover changes as this country lies in a tropical region; thus, rainfall is very high, and it is more pronounced than the annual cycle of estimation. The winter season is very dry, and accounts for only 2%–4% of the total annual rainfall causing brutal conditions for the croplands, facing mild drought and splash erosion, which carries large amounts of silt to the river and the low flow of water causes it to be deposited onto the bed; immediately after this season there is rainfall overflowing the river banks.
- Shrimp cultivation and human interventions, a conflicting issue occurring in the coastal area of Bangladesh between fisherman and crop farmers, where the farmers concerned about crop cultivation and fisherman about the fish. However, climate change is a very harsh indicator here as coastal farmers are forced to cultivate shrimp. The real field scenario is that the salinity of the winter season soil reaches its peak, which is prevent maximum crop cultivation and poor farmers cannot survive only annual rainy season crops; thus, the remaining option of shrimp cultivation is more profitable and this finally leads to them becoming fishermen.

2.2.5 Climate Change-Induced Major Hazards and Disasters and the Implications

The coastal area of Bangladesh usually suffers from different natural, quasi-natural, and other man-made hazards and disasters. Southern coastal communities usually suffer from cyclones, storm surges, salinity intrusion, a strong back water effect, flood, heavy rainfall, tidal surge, coastal high tides, waterlogging, short-term drought, prolonged high tides, erratic rainfall, and so on. Previous perceptions and seasons of occurrence are changing because of the climate change, field-based perception, and the period of suffering presented in the seasonal calendar of hazards experienced in this region. Details of the implications of the natural and quasi-natural disasters are presented in Sect. 3.3.4 (Fig. 2.5).

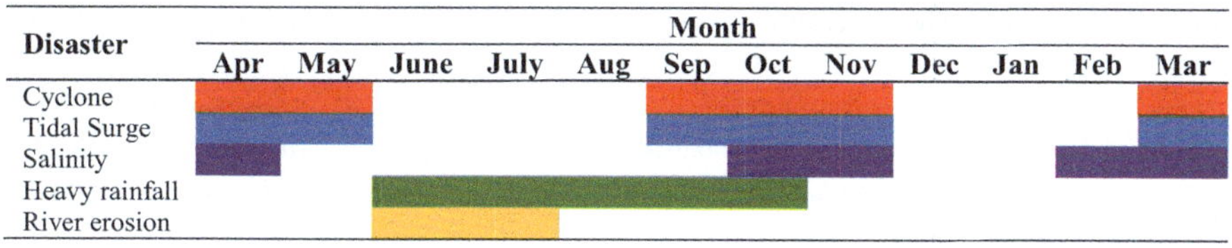

Fig. 2.5 Seasonal calendar of hazards of the coastal area

2.2.5.1 Cyclones

Cyclones are the main devastating disaster in the coastal area of Bangladesh due to its terrain features, geographical location, and funnel shape. Almost every year this area suffers from cyclones, but in the past couple of years its magnitude and intensity have both increased. Although this study is not focused on cyclone formation but scientifically on temperature rise means that global warming has a triggering effect for cyclone formation, which is also causes damage to pavement and road structure. Several devastating cyclones that have hit the different parts of the country in the last decades are presented.

Table 2.4, which also alternatively presents the effective measures that reduced casualties owing to precautions. During cyclone "Amphan" around 2.4 million people and 0.5 million livestock were evacuated to 15,000 cyclone shelters (Shahin et al. 2020a) within 2 days – a unique example of rapid evacuation during a disaster (Rahman 2020).

2.2.5.2 Sea Level Rise, Inundation, and Waterlogging

Planet earth has been warming to a significant level in the last couple of years, which is undeniable now if the community visits the coastal area of Bangladesh. After illegal withdrawal of upstream water from the GBM basin by India, the flow of water of these lowline rivers is significantly reduced (Rahman et al. 2020), which leads to soil deposition into the river bed of around 6 mm/year (Brammer 2014b), reducing the carrying capacity of the river channel and even the Bay channel. For this reason, almost every year Bangladesh faces flash floods and prolonged periods of waterlogging into the north west, such as the Rangpur and Dinajpur districts, to the south-central region (Bhattacharjee 2011). The fate of coastal people is broken, such as the washout and broken earthen embankment shown in Fig. 2.6, damaged during the most recent cyclone, Amphan, which occurred on 21 May 2020. The image was captured during the low tide from 22° 23′ 17.818″ N latitude and 89° 11′ 9.613″ E longitude, local name "Ghola Bandh, Shymnagar, Satkhira," dated 10 June 2020, 18.33 BST. This devastating cyclone and the height of the associated storm surge damaged most of the earthen embankments and inundated associated areas, with around 3 million people suffering because of it. Therefore, there is no doubt that this is type of bag (shown in Fig. 2.6) are no longer applicable for the protection of embankment in the coastal areas of Bangladesh.

Table 2.4 List of some devastating cyclones (with casualties) that also made landfall over Bangladesh

Date and description of cyclones/cyclone name	Max. wind speed (k/h)	Surge height (m)	Deaths
1867: cyclone and severe cyclonic storm with hurricane winds	–	3–13.7 m	200,000
28 May 1963: severe cyclonic storm	200	6.0 m	11,520
12 November 1970: severe cyclonic storms with hurricane winds	224	10 m	300,000
25 May 1985: severe cyclonic storm	154	4.6 m	4264
29 November 1988: severe cyclonic storms with hurricane winds	160	4.5 m	5708
29 April 1991: severe cyclonic storm with hurricane winds	225	6–7.6 m	138,000
2 May 1994: severe cyclonic storm with hurricane wind	220	3.6–4.8 m	188
19 May 1997: severe cyclonic storm with hurricane wind	220	4.55 m	155
15 November 2007: severe cyclonic storm with hurricane wind (Sidr)	)223	6.02 m	3363
25 May 2009: cyclonic storm (Aila)	92	2.50 m	190
17 May 2013: cyclonic storm (Mahasen)	85	1 m	11
31 May 2017: severe cyclonic storm (Mora)	150	1.5 m	135
17 December 2018: very severe cyclonic storms (Titli) (right)	175	Undefined	343
5 May 2019: extremely severe cyclonic storm (Fani)	250	5–8 m	17
11 November 2019: severe cyclonic storm (Bulbul)	155	2–3 m	41
21 May 2020: super cyclonic storm with hurricane wind (Amphan)	260	3–5 m	128

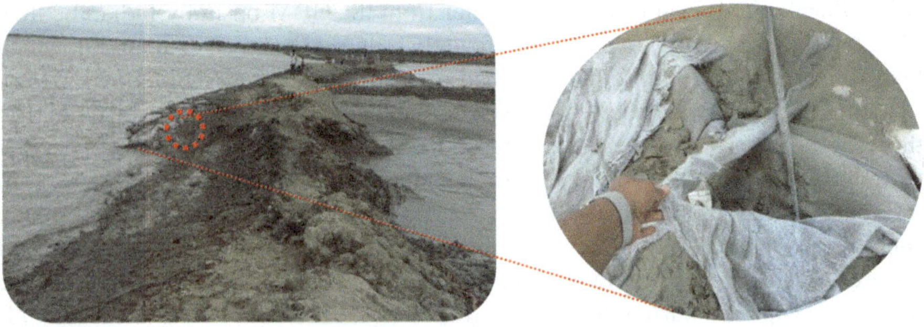

Fig. 2.6 Super cyclone Amphan washed out earthen embankments in Satkhira district (field visit: 10 June 2020)

2.2.5.3 Coastal High Tides

Another major hazard in the coastal area is the high tide, which specifically attacks every year those low-lying areas where cyclone Sidr and other successive cyclone/storm surges have attacked. Dams, polders, and other structures have been constructed to save these coastal communities, but almost every year, riverbank erosion significantly deteriorated these structures. As it is low and adjoining the Bay of Bengal community life, crops, fish resources, forest, domestic animals, houses, industry, and infrastructures are fatally affected by the regular coastal high tides in the district.

2.2.5.4 Salinity

Salinity is the malicious friend of the coastal area, as it has great negative health effects, but makes a significant contribution to income-generating activities. For instance, every year, Bangladesh earns 40% of its foreign currency by exporting shrimp, which is not possible to cultivate without acute saline water (Azad et al. 2009). By then, it is well known that this saline water and soil salinity alter their lifestyle, but they intentionally channelize for shrimp cultivation as the only high profitable business (Islam and Tabeta 2019). Although it has increased conflicts between agrarian farmers and fishermen, the final profit margin makes solutions beyond health crises. It is also creating a very bad impact on crops, forests, fish resources, and infrastructure, as well as the whole lives and livelihoods of the people. Saline water is easily passed to the lowland by the dams, roads, and sluice gates that are damaged during a sea surge and water cannot go back into the sea or river from the affected lands.

2.2.6 Protection Measures for Coastal Landscape

Coastal districts are facing more intensified and a greater magnitude of natural calamities over time, but measures taken by government or international organizations are insufficient and not up to the mark (Brammer 2014a). It is obvious that resources and costs are not the root cause of not taking development measures for coastal areas in Bangladesh, but a huge lack of smart planning, expert/researchers' work-based outcome, corruption, and timing, and many more, as recently Bangladesh has been successfully running several mega projects. Although Government has taken measures to help the coastal community, most are related to the construction of flood protection embankments (Ministry of Environment and Forests 2008) and not constructing earthen embankments but rather reinforced cement concrete (RCC) structures. Recent study shows high-tide water flow above the existing flood dam, a clear sign that the rise in sea level was inundating exposed sub-district of Barguna, Patuakhali district in 2018. Some of the protection measures where detailed structural engineering and bio-engineering measures are described in Sect. 3.6.

- First and foremost: realistic water distribution of the transboundary rivers between India and Bangladesh needed to be resolved to reduce riverbed siltation, flash floods, erosion, inundation, and waterlogging.
- Newly settled land needs to stabilize by taking appropriate measures like planning developed rooted plants and immediately after settlement, it also needs to be prepared for cultivation to meet increasing demands.
- Cultivating saline-tolerant varieties and shrimp cultivation in acute saline zones.
- Existing flood dam height and coastal embankment height needed to re-fix, considering current hazards such as the wind speed of cyclones and the height of storm surges.

2.3 Coastal Roadway Communication and Disasters

Landscape changes and erosion procedures impact another substantial sector, the communication sector, which is an extensively required livelihood and development. In the coastal area of Bangladesh the roadway is usually the main transportation system, although water and air ways are available in a few districts but not enough compared with what is needed. Studies show that road transport shares both passenger and freight traffic that is greater than the combined share of rail and water transport (Quium and Hoque 2002). Every year, there are many road crashes due to high road traffic and damage to roadways in the coastal areas of Bangladesh (Quium and Hoque 2002). Bangladesh is one of the most vulnerable countries to climate change in the world and climate change triggering natural and quasi-natural disasters contributes to damage to the road network along with the quality of construction (Shahin et al. 2020a). In the event of emergency evacuation and relief, the road infrastructure is highly essential not only for volunteers but also for survivors to access emergency medication (Amin et al. 2020). In recent decades, coastal communities have faced several high-intensity natural disasters, for example, super cyclones Sidr (2007), Aila (2009), Mohasen (2011), Bulbul (2019), and "Amphan" hit on 20 May 2020. All these devastating cyclones with storm surges, floods, and erosions, caused severe damage to roadways.

To address substantiate road network failures in the coastal area, an analysis of the current situations and historical trends is required. Thus, the study investigates disaster-induced road structure damage, resilience measures of the road networks against various climate change-induced disasters and exploring lacks and gaps in resilience measures for road network sustainability along with the constructing materials, such as the strength of the bituminous concrete is satisfied up to 4% of salt (Shahin et al. 2015). For effective functioning of these roadways, it is high time that the constructing materials, strengths, and manuals were checked and revised taking into consideration climate change impacts and issues.

2.3.1 Existing Road Networks and Study Regions

The types and functions of the road network for coastal districts are almost the same; therefore, two coastal districts from the mid-central coastal zone named Barguna and Patuakhali are selected owing to their great homogeneity with other coastal districts, and the availability of sufficient field-based primary data, high-density population, and homogeneity of natural calamities. Geographically located between 21°48′ and 22°29′ north latitudes and 89°52′ to 90°41′ east longitudes (Fig. 2.1) with a total area of 3220.15 km^2 and 1831.31 km^2 respectively. Neighboring districts are Barisal on the north side, the Bay of Bengal in the south, Bhola and the Tetulia River are in the east of Barguna district, and Pirojpur and Bagerhat districts in the west (RHD 2009). Types of roads in the study area are shown in Fig. 2.1.

Cross section of a typical road section is presented in Fig. 2.7, where the pavement is constructed on the top for traffic movement, and the bottom of the pavement is called the subgrade embankment and is built with earth. A base course is placed on top of it in two or more layers and the surface course is placed on top of the pavement. The shoulder on both sides of the pavement is for the pedestrian, light vehicle movement, and car parking, and the earth is filled and compacted up to the required height from the earth with compacting materials. The pavement has slopes on both sides from the center of the road called the camber and has its own camber design, which is also called crossfalls. Each embankment has slopes on both sides according to the design at the end of the toe of the slopes. To make the embankment stable the part called the "berm" is prepared (RHD 2009).

However, in the coastal area of Bangladesh, four main types of road networks are available, i.e., earthen road, flexible (bituminous) pavement, brick pavement, and

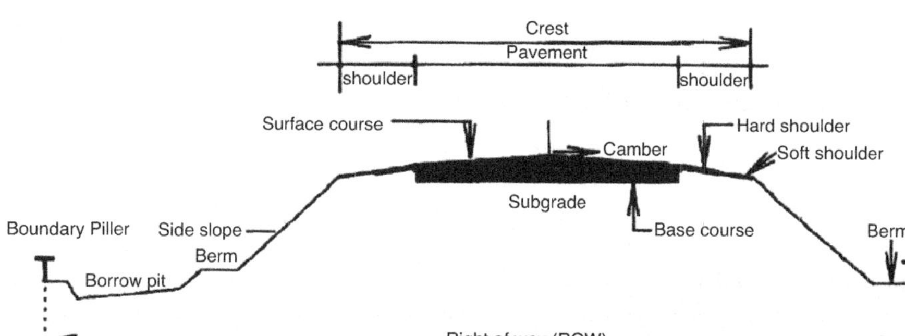

Fig. 2.7 Typical cross section of a road structure (Source: RHD, Bangladesh)

Table 2.5 Types of road structure in the coastal districts of Barguna and Patuakhali

	Upazila	Earthen (km)	Pavement (BC) flexible (km)	Pavement brick (km)	Pavement rigid (km)	Total
Patuakhali District	Patuakhali Sadar	2,313.05	287.20	84.60	16.80	2,701.65
	Dashmina	873.98	85.71	22.85	1.48	984.02
	Bauphol	1894.63	377.95	87.41	14.82	2374.81
	Rangabali	732.31	34.70	34.22	34.79	836.02
	Dumki	366.70	73.47	34.01	5.52	479.70
	Galachipa	1813.05	182.92	53.10	–	2049.07
	Kalapara	1713.82	187.87	79.50	3.97	1985.16
	Mirzagonj	761.62	90.09	72.70	2.51	926.92
	Total	**10,469.16**	**1319.91**	**468.39**	**79.89**	**12,337.35**
Barguna District	Barguna sadar	1249.77	228.89	55.43	1.80	1535.89
	Amtali	1111.95	207.04	27.06	3.37	1349.42
	Bamna	252.15	108.32	26.42	3.50	390.39
	Betagi	1288.07	126.83	47.43	1.98	1464.31
	Patharghata	712.66	125.60	73.97	3.88	916.11
	Taltali	556.04	108.79	17.97	4.67	687.47
	Total	**5170.64**	**905.47**	**248.28**	**19.20**	**6343.59**

Source: LGED, Bangladesh

rigid pavement. In Patuakhali, 84% (10,469 km) of roads are earthen, and 11% (1320 km) of the roads are flexible pavements. as shown in Table 2.5. In Barguna, earthen roads cover 81% (5,171 km) and 14% (906 km) of the roads are flexible pavements. The substantial part of the road network, earthen roads, faces washout during storm surges in cyclones. The flexible pavements are mostly damaged by different types of disasters such as salt intrusion, washout, and flood water intrusion in the subgrade. Rigid pavements, a small part of the road network in the town and markets, are damaged by water intrusion in the subgrade (mud-pumping), and salt accumulation.

2.3.2 Damage to Coastal Roads and Highways

Pavement damage is harmful, but not all types. Some of these are severe whereas others need only repair rather than reconstruction. The most frequent types of pavement damage are shown in Fig. 2.8. Longitudinal and transverse cracking are the most common and substantial phenomena in high traffic areas. Potholes are major defects of village roads, usually where they have been waterlogged with rain for a few days. The waterlogging weakens the bond strength of the bituminous pavement, as water is an enemy to bitumen. Rutting is found as a massive and frequent defect

Fig. 2.8 Most frequent pavement damage types in the coastal area

that is triggered by flood water intrusion in the subgrade of roads. Edge erosions are found as a serious surface deterioration in the coastal areas, although it is also a problem in other regions.

During a massive flood, the roads are mainly damaged owing to the looseness of the subgrades (Shahin et al. 2020b). Major impacts of the disasters on roadways are washout, subgrade and surface deformation, surface texture losses, displacement failure, stripping, surface deformation, rutting, potholes, road edge erosion, bleeding, material degradation, and cracks (Sarie et al. 2015). As is known, stripping, subgrade looseness, washout, and potholes are attributable to the water-induced deterioration, which destroys the roads very frequently. 93.33% of duty engineers (respondents) claimed that washout of the roads has major impacts, whereas 13.33% opined that rutting was a major impact of disasters on the roads. Occasionally, binder from within the pavement migrates to the pavement surface, resulting in flushing or bleeding (Zumrawi 2016).

2.3.3 Implications of Climate Change, Hazard, and Disaster for the Road Network

Bangladesh is one of the most vulnerable countries to climate variability and change owing to its geographical location, low deltaic flood plain, and hydro-meteorological influence of erratic monsoon rainfall (Shahin et al. 2015, 2020a; Minar et al. 2013; Nateque Mahmood et al. 2014; Sikder 2010). Increased risks for severe flooding, more frequent extreme weather events, salinity intrusion, erratic rainfall pattern, increased temperature, and a potential sea level rise pose new risks to millions of lives, as well as damage to properties/assets and infrastructures, including the road networks (Haque and Jahan 2016) particularly in coastal districts. Prime factors for roadway deterioration are heavy trucks, and business class buses that carry loads beyond the road's capacity, for instance, inter-country cargo trucks between Bangladesh and India operate on several coastal roads. The subgrade of coastal roads is weak owing to water intrusion that accelerates subgrade failure and surface deformation. Second, in recent decades, Bangladesh has suffered from high magnitude natural disasters including cyclones with storm surges, flood, salinity intrusion, waterlogging, coastal erosion, and drought (De Carteret et al. 2010; Shahin et al. 2015). Salinity triggers the formation of potholes in granular pavement, rutting, shape loss, crumbling of concrete kerbs, subsequent spalling of concrete, and corrosion of steel reinforcement.

2.3.3.1 Environmental Parameters

The environmental parameters that have been significantly changed in recent years are attributable to the deterioration of road infrastructures. Increased temperature, rainfall, floods, frequent freeze–thaw phenomena, storm surges, and sea level rise affect the performance of road infrastructure, maintenance, and operations. Specific impacts of climate change on road transport infrastructures include thermal expansion of bridge joints, the displacement of flexible pavement layers, overflow of side drains and cross drainage works, inundation of coastal roads due sea level rise and flooding, and road closures due to landslides (Chai et al. 2014). Unpredicted precipitation causes overflow of the drainage system, water infiltration into the subgrade and shoulder, which cause surface course deterioration. The coastal areas in Bangladesh are low lying and flat in nature, which means that in the coastal areas situated below sea level this is caused by temperature rise, as the height is less than 3 m. The major environmental parameters, which have abruptly changed in the last couple of years are temperature and precipitation. The parameters are discussed in the following subsections with their impacts, which are summarized in Table 2.6.

2.3.3.2 Precipitation

Barguna and Patuakhali districts are situated in tropical climatic zones with fluctuating erratic precipitation. Average annual precipitations are 2516 mm in Barguna and 2654 mm in Patuakhali. The fluctuation in precipitation of ~576 mm per year is shown in Fig. 2.9. The wettest months from May to September face high levels of precipitation, which cause overflow of the existing drainage system, and roads are inundated, which is the preliminary cause of the stripping of flexible pavements.

2.3.3.3 Temperature Rise

Barguna and Patuakhali districts face unpredictable fluctuations of temperature. Average annual temperatures in Barguna and Patuakhali districts are 26.0 °C and 25.9 °C respectively. May is the hottest month, with average temperatures of

Table 2.6 Potential impacts of climate change and types of distress

Climatic events	Potential impacts	Distressed type
Sea level rise	• The rise in sea level affects coastal roads, leading to realignment or abandonment of roads in affected areas by increasing soil salinity levels	• Failure of the strength of bituminous concrete, blistering, cracking, and potholes
Cyclone and storm surge	• Rainfall and winds associated with storm/cyclone create flooding, inundation of embankments, and affect road transport • Disrupt traffic safety and emergency evacuation operations, affect traffic boards and information signs	• Holes, road edge erosion, washout, ditches, embankment damage, surface defects, cracks, layer movement, interlayer bonding loss, surface texture loss, degradation of layer materials
Rainfall	• Increased intensity of precipitation creates floods, affects drainage, road pavement, bridges and culvert waterways and clearance, damaged bridges and culvert foundation owing to scouring • Erosion and soil instability	• Cracks, ruts, failures, holes, surface deformation, surface defects, patches, stripping, erosion of the subgrade, edge joint cracks

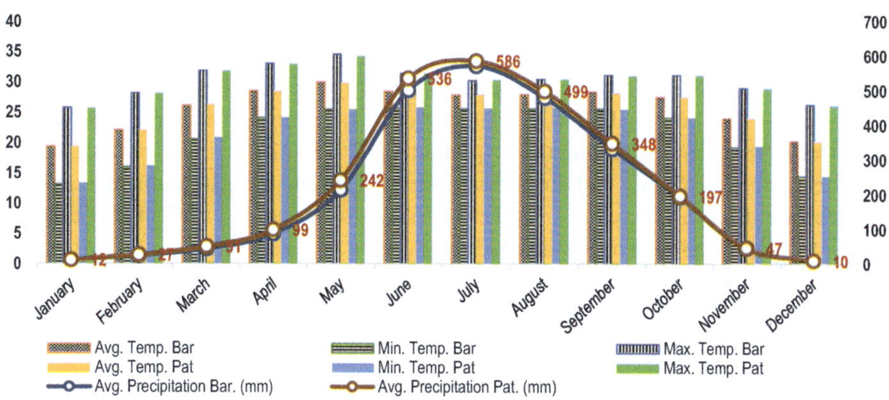

Fig. 2.9 Monthly average temperature and precipitation in coastal districts in 2018 (Barguna and Patuakhali)

29.9 °C, 30.0 °C in Barguna and Patuakhali where December and January are the driest months. Temperature fluctuations of ~10.4 °C per year are shown in Fig. 2.9. Extreme temperature is likely to have several adverse impacts on road pavements. Extreme temperature is the main cause of generating cyclones, as the scientist McCarthe suggested 26 °C as the temperature that create cyclone seeds/eye. The average maximum temperature is increasing at a rate of 0.03 °C per year but the annual minimum temperature is decreasing at a rate of 0.003 °C per year, resulting in climatic variability in Bangladesh (Neema 2014). Successive extreme hot weather events such as stronger and more frequent cyclones, storm surges, and sea level rise wash out the roads, and inundate some road networks. A 1-m rise in sea level would submerge 18% of the total land area in Bangladesh (Minar et al. 2013), which could be accelerated by the extreme temperature. The inundation-induced waterlogging causes accumulation of salinity that reduces the foundation, basement structural strength, and lifespan of the pavements. The excessive flood water causes the drainage systems to overflow and water intrusion into the base course and subgrade, leading to deformation while vehicles are moving on those roads/embankments.

Extremely hot temperature leads to more evaporation and salt accumulation in the riverine humid climate areas, increasing erratic precipitation and rusting of the reinforcement in concrete structures, requiring rural infrastructural investment, especially in coastal districts, which must be resilient to climate change associated with extreme events (ADB 2011). Temperature (both an increase in the annual average temperature and the seasonal variation) has a significant impact on pavement distress, including longitudinal cracking, fatigue cracking, and asphalt course rutting (Qiao et al. 2013). The potential impacts are shown in Table 2.6.

2.3.3.4 Impact History of Natural Disasters on the Road Network

As has been discussed in the previous sections, coastal environment and climatic variability is unique in some sorts of aspects in this coastal region; in this regard, real field scenarios and community perception of historical extent analysis is most important as they are dealing with long periods. As community people are the first to suffer from disasters, among the respondents, 87% think that the consequences of disasters are destructive and the remaining 13% think that the consequences of disaster are harmful. Here, harmful refers to the consequences of disasters that are likely to cause dysfunction of the road over a long period of time, for example, the effects of salinity on the pavements. "Destructive" means consequences of disasters that are likely to cause dysfunction of the road immediately after the events or within a very short time of the events, for example, washout of roads by the storm surges.

Disasters have adverse impacts on road networks all over the world and the coastal area is no exception. Coastal road transport infrastructure and its functions are extremely affected by climate change events such as erratic rainfall, extreme temperature, sea level rise, and storm surge (Regmi and Hanaoka 2011) and 93.33% respondents opined that flood causes severe damage by creating holes, road edge erosion, washout, ditches, embankment failure, surface defects, cracks, layer movement, cohesion and adhesion loss, surface texture loss, material degradation, whereas 73.3% of respondents agreed on the road network damage by cyclones. Coastal erosion is attributable to the pavement destruction, as said by 53.3% of respondents (Fig. 2.10). The moisture, a great enemy to the flexible pavement, that comes from flooding, erratic rainfall, and storm surges, destroys the pavement, as was suggested by 80% of duty engineers in the study area based on their field experiences. Major impacts include rapid road asphalt deterioration due to increased

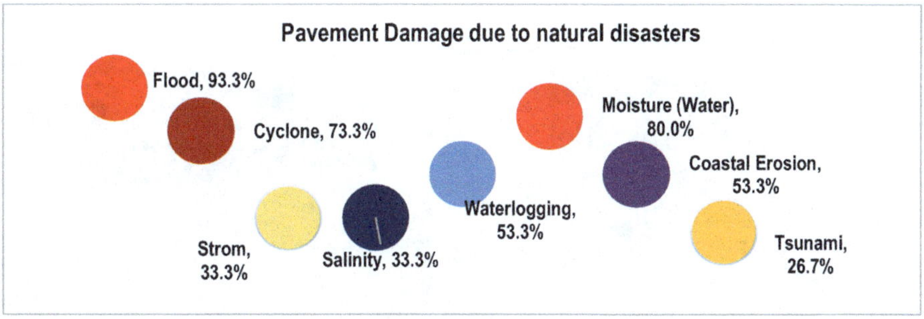

Fig. 2.10 Percentage of pavement damage due to natural disasters

flooding of roadways, prolonged heat waves, erosion of the road base, and damage to the upper infrastructure (USAID 2012). Increased frequencies of high-intensity cyclones, storm surges, and floods cause all types of pavements to deteriorate in various ways including stripping, potholes, displacement, corrugations, subgrade deformation, surface failure, and cracking. The damaged roads incur higher costs of maintenance and increased road crashes. In addition, road damage also increases the costs of vehicle operation and maintenance, travel time, and hinders access to markets and social and economic facilities.

2.3.3.5 Cyclone and Storm Surge

Cyclones and storm surges are devastating in the coastal areas, and cause severe damage to road structures. 73.3% of respondents blamed cyclones with storm surge for washing out the road embankment fully or partially. In the last couple of years, the frequency and intensity of cyclones and storm surges has increased significantly. Cyclones along with storm surges cause severe damage to the road network, prolonged inundation of pavements from flooding causing the entry of moisture into pavements, accelerating adhesion and cohesion loss, rutting, and cracking. Cyclones uprooted most of the roadside trees, which causes slope failure of the road embankment, whereas storm surges inundate and wash out the road embankment. Increased damage to pavements from flooding results in rapid deterioration of pavements, and reduction in the life of a pavement. In the super cyclone Sidr, most of the road embankments were washed out by the storm surges in the affected coastal areas of Bangladesh.

Figure 2.17 illustrates major damage due to the cyclone, such as lifting roadside trees and bridge decks, washout, and cracks. Some devastating cyclones with storm surges (with casualties) that made landfall in coastal areas of Bangladesh are presented in Fig. 2.17 (Khatun et al. 2016). The study shows that the cyclone and storm surge lifted roadside trees in 86.7% of cases; cracks and bridge decks where surge water mostly washed out and damaged basements occurred in 60% and 73.3% respectively.

Fig. 2.11 Super cyclone "Amphan" washes out earthen embankment in Satkhira District (Field Survey: 10 June 2020)

Earthen Road Network Status After Recent Cyclone "Amphan"

Storm surges in cyclone Amphan that partially or fully washed out the embankment in the affected area in Satkhira district of Bangladesh are presented in Fig. 2.11. There may have several causes of the damage, but most visible causes were weak earthen embankment (saturation of foundation soil), lack of proper compaction, inefficient protection works (i.e., geotextile bags), and overflow of storm water. Most of the storm surges have similar impacts on the embankments in Bangladesh. A witness said that "super cyclone Sidr washed out some roads and embankments in such way that you cannot imagine the existence of those infrastructures in Barugan district."

The existing protection works for the embankments are inadequate to cope with the current high-intensity storm surges, cyclones, and floods. Advanced methods of embankment construction for the protection of coastal communities and road pavement should be adopted. The repair of the remnants is costlier, and almost every year such type of events take place in the coastal areas, causing devastation to individual's properties, lives, infrastructures, and social systems. The coastal people have to rebuild their house almost every year, which makes them poorer year by year. Therefore, advanced technologies (e.g., GIS, RS, Radar signal), and construction methods (e.g., high degree of compaction, high strength materials, full height rigid embankments) should be emphasized instead of the costs of construction giving priority to lives and infrastructures.

2.3.3.6 Salinity Intrusion

Salinity intrusion in the coastal area is a common phenomenon that has increased in the last 40 years owing to illegal stream water withdrawal by India (Rahman and Rahaman 2018), high interest profit margin shrimp cultivation rather than crop cultivation, strong backwater effect due to sea level rise, and so on. Saline water intrusion is caused by storm surges, sea level rise, and flooding in the coastal regions.

Saline water intrusion into rivers and canals leads to salinity intrusion into the ground water. Subsequent inundation of coastal areas increases the salinity level by the gradual accumulation of salt contained in sea water. In the coastal areas of Bangladesh, salinity comes in contact with the pavements in several ways, such as waterlogging, inundation, storm surges, earth filling with saline soil, and water table rise. Shahin et al. (2015) found that 4% salt could be tolerated by the flexible pavement with substantial stability loss, but still above the required stability. Beyond the limit, salt damages the pavement in several ways. The moisture content of saline soil is subjected to intrusion into the road structural materials, which causes rutting in granular pavements, differential shape loss resulting in rough pavements, sealant blister leading to loss of sealant, water infiltration, and potholing (Pirlea and Burlacu 2013). In around 72.73% of cases it is causing strength failure, in 63.6% of cases responding blisters, and 45.5% and 36.40% making cracks and potholes respectively. Salinity also damages the soil quality, which influences the road edge erosion and surface deformation. Osmosis occurs in the presence of salts or salt solutions in aggregate pores and creates an osmotic pressure gradient that actually sucks water through the asphalt film (Little and Jones 2003). It is known that the salt is detrimental to flexible and rigid pavements, and the effects of salt have been observed in the Barguna and Patuakhali districts of Bangladesh.

2.3.3.7 Moisture

Moisture in the form of water is mainly responsible for the early deterioration of pavement foundations. Moisture accumulates and is retained owing to the frequent flooding at high tides, waterlogging, and intrusion of water into the soil of the road embankment. Moisture is also trapped on the uneven and wavy surface of the road, which is retained for a long time because sunlight is prevented from penetrating owing to the roadside trees that are planted to prevent the slope failure. Loss of strength and durability due to the effects of water are caused by loss of cohesion of the asphalt film, failure of the adhesion between the aggregate and asphalt, and degradation of the aggregate particles subjected to freezing (Little and Jones 2003). Moisture damages pavements because it permeates and weakens the mastic, making it more susceptible to moisture during cyclic loading. The stripping of flexible pavement is attributed to this moisture. The literature refers to five mechanisms of moisture damage (stripping) for the flexible pavements, i.e., displacement, film rupture, spontaneous emulsification, pore pressure, and detachment.

According to the displacement theory, the binder aggregate function, in the presence of water, becomes thermodynamically unstable and retracts to a more stable position (Caro et al. 2008; Frazier Parker 1989; Asphalt Institute 1981). The stripping by displacement results from the penetration of water to the aggregate surface through a break in the asphalt film (Asphalt Institute 1981; Taylor and Khosla 1983). In the film rupture theory, dynamic loading and the kneading action of traffic induce rupture at the sharp corner of the aggregate where the binder film is thinnest, while water may easily displace bitumen (Caro et al. 2008; Thelen 1958; Emery and

Seddik 1997). Spontaneous emulsification is an inverted emulsion of water droplets in asphalt cement. Once the emulsion formation penetrates the substrata, the adhesive bond is broken (Fromm 1974). The formation of such emulsions is further aggravated by the presence of emulsifiers such as clays and asphalt additives (Asphalt Institute 1981; Fromm 1974). The build-up of pore pressure in a mixture of high void content may result in stripping phenomena (Halberg 1950; Yilmaz and Sargin 2012). The detachment theory attributes the adhesion to a thermodynamic replacement of the bitumen by a thin film of water that may come from either outside or from within the aggregate while the bitumen coating remains intact (Taylor and Khosla 1983; Hughes et al. 1960; Terrel and AI-Swailmi 1994; Horgnies et al. 2011; Abo-Qudais and Al-Shweily 2007).

Water also infiltrates pavements through the shoulders, the surface, or the sides of lower layers. The water that infiltrates into the foundation erodes and weakens the soil. Water trapped in the unbound flexible base and subbase layers can increase the tensile stresses at the bottom of the asphalt layers, leading to cracking or rutting. The moisture in the form of water damages the subgrade along with the surface courses. There is deformation in the subgrade while the cyclic wheel load moves on it. Failure of subgrade leads to the formation deformation on the surface course, potholes, cracking, and rutting. Surface raveling or a loss of surface aggregate can also occur, especially with chip seals. Water beneath jointed RCC pavements can cause erosion of base materials leading to accelerated faulting, corner breaking, and transverse and diagonal cracking (Parsons and Pullen 2016). In the operational phase, roads are wetted for long periods, specifically during the rainy season and immediately after rain evaporation and draining moisture is reduced to its optimal level and by this continuous freeze–thaw causes stress in the pavement, resulting in cracking, or in less intense pavement failure, damage to the road by creating cracks, holes, surface deformation, surface defects, patches, strip, erosion of the subgrade, and edge joint cracks (Fig. 2.12).

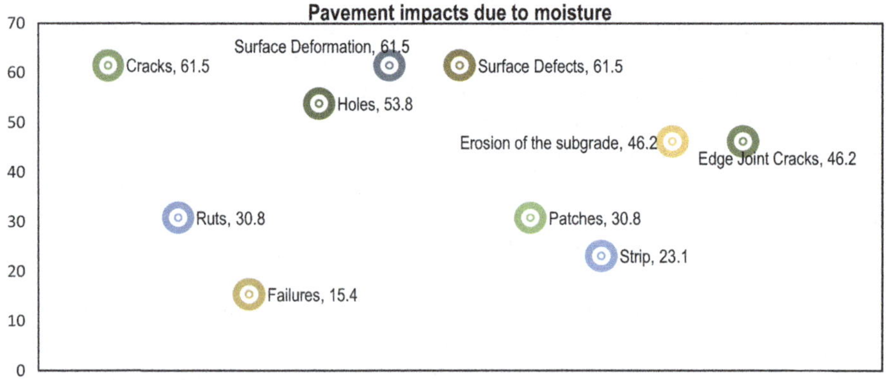

Fig. 2.12 Professional clusters percept: percentage of moisture contributing to different types of damage

2.3.3.8 Inefficient Drainage Facilities

Adequate drainage systems are very important for draining out the water that comes from rain, flooding, high tides, and storm surges. The pavements have to face the stripping and related failure in the presence of water. Faulty drainage systems accelerate the accumulation of moisture on the surface and foundation that deteriorates the pavements and foundations. Drainage systems include a water handling system, with surface slope, shoulders, roadside drains, and culverts. In the coastal areas, inadequate drainage is mostly attributable to a roadway failure. Poor design can direct water back onto the road or prevent it from draining away. Too much water remaining on the surface combined with traffic action may cause potholes, cracks, and pavement failure (Abhijit 2011). Therefore, poor drainage causes the premature failure of the pavement. The major causes that make the existing drainage system faulty or inefficient are as follows: natural causes including floodwater, short duration of heavy rainfall, whereas manmade causes include seizing of canals and ditches, poor maintenance of culverts, blockage in drainage system, exceeding manhole capacity, lack of good drainage, and lack of public awareness (Mahmud et al. 2015).

2.3.3.9 Inundation and Waterlogging

Geo-materials provide the bulk of the structural strength or the foundation of the base, subbase, and subgrade of most pavements. The inundation of coastal areas has occurred because of the sea level rise, excessively heavy rainfall, high tides, storm surges, and floods. The water that causes inundation is trapped for a long time, causing waterlogging. Frequent inundation that is a common phenomenon in the coastal areas saturates the foundation and deteriorates the pavements by stripping and corrugation. Floodwater can reduce the strength of foundations significantly owing to a long-term increase in saturation, which leads to excessive strains and deformations under traffic loading and consequently premature failure of the pavement. The extent of damage depends on the vulnerability of the pavement to the ingress of surface water, the amount of water, the inundation period, and the drainage features of the pavement. The inundation cycles cause the deposition of salt that leads to pavement damage by the saline action. In addition to cyclones and storm surges, salinity intrusion causes damage to the pavement in several ways.

Waterlogging is responsible for the penetration of water into the foundation of road structures and water table rise, which increase the pore water pressure leading to cracking of the surface when wheel loads are passing. Moreover, waterlogging causes salt intrusion, which causes premature failure of the pavements. Waterlogging, heavy rainfall, seasonal high tide, and saline water intrusion into the bituminous pavement can lead to the isolation of asphalt binder from the aggregate surface (stripping), creating loose aggregate particles, and subsequently leading to the formation of potholes and disintegration of the layer. In addition, fast-flowing water

Fig. 2.13 Super cyclone Amphan creates waterlogging in Satkhira District

near streams can cause erosion of materials at different layers, such as base courses, and weaken the foundation of the top layers, thus causing the overall failure of the pavement structure. Water in RCC pavements can lead to an acceleration of cracking, leaching of chemicals, and corrosion of reinforcements and dowels.

The super cyclone "Amphan" hit Bangladesh on 20th May 2002, and created waterlogging in the affected area in Satkhira district, as shown in Fig. 2.13. This secondary impact of storm surge will further cause water intrusion into the foundation of the embankment, which will weaken the embankment. The embankment will gradually be deformed, and slope failure will take place at several weak points. The trapped sea water is also responsible for the salinity intrusion. Most cyclones with storm surges caused waterlogging in the affected areas, e.g., Sidr, Mohasen, Aila, and Bulbul.

2.3.4 Mitigation Measures

Nowadays, wheel loads are not only the miscreants for the deterioration of the pavements; several environmental load factors affecting the stability and durability of pavements due to global climate change should also be considered for the design, construction, and maintenance. The changes in the rainfall, temperature, and evaporation patterns can alter the moisture balances in the pavement foundation, which can lead to the reduction of the structural strength of the pavement. As the frequency of the extreme rainfall events has increased over the past few years, it can influence the structural strength and surface condition of the pavements (Sultana et al. 2015). In order to protect the road from early deterioration, the concept of cost-based analysis for holistic resilience measures needs to be undertaken during the design and construction of the road network (Manyena 2006), which may reduce accidental death, protecting the environment by creating less dust, reduce overall costs, and increase the durability of the road network. The mitigation measures refer to the practiced resilience measures in the study areas and suggested measures that are required to be adopted.

2.3.4.1 Institutional Framework for Disaster Resilience

Disaster resilience is the capacity of a system, community or society potentially exposed to hazards to adapt, by resisting or changing in order to reach and maintain an acceptable level of functioning and structure (UNISDR 2005). Disaster resilience is also the ability of countries, communities, and households to manage change, by maintaining or transforming living standards in the face of shocks or stresses – such as earthquakes, drought or violent conflict – without compromising their long-term prospects. Community resilience is defined as the ability of a community to withstand and mitigate the stresses of disasters (Chandra et al. 2011). Engineering resilience for both physical and social systems consists of properties including robustness, redundancy, resourcefulness, and rapidity (Bruneau and Reinhorn 2006). To make the road resilient to disaster, particularly in the coastal area, several measures can be taken to reduce the risk to roads of erosion from extreme waves including constructing earth levee banks with rip-rap protection and embankments, and installing larger drains and additional culverts to accommodate excessive runoff (ADB 2013a). In Bangladesh, the Local Government Engineering Department (LGED) and the Roads & Highway Department (RHD) mainly work on the transport infrastructural development with some international partners such as ADB, The World Bank, KfW, JICA, UNISDR, IFAD, and so on. The international organizations fund the disaster resilience projects, for example, construction of cyclone shelters in coastal districts to increase the community resilience with regard to cyclones, storm surges, and floods.

Making infrastructure resilient calls for engineering and non-engineering measures that consider the links between built and natural environments and among institutional frameworks. Careful consideration of development goals, prevailing situations, resources, and opportunities is needed to push the resilience agenda forward. Given the multifaceted dimensions of resilience, coordinated action from various sectors and stakeholders is imperative to achieve a safer future (ADB 2013b). The National Disaster Management Authority (NDMA) of India, in collaboration with UNISDR, organized a Conference including a featured event on "Disaster Risk Resilient Infrastructure for Sustainable Development," which highlighted the need for stronger collaboration and co-operation in the area of disaster-resilient infrastructure. Longevity and sustainability of infrastructures can be achieved by taking disaster resilience measures, especially in coastal districts associated with extreme events.

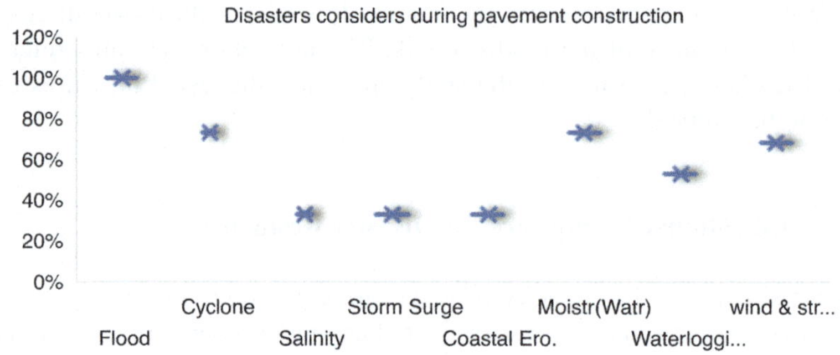

Fig. 2.14 Disasters considered during pavement construction in coastal areas

2.3.4.2 Factors That Need to Be Taken into Consideration

Climate change poses a major challenge to the policy makers in developing the most appropriate guidelines for the design, construction, and maintenance of road infrastructure.

The increase in the temperature and rainfall has a substantial impact from the planning stage through to the operation stage of the roads. Temperature fluctuation escalates the instances of thermal cracks on the roads (Qiao et al. 2013). and heavy precipitation leads to stripping and deformation. The poor road conditions resulting from natural calamities increase the chances of accidents and delay the transportation of necessities in the affected communities. Climatic factors, especially temperature and moisture, directly affect the performance of flexible pavements, which is considered during construction of roads, are presented in Fig. 2.14. These factors have long been a consideration in pavement design and practice because they may alter the deterioration of pavement materials, thus affecting pavement performance. Several factors need to be considered, including temperature, precipitation, height of storm surges, historical and predicted flood height, salinity, waterlogging potential, coastal erosion, and historical disasters (e.g., cyclones, Nor wester; Fig. 2.14).

Flood and storm surge height consideration during pavement construction is a resilience measure for increasing the lifespan of the pavement. Professionals involved with the design, construction, and maintenance of road pavements and drainage systems have already taken several measures to make the pavement resilient to disasters. Current resilience measures take into consideration 3–5 ft flood height and in maximum cases a 10-year return period of severe flooding, which is only 3–5 years for cyclones, construction of drainage, as well as structural and non-structural measures during the design and construction of the road network. In response to storm surge risk, 60% of cases report that they consider storm height up to 1.5 m, 20% consider 2.5 m, and the other 20% consider storm height at 1 m.

Fig. 2.15 Current structural and nonstructural measures. (**a**) Roadside drainage. (**b**) Excavation channel

2.3.4.3 Current Practice of Resilience Measures for Pavement Construction

Road pavements are required to be resilient to climate change-induced natural disasters such as flooding, storm surges, sea level rise, inundation, cyclones, and saline water. To accomplish the resilience of pavements, various structural and non-structural measures have been practiced in the study area. Structural measures include construction of embankment, dam, sluice gates, block damping, and road-side drainage, where geotextile, tree plantation, and vetiver grass plantation are the common non-engineering measures. Figure 2.15 shows some representative structural and nonstructural measures that mitigate the adverse impacts and increase the resilience of the road pavement to climate change-induced disasters.

Roadside Drainage Roadside drainage allows the free flow of water, which can lead to flooding. Drainage systems also prevent the accumulation of stagnant water to keep the fundamental road structure as dry as possible. Figure 2.15a shows the construction of roadside drainage for water discharge in the study area. This helps to remove flood, rain, and storm water from the road, which prevents the road from failures, holes, washout, etc. Findings suggest that the important roadside drainage is built at the last stage of road construction, and that the construction quality is very poor and inadequate regarding the required cross-sectional area. In some cases, drainage works are left owing to corruption. Moreover, many roads do not have roadside drainage. Inadequate roadside drainage reduces the pavement performance by creating ditches, embankment damage, road edge erosion, holes, cracks, wash-out, rutting, surface deformation, adhesion, and cohesion loss. Respondents (i.e., executive engineers, senior assistant engineers, Upazila engineers, and sub-assistant engineers) opined that the construction of drainage is poor in quality, required depth, discharge rate, and durability.

Fig. 2.16 Current resilience measures. (**a**) Planting vetiver grass. (**b**) Sluice gate for water regulation (Patuakhali)

Canal Excavation For flood and stormwater discharge Bangladesh Water Development Board (BWDB) and LGED excavate the canal, which helps to prevent the road from inundation, thus reducing road damage (Fig. 2.15b). A canal can remove the excess water and prevent the road from washout, holes, road edge erosion, ditches, etc. Canals are mainly re-excavated before the rainy season. The canals are insufficient and existing canals are gradually converted into the built area. Many canals are filled for construction, and buildings are constructed over the canals which reduces their capacity and increases sediment deposition; ultimately the canals stop discharging. An example of a witness, "re-excavation of canals is only the cleaning of both sides on the slope without reclaiming the depth and width of the canals".

Vetiver Grass Vetiver grass is planted on the roadside, which helps to prevent the road from slope protection as shown in Fig. 2.16a. Soil and climatic conditions in the study area indicate that vetiver grass can grow and sustain that region. Vetiver grass stabilizes the slope of coastal roads and embankments against the tidal surge. The special characteristics of vetiver grass are its longer life, high resistance to extreme climatic variation, finely structured root system, cost-effectiveness and environmental friendliness. The vetiver grass is effective to some extent, but investigation of its performance regarding the storm surges and frequent inundation in the coastal areas is required. It is recommended to study the performance of the vetiver grass against the storm surges, cyclones, frequent inundation, and floods.

Sluice Gates In order to regulate flood, rain, and storm water sluice gates are constructed, and are shown in Fig. 2.16b. Sluice gate is designed to adjust flow rates in sluices and canals. Sluice gates commonly control water levels and flow rates in rivers and canals. A sluice gate can regulate the excessive water inside an embankment; thus, it protects the road pavements from deterioration such as deformation, slope failure, rutting, holes, etc. The sluice gates are essential for discharging the flood and rain water. The number and location of sluice gates demand rigorous study because in most of the cases sluice gates are misplaced owing to the availability of land and social acceptance. In addition, the durability of the sluice gate is reduced as the suppliers provided very low-quality materials. The operation and maintenance of this infrastructure should involve the community people, who are overlooked in the existing policy (Table. 2.7).

2.3.5 Resilient Pavement Design Considering Changing Climate

It is evident that pavements are at an increasing and critical risk from climate-driven stressors; anticipated changes could alter the frequency, duration, and severity of road failures, as well as the time and cost of reconstructing the pavement systems. This study was conducted for the purposes of better understanding the potential impacts of climate change on pavement design and maintenance in the future and to identify limitations in the sector's knowledge. Findings indicate that pavements face failure from a combination of poor drainage and water impacts. Increased precipitation, flooding events, and/or sea level rise will increase the moisture content of the granular (soil) sub-layers under the pavement surface. The increase could be permanent (in the case of higher groundwater tables) or temporary (longer drainage times). Coastal roadways, especially local and secondary roads, will be flooded more frequently as the limit of high tide moves landward and storm surge and wave effects reach farther inland than in the past. Higher moisture content weakens the pavement base and subgrade; this can lead to increased cracking or rutting of the pavement surface owing to loss of underlying support, hastening the failure of the entire pavement structure. Flooding can also lead to sudden catastrophic failure (washout) if the event is sufficiently large. Impacts from temperature and water ultimately will require new pavement designs, roadway elevation or re-routing in order to accommodate expected changes (Daniel et al. 2014). Thus, it is mandatory to design and construct climate-resilient road infrastructure for the purposes of protecting the road from early deterioration as well as reducing maintenance costs.

2.3.5.1 Issues and Improved Resilience Measures

Historical Events

The policy makers consider a storm height of 1.5 m for the infrastructural development in the study area. The literature (Shahin et al. 2020a; Harwood 2012) suggests that the study area might have experienced a storm height of 6 m during the cyclone Sidr. (Table 2.4). Therefore, the design and construction standard should be revised based on current knowledge. In the last 20 years, Bangladesh has faced more than

Table 2.7 Lacks and gaps in current disaster resilience measures

Gaps	Impacts	Current measures/need to improve
Insufficient roadside drainage	• Creating cracks, ruts, failures, holes, surface deformation, surface defects, patches, stripping, erosion of the subgrade, edge joint cracks	• Construction of drainage for flood and rainwater movement • Constructing sluice gates for water movement • Excavation of canal for rapid rain/flood water movement
Not considering historical disasters	• A historical cyclone in the southern part causes storm levels to raise up to 6 m (19 ft) but most of the road is constructed by considering a storm height of 1.5 m • Flooding of coastal roads	• Revised flood and storm height needs to be considered based on real field conditions • Consider return period of flood, cyclone, storm surge, and salinity
Dumping waste on road	• Seepage and moisture making holes	• Proper maintenance and operation • Awareness among the villagers
Coastal erosion not properly managed	• Pavement is damaged with scouring, which needs to improve	• Planting vetiver grass and trees • Protecting coastal roads, common existing measures are geotextile, block damping, tree plantation, protective wall
Improper measures for moisture control	• Loss of strength and durability	• Several drainage systems have been constructed for rain, flood, and storm water movement
Inadequate dam height	• Not protecting saline water intrusion	• Dam height needs to be revised based on the previous and current tide, storm surge height, and sea level
Flexible pavements with lower grade materials	• Flexible pavements are more sensitive to climate change than rigid pavements	• A rigid and strong pavement with a high standard is the most disaster resilient
Lack of signposting/mirror in roads with maximum bends	• Accidents are frequently occurring, causing human and economic losses • Reduce the service life, trigger maintenance earlier, and eventually incurs additional costs	• To avoid preventable accidents, it is required to install signposting/mirror early embankment can be constructed along the coastal belt of Bangladesh to protect from tropical cyclones in Bay Bengal Areas deterioration of the pavement
Lack of documentation and study on the subject matter	• Practitioners suffering owing to proper guidelines to consider all factors	• The limited literature on road transport and climate change studies in Bangladesh indicates that further research is needed for assessment and quantification of impacts and adaptation strategies for developing resilient and sustainable road transport infrastructure

six cyclones, which caused severe damage because of the unusual storm height. Therefore, it is mandatory to consider storm height based on historical storm data to avoid further road damage.

Corruptions

Although the policy makers adopted few resilience measures in the study area such as raising road height above the flood level, introducing sand drains, and planting vetiver grass in the road slope, the contractors and suppliers provided low-quality material and did not maintain the quality of construction of the road pavements because they are politically privileged. The politically privileged win the bidding for the construction of roads and they implement nominal works to show to the authorities. The field officials and the consultants supervise the work regularly and carefully, but they sometimes face pressure to manage the politically privileged contractors. The noncooperation of such politically backed contractors may lead to implementation of sub-standard works, as well as failure to achieve the goal of resilient construction for sustainable development. Corruption is the main barrier to the development of resilience measures for sustainable road pavements.

Slope Protection Works

Tree plantation is a low-cost solution to the slope protection in the coastal areas of Bangladesh. Trees are planted on the slope to tie the soil on the slope. Cyclones uproot many of the trees on the slope, which partially damages the slope. The uprooted trees fall on the road pavement causing damage to the pavements. Wood trees should be avoided because their adverse effects outweigh the advantages. The uprooted trees block the roads during cyclones, which prolongs the rescue and relief work because vehicles cannot enter the area with emergency foods, drinks, water, and other supplies (Fig. 2.17). The roots of trees sometimes enter beneath the pavement and cause cracking. The roots of the trees reduce the lifespan of the

Fig. 2.17 Roadside trees are uprooted in cyclone Amphan (Source: Dhaka tribune, online)

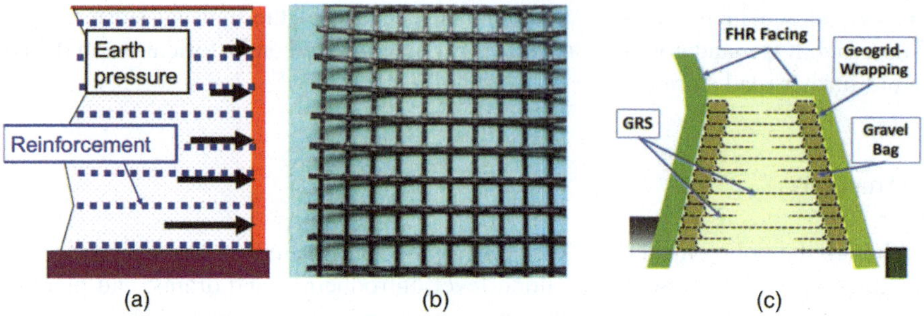

Fig. 2.18 Typical full-height rigid facing (FHR) embankment can be constructed along the coastal belt of Bangladesh to protect from tropical cyclones in Bay Bengal Areas (Tatsuoka et al. 2012)

pavements as has been found in several cases of tree plantation along the roadside. Herbs with deep spread roots could be planted on the slope to protect it, for example, vetiver grass. Full-height, rigid-facing, geosynthetically reinforced soil can be applied for slope protection against cyclones, storm surges, and floods. The use of jute/geotextile bags is a conventional method of slope protection of the embankment.

GRS Embankment

In addition to the existing resilience measures, it is recommended to construct full-height rigid (FHR) facing road embankment by introducing geosynthetic-reinforced soil (GRS) (Tatsuoka et al. 2012, 2014). The embankments that withstand storm surges, floods, and heavy rain can be constructed with GRS. GRS embankments are much stronger than the conventional embankments that had been applied in Japan for similar purposes (Tatsuoka et al. 2014). The FHR facing embankment can be constructed along the coastal belt of Bangladesh, as shown in Fig. 2.18. The advantages of the FHR facing embankments are a small force on the face, small overturning moment, and the lateral force at the bottom, no need for pile foundation, cost-effectiveness, mass narrower space, and self-supporting structures. There is a great deal of friction between the geogrid and soil (compaction), where the geogrid has high extension resistance. A GRS integral bridge is a probable solution for the abutments and backfill failure of the bridge in the case of storm surges and flooding. Along the coastline, FHR facing GRS embankment/dikes can be constructed for the protection of properties and lives, whereas a geogrid-wrapped GRS embankment can be a good solution for the construction of disaster-resilient road embankments inside the coastline in the study area.

2.3.6 *Adaptation Strategies for Climate Change*

Climate change-resilient roads comprise a set of interventions such as structural and nonstructural interventions. Structural interventions provided enough strength to cope with the loads imposed due to climate change and disasters, e.g., higher compaction, FHR facing GRS embankment, slope protection, grass plantation. Nonstructural interventions such as climate change policy could reduce the greenhouse gas emission, thus reducing the frequency and impacts of disasters. Technological interventions are also important to reduce the severity of the impacts of climate change. Road surface defects can be identified using GIS, remote sensing, image processing, and radar signals to take rapid actions during initial stages of formation of potholes, deformation, rutting, and cracking (Xing et al. 2019; Murdzek et al. 2018). This strategy has been used in many countries to identify the defects on the pavements. The cost of road construction is the highest in the world, estimated to be about BDT 1610 million/km of a two-lane road in Matarbari, Chittagong. FHR facing GRS embankment is cost effective, as was suggested by Tatsuoka et al. (2014). Therefore, GRS embankment can be proposed in the coastal areas of Bangladesh for resilient road pavements to adapt to climate change. The adaptation strategies can be divided in two categories, i.e., structural and bioengineering measures.

2.3.6.1 Structural Measures

Several structural developments and improvements have been discussed in the previous sections. Even though there are many measures, this study emphasizes the following structural measures to focus on the climate change-resilient road pavements in the coastal area of Bangladesh based on the field investigations and literature search.

- Policy: climate change adaptation policy development to ensure the implementation of disaster resilience measures for infrastructural development. Corruption control strategy is required to be revised. An example, e-tendering is adopted to control corruption of authority, but e-tendering sites are kept disabled during the tender submission period in most cases to privilege some preferred bidders. The system of e-tendering should be automated and controlled to avoid corruption.
- Planning: planning of roads with a disaster-resilient cross-section and dimensions is of utmost importance to attain disaster resilience. The tree plantation should be avoided on the slope. Herbs (vetiver grass) can be planted to tie the soil o the slope to some reported areas, but not at all. Alignment of new roads should avoid vegetative cover. Historical events must be considered for the planning of new roads, e.g., the height of storm surges.

- Stability: high strength road embankments and pavements, e.g., FHR facing GRS embankment. Slope stabilization with FHR facing structure to withstand persistent impacts of storm surges, floods, and cyclones. The performance of the conventional slope stabilization structures such as a dry-stone wall, gabion wall, and jute bag wall could be modeled against high-intensity storms. The degree of compaction should be strictly controlled (Tatsuoka et al. 2012), to ensure that additives can be mixed with sand for filling (Shahin et al. 2020b).
- Drainage: sufficiently improved drainage facilities should be installed to discharge surface runoff, infiltration, and storm water. Improved drainage systems to avoid erosion of road materials and discharge of runoff. The drainage system includes drainage and crosses drainage structures such as cascades, small check walls, culverts, and causeways.
- Materials: high-strength materials are recommended for the pavement construction to withstand the weather of climate change, for example, low-density polyethylene is a good additive that increases the strength of asphalt concrete 198% compared with pure bitumen (Shahin 2015). Durability of paving materials should be emphasized to cope with salinity.

2.3.6.2 Bioengineering Measures

Bioengineering is widely practiced in slope protection in Bangladesh, although its performance is questionable. Bioengineering refers to the use of vegetation, either alone or in conjunction with civil engineering structures such as embankments, retaining walls, dams, and drainage to manage water and debris, thereby reducing instability and erosion on slopes. Bio-engineering measures are also taken during earthwork of road construction. Bioengineering includes broadcasting seeds, seedling of local plants. It may be suitable for the area where inundation, flooding, and severe storm surges are absent. Typical bioengineering includes the following:

- Grass seed: generally grass seed is spread, or, alternatively, the grass is planted manually in lines across the slope of the road embankment. The growing grass root ties the soils on the slope, thus resulting in slope stabilization by armoring and reinforcing of slopes. The performance of the grass roots against the storm surges is required to be justified. The LGED recently started slope stabilization using vetiver grass plantation. There are many roads along the canal sides, where the water current directly hits the slope of the embankment. Grass plantation is ineffective in such cases.
- Tree plantation: shrubs or trees are planted at regular intervals on the slope, which later creates a dense network of roots in the soil supporting the slope. This a very cheap method of slope stabilization; even these trees are uprooted in the event of a cyclone and damage to the slope thus triggers the failure of road embankments.

- Brush layering, palisades, and fascines: in this system, logs of wood are cut to required lengths and are placed across the slope, usually following the contour, which forms a strong barrier, preventing the development of rill and trap soils from moving down the slope. The system catches debris and armors and reinforces the slope. This method is a temporary solution for some special cases, e.g., slope failures without surge load owing to loosened soil.

Adaptation to climate change for the disaster-resilient road embankment was summarized in Table 2.8. Along with the available resilience measures, the proposed adaptation measures shown in that table will be effective in making the road climate change resilient.

2.4 Climate Change and Health Interface Threatens Humanity

In this era of pandemic, "human health" is the most frequently pronounced term ever, although its homogenous words have been used for a long time. This term usually refers to a state in which a person is able to work well physically, mentally, socially, economically, and spiritually within the environment where he/she belongs. People have been dealing with various disasters in this world for centuries, but currently they are living under the greatest stress in the last 50 years due to multidimensional dangers and disasters. Climate change, global warming, and sea level rise are not only environmental terms now but also concerning issues for current health science. It is evitable that every implication that has occurred on this planet or on its outer surface has a health inference for humans and animals. For instance, salinity level increase leading to premature birth/miscarriage in coastal communities, temperature increase leads to heat stroke, erratic rainfall leads to seasonal flu, sea level rise destroys mangrove habitats, and many more. However, these climatic issues combined with intensified natural hazards and disasters that demand special care are identified in this section.

Bangladesh is a home for climate-induced disasters and almost every year a significant number of the population die because of these calamities; furthermore, the relatively low provision/budget for the health sector makes further development complex. The highest concentrations of poor people and malnourished children are in the coastal area, with a lower capacity to take any adaptive action against adverse situations, which leads to socio-economic vulnerability (Yu et al. 2010). This unique topographic feature combined with a high population density, rapid urbanization and widespread poverty, food insecurity, gender discrimination and poor health infrastructure, exacerbates the impacts of climate change on human health in Bangladesh (United Nations Population Division 2018). Without a healthy generation, no development work may be sustained; therefore, combined approaches such as health issues with environmental (i.e., coastal erosion, transportation network

Table 2.8 Climate change adaptation techniques for road embankments in the coastal area

Potential climate change issues	Impacts on roads	Adaptation options
Sea level rise	• Inundation of roads in coastal areas • More frequent or severe flooding of underground tunnels and low-lying infrastructure • Erosion of road base and bridge supports • Bridge scour • Loss of coastal wetlands and barrier shoreline • Land subsidence	• Protection of high-value coastal real estate with GRS integral levees, seawalls, and dikes • Strengthening and heightening of existing levees, seawalls, and dikes • GRS integral bridge construction • Relocation of some sections of roads inland • Restriction of most vulnerable coastal areas from further development
Rainfall	• Increases in flooding of roads and underground tunnels • Overloading of drainage systems, causing backups and street flooding • Increases in road scouring, road washout • Impacts on soil moisture levels, affecting the structural integrity of roads, bridges	• Construction of GRS integral road embankment • Protection of critical evacuation routes • Upgrading of road drainage systems • Protection of bridge piers and abutments with riprap • Increases in culvert capacity • Increases in the standard for drainage capacity for new transportation infrastructure and major rehabilitation projects (e.g., assuming a 500-year rather than a 100-year storm)
Cyclone (category 4–5)	• Greater probability of infrastructure failures • Increased threat to stability of bridge decks • Increased damage to signs, lighting fixtures, and supports	• Construction of GRS integral bridges that is recommended by Tatsuoka et al. (2014) • Changes in bridge design to tie decks more securely to substructure and strengthen foundations • Increases in drainage capacity for new transportation infrastructure or major rehabilitation projects (e.g., assuming more frequent return periods) • Development of modular traffic features and road sign systems for easier replacement • Adoption of modular construction techniques where infrastructure is in danger of failure
Flood	• Overloading of drainage systems • Greater probability of dam, embankment failures	• Increases in GRS integral culvert construction • GRS integral road embankment construction • Canal excavation • Enough drainage facilities and increased capacity

GRS geosynthetic reinforced soil

development), social, and economic, might be a suitable solution. The added burden of increased health problems due to climate change will worsen the situation and push back its developmental achievements (Rahman 2008).

Climate variability is already threatening to reverse some of Bangladesh's significant gains in improving the status of maternal and child health nutrition across the country (Sinharoy et al. 2018). Increase in maternal health complications due to climate change has been linked to a number of adverse health effects and pregnancy outcomes, including vertical transmission to the fetus, preterm birth, low birth weight, pre-eclampsia and eclampsia, and perinatal and maternal death (Rylander et al. 2013). In contrast, low birth weight in infants can in turn lead to malnourishment in children under 5 years (Shammi et al. 2019; Rahman et al. 2016a). Besides, along with extreme weather events, the outbreak of different water- and vector-borne diseases, air pollution, heat wave, and salinity intrusions have increased the death toll gradually in different regions of the country. Therefore, climate changes affect human health directly and indirectly in an unequivocal way in Bangladesh.

2.4.1 Climate Change Impacts on the Health Sector

Climate variability impacts health in numerous ways, including through increased heat, salinity intrusion poor air quality, and extreme weather events (floods, storms, cyclones), as well as through changes in temperature and precipitation that alter vector-borne disease, reduce water quality, and decrease food security (Sorensen et al. 2018). There are three basic pathways by which climate change affects human health, which appear below.

Direct impact: direct health consequences of climate change primarily relate to changes in the frequency of extreme weather, including heat, drought, heavy rain, cyclones, and storms.

Effects mediated through natural systems: environmental changes and ecological disruption increase the frequency of maternal and child health illness and different vector-borne and water-borne diseases.

Effects heavily mediated by human systems: diverse health consequences, e.g., traumatic, infectious, nutritional, psychological, etc. occur in demoralized and displaced populations in the wake of climate-induced economic dislocation, environmental decline, and conflict situations (Smith et al. 2017; Haider 2007).

2.4.1.1 Vulnerable Group to Climate Change

Climate variability and change has dire implications for every aspect of life, particularly human health, while children, the elderly, and women are the most vulnerable. Adverse effects of climate change are presently concentrated amongst children for reasons of physiological susceptibility, small body mass to surface area ratio, and a higher infection rate in infants (Basu and Ostro 2008). Older people are at greater risk because they tend to be less mobile in extreme weather events such as storms, floods, and heat-waves. (Hansen et al. 2012). The health risks associated with exposure to climatic stressors differ substantially between men and women. Women are more affected than men by a range of climate hazards, owing to differences in the prevalence of poverty, undernutrition, and exposure to water-logged environments (Sharmin and Islam 2013). Pregnancy is a period of increased vulnerability to a wide range of environmental hazards, including extreme heat and infectious diseases (Strand et al. 2012). In fact, women and children, especially of low socioeconomic status, are disproportionately impacted by climate-related shocks (Balasubramanian 2018).

2.4.1.2 Maternal and Child Health Impact

Miscarriage Increasing salinity of natural drinking water sources is one of the many problems in coastal areas of many countries, such as Bangladesh, China, Vietnam, California, Brazil, and the Netherlands (Vineis et al. 2011). Water salinity caused by rising sea level negatively affects the health of women living in the coastal zone of Bangladesh. It was recently proven that women living within 20 km of the coastline and 7 m above sea level were 1.3 times more likely to miscarry than women who live inland. For this reason, a higher percentage of pregnancies end in miscarriage in Chakaria (11%) costal area than in Matlab (8%) sub-district (BBC 2018; Kamenova 2019). In addition, heat stress during pregnancy period increases the risk of congenital defects, spontaneous abortion, and miscarriage. A pilot study in Dhaka city found elevated body temperature among outdoor women workers where maximum core body temperature was up to 37.5 °C or more at many points of work. Consequently, outdoor women workers who are pregnant are at a higher risk of anomaly than women who work indoors (Rahman et al. 2016b).

Pre-Term Birth Drinking saline water increases the risk of both (pre)eclampsia and gestational hypertension, which are the risk factors for stillbirth and pre-term delivery. In Khulna district, seasonal hypertension is observed among pregnant women, especially during the dry season when less precipitation increases drinking water salinity (Rahman et al. 2016a). Moreover, salinity intrusion correlated significantly with infant mortality during the last stages of gestation in coastal Bangladesh (Shammi et al. 2019).

2.4.1.3 Heat Stress

Increasing heat and extreme temperature is the climate-related illness of greatest concern in many countries, particularly in the USA and Australia (Bambrick 2016; Dlugokencky and Tans 2014). High temperatures adversely impact the human body by interfering with its ability to dissipate heat and thermoregulate, leading to heat-stroke in heat-susceptible individuals such as poor, elderly, young children, minority groups, and outdoor workers (Martiello and Giacchi 2010). Heat exhaustion or hyperthermia is also a major concern to the people living in the urban areas of Bangladesh. Raise of temperature increases hospital admission for renal diseases and mental disorders including dementia, neurotic and somatoform disorders among susceptible individuals, especially the elderly people in this country as a consequence of heat stress (Hansen et al. 2008). In addition, the incidence of heatstroke among rickshaw pullers and people working in industries has increased in recent years (Rahman 2008).

2.4.1.4 Climate Change and Vector-Borne Diseases

In Bangladesh, changing climate conditions are leading to increased rates of different vector-borne diseases such as *Leishmaniasis* (locally known as kala-azar), malaria, dengue hemorrhagic fever (DHF), and Chikungunya (Costello et al. 2009). Variations in climate such as changes in temperature and more frequent and intense rainfall have consequences for waterlogging conditions, which emerged in floods in urban areas owing to poor drainage, and an inefficient sewerage system. Thus, climate change has a detrimental effect on malaria/dengue through its influence on suitable vegetation and vector-breeding sites (Kumar et al. 2011).

Although dengue fever has been prevalent in many Asian, African, and North American countries since the 1780s, it was not dominant in Bangladesh until 2000 (Bhatia et al. 2013). Thereafter, owing to tremendous changes in weather patterns, dengue fever first attacked Bangladesh in 2000 with the principal vectors *Aedes aegypti* and *A. albopictus* (Hasib and Chathoth 2016). In 2019, dengue outbreaks surpassed all previous records in the capital city of Dhaka and peaked with 70,188 cases with 67 dengue-related deaths (Hsan et al. 2019). Scenarios of severe the dengue attack in Dhaka city and two dengue patients are shown in Fig. 2.19. Incidences of vector-borne diseases are higher in Dhaka, Chattogram, and Khulna districts of Bangladesh during the rainy season, especially from May to October (ProMED 2019). Latterly, dengue fever spread severely in many Asian countries such as Indonesia, Singapore, Pakistan, and China, whereas Sri Lanka was facing an extreme dengue epidemic, with at least 301 patients dead in 2017 (Tissera et al. 2020).

Alongside dengue, a major outbreak of Chikungunya by *Aedes* vector was reported in Bangladesh, with 984 cases confirmed and more than 13,176 clinically

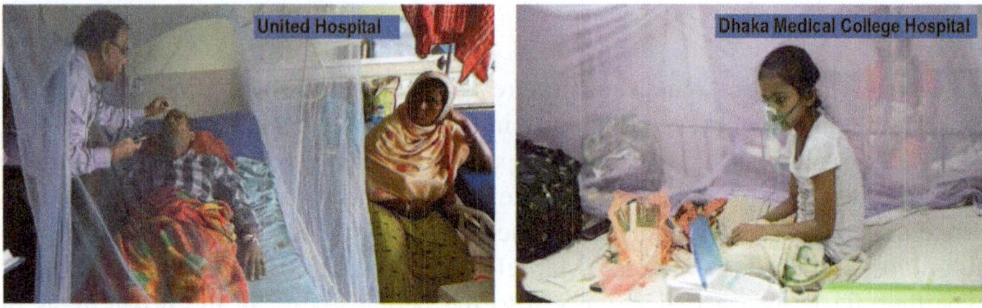

Fig. 2.19 Dengue patients admitted to a dengue unit in 2019 (Picture captured: Ashraf Uddin Sohag)

confirmed cases in 17 out of 64 districts, including the capital Dhaka in 2017 (Kabir et al. 2017). Moreover, malaria caused by *Plasmodium falciparum* is also endemic in the country in 13 eastern and northeastern bordering districts of Bangladesh. Malaria trends do show a decline in the percentage of population exposed, but approximately 34% of the population of these regions are still at risk (Islam et al. 2013). Although malaria is endemic in Kurigram, Sherpur, Mymensingh, Habigonj, Sylhet, Maulovibazar Cox's Bazar, and Chittagong district, Chittagong Hill Tracts districts have a very high prevalence of malaria where pregnant women are at a higher risk of asymptomatic *P. falciparum* infection (Hasib and Chathoth 2016; Khan et al. 2011)

2.4.1.5 Water-Borne Disease

The combination of higher temperatures and potential increases in precipitation creates the conditions for greater intensity or spread of many water-borne infectious diseases in Bangladesh (Shahid 2010). The seasonal peak of *Escherichia coli*, *Vibrio cholerae* and rotavirus diarrhea in Bangladesh coincides with the time when food is most contaminated owing to higher bacterial growth caused by high temperatures (Huq et al. 2005). For instance, morbidity due to rotavirus gastroenteritis created a burden on health care systems as 3783 children were hospitalized in Dhaka for gastroenteritis from July 2012 to June 2015 (Satter et al. 2017). On the other hand, noncholera diarrhea cases in Dhaka increase with higher temperatures, particularly in those individuals of a lower socio-economic and sanitation status. Many scientists argued that cholera in Bangladesh increases with the increase in sea surface temperature (Hashizume et al. 2007). Therefore, it is evident that the rise in temperature due to global warming increases the frequency of diarrheal diseases in Bangladesh.

2.4.1.6 Noncommunicable Diseases

Climate change and its rapid emergence in past decades have altered the distribution of some noncommunicable diseases in Bangladesh. Salinity intrusion is one of the major climate-changing variables, causing numerous health problems such as cardiovascular diseases (CVD), abdominal pain, and skin diseases in different coastal regions, particularly in Pirojpur, Bagerhat, and Satkhira districts (Chakraborty et al. 2019; Moumita et al. 2015). On the other hand, the prevalence of dehydration and respiratory disorders also increased during extreme heat waves and contribute up to 10% of the total burden of diseases (World Health Organization 2003). Outdoor air pollution is another major environmental threat to this country, especially to the urban areas such as Dhaka. As a result of extreme levels of air pollution, lead poisoning is also observed in several chemical and plastic manufacturing industries in Bangladesh (Sharmin and Islam 2013).

2.4.1.7 Climate Change and COVID-19

Coronavirus disease (COVID-19), a highly infectious communicable disease, is the major public health concern of the twentieth-first century. The disease is caused by a newly identified severe acute respiratory syndrome coronavirus 2 (SARS-CoV-2) virus that originated in bats. The outbreak was first identified in a seafood market of Hubei province, Wuhan, China, in December 2019. Thailand confirmed the first COVID-19 positive case outside of China on 13 January 2020. Owing to its rapid spread throughout the country, the World Health Organization declared a Public Health Emergency of International Concern on 30 January, and finally announced the disease as a pandemic on 11 March. Meanwhile, Bangladesh reported its first confirmed COVID-19 case on 8 March 2020. This pandemic situation is still ongoing and had infected 84,379 people with 1209 deaths in Bangladesh by 15 June 2020. The virus also spread to 213 countries of the world, infecting 7,810,339 people with a heavy death toll of 430,133 deaths. The virus spreads from close person-to-person contact through respiratory droplets from coughing and sneezing, touching the mouth, nose, or eyes after touching any infected area, and may also spread through air, as tiny droplets remain in the air for some time. As the mouth and nose are the main routes of transmission, wearing a mask, using hand sanitizer frequently, and maintaining physical distance are taken as preventive measures.

However, even though it seems that the virus was transmitted from bats, it is not the actual proven source. For instance, during the Influenza A virus pandemic in 1918, researchers opined that the virus originated from birds. Latterly, it was proven that the actual source of the virus was the ice and water of lakes situated in northeastern Siberia. As sea ice forms, brine channels and pockets enriched in salts and nutrients are created and remain in a liquid state inside the ice, which supports microbial growth. For this reason, bacteria can grow in Antarctica and the Arctic Ocean and when the snow melts because of temperature rise caused by climate change, the virus bacteria leaks from there and is transmitted throughout the world via biotic and abiotic reservoirs. Similarly, when snow melted from the lakes of northeastern Siberia, many migratory birds passed the area and carried the virus to Asia, North America, Europe, and Africa. This theory indicates the outbreak of a new virus as a long-term consequence of climate-changing events, including increased temperatures, sea level rise, and changing weather patterns. Thus, the outbreak of COVID-19 also may be a long-term consequence of climate change, but it needs proven research to justify the argument (Yau and Seth-Pasricha 2019; Zhang et al. 2007).

2.4.1.8 Extreme Weather Events and Consequences

Bangladesh was ranked sixth on Germanwatch's 2015 Global Climate Risk Index because it is significantly affected by extreme weather events including floods, cyclones, and droughts (Kreft et al. 2015). An extreme climatic phenomenon causes destruction of physical and social infrastructure, assets, crop production, and loss of lives. An economic summary and human health effects are presented in Table 2.9. Apart from this, natural disasters also have indirect effects on socio-economic conditions and the well-being of women, girls, and young children.

Poverty Aftermath of disasters such as frequent flooding, cyclones, and river erosion means that many families have to live with the constant threat of insecurity and increased poverty that adversely affect their livelihood (Kamal et al. 2015).

Sexual Abuse and Child Marriage Adolescent girls face high levels of sexual harassment and abuse owing to the lack of privacy in emergency shelters and some are pushed into early marriage (Bartlett 2008).

Educational Determination In disaster-prone areas, particularly in the Khulna and Barishal division, schooling is often disrupted when educational infrastructures are destroyed and used for shelters, which forces young children into child labor (UNICEF 2016).

Psychiatric Disorder In a conflict situation, mental health problems also grow among climate-affected people, which was frequently observed in the post-Sidr and post-Aila periods (Kabir et al. 2016).

Population Displacement and Communicable Disease Population displacement due to an extreme climatic event leads to increases in communicable diseases and poor nutritional status resulting from overcrowding, and a lack of safe water, food, and shelter in Bangladesh (Kouadio et al. 2012).

Table 2.9 Overview of extreme weather events and its overall impact in Bangladesh

Weather event	Year	Most affected area	Provoked loss
Cyclone	2020 (Amphan)	26 districts	• 10 people died from accidents and mental instability
	2019 (Bulbul)	7 districts	• 15 were injured and 13 people died • More than 108,000 houses and 117,000 ha of crops were affected
	2013 (Viyaru)	Patuakhali, Bhola, Barguna	• 13 people died • Approximately 49,000 houses were destroyed and 45,000 houses partially destroyed
	2009 (Aila)	Satkhira, Khulna	• Total death 190 • Affected nearly 4 million people • Over 350,000 acres of cropland were destroyed
	2007 (Sidr)	Barguna Patuakhali	• Killed 4000 people • Economic impact of US$1.7 billion
	1970 (Bhola cyclone)	Bhola	• Estimated to have killed 500,000 people • Economic loss of US$1.7 billion
Flood	2019	15 districts	• 114 people have died • 14,781 people have fallen ill of different diseases
	2017	31 districts in the northern part	• A total of 121 people died • 15,529 ha of land fully inundated
	2007	Dhaka	• Most of the deaths were caused by snake bites • Huge losses in the crop, dairy, poultry, and fisheries sectors • Economic losses US$1.06 billion
	2004	Comilla, Feni, Jessore, Satkhira, Magura	• Total deaths 747 • US$2.2 billion in economic losses
	1998	75% of the total area of the country	• 69% of Aush rice production, 82% of deep water Aman and 91% of transplanted Aman were lost • Economic impact of US$2.3 billion
Drought	2006	Northwestern region (Rangpur)	• Reduced Aman crop production of about 25–30% • 59% of drought victims left the area in 2006
	2000		• Severe damage of crop production • Population displacement
	1995		• Damage to rice, jute, bamboo, and other crops
	1994		• 84% of drought victims migrated from the drought-prone area
	1973		• Widespread damage to crops was in part responsible for the famine in northwestern Bangladesh in 1974

Majumder (2013), Islam et al. (2014), Kabir and Hossen (2019)

Thereafter, it implies that climate change not only takes human lives, it also has a long-term adverse effect from the perspective of nutrition, food security, education, and finance that delays the ultimate development of the country for decades.

2.4.2 Relationship Among Climate Change, Food Security, and Malnutrition

Climatic events and food security issues are multi-faceted and transcend national boundaries, which are further linked to malnutrition and hunger. Although tidal surges, floods, drought, and heavy rainfall are primarily responsible for damaging the cropping system, nevertheless, salinity intrusion is the principal source of food insecurity in the coastal region of Bangladesh, indirectly affecting the nutritional status of the coastal people (Moumita et al. 2015).

Climatic fluctuations destroy arable land and reduce food production, which in turn increases the prices of essential goods (Nahar 2016). Conversely, people are turning to shrimp farming as a way to manage the increased salinity of land that had undermined the viability of their previous agricultural activities (Vineis et al. 2011). Along with this, homestead plantations, particularly sweet water-loving fruit species have been declining in recent years owing to increasing soil salinity (Alam et al. 2017). On the other hand, erratic and irregular rainfall, as well as high temperature, reduced the production of target species in marine and freshwater systems by affecting the readiness, maturity and gonad development of fishes in the breeding season. In addition, increased salinity and change in water quality can instigate a change in species composition and distribution, especially in coastal areas, that indicates the change in the seasonal abundance of individual fish in communities (Dasgupta et al. 2017) (Fig. 2.20).

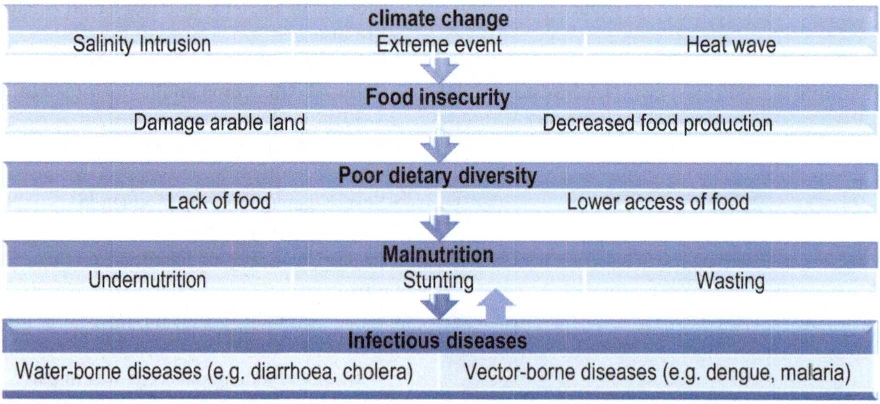

Fig. 2.20 Inter-relationship among climate change, food insecurity, and malnutrition

Moreover, livestock are more vulnerable to climate change because they have no protective structures and supplemental feed. Drought events also lead to a decline in meat, egg, and milk production from the livestock sector by altering the metabolic system and creating heat stress (Chowdhury et al. 2016). Thus, it implies that climate change precedes food insecurity by affecting the availability and accessibility of food at the community level. As a result, people consume a less diverse diet, and have inadequate food and nutrition, leading to both short-term (e.g., underweight, wasting) and long-term malnutrition (stunting). People, especially children, who are suffering from malnutrition are more likely to be affected by infectious diseases, such as malaria, diarrhea, pneumonia, etc. Conversely, illness due to infectious diseases increases the risk of becoming malnourished (Saha et al. 2020). Therefore, climate changes induce human health and nutritional status by impacting food security, which leads the country to prolong malnutrition.

2.4.3 Health Adaptive Measures by the Government of Bangladesh Responding to Climate Change

The health consequences of climate change are complex and interrelated with many factors. Although Bangladesh is a virtually zero contributor to greenhouse gas emissions, it has to suffer disastrously the undeniable and unequivocal effects of climate change (Ali 1999). However, the country is well positioned to address these challenges, as there is a political will to deal with climate adaptation, and progress is being made (Zamudio and Parry 2016). To address the health concerns due to climatic fluctuation, the government of Bangladesh established the Bangladesh Climate Change Strategy and Action Plan (BCCSAP) in 2009, which included six pillars, with health belonging to the first pillar. In 2012, the government of Bangladesh took the further initiative of developing the National Adaptation Plans (NAPs) intended to provide support for medium- and long-term adaptation planning.

Finally, to establish a climate-resilient health system, the World Health Organization operational framework, the Health component of the National Adaptation Plan (HNAP) needed to be developed in 2018. Therefore, to mainstream climate change into existing health programs and operational plans, the HNAP Action Plan outlines key short- and medium-term health adaptation activities with an estimated budget of US$ 1,038,000 to be carried out over a 5-year period (2018–2023). Major activities of the HNAP Action Plan includes strengthening the Climate Change and Health Promotion (CCHP) Unit, assessing vulnerability, providing in-service training for the health workforce, developing a knowledge-sharing platform, implementing "green hospitals," and broadening medical and disaster response supply systems during emergencies.

Along with the Bangladeshi government, some prominent organizations are contributing to the combating climate change issue, including the Bangladesh Centre for Advanced Studies (BCAS), the Centre for Natural Resource Management (CNRS), IUCN Bangladesh, Action Aid Bangladesh, CARE Bangladesh, Oxfam Bangladesh, Practical Action Bangladesh, Bangladesh Red Crescent Society, Concern Worldwide, and Caritas (World Health Organization 2017). Besides, researchers and policy makers are thinking about climate-smart agriculture and community-level adaptation strategies to mitigate the severe health impact of climate change. Above all, increasing awareness among people and collaboration between all sectors, both national and international, are highly required to address climate change issues in countries like Bangladesh.

2.5 Conclusion

Low- to middle-income countries such as Bangladesh, are trying to hold them to a (top level achievements), but the "sins" of "industrialized" or "developed" countries are some of the main hindrances to not achieving this, as they are leaders for changing global climate. These unbalancing practices are threatening the civilization by melting the polar ice, increasing waterbodies, and sinking low-lying countries such as Bangladesh. Such predictions are the harsh reality in the coastal areas of sea-facing countries. Almost every year, coastal districts are facing high-intensity and -magnitude natural calamities that are breaking the previous records and if this continues, it might be that the earth does not have much time to hold its habitats. The very recent cyclone "Amphan" severely affected the coastal areas of Bangladesh and the houses of about three million people, cultivated lands, and fish-farms have gone under water, and this is occurring almost every year. However, natural calamities are neither preventable nor avoidable but it is possible to obey some measures to make a resilient community. Therefore, temporally, ANN, and CA-Markov predicted erosion-prone areas will be required to protect the landscape by implementing appropriate engineering measures. For the road networks, a perception of professional clusters has been presented and found that the washed out base structure, surface deformations, and potholes are the main problems, but the measures taken are insufficient and need to be revised based on real current scenarios where the proposed structural bioengineering measures, specifically GRS embankment, might be a sustainable solution. Finally, Bangladesh has water everywhere, but a great lack of drinking water sources, which is a leading cause of human health deterioration specifically in the coastal area. Health care is not up to scratch yet, owing to budget shortages and "corruption," which needs to be eliminated. This study, a door to further investigation, research, and development for improving the life status of the coastal communities, who are leading a very risky life and who are field warriors for any kind of hazard and disaster.

References

Abhijit P (2011) Effects of bad drainage on roads. *Environ Res* 1(1):1–8

Abo-Qudais S, Al-Shweily H (2007) Effect of aggregate properties on asphalt mixtures stripping and creep behavior. *Constr Build Mater* 21(9):1886–1898

ADB (2011) Annual report 2011, ISSN 306–8370 ISBN 978–92–9092-628-3 (print), 978–92–9092-629-0. Asian Development Bank, Mandaluyong City, Philippines, p 2012

ADB (2013a) Annual report 2013: nonsovereign approvals by developing member country, 1983–2013. Asian Development Bank, 2013, Mandaluyong City, Philippines

ADB (2013b) Annual report 2013: promoting environmentally sustainable growth in Asia and the Pacific. Asian Development Bank, 2013, Mandaluyong City, Philippines

Admiral R, Alam MK (2016) Ocean/Blue economy for Bangladesh

Ahamed F, Hossain MY, Fulanda B, Ahmed ZF, Ohtomi J (2012) Indiscriminate exploitation of wild prawn postlarvae in the coastal region of Bangladesh: a threat to the fisheries resources, community livelihoods and biodiversity. *Ocean Coast Manag* 66:56–62

Ahmed A, Drake F, Nawaz R, Woulds C (2018) Where is the coast? Monitoring coastal land dynamics in Bangladesh: an integrated management approach using GIS and remote sensing techniques. *Ocean Coast Manag* 151:10–24

Ahsan ME (2013) Coastal zone of Bangladesh fisheries resources and its potentials. Lap LAMBERT Academic Publishing, Saarbrücken, p 69

Alam MZ et al (2017) Effect of salinity intrusion on food crops, livestock, and fish species at Kalapara Coastal Belt in Bangladesh. *J Food Qual* 2017:1

Ali A (1999) Climate change impacts and adaptation assessment in Bangladesh. *Clim Res* 12(2–3):109–116

Amin MSR, Tamima U, Amador L (2020) Towards resilient roads to storm-surge flooding: case study of Bangladesh. *Int J Pavement Eng* 21(1):63–73

Asphalt Institute (1981) Cause and prevention of stripping in asphalt pavements, Educational series No. 10. Asphalt Institute, College Park

Azad AK, Jensen KR, Lin CK (2009) Coastal aquaculture development in Bangladesh: unsustainable and sustainable experiences. *Environ Manag* 44(4):800–809

Balasubramanian M (2018) Climate change, famine, and low-income communities challenge sustainable development goals. *Lancet Planet Health* 2(10):e421–e422

Bambrick (2016) Getting better faster: a 90 day plan for coaching new teachers, San Francisco, CA, Jossey-Bass, 2016. Print

Bartlett S (2008) Climate change and urban children: impacts and implications for adaptation in low-and middle-income countries. *Environ Urban* 20(2):501–519

Basu R, Ostro BD (2008) A multicounty analysis identifying the populations vulnerable to mortality associated with high ambient temperature in California. *Am J Epidemiol* 168(6):632–637

BBC (2018) How climate change could be causing miscarriages in Bangladesh – BBC News

BBS, SID (2011) *Population and housing census 2011 (Bangladesh)*. Bangladesh Bureau of Statistics, Dhaka

Bhatia R, Dash AP, Sunyoto T (2013) Changing epidemiology of dengue in South-East Asia. *WHO Southeast Asia J Public Health* 2(1):23

Bhattacharjee S (2011) Uncertainties of flash flood in context of climate change in Sunamganj District, Bangladesh. In: *World Environmental and Water Resources Congress 2011: bearing knowledge for sustainability – Proceedings of the 2011 World Environmental and Water Resources Congress*, pp 4564–4571

Brammer H (2014a) Climate Risk Management Bangladesh's dynamic coastal regions and sea-level rise. *Clim Risk Manag* 1:51–62

Brammer H (2014b) Bangladesh's dynamic coastal regions and sea-level rise. *Clim Risk Manag* 1:51–62

Bruneau M, Reinhorn A (2006) Overview of the resilience concept. In: *The 8th U.S. National Conference on earthquake engineering April 18-22, 2006, San Francisco, California, USA. paper no. 2040*, p 3

Byard R, Payne-James J (2015) *Encyclopedia of forensic and legal medicine*. Academic Press, Amsterdam

Caro S, Masad E, Bhasin A, Little DN (2008) Moisture susceptibility of asphalt mixtures, part 1: mechanisms. *Int J Pavement Eng* 9(2):81–98

Chai G, van Staden R, Guan H, Kelly G, Chowdhury S (2014) The impacts of climate change on pavement maintenance in Queensland, Australia. In: *Transport research arena*. Wiley, Hoboken

Chakraborty R, Khan KM, Dibaba DT, Khan MA, Ahmed A, Islam MZ (2019) Health implications of drinking water salinity in coastal areas of Bangladesh. *Int J Environ Res Public Health* 16(19):3746

Chandra A et al (2011) Building community resilience to disasters. RAND Health, Santa Monica

Chowdhury QMMK, Hasan M, Ahmed J, Shykat CA, Islam MS, Hossain M (2016) Impact of climate change on livestock in Bangladesh: a review of what we know and what we need to know. *Am J Agric Sci Eng Technol* 3(2):21–31

Costello A et al (2009) Managing the health effects of climate change: lancet and University College London Institute for Global Health Commission. *Lancet* 373(9676):1693–1733

Daniel JS, Jacobs JM, Douglas E, Mallick RB, Hayhoe K (2014) Impact of climate change on pavement performance: preliminary lessons learned through the infrastructure and climate network (ICNet). Clim Eff Pavement Geotechnol Infrastruct – Proc Int Symp Clim Eff Pavement Geotech Infrastruct 2013 5(November 2015):1–9

Dasgupta S, Huq M, Mustafa MG, Sobhan MI, Wheeler D (2017) The impact of aquatic salinization on fish habitats and poor communities in a changing climate: evidence from Southwest Coastal Bangladesh. *Ecol Econ* 139:128–139

De Carteret S, Buzzi RSO, Fityus (2010) *A review of the effects of salinity on road pavements and bituminous surfacings*. Kasetsart University, Bangkok

Dlugokencky E, Tans P (2014) Trends in atmospheric carbon dioxide. National Oceanic & Atmospheric Administration, Earth System Research Laboratory (NOAA/ESRL)

Emery J, Seddik H (1997) Moisture-damage of asphalt pavements and antistripping additives: causes, identification, testing and mitigation. Canada Transport Association, Ottawa

Frazier Parker J (1989) A field study of stripping potential of asphalt concrete mixtures, Project Number ST 2019-6. Highway Research Center, Auburn University, Auburn

Fromm HJ (1974) The mechanisms of asphalt stripping from aggregate surfaces. *Assoc Asph Paving Technol* 43:191–223

Habiba U, Abedin MA, Shaw R, Hassan AWR (2013) Salinity-induced livelihood stress in coastal region of Bangladesh. *Community Environ Disaster Risk Manag* 13:139–165

Haider MR (2007) Climate change and its impacts on public health. New age living climate change

Halberg S (1950) The adhesion of bituminous binder and aggregate in the presence of water. Medd Statens Vaginst Stock Se 78

Hansen A, Bi P, Nitschke M, Ryan P, Pisaniello D, Tucker G (2008) The effect of heat waves on mental health in a temperate Australian city. *Environ Health Perspect* 116(10):1369–1375

Hansen J, Sato M, Ruedy R (2012) Perception of climate change. *Proc Natl Acad Sci* 109(37):E2415–E2423

Haque A, Jahan S (2016) Natural disasters in Bangladesh: impact analysis through input-output model. *J Bangladesh Inst Planners* 8(December):15–21

Harwood P (2012) Cyclone Sidr (2007). [Online]

Hashizume M et al (2007) Association between climate variability and hospital visits for non-cholera diarrhoea in Bangladesh: effects and vulnerable groups. *Int J Epidemiol* 36(5):1030–1037

Hasib E, Chathoth P (2016) Health impact of climate change in Bangladesh: a summary. *Curr Urban Stud* 4(1):1–8

Hoq ME (2003) Sustainable use of mangrove fisheries resources of Sundarbans, Bangladesh. *Trop Agric Res Ext* 6:113

Horgnies M, Darque-Ceretti E, Fezai H, Felder E (2011) Influence of the interfacial composition on the adhesion between aggregates and bitumen: investigations by EDX, XPS and peel tests. *Int J Adhes Adhes* 31(4):238–247

Hsan K, Hossain MM, Sarwar MS, Wilder-Smith A, Gozal D (2019) Unprecedented rise in dengue outbreaks in Bangladesh. *Lancet Infect Dis* 19(12):1287

Hughes RI, Lambs DR, Prodes O (1960) Adhesion in Bitumen Macadam. *J Appl Chem* 10(11):433–444

Huq A et al (2005) Critical factors influencing the occurrence of Vibrio cholerae in the environment of Bangladesh. *Appl Environ Microbiol* 71(8):4645–4654

IPCC (2007) Mitigation of climate change: contribution of working group III to the fourth assessment report of the Intergovernmental Panel on Climate Change. Cambridge University Press, Cambridge

Islam MR, Tabeta S (2019) Shrimp vs prawn-rice farming in Bangladesh: a comparative impacts study on local environments and livelihoods. *Ocean Coast Manag* 168:167–176

Islam N, Bonovas S, Nikolopoulos GK (2013) An epidemiological overview of malaria in Bangladesh. *Travel Med Infect Dis* 11(1):29–36

Islam A, Tasnuva A, Sarker SC, Rahman MM, Mondal MSH, Islam MMU (2014) Drought in Northern Bangladesh: social, agroecological impact and local perception. *Int J Ecosyst* 4(3):150–158

Islam MM, Islam MH, Shahin M (2016) Communities strength and weakness for responding cyclone. In: Proceedings of the 3rd International Conference on Civil Engineering for Sustainable Development (ICCESD 2016), 12~14 February 2016, KUET, Khulna, Bangladesh (ISBN 978-984-34-0265-3), no. February, pp 978–984

Islam MM, Rahman MS, Kabir MA, Islam MN, Chowdhury RM (2019) Predictive assessment on landscape and coastal erosion of Bangladesh using geospatial techniques. *Remote Sens Appl Soc Environ* 17(August):100277

Islam MM, Mohaiman R, Mostafa CAKM, Saifur Z (2020) Spatiotemporal mapping mangroves of Tengragiri wildlife sanctuary under Barguna district of Bangladesh using freely available satellite imagery. *Model Earth Syst Environ* 6(Mathbor 2007):917

Kabir MH, Hossen MN (2019) Impacts of flood and its possible solution in Bangladesh. *Disaster Advances* 12:48–57

Kabir R, Khan HTA, Ball E, Caldwell K (2016) Climate change impact: the experience of the coastal areas of Bangladesh affected by Cyclones Sidr and Aila. J Environ Public Health 2016

Kabir I, Dhimal M, Müller R, Banik S, Haque U (2017) The 2017 Dhaka chikungunya outbreak. *Lancet Infect Dis* 17(11):1118

Kamal SMM, Hassan CH, Alam GM, Ying Y (2015) Child marriage in Bangladesh: trends and determinants. *J Biosoc Sci* 47(1):120–139

Kamenova D (2019) For some elements of the modern educational collage. J Pedagogy Educational Manag 3:2–4

Kermack WO, McKendrick AG (1927) A contribution to the mathematical theory of epidemics. *Proc R Soc Lond Ser A Contain Pap Math Phys Character* 115(772):700–721

Khan AE et al (2011) Drinking water salinity and maternal health in coastal Bangladesh: implications of climate change. *Environ Health Perspect* 119(9):1328–1332

Khatun MA, Rashid MB, Hygen HO (2016) Climate of Bangladesh

Kim ST, Yu JY, Lu MM (2012) The distinct behaviors of Pacific and Indian Ocean warm pool properties on seasonal and interannual time scales. *J Geophys Res Atmos* 117(5):D05128

Kirui KB et al (2013) Ocean & coastal management mapping of mangrove forest land cover change along the Kenya coastline using Landsat imagery. *Ocean Coast Manag* 83:19–24

Kothari M, Poongavanam S (2019) Effective role of bimstec in the governance of the Bay of Bengal. *Int J Innov Technol Explor Eng* 8(6 Special Issue 4):1295–1298

Kouadio IK, Aljunid S, Kamigaki T, Hammad K, Oshitani H (2012) Infectious diseases following natural disasters: prevention and control measures. *Expert Rev Anti-Infect Ther* 10(1):95–104

Kreft S, Eckstein D, Melchior I (2015) Global climate risk index 2017: who suffers most from extreme weather events, Weather loss events. Germanwatch, Bonn

Kumar U, Khan MSA, Rahman R, Mondal MS, Huq H (2011) Water security in Peri-Urban Khulna: adapting to climate change and urbanization, Peri Urban Water Security Discussion Paper Series. Oxford University Press, Delhi
Laurance WF et al (2012) Averting biodiversity collapse in tropical forest protected areas. *Nature* 489(7415):290–293
Little DN, Jones IV (2003) Chemical and mechanical processes of moisture damage in hot-mix asphalt pavements. In: *Moisture Sensitivity of Asphalt Pavements-A National Seminar, Transportation Research Board*, pp 37–74
Liu J, Pu H, Song S, Du M (2018) An adaptive scale estimating method of multiscale image segmentation based on vector edge and spectral statistics information. *Int J Remote Sens* 39(20):6826–6845
Mahmud SMN, Islam MT, Rabbi MF (2015) Typical pavement distresses of Dhaka City roads. In: *International Conference on Recent Innovation inCivil Engineering for Sustainable Development*, pp 763–768
Majumder S (2013) The economics of early response and resilience: Bangladesh country study. UK Government Department for International Development, London
Manyena SB (2006) The concept of resilience revisited. *Disasters* 30(4):434–450
Martiello MA, Giacchi MV (2010) High temperatures and health outcomes: a review of the literature. *Scand J Public Health* 38(8):826–837
Mascarenhas A (2011) Response of coastal vegetation and the need for green belts along the Tamil Nadu coast, India: the December 2004 tsunami experience. In: *Tsunamis: causes, characteristics, warnings and protection*, pp 131–147
Minar MH, Hossain MB, Shamsuddin MD (2013) Climate change and coastal zone of Bangladesh: vulnerability, resilience and adaptability. *Middle-East J Sci Res* 13(1):114–120
Ministry of Environment and Forests (2008) *Climate Change Strategy and Action Plan*. Ministry of Environment and Forests, Dhaka
Mirza MMQ (2002) The Ganges water-sharing treaty: risk analysis of the negotiated discharge. *Int J Water* 2(1):57–74
Moumita C, Biswas BK, Muktadir H (2015) Nutritional status of women living at South-west Coastal Belt of Satkhira Bangladesh. *J Environ Sci Nat Resour* 8(2):41–46
Mozumder MMH, Uddin MM, Schneider P, Islam MM, Shamsuzzaman MM (2018) Fisheries-based ecotourism in Bangladesh: Potentials and challenges. *Resources* 7(4):61
Murdzek R, Malik H, Leśniak A (2018) The use of the DInSAR method in the monitoring of road damage caused by mining activities. *E3S Web Conf* 36:02005
Nahar MA (2016) The impact of climate change in Bangladesh on the rice market and farm households. University of Arkansas, Fayetteville
Nateque Mahmood M, Prasad Dhakal S, Keast R (2014) The state of multi-purpose cyclone shelters in Bangladesh. *Facilities* 32(9/10):522–532
Neema MN (2014) Spatial and temporal changes of climatic variables and its impact on natural disasters in Bangladesh. *Am J Environ Prot* 3(3):122
Parsons TA, Pullen BA (2016) Relationship between climate type and observed pavement distresses. In: *International Conference on Transportation and Development 2016*, pp 88–102
Pirlea O, Burlacu A (2013) The influence of sea water on promenade pavement. In: Recent advances in civil and mining engineering, pp 268–273
ProMED (2019) Brookline, MS, USA; International Society for Infectious Diseases
Qiao Y, Flintsch GW, Dawson AR, Parry T (2013) Examining effects of climatic factors on flexible pavement performance and service life. *Transp Res Rec J Transp Res Board* 2349(1):100–107
Quium ASMA, Hoque SAMA (2002) The completeness and vulnerability of road network in Bangladesh. In: Ali MA, Seraj SM, Ahmad S (eds) *Engineering concerns of flood*, pp 59–75
Rahman A (2008) Climate change and its impact on health in Bangladesh. *Reg Health Forum* 12(1):16–26

Rahman E (2020) 2.4m people, over half a million livestock moved to some 15,000 cyclone shelters in Bangladesh. Bangladesh Red Crescent reports first death from cyclone Amphan Bangladesh moves hundreds of Rohingyas to shelters as cyclone Amphan nears, pp 14–15

Rahman MM, Rahaman MM (2018) Impacts of Farakka barrage on hydrological flow of Ganges river and environment in Bangladesh. *Sustain Water Resour Manag* 4(4):767–780

Rahman MS, Howlader T, Masud MS, Rahman ML (2016a) Association of low-birth weight with malnutrition in children under five years in Bangladesh: Do mother's education, socio-economic status, and birth interval matter? *PLoS One* 11(6):e0157814

Rahman J, Fakhruddin SHM, Rahman AKMF, Halim MA (2016b) Environmental heat stress among young working women: a pilot study. *Ann Glob Health* 82(5):760–767

Rahman MB, Islam MM, Ruma M, Rahman MM (2020) Effects of environmental condition on spatial-temporal changes of fish diversity and morphology of Shitalakshya River. *Ecología* 10(1):38–49

Ralph TU, Gough WA (2009) The influence of sea-surface temperatures on Eastern North Pacific tropical cyclone activity. Theor Appl Climatol 95(3–4):257–264

Regmi MB, Hanaoka S (2011) A survey on impacts of climate change on road transport infrastructure and adaptation strategies in Asia. *Environ Econ Policy Stud* 13(1):21–41

RHD (2009) Survey manual. *Minist Commun Roads Highw Dep Bangladesh* 3(20):3–7

Rubio-Cisneros NT et al (2019) Poor fisheries data, many fishers, and increasing tourism development: interdisciplinary views on past and current small-scale fisheries exploitation on Holbox Island. *Mar Policy* 100:8–20

Rylander C, Øyvind Odland J, Manning Sandanger T (2013) Climate change and the potential effects on maternal and pregnancy outcomes: an assessment of the most vulnerable–the mother, fetus, and newborn child. *Glob Health Action* 6(1):19538

Saha UR, Chattopadhyay A, Richardus JH (2020) Correction: trends, prevalence and determinants of childhood chronic undernutrition in regional divisions of Bangladesh: evidence from demographic health surveys, 2011 and 2014. *PLoS One* 15(2):e0229677

Sarie F, Bisri M, Wicaksono A, Effendi R (2015) Types of road pavement damage for road on peatland, a study case in Palangka Raya, Central Kalimantan, Indonesia. *IOSR J Environ Sci* 9(12):53–59

Sarwar MGM, Woodroffe CD (2013) Rates of shoreline change along the coast of Bangladesh. *J Coast Conserv* 17(3):515–526

Satter SM et al (2017) Hospital-based surveillance for rotavirus gastroenteritis among young children in Bangladesh: defining the potential impact of a rotavirus vaccine program. *Pediatr Infect Dis J* 36(2):168

Shahid S (2010) Probable impacts of climate change on public health in Bangladesh. *Asia Pac J Public Health* 22(3):310–319

Shahin M (2015) *Effect of water on the compressive strength of bituminous mixes*. LAP LAMBERT Academic Publishing, Saarbrücken

Shahin M, Khokon ZH, Sobhan MA, Ahmed TU (2015) Salt tolerance limit of bituminous pavement. *SSRG Int J Civ Eng* 2(4):1–7

Shahin M, Billah M, Islam MM, Parvez A, Zaman AKMM (2020a) Cyclone shelters need sustainable development. Int J Disaster Resil Built Environ. https://doi.org/10.1108/IJDRBE-12-2015-0061

Shahin M, Khan MMH, Bari MN (2020b) A disaster resilient road: effects of fines on density and shear strength of sands. *Int J Transp Eng Technol* 6(2):38

Shammi M, Rahman M, Bondad SE, Bodrud-Doza M (2019) Impacts of salinity intrusion in community health: a review of experiences on drinking water sodium from coastal areas of Bangladesh. *Healthcare* 7(1):50

Shamsuzzaman MM, Islam MM, Tania NJ, Abdullah Al-Mamun M, Barman PP, Xu X (2017) Fisheries resources of Bangladesh: present status and future direction. *Aquac Fish* 2(4):145–156

Sharmin Z, Islam M (2013) Consequences of climate change and gender vulnerability: Bangladesh perspective. Available SSRN 2200116

Sikder MT (2010) The impacts of climate change on the coastal belt of Bangladesh: an investigation of risks & adaptations on agricultural sector. In: *International Conference on Environmental Aspects of Bangladesh (ICEAB10), Japan*, pp 26–28

Sinharoy SS, Waid JL, Haardörfer R, Wendt A, Gabrysch S, Yount KM (2018) Women's dietary diversity in rural Bangladesh: pathways through women's empowerment. *Matern Child Nutr* 14(1):e12489

Smith KR et al (2017) Human health: impacts, adaptation and co-benefits. In: Climate change 2014 impacts, adaptation, and vulnerability: Working Group II Contribution to the Fifth Assessment Report of the Intergovernmental Panel on Climate Change. Cambridge University Press, Cambridge

Sorensen C, Murray V, Lemery J, Balbus J (2018) Climate change and women's health: Impacts and policy directions. *PLoS Med* 15(7):e1002603

Strand LB, Barnett AG, Tong S (2012) Maternal exposure to ambient temperature and the risks of preterm birth and stillbirth in Brisbane, Australia. *Am J Epidemiol* 175(2):99–107

Sultana M, Chai GW, Martin TC, Chowdhury SH (2015) A study on the flood affected flexible pavements in Australia. In: 9th International Conference on Road Airfield Pavement Technology, December 2010, pp 9–13

Summit WF (1996) Rome declaration on world food security and world food summit plan of action. FAO, Rome

Tatsuoka F, Tateyama M, Koseki J (2012) GRS structures recently developed and constructed for railways and roads in Japan. In: *Advances in transportation geotechnics 2*. CRC Press, Boco Raton, pp 63–85

Tatsuoka F, Tateyama M, Koseki J, Yonezawa T (2014) Geosynthetic-reinforced soil structures for railways in Japan. Transp Infrastruct Geotechnol 1(1):3

Taylor MA, Khosla NP (1983) Stripping of asphalt pavements: state of the art. *Transp Res Rec* 911:150–158

Terrel R, Al-Swailmi S (1994) Water sensitivity of asphalt-aggregate mixes: test selection, National Search Council Strategic Highway Research, Oregon state Univ. Corvallis. Dept. of civil engineering. Strategic Highway Research Program, Washington, DC

Thelen E (1958) Surface energy and adhesion properties in asphalt – aggregate systems. *Transp Res Board* 192:63

Tissera HA et al (2020) Severe Dengue Epidemic, Sri Lanka, 2017. *Emerg Infect Dis* 26(4):682

Torikul MH (2015) Climate change, natural disaster and vulnerability to occupational changes in coastal region of Bangladesh. *J Geogr Nat Disasters* 5(1):1000134

Uddin AMK, Kaudstaal R (2003) Delineation of the coastal zone. Program Development Office for the Integrated Coastal Zone Management Plan, Dhaka, pp 1–42

UNICEF (2016) Learning to live in a changing climate: the impact of climate change on children in Bangladesh. Bangladesh United Nations Children's Fund, Dhaka

UNISDR (2005) Hyogo framework for action 2005–2015. In: *World Conference on Disaster Reduction*, January 2005, pp 1–22

United Nations Population Division (2018) World urbanization prospects: the 2018 revision. United Nations, New York

USAID (2012) Addressing climate change impacts on infrastructure preparing for climate change. USAID, Washington, DC

Vineis P, Chan Q, Khan A (2011) Climate change impacts on water salinity and health. *J Epidemiol Glob Health* 1(1):5–10

World Health Organization (2003) *Climate change and human health: risks and responses*. World Health Organization, Geneva

World Health Organization (2012) The global action report on preterm birth. Geneva World Health Organization, Geneva

World Health Organization (2017) Status of the development of health national adaptation plan for climate change in South-East Asia. World Health Organization, Regional Office for South-East Asia, New Delhi
Xing X, Chang HC, Chen L, Zhang J, Yuan Z, Shi Z (2019) Radar interferometry time series to investigate deformation of soft clay subgrade settlement-A case study of Lungui Highway, China. *Remote Sens* 11(4):429
Yau S, Seth-Pasricha M (2019) Viruses of polar aquatic environments. *Viruses* 11(2):189
Yilmaz A, Sargin Ş (2012) Water effect on deteriorations of asphalt pavements. *Online J Sci Technol* 2(1):1–6
Yu W et al (2010) *Climate change risks and food security in Bangladesh.* Routledge, London
Zamudio AN, Parry J-E (2016) Review of current and planned adaptation action in Bangladesh. International Development Research Centre, Ottawa
Zhang G, Shoham D, Gilichinsky D, Davydov S, Castello JD, Rogers SO (2007) Evidence of influenza a virus RNA in Siberian lake ice. *J Virol* 81(5):2538
Zumrawi MME (2016) Investigating causes of pavement deterioration in Khartoum State. *Int J Civ Eng Technol* 7(2):203–214

Chapter 3
Climate Change Impacts on Human Health Problems and Adaptation Strategies in the Southeastern Coastal Belt in Bangladesh

Kazi Md. Barkat Ali, Umme Kawser, and Md. Nazrul Islam

Abstract Climate change causes extreme natural events and at the same time, increases social vulnerabilities such as health impact. Changing climatic patterns intensify the frequency of the disasters and magnitude, which bring suffering as well as creating difficulties in human health. The effects of climate change hit St. Martin's Island, Bangladesh. The increasing frequency of disasters, along with changing climate, has generated pressure on the health conditions of the inhabitant. Thus, the present study strives to identify impacts of climate change on disasters and associated health difficulties, as well as the coping strategies of the people under these changing conditions. The study also emphasizes the available treatment facilities on the St. Martin's Island. Data were collected following a semi-structural questionnaire survey by using a random sampling of 390 inhabitants. About 78.72% of respondents observed climatic change on the Island in the last few years. This changing climate has increased the frequency and severity of disasters such as excessive rainfall during the rainy season (19.06%), the transgression of the shoreline (17.95%), drought in summer (7.18%), shortage of fresh water in summer (14.11%), increasing thunderstorms (5.45%), and suffering from diseases (13%). Changing temperatures have created heat-related stress where high salinity and post-disaster shortage of drinking water cause the breakout of waterborne diseases among the inhabitants, which has worsened owing to the deplorable treatment facilities. The low level of adaptation strategies to cope with the climate-induced disasters and related health problems make the community extremely vulnerable. This

K. M. Barkat Ali (✉) · U. Kawser
Department of Geography and Environmental Studies, University of Chittagong, Chittagong, Bangladesh

M. N. Islam
Department of Geography and Environment, Jahangirnagar University, Savar, Dhaka, Bangladesh
e-mail: nazrul_geo@juniv.edu

M. N. Islam, A. van Amstel, *Bangladesh II: Climate Change Impacts, Mitigation and Adaptation in Developing Countries*, Springer Climate,
https://doi.org/10.1007/978-3-030-71950-0_3

assessment will help to understand the vulnerability of communities locally and will try to give new insights into policy approaches in the region.

Keywords Climate change · Disasters · Human health · Adaptation strategies · St. Martin's Island · South-Eastern Coastal Belt · Bangladesh

3.1 Introduction

Climate is the fundamental environmental influence on human health and well-being. Henceforth, any major change in climatic settings is likely to have significance for the health of the human population. Anthropogenic climate change is likely to increase health risks in most regions of the world. One of the single most important challenges of this century is climate change (Elahi 2016). Interpretations of rises in global temperature and global sea level in recent eras demonstrate the undeniable warming of the Earth's climate (IUCN 2011). Climate change-associated disasters including flood, drought, sea-level rise, salinity, temperature, and rainfall variations, etc., have become key anxieties for many countries of the world because of longstanding allegations along with adverse impacts on development events (CCC 2009). Both developed and developing countries of the world experience natural calamities in the form of drought and floods, though on diverse scales (Ahmad et al. 2010). Inhabitants of arid or semi-arid regions, in low-lying coastal zones, in water-scarce or flood-prone parts, or on small coastal islands are predominantly susceptible to climate change. In general, small islands are considered to be very vulnerable to climate change and particularly sea-level changes (Watson et al. 1998; CCKN 2001; Banglapedia 2015). Estimation shows that millions of coastal inhabitants of southern Asia will be disturbed by sea-level rise and intensification in the strength of tropical cyclones (Brammer 2002; Water Resources Planning Organization 2006; Cruz et al. 2007; BWDB 2013; Laila 2013). Globally documented data show the impacts of climate change on human health (Luber et al. 2014; Petkova et al. 2015). Human health in developing countries of the tropical regions will be affected by climate change (IUCN 2011; Tsurita et al. 2013). Global concerns are increasing regarding both the water and human health impacts of climate change (Rabbani and Huq 2010). It is considered that Bangladesh is one of the most vulnerable countries in the Global Climate Risk Index (CRI), which is developed by Germanwatch. From 1992 to 2011, Bangladesh has been listed as one of the most climate change-affected countries (Germanwatch 2013; Laila 2013). Bangladesh is extremely vulnerable to climate change because of its low lying topography (Minar et al. 2013). The country has been faced with tropical cyclones, storm surges, floods, droughts, and river erosion, which are the major natural disasters (Kabir et al. 2016). Recent studies show that frequencies of natural hazards have increased in Bangladesh, especially affecting coastal areas and their communities (BCCASP 2009). These have killed hundreds of thousands of people and destroyed homes and livelihoods, which cost approximately US$16 billion in damages (Asia Foundation 2012). In terms of health disasters, because of climate change, Bangladesh has appeared to be one of the world's projected hot spots for prospective tropical diseases (Elahi 2016). Evidence for the effects of climate change on vector-borne and other infectious diseases has been found in research (Haines et al. 2006; Kabir et al. 2016). Extreme climatic events such as floods, droughts, and cyclone. directly and indirectly affect the health of the people of Bangladesh (CCC 2009), particularly in the coastal region (Rabbani and Huq 2010). In the coastal area, to sustain livelihoods, domestic use, and health services, people are struggling to access water resources. Besides, in the coastal zones, salinity intrusion in both surface and groundwater and due to drought and lack of rainfall, the inadequacy of water in the northwestern part of the country makes the health of local communities vulnerable (Rabbani and Huq 2010). Diarrhea, dysentery, etc., are also proliferating, particularly in summer. It has been anticipated that the spread of many infectious diseases in summer might be the combined effect of higher temperatures and probable increase in precipitation (MoEF 2005). Additional health stresses such as dehydration, malnutrition, and heat-related morbidity, especially among children and the elderly of the country, are also brought about by climate change (CCC 2009). Many areas in Bangladesh are clearly under threat of climate change, and people in the coastal region are more vulnerable, caused by relatively higher poverty than the other parts of the country (Hossain et al. 2010). St. Martin's Island, the only coral island of Bangladesh, located in the southerly part of Cox's Bazar district is most vulnerable to climate change owing to its geographical location, natural disasters, the impact of communities around the Island, and at the same time, inhabitants are confronting significant health problems. Many climate change studies have been conducted on the southeastern coastal region, especially on St. Martin's Island, such as Minar et al. (2013), Hasan et al. (2008), and Alam et al. (2015). Climate change has occurred as the greatest threat to humankind. Climate change will have an adverse impact on all aspects of human development, including livelihoods, safe water and sanitation, food security, health care, shelter, etc. Disadvantaged groups are the most affected by the climate extremes and have a very small capacity to cope with the risks. Bangladesh already feels the impacts of climate change through irregular rainfall pattern, floods, flash floods, cyclones, saline intrusion, drought, sea-level rise, tidal surge, and waterlogging. In the coastal areas of Bangladesh, poor groups are the most exposed to the impacts of climate change along with extreme climatic events and environmental degradation. However, studies have addressed climate change-related vulnerability, effects, and response. Still, very few studies have been directed on the assessment on social vulnerability at the local level of the coastal zone, especially on St. Martin's Island, Bangladesh. St. Martin's Island, which frequently faces a major threat owing to climate change-induced disasters and health problems, no prior specific study on the disaster vulnerability of St. Martin's and the related health impact has yet been conducted. By understanding the poor socio-economic conditions and research gap, this study is focused on assessing the impacts of climate changes on disasters and associated climate change impacts on the health of the southeastern coastal community of St. Martin's Island. Thus, important aspects of coping with climate change can be identified for St. Martin's Island, Bangladesh, by this assessment (Huq et al. 2003; IPCC 2007; Hughes and McMichael 2011).

3.2 Coastal Region of Bangladesh

The coastal region of Bangladesh is densely populated, with 28% of the total population of the country living here, which is expected to increase from 36.8–43.9 million in 2015 to 60.8 million by 2050 (PDO-ICZMP 2003). Amounting to an area of 47,201 km^2, it consists of 19 districts, which is about one-third of the total area of the country. About 62% of the land of the coastal zone has an average elevation of up to 3 m and 86% up to 5 m (MoWR 1999; Coastal Zone Policy 2005; Nazem and Mahboob 1992). Major disasters in the coastal zone included floods, cyclones, and waterlogging. There are more than 300 islands in this country, of which 186 are documented. These islands are inhabited by thousands of people whose livelihood solely depends on the coastal environment and is influenced by climate change (Islam 2001). The coastal zone of Bangladesh is highly affected by climate change, which has an impact on the lifestyle and livelihood of the people living in the coastal areas (Islam 2004; Luber and Prudent 2009; Rahman et al. 2010). The disasters and hazards faced by these people are inundation of low-lying coastal settlement, coastal wetlands and mangrove forest, waterlogging, drainage congestion, backwater effect, prolonged flooding, land erosion, saltwater intrusion, etc. (UNFCCC 2007). Nineteen coastal districts have three geophysical characteristics that distinguish the coastal zone from rest of the country: the interplay of the tidal regime, salinity in soil and water, and cyclones and storm surges, with economic and social implications for the population (ICZMP 2001). The coastal zone in Bangladesh has its own specific coastal landform and distinct characteristics. The coastal area of Bangladesh is broadly divided into three regions: western, central, and eastern (Fig. 3.1) (Warrick and Ahmed 1996) (Map 3.1).

3.3 Eastern Coastal Zone of Bangladesh

The eastern coastal zone of Bangladesh starts from Bodormokam, the southern tip of the mainland to the Feni river estuary. It is also the part of the GBM (the Ganges–Brahmaputra–Meghna) delta system. The characteristic features of this zone is that it is a very narrow, unbroken, straight shoreline, and is less dynamic (Islam 2004; Luber and Prudent 2009; Rahman et al. 2010). Parallel to this zone, there runs a series of small islands. Karnafully, Sangu, and Matamuhury are significant rivers of the zone. Dividing Bangladesh from Myanmar, the Naf River falls to the Bay of Bengal. Soils of the eastern coastal site are dominated by submerged sands and mudflats. About 145 km long, a sandy beach from Cox's Bazar headed for Teknaf is formed by the submerged sand of the zone (Islam 2001; Rasheed 2016). Patenga and Cox's Bazar, two of the country's most important sandy beaches from tourists' perspective, are located on this coast. Core economic

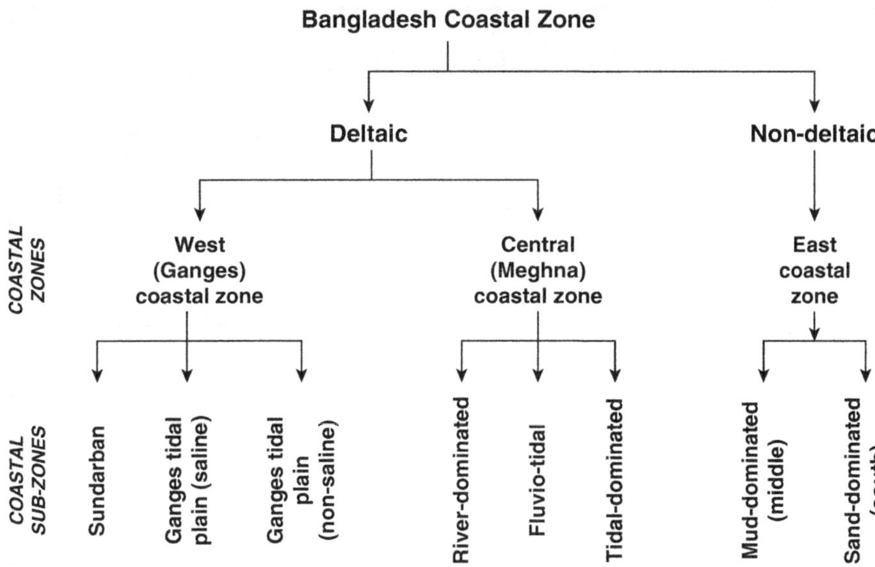

Fig. 3.1 Bangladesh coastal zone sub-divisions. (Source: Warrick and Ahmed 1996)

activities of the zone are fish farming, fishing in the bay, salt production, and tourism. Sandweep, Kutubdia, Maishkhali, and St. Martin's are the main islands in the eastern coastal zone of Bangladesh. St. Martin's, the only coral island of the country, is located in the Bay of Bengal in the eastern coastal area of Bangladesh (Map 3.2).

3.4 Geography of St. Martin's Island

St. Martin's Island is situated in the northeastern part of the Bay of Bengal and also the southeastern part of Bangladesh. It is located 228 km from Chattogram city and about 9 km south of the Cox's Bazar–Teknaf peninsular tip and at the mouth of the Naf River about 8 km west of the northwestern coast of Myanmar. As the southeastern-most point of Bangladesh, the Island lies between a latitude of 20°34′ and 20° 39'N, and a longitude of 92°18′ and 92° 21′E. The entire area of the Island is about 12 km^2, including the rocky platforms extending into the sea, and the land area of the Island is 5.9 km^2. Depending on the tidal level, the surface area is about 8 km, and the beach is about 14 km long. The Island is almost plain land and is 3.6 m above the mean sea level. The 9.66 km-wide channel between the mainland and the reef is much shallower than the open sea southwest of the Island (Feeroz 2009) (Map 3.3).

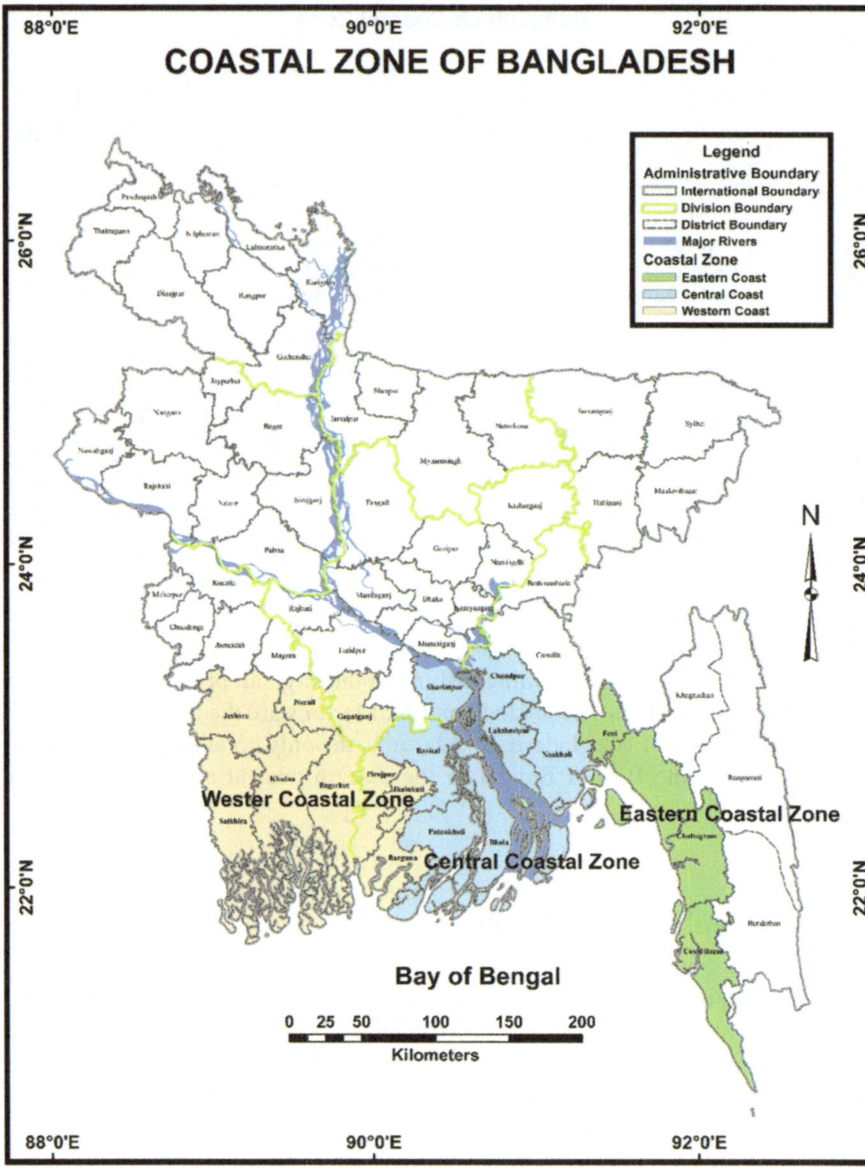

Map 3.1 **Coastal** zone of Bangladesh. (Source: Prepared by author)

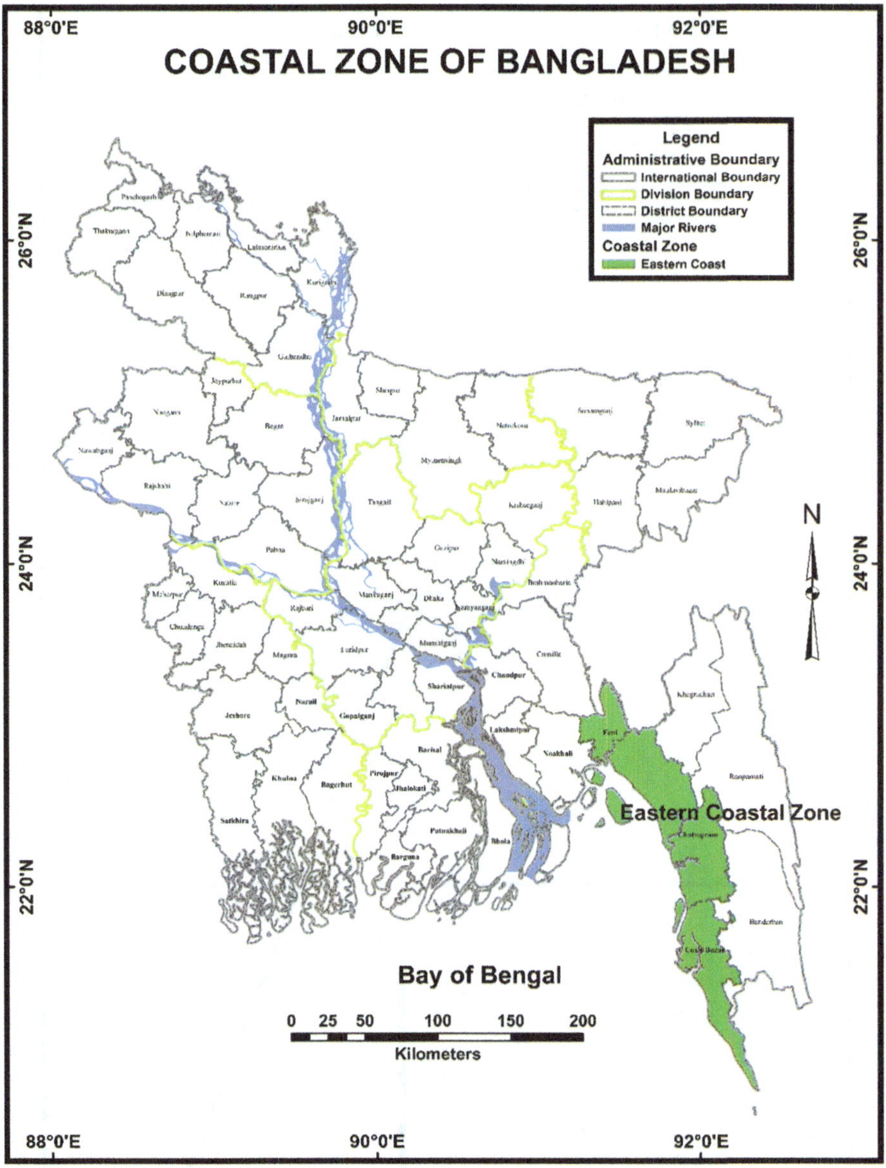

Map 3.2 Eastern coastal zone of Bangladesh. (Source: Prepared by the author)

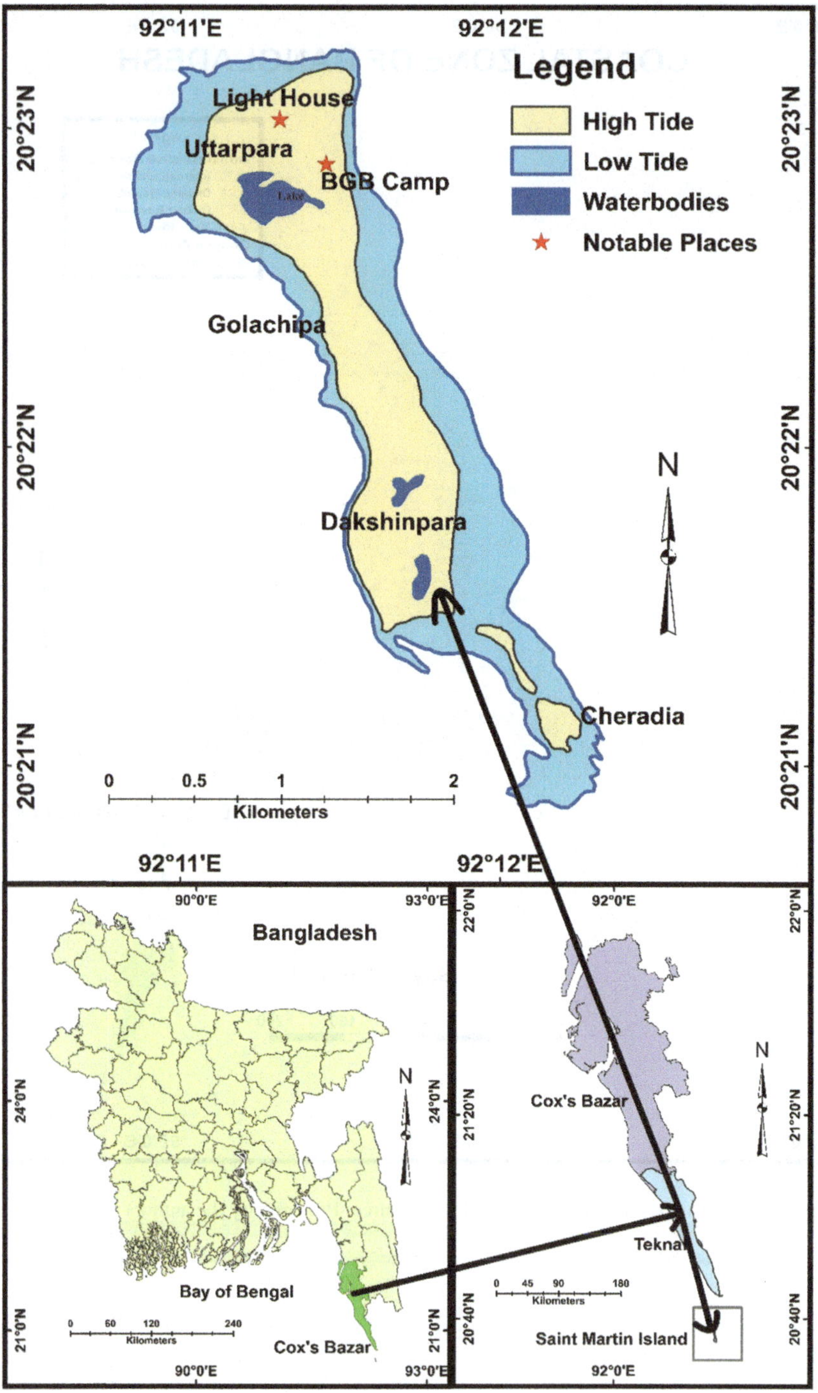

Map 3.3 Map of the study area. (Source: Prepared by the author)

3.5 Climate Characteristics of Saint Martin Island

Subtropical monsoon has heavily influenced the environment of the Island. The summer season extends from March to May, the monsoon from June to September, when much of the rainfall occurs, and from October to February is winter, with calm weather and tropical rain (Thompson et al. 2010). Depending on weather data from Cox's Bazar station, maximum monthly temperatures vary from 27.6 °C (January) to 32.7 °C (May), and the minimum monthly air temperatures fluctuates between 15.8 °C (December) to 26.2 °C (May).

With an average of 79.7% at Cox's Bazar, humidity remains comparatively high throughout the year. (MoEF 2001b; Thompson et al. 2010). Winds flow from the northwestern direction from November to February on the Island, whereas it blows in the southwestern direction from March to May and from the southeast from June to September. The sea becomes very rough in the rainy season. The Island usually has pleasant and calm weather between November and March, which makes these months the peak tourist season (Thompson et al. 2010).

3.6 Life and Livelihood of St. Martin's Island

St. Martin's Island is small and the only coral island of Bangladesh. It is the smallest union (local administrative unit) of Bangladesh, with a population of about 4000 (BBS 2013). About 250 years ago, the first habitation was started by Arabian sailors who named the Island "Jazeera." During the British regime, the Island was named after the DC (Deputy Commissioner) of Chittagong, Mr. Martin, as St. Martin's island (Wikipedia 2012). The local name of the Island is "Nairkel Jinijra," which means "coconut island". The income and occupation pattern of households is also different than other parts of the country, because of the identical character of the location and physical nature. Fishing, agriculture, and tourism are the main economic activities of the people on this Island. Most of the inhabitants are at the middle-income level. Education and health facilities are not available here. Water transport is the prime transport system of the Island. Owing to increases in climatic disaster, poverty, and lack of employment opportunities, many people are migrating from this Island to other places in the country such as Dhaka or Chattrogram.

3.7 Materials and Methods

Both primary and secondary sources of data were used in this analysis. Preliminary data were collected through a questionnaire survey, direct observation, in-depth interview techniques and mixed types of a questionnaire survey with open-ended and closed questions. A total of 390 samples were selected using a random sampling technique to collect field data on household levels. Direct observation was used to perceive the island position, potentiality of disaster vulnerabilities, and coping practices of local people to reduce the risk. In-depth interviews were used to collect details and erudite information from the particular respondent groups such as social elite groups (community leaders, Imam of the mosque, local honorable aged dwellers), local political leaders (Paurashava and Union chairman and Union members), as well as relief and disaster volunteer groups, and an employee of the island meteorology office. All types of respondents were selected based on age and duration of living in the study area. Climatic data were collected from the DMB, BMD, and BBS. Data were analyzed using Statistical Package for Social Science (SPSS version 16.0), Microsoft Excel. Mathematical tools such as averages, percentages, tables, and graphs were used to present the analyzed data.

3.8 Demographic and Socio-economic Characteristics of the Respondents

Impacts of natural disasters to a community vary based on the status of the household and resources, economic, social, and environmental perspectives of the community (Laila 2013). More than half of the inhabitants participating in the study were male (66.93%), and rest were female (33.07%). It was found that the highest age range of the participants was 36–45 (27.18%). The second highest age category was the 46–55 age group (15.64%). However, the smallest age group was 86–90, which comprised only 0.51% of our research group. Table 3.1 shows the demographic and socio-economic conditions of the study area.

Table 3.1 Demographic and socio-economic characteristics of the respondents

Age	16–25 (10.69%), 26–35 (35%), 36–45 (27.18%), 46–55 (15.64%), 56–65 (4.87%), 66–75 (4.4.36%), 76–85 (1.54%), 86–95 (0.51%)
Gender	Male (66.93%) and female (33.07%)
Occupation	Fisherman (14.87%), farmer (6.66%), businessman (26.15%), government job (8.98%), unemployed (2.82%)
Houses	Katcha (42.82%), semi katcha (14.87%), pucca (11.03%), semi pucca (23.59%), slum (7.7%)
Education	Illiterate (32.56%), primary (30%), secondary (27.44%), higher secondary (3.08%), graduate (5.39%), and postgraduate (0.51%)
Duration of living	From birth = 41%, 20–30 years = 35% and > 30 years = 24%
Monthly income	<BDT5000 (18.25%), BDT5001–10,000 (35.25%), BDT10,001–15,000 (24.75%), BDT15,001–20,000 (9.50%), BDT20,001–25,000 (6.50%), above BDT25,000 (6.75%)
Sanitation conditions	Open space (13.25%), healthy (76.50%), neighbor's latrine (2.75%), open latrine (7.00%), drain (0.50%)
Source of drinking water	Pond/lake (11.87%), rainwater (18.35%), collect from Teknaf (8.25%), Well/ tube well (58.58%), others (2.97%).

Sources: Field Study (2018)

Major occupations of the respondents are businesses related to fisheries and tourism (26.15%) and the second most popular occupation is fisherman (14.87%), which is the reflection of surrounding open sea and the coral as well as the aesthetic beauty of the Island, which attracts the inhabitants to these occupations. According to Feeroz (2009), throughout the tourist period (November to February), 3000 people typically visit the Island. Owing to limited land resources, agriculture does not flourish on the Island (6.66%). From the housing pattern, it is clear that a significant amount is mainly katcha (42.82%) and the proportion of built infrastructure is deficient, only 11.03%. Sanitation conditions are satisfactory; around one-fourth of the population is blessed with health sanitation. For more than half of the respondents, the drinking water source is mainly a well/tube well.

During natural disasters, community characteristics such as age and economic status have a substantial effect on the capacity for the community preparedness, response, and recovery with regard to the disasters (Oxfam America 2009; Petkova et al. 2015). Previously vulnerable groups of the communities, including women, children, and the elderly will be disproportionately affected by climate change, especially those with a lower socio-economic status and with pre-existing health issues (Luber et al. 2014). It is evident from the analysis of demographic and social structures of the Island that the socio-economic condition is relatively weak, which marks the high demographic and social vulnerability of the Island to disasters. The housing conditions, occupation and literacy conditions of the Island indicate that most of the inhabitants live in katcha houses, which are mainly built by bamboo and straw, which would not give them any protection during disasters. At the same time, it is a reflection of the low status of the income condition of people (Watson et al. 1996; Rashid et al. 2009). Again, the low literacy rate affects the understanding of the people of the climate change, adaptation, mitigation knowledge, and information. A significant portion of the respondents were women, and most of them are illiterate, a few of them hardly having primary education. According to the UNDP (2010), women are most vulnerable to climate change because of social and cultural restrictions viz. limited access to resources, lack of education and access to information, limited mobility, and limited roles in decision making. Considering that most of the women are illiterate in the study area, this affects their ability to understand climate change risk, adaptive capacity, and knowledge. They have limited access to mobility and decision making, cannot follow the early warning system, and have an inadequate ability to save themselves from disasters and post-disaster health impacts and get proper health care, being somewhat willing to depend on local empirics for treatment.

3.9 Climatic Scenario on the Southern Coast of Bangladesh

According to a study by Huq and Alam 2003, climate change is projected to increase the sea level to 43 cm by 2050, with a growing frequency and magnitude of cyclones and storm surges also predicted. In Bangladesh, an average 0.5 °C warming is comparable in extent to the global observed mean warming (Warrick and Ahmed 1996; Ali et al. 2015). To understand the present climatic condition and identify changes

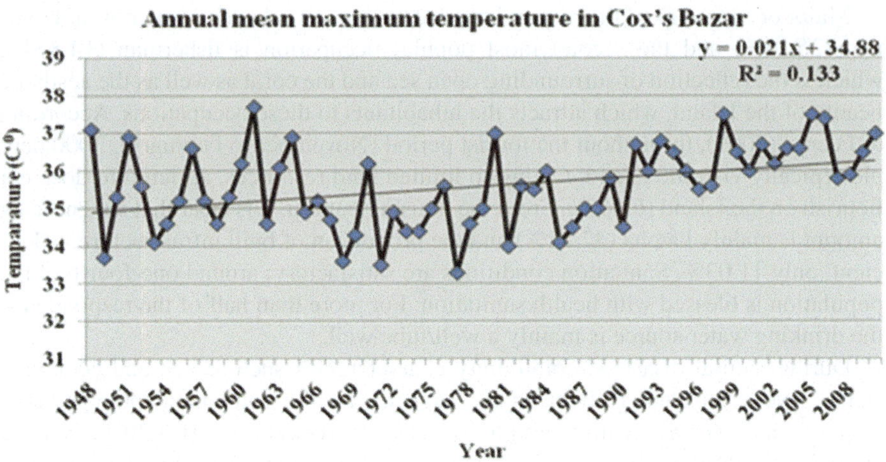

Fig. 3.2 Annual mean maximum temperature in Cox's Bazar area. (Adapted from Ali et al. 2015. Source: DMB 2012; Prepared by the author)

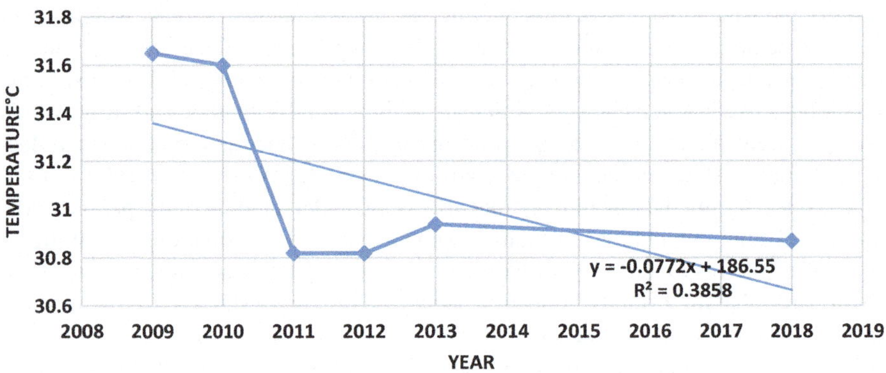

Fig. 3.3 Temperature changes in Cox's Bazar (2008–2018). (Source: BMD 2017 and BBS 2014 & 2019; Prepared by the author)

in the climate in the southeastern coastal belt of Bangladesh, temperature as well as rainfall data have been scrutinized. The study by Ali et al. (2015) found that there has been a strong positive correlation of changing temperature (1948–2008) and rainfall (1970–2010) in Cox's Bazar region. According to Ali et al. (2015), the temperature was rising by 0.021 times every year or 2.1% more compared with its former equivalent, and Figs. 3.2, 3.3, 3.4, and 3.5 show the results.

From 2008 to 2018, the temperature changed by −0.78 times every year compared with its previous counterpart. $R^2 = 0.38$ means that the given model can explain that the total variation of temperature was 38.58%.

Figures 3.3 and 3.4 show rainfall deviation in the Southern region based on data from Cox's Bazar station. According to Ali et al. (2015), the rainfall increased 0.092 times every year or 9.2% more compared with its previous counterpart.

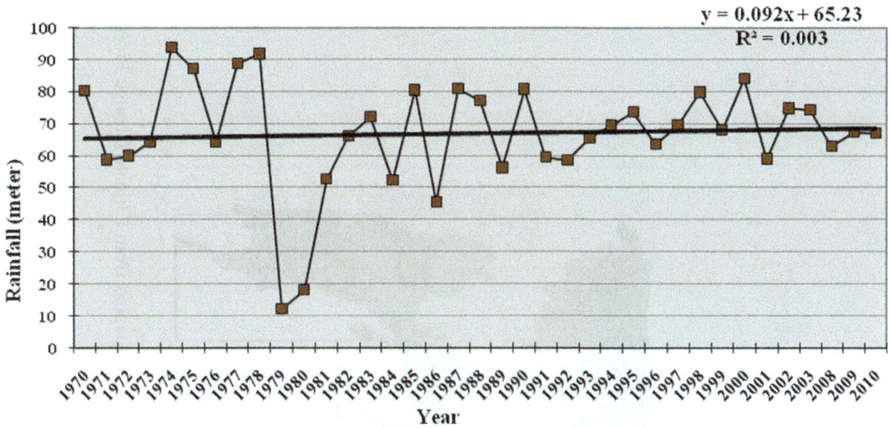

Fig. 3.4 Annual mean rainfall in Cox's Bazar (1970–2010). (Adapted from Ali et al. 2015. Source: DMB 2012; Prepared by the author)

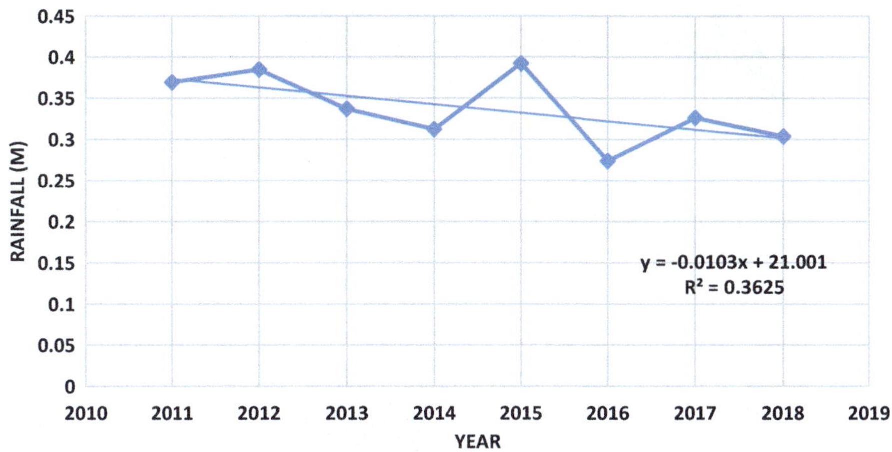

Fig. 3.5 Rainfall changes in Cox's Bazar (2010–2018). (Source: BMD 2017 and BBS 2014 & 2019)

The present study discovered that the rainfall changed −0.01 times more every year compared with its previous counterpart.

Again, according to the study of Ministry of Foreign Affairs (2018), the Climate Change Profile: Bangladesh, supports the same trend of climatic variations as a decrease in temperature and increase in rainfall on the southern coast of Bangladesh from 1950 to 2000.

Although the previous study showed the decadal intensification of both temperature and rainfall, in recent years, both temperature and precipitation have decreased. There are temperature and rainfall variabilities with rising and declining trends in the southern coastal zone of Bangladesh (Map 3.4).

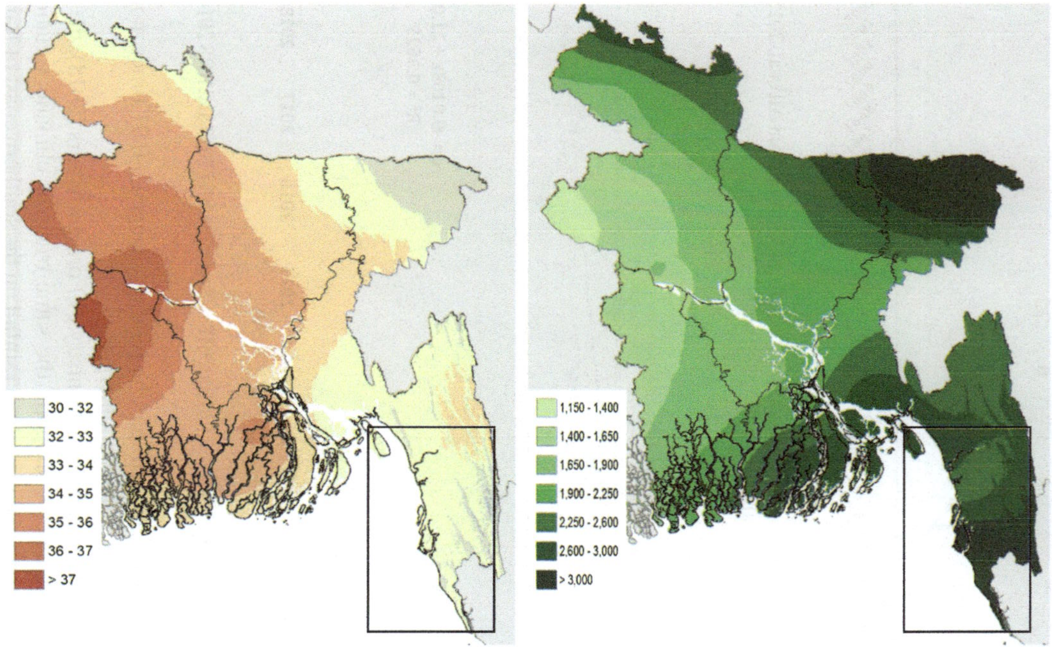

Map 3.4 Current average maximum temperature and current average annual rainfall of Bangladesh (1950–2000). (Adapted from Climate Change Profile: Bangladesh, 2018. Source: Thomas et al. 2013)

St. Martin's is surrounded by the open ocean and located at a more southern tip than Cox's Bazar, and is supposed to experience high lowest temperatures and lower extreme temperatures than Cox's Bazar (Tomascik 1997), which has already been proven by the data of previous studies in this region.

3.10 Knowledge and Understanding of Climate Change of Respondents

To identify the depth, magnitude, and respondents' knowledge, the FGD and questionnaire reconnoitered information on their ideas of climate change. The study shows that respondents did not have an explicit understanding of climate change in the study area. Experience of the term "climate change" among the households in the study area was not very satisfactory. Although 63% of respondents said that they have an idea about climate change, they think only rising temperature means climate change (Fig. 3.6).

This is very surprising that besides living in a disaster-prone area, which has the effects of climate change imposed upon it, the inhabitants had minimal knowledge of climate change; in particular, female respondents had nearly zero experience on this aspect. The percentage of 7.28% who had no answer to the question reflects the situation that is affecting the ability and knowledge of understanding the importance of the adaptation and mitigation measures of climate change of the inhabitants and eventually increasing the vulnerability of the illiterate as well as poor respondents to climate change. Although they have insufficient knowledge on climate change, they observe changes in temperature, rainfall, and the frequency and severity of the disasters on St. Martin's Island (discussed below).

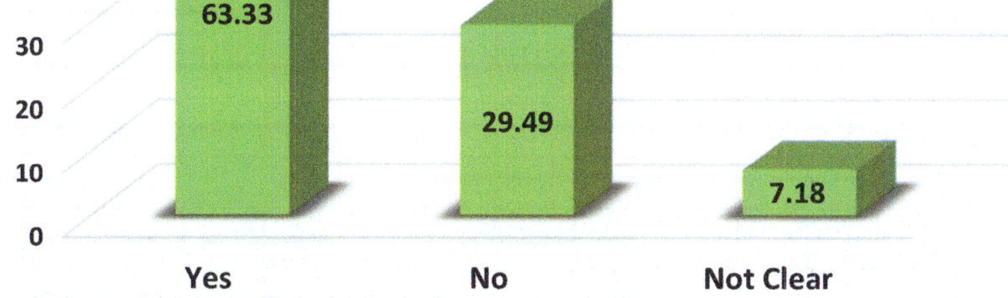

Fig. 3.6 Respondents' knowledge on climate change. (Source: Field Study 2018)

3.11 Major Disasters Along with the Seasons in the Study Area

In Bangladesh, heavy rainfall occurs during the season, and most of the disasters such as cyclones, storm surges along with high tides, and flash floods have occurred in pre-monsoon and post-monsoon seasons. Cyclonic storms in the southeastern belt region are frequent and are usually associated with storm surges. Cyclones develop in the Bay of Bengal, generally in April–May and October–November, and those that make landfall cause severe damage to human settlements and vegetation (Thompson et al. 2008). During the summer, owing to high temperatures, sudden low pressure is created in the Bay of Bengal, which in turn creates tropical cyclones and storm surges in the study area.

Cyclones, bay bank erosion, storm surges, sea-level rise, salinization, thunderstorms, disaster-borne disease, irregular rainfall, temperature increase, etc., frequently occur in the study area. From the study, commonly occurring disasters in the study area are divided into two classes, i.e., generally occurring hazard-related vulnerabilities and climate change-induced disaster-related vulnerabilities. Table 3.2 provides an insight into the disaster periods of the study area throughout the year.

Since 1970, a total of 14 severe cyclones have affected the Cox's Bazar area, with four significant cyclones occurring since 1991 (DMB 2008) and since 2000, five cyclones have attacked Chittagong Coast and the magnitude is increasing. From the current study, the majority of the disasters occurred in the rainy season (65.10%) and at the beginning and end of the summer (34.90%). It is found that on St. Martin's, cyclones (34.97%) bundling with storm surge (12.24%) is the highest-ranking disaster, whose frequency is increasing. Other disasters such as bay bank erosion (15.38%), salinity (12.59%), sea-level rise (8.15%), and flash floods (5.93%) mainly appear in the rainy season from March to November in the study area, of which storm surge is the most devastating (Fig. 3.7).

Table 3.2 Disaster calendar of the study area

Month / Disasters in the study area	Jan	Feb	Mar	Apr	May	Jun	Jul	Aug	Sept	Oct	Nov	Dec
Cyclone				←	→				←	—	—	→
Storm Surge									←	—	—	→
Bay Bank Erosion	←	—	—	—	—	—	—	—	—	—	—	→
Sea Level Rise	←	—	—	—	—	—	—	—	—	—	—	→
Salinization	←	—	—	—	—	→					←	→
Flash Flood				←	—	—	—	—	—	—	→	
Thunderstorm				←	—	—	—	—	—	—	→	

Prepared by the author (Source: Field Study, 2018)

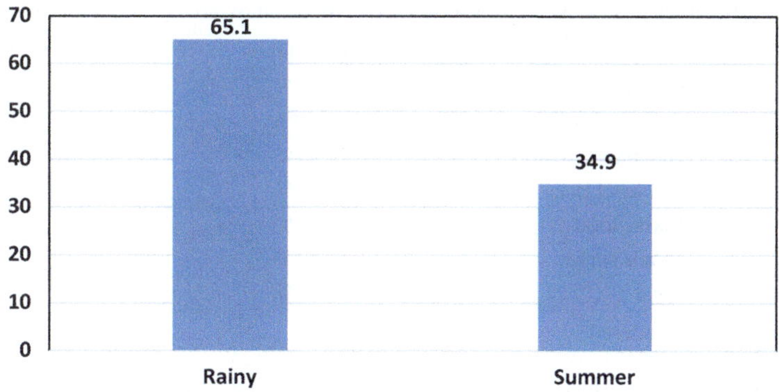

Fig. 3.7 Seasons of disasters in the study area. (Source: Field Study 2018)

Disasters such as landslides (1.96%) and thunderstorms (8.13%) are some of the everyday calamities. According to the study, 27.69% of respondents said that three to four storms have occurred every year for the last 30 years, whereas it was only one to two before, increasing the Island's vulnerability to disaster. In the words of the respondents, disasters are occurring almost suddenly. Although 57.44% of respondents answered that before disasters they usually do not see any signs of natural disturbances, 36.92% said that they sometimes saw signs such as a strong wind flow, cloudy sky, a sudden change in temperature, and high tide-like symptoms.

3.12 Climate Change-Induced Disasters in the Study Area

In the words of Trenberth (2005), in the areas where tropical storms occur, global warming is already altering the environmental conditions, providing more energy to fuel the showers, which make the storms more intense (in terms of rainfall and wind speed) (Aalst 2006). In parts of Africa and Asia, over the past few decades, the incidence and intensity of droughts have amplified (Aalst 2006).

Referred to as one of the most vulnerable countries to climate change, the topography and geographical location of Bangladesh marks it as predominantly susceptible to extreme weather events such as tropical cyclones, storm surges, and floods (CCPB 2018). Disasters are prevalent and familiar in the study area, but vicissitudes in the climate are deteriorating the condition of the disasters of the Island. Owing to deviations in the climatic variables, fluctuations occur in the pattern and frequency of disasters. At the time of the study, it came out of the analysis that changes in the climatic variables have forced the increasing severity of the climatic events in recent years. Table 3.3 shows the rank-wise varying severity and rate of disasters due to the controls of shifting climatic variables in the study area.

From the field study 2018, more than three-quarters (78.72%) of respondents said that they had observed climatic change in the Island in last several years,

Table 3.3 Climate change-induced disasters (*CCID*) in the study area

SL	CCID	Percentage	Rank[a]
1	Cyclone	51.11%	1
2	Salinization	12.59%	2
3	Bay bank erosion	11.11%	3
4	Sea level rise	8.15%	4
5	Flash flood	5.93%	5
6	Storm surge	11.11%	0
	Total	100.00%	

Multiple answers considered
Source: Field Study (2018)
[a]Rank was demarcated considering the percentage of a questionnaire survey, 1 is the highest severity and frequency, and 5 is the lowest
SL sea level

including excessive rainfall during the rainy season (19.06%), the transgression of the shoreline (17.95%), drought in summer (7.18%), shortage of freshwater in summer (14.11%), increasing thunderstorms (5.45%), and suffering from diseases (13%).

The United Nations Development Programme (UNDP) has ranked Bangladesh top among all countries around the world in terms of vulnerability to tropical cyclones. On average, the country is hit by a severe cyclone every 3 years. (MoEF 2009; CCPB 2018). In the existing research, it was observed that cyclone is the highest-ranking disaster both generally occurring and climate change-induced. The cyclone is more severe as an impact of climate change. Although bay bank erosion, storm surges, sea-level rise, salinization, and flash floods are generally occurring disasters on the Island, their magnitude and occurrence have been very high in recent years.

It is evident that owing to changing climate, salinization is another severe problem for the Island. As a result of drought, the freshwater sources have dried up and become salinized, which will, in turn, create a scarcity of freshwater for the inhabitants to drink, which is the most hazardous effect of climate change. If this situation continues, the Island will start to suffer from a lack of drinking water and irrigation water availability. As the Island is located in the open sea and the nearest mainland located 49.6 km away, it is not possible to manage drinkable freshwater for the inhabitants. As a result, the area will be unsuitable for living very soon.

3.13 Impacts of Climate Change Induced Disaster in the Study Area

St. Martin's Island, the only coral island of Bangladesh, is gifted with immense marine as well as land resources and having exceptional biodiversity (Feeroz 2009). As a coastal island, it faces disasters such as cyclones, sea-level rise, bay bank erosion, salinity intrusion whereby the disaster situation is deteriorating owing to the

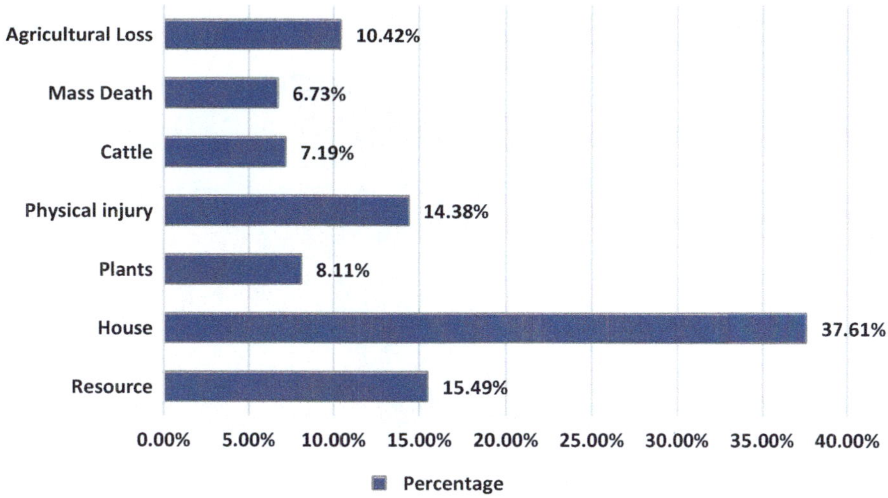

Fig. 3.8 Effects of cyclones in the study area. (Source: Field Study 2018)

swelling rate of recurrence and extents due to climatic changes. That brings massive destruction of life, property, habitats, agriculture, and biodiversity. About 55.64% of the respondents of the study area agreed that they are suffering from recurrent occurrence and growing rigorousness, particularly of cyclones. Losses from cyclones in the study area are given in Fig. 3.8.

The figure shows that loss of houses and other properties such as tourism-related infrastructure is the significant destruction from the cyclone. In addition, physical injuries, agricultural failure, mass deaths are other forms of considerable suffering that happen after cyclones. Owing to the growing rates of the cyclone, these destructions have become day-to-day incidences.

Salinity is another problem that is becoming more and more complicated day by day owing to climate change. As a result of increasing salinity intrusion in the study area, the inhabitants lack drinking and cooking water (69.02%) and a lack of freshwater availability in households (21.60%). In addition to this, the lack of freshwater for irrigation (9.39%) is another issue that is emerging as an alarming problem on the Island.

Along with cyclone, sea-level rise and salinity intrusion drought in the dry season is another evolving problem due to changing climate in the study area. As a result of a prolonged dry summer season inhabitants on the Island faces glitches, especially in cultivation in crops (7.27%) and drinking water (25.45%). Besides, this also leads to temperature-related illness (30.91%) among the inhabitants.

Another impact of climate change is bay bank erosion, which causes migration of people (25.88%) from the Island to the mainland or other parts of the Island, losses of arable land (35.59%) and losses of houses (38.2%).

As a result of rapid sea-level transgression, inhabitants face the loss of a considerable amount of landmass, besides forest, erosion of the coastline, salinity

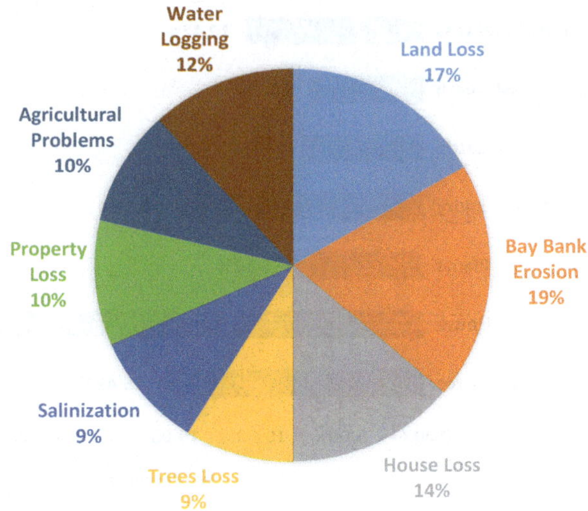

Fig. 3.9 Losses from sea-level rise in the study area. (Source: Field Study 2018)

intrusion, salt waterlogging, which causes agricultural land to become unsuitable for cultivation and creates a problem for farming production. The impacts of climate change-induced sea-level rise are shown in Fig. 3.9.

As a tiny island on the open coast, this area is highly disaster-prone. Along with this, climate change adds further difficulties by accelerating the regularity and magnitude of disasters. Analysis of the present study data indicates that the sea level transgression is happening on the Island, which causes considerable loss of the limited land mass of the Island, as a result of which, according to the information of the inhabitants during the study, coastal boulders, known locally as "Tila," have already been lost to the sea along with coastal forests, and saline water is intruding more inland from the coast. This results in a lack of fresh drinking water and water availability for irrigation. In addition to this, tourism creates immense pressure on the freshwater resource of the Island. If the present sea level transgressions continue, it is easily predictable that very soon, the Island will lose its habitat and be lost in the ocean.

3.14 Climate Change and Human Health

The connection between climate change and human health is multidimensional. The health and well-being of human populations are sensitive to shifts in weather patterns and other aspects of climate change (IPCC 2014). There are three essential pathways by which climate change affects health, which can be classified as primary or direct impacts, secondary or indirect impacts, and tertiary or long-term impacts respectively (Butler 2014). Most humans, as well as natural systems, are sensitive to climatic changes. Nevertheless, human health differs from other impacts which dominates in various ways. The pathways of climate change effects on human health are often long and complicated, and uncertainties accumulate along the way. The majority of health impacts are strongly influenced by behavioral factors and the socio-economic characteristics of the community such as the overall level of economic development, the state of sanitation and public health systems, and building standards. Meanwhile, climate is but one of many factors that govern the status of population health, and judging the role of climate in disease incidence requires careful analysis. In calculation, obtaining data is often more involved in human health than in other impact spheres because it usually depends on human co-operation, and there is limited scope for controlled trials. Human health is distinct from other climate-sensitive domains, but anthropogenic climate change is also distinct from other environmental extortions of human health. The most important differences are the large spatial scale of the problem, the very long time span to be measured, the improbability related to future scenarios of climatic hazards, and the complexity of the relationship between climatic factors and health outcomes.

3.15 Health Impacts of Climate Change-Induced Disaster in the Study Area

"A state of complete physical, mental and social well-being and not merely the absence of disease or infirmity" (WHO 1948). Human health and environment are intimately related. Climate is one of the critical environmental factors that influence human health (Paul 2010). Temperature-related mortality and morbidity are the most easily measurable impacts of changing climate on human health. From the very young to the elderly as well as individuals of low socio-economic status, and individuals with previous medical conditions, particularly cardiovascular and respiratory, are at a high risk (Basu and Samet 2002, Basu 2009; Petkova et al. 2015). According to the Intergovernmental Panel on Climate Change (IPCC) (2001) human health is directly affected by climate change directly (e.g., impacts of thermal stress, death/injury in floods and storms) and indirectly by changes in the varieties of disease vectors (e.g., mosquitoes), waterborne pathogens, water quality, air quality, and food availability and quality. The pathways of climate change impacts on health according to the IPCC are shown in Fig. 3.10.

Although it is instinctive that extreme events (for example, drowning during floods), can have health impacts such as death or injury during a case, it can also occur before or after an extreme event because individuals may be involved in activities, for instance, disaster preparation and post-event cleanup, that put their health at risk (USGCRP 2016). In Bangladesh, the health of the people is affected directly and indirectly by extreme events, for example, cyclone, floods, and drought, etc., every year (CCC 2009). Rodó et al. (2002) identified, with the escalating of El Niño events, increases in cholera cases in Bangladesh (Ahmed 2006). Changing climate conditions are leading to amplified rates of communicable and noncommunicable

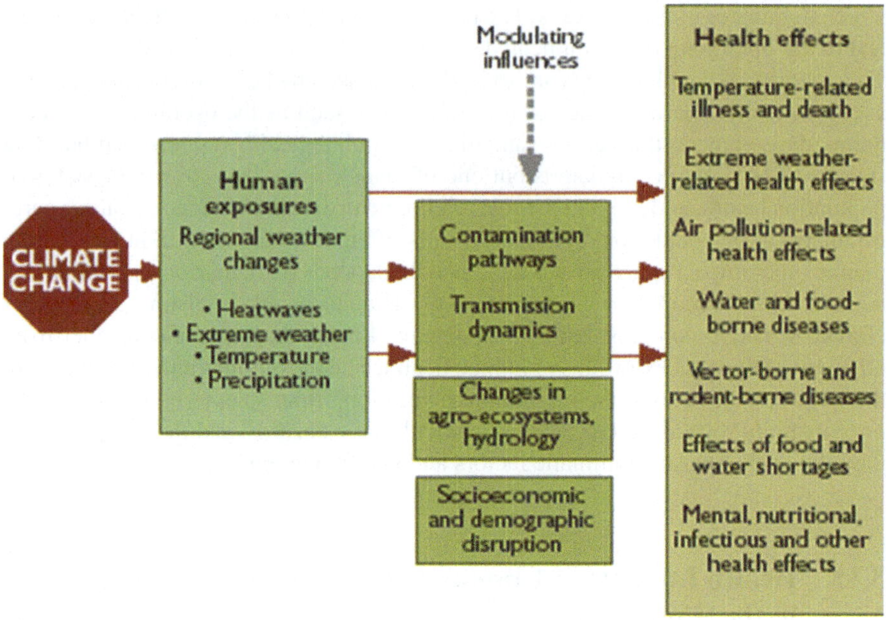

Fig. 3.10 Pathways of climate change affect human health. (Modified from Patz et al. 2000, adapted from IPCC 2001)

diseases in children, which include vector-borne infections, waterborne diseases, and diarrhea in Bangladesh. In vulnerable groups in urban areas, more recurrent and intense heat waves will cause mortality and morbidity to rise (UNICEF 2016).

3.16 Temperature-Related Health Problems

As a result of significant changes in the climate system on the Island, inhabitants have been facing a wide range of health problems. Inhabitants experience adverse health problems owing to rising heat. The majority of our respondents (61.03%) said that the temperature had increased on the Island. Over half were suffering from a temperature-related illness (59.24%). Figure. 3.11 shows the temperature-related health problems of the inhabitants-.

Owing to increasing temperature, people around the Island mainly suffer from diseases such as diarrhea (43.75%), typhoid (6.25%), and skin diseases (19.27%). They also suffer from fever (29.07%), headache (9.25%), high blood pressure (1.32%), and digestion problems (1.32%) as a consequence of high temperature.

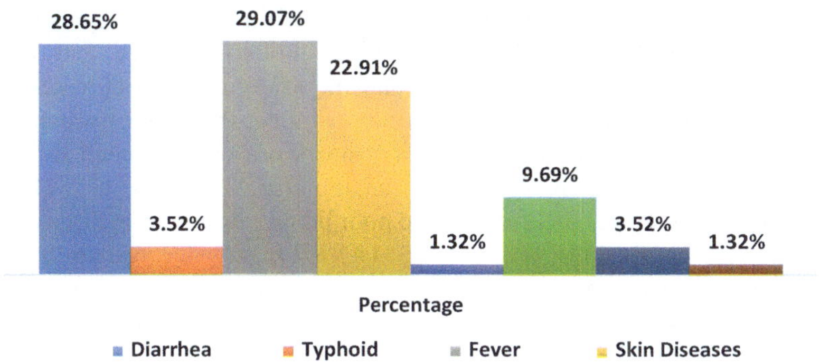

Fig. 3.11 Temperature-related health problems of the respondents. (Source: Field Study 2018)

Fig. 3.12 Waterborne diseases in the study area. (Source: Field Study 2018)

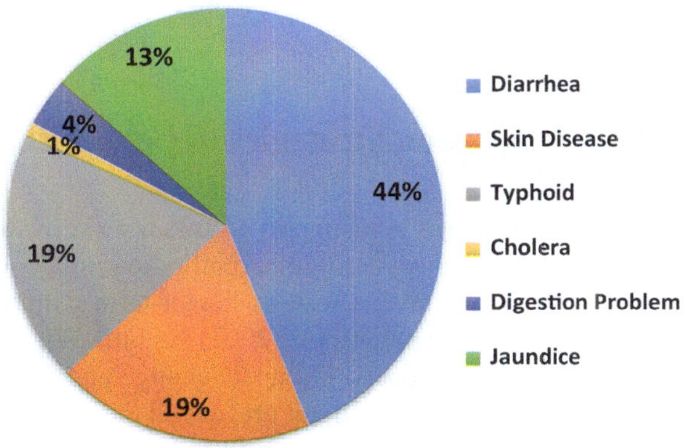

3.17 Waterborne Diseases

Changing climate has direct implications for health. Fecal–oral transmission is one of the focal origins of waterborne diseases in Bangladesh (Biswas et al. 2015). According to the study of Kabir et al. 2016, during and after Sidr, people suffered from waterborne diseases such as diarrhea, dysentery, viral fever, and skin diseases. In the present study, respondents also said that after any storm, waterborne diseases spread on the Island because of pollution and high levels of saline in the drinking water. Again, owing to the scarcity of freshwater during summer, the inhabitants of the Island drink contaminated water. As a result, they suffer from various waterborne diseases. Figure 3.12 shows the waterborne diseases affecting St. Martin's Island.

Data analysis from the present study shows that owing to the impacts of changing climate and the resultant disasters, waterborne diseases are prevalent on the

Island like other climate-affected, coastal disaster-hit areas. Respondents were suffering from diarrhea, skin diseases, typhoid, cholera, jaundice, and digestion problems as a result of climate change and the aftermath of a climate-induced disaster. According to respondents, after any cyclone, the surface water sources become highly saline, and surface water sources are contaminated by disaster debris. Owing to unhygienic conditions in the shelter centers, as well as the destruction of sanitation conditions of the poor people, waterborne diseases break out after cyclones. Besides, owing to the lack of freshwater availability due to drought and extreme summer temperatures, people are forced to drink contaminated water, as well as saline water, which creates other waterborne diseases among the inhabitants.

It is clear from the analysis that climate change and related disasters cause massive health problems for the inhabitants of the Island. Although it is a tiny island, it has all types of health-related climatic effects, like other climate change-affected coastal zones.

3.18 Adaptation Strategies to Cope with the Problems

"Adaptation involves adjustments to enhance the viability of social and economic activities and to reduce their vulnerability to climate, including its current variability and extreme events as well as longer-term climate change" (Smit 1993). The IPCC Third Assessment Report (2001) says that adaptation has the potential ability to reduce adverse impacts of climate change and to enhance the beneficial effects, but will incur costs and will not prevent all damage.

Based on the time scale prepared by the Committee on Approaches to Climate Change Adaptation 2010, adaptation can be classified as-

- Short-term: less than 10 years
- Medium- and long-term: 10 to 100 years
  - Medium-term: 10–30 years (inclusive)
  - Long-term: more than 30 years to 100 years
- Short-term adaptation: it is essential to initiate and inspire urgent action to preclude or mitigate short-term impacts, which are already stirring and likely to arise from climate change. It is also called a reactive process.
- Medium- and long-term adaptation: response actions are necessary to enhance adaptive capacity to prevent and mitigate possible impacts, by assessing the risks of results that may occur in the medium and long term, and by controlling the effects, reducing vulnerability, and strengthening resilience. It is also termed a proactive process.

There are some steps to the adaptation process (Fig. 3.13).

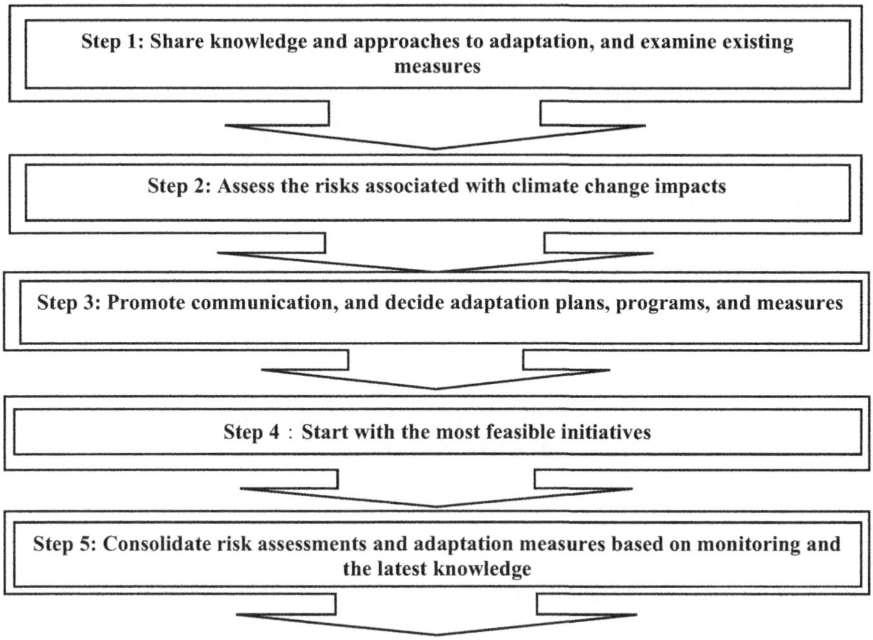

Fig. 3.13 Five steps for adaptation planning and implementation

3.19 Adaptation Measures in the Pre-Disaster Period to Mitigate Loss

3.19.1 Pre-Disaster Warning

To mitigate the impacts of disasters on life and health and on proper adaptation, a pre-disaster warning is a pre-requisite. St. Martin's is a very miniature island. Here, in times of disasters, appropriate measures are taken to warn the local people. Although most of the people are poor, they manage to get early warning. About 68.41% of respondents claim that they get warning from local government. Others get information from the radio (11.49%), television (10.77%), and personal contact from the meteorological department office on the Island (10%). The above statistics show that the pre-disaster warning system on the Island is quite developed and people get enough time to take proper shelter to save themselves from avoidable death.

3.19.2 Safety Measures

Safety decisions are mainly taken by the head of the family or more precisely, the male members of the family. About 85.13% of respondents said that the head of the family makes the decision on taking shelter at the shelter center. In the pre-disaster period after getting the signal, 69.23% people take shelter at government shelter centers. Still, a large portion of people prefer to stay home (30.77%) because of the unhygienic conditions and congested situation of the shelter centers. The majority of the indigent respondents take shelter during and after the disaster because of a lack of supporting settlement system for security during a disaster. Respondents who had a supporting settlement system did not leave their home. There were fewer shelter facilities than needed. Mainly primary schools, hospitals, and Upazila complexes were used as shelter centers. Sanitation facilities were not sufficient to support huge numbers of people at a time. There was also a lack of safety for girls and women.

3.20 Adaptation Strategies to Deal with Climate Change-Related Health Problems

A large number of respondents suffered from waterborne diseases owing to the unhygienic situation in the shelter during and after a disaster. Again, there were no measures for providing health care, especially for pregnant women, children, or older people. According to the survey, respondents were not taught any particular coping strategy to deal with health impacts owing to climate change.

3.20.1 Treatment Facilities in the Study Area

The study shows that climate change causes severe waterborne diseases, and inhabitants of the Island greatly suffer from temperature-related illness owing to changes in the climatic pattern and post-disaster period. But treatment facilities on the Island are not very satisfactory for dealing with this problem or mitigating it. Existing treatment facilities and conditions of the Island are shown below in Fig. 3.14.

To adapt to health problems related to changing climatic patterns and during as well as post-disaster health difficulties, respondents depend on their ability, situation, and knowledge. About 24% of the respondents directly take medicine from a local dispensary, and 17.69% take medication according to their expertise without any specification by the doctor. During the study, it was found that there was only one hospital on the Island, but doctors, nurses and no one related to the hospital was found at the station. There was only one qualified doctor in the health complex, which is highly inadequate (Fig. 3.15).

The hospital also lacks treatment instruments and medicines. Some respondents were capable of going to Teknaf (9.49%) for better treatment, but this percentage is meagre owing to the high cost. Although the Island is one of the most attractive tourist spots in the country, there are no treatment facilities for tourists. Again, most of the low-income inhabitants favor taking treatment from local quacks and

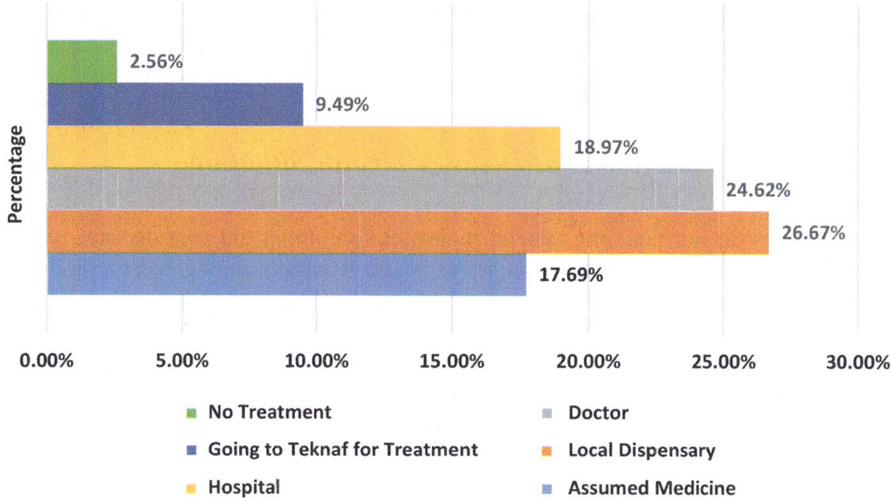

Fig. 3.14 Existing treatment facilities in the study area. (Source: Field Study 2018)

Fig. 3.15 The only treatment center of St. Martin's Island. (Source: Field Study 2018)

consultation with the pharmacy man instead of going to the hospital. This is because the treatment facilities are meager, and the doctor is not available when necessary.

3.20.2 Adaptation to Post-Disaster Health Situation

Respondents do not take any special measures to combat the post-disaster health conditions. To cope with the post-disaster health situation, they preserve food and medicine (67.44%). But in the case of drinking water, they mostly use the untreated water after disasters, which is the major reason for the break out of post-disaster waterborne diseases. Only 21.79% of respondents use treated water for drinking and household purposes under post-disasters conditions.

3.20.3 Adaptation to the Salinity Problem

Salinity is one of the significant climatic disasters on the Island. During the dry summer months, high tides and cyclones with storm surges create this problem. But adaptation to this problem to get fresh water is very low. However, they use some process to get fresh water during salinity problems. Figure 3.16 shows the adaptation strategies to cope with salinity problems on the Island.

From the analysis of data of the study, it was found that in times of severe salinity problems, deep tube well water is the primary source of freshwater. Some are used to boiling water or filtering. People with good economic conditions buy drinking

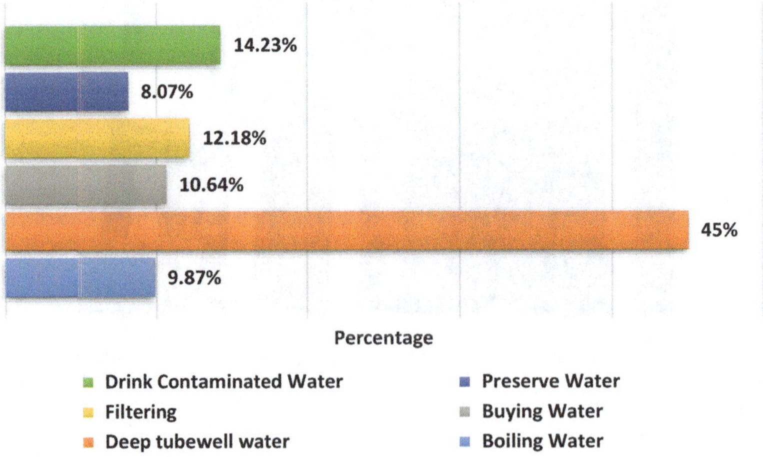

Fig. 3.16 Freshwater sources during the salinity problem. (Source: Field Study 2018)

water. However, a large number of the inhabitants drink this contaminated water, which is very injurious to health.

The above discussion makes it evident that there is an increasing rate of climate change impacts on the health conditions of the local inhabitants of the Island. Still, the health care facilities were inferior beforehand, and there are almost no health care facilities available, which can be taken into consideration, especially no health care measures for increasing disease situations. Adaptation to coping with the climate change-induced disasters and associated health problems is inferior on St. Martin's Island. Either the inhabitants or the government takes any adaptation strategy to cope with the condition.

3.21 Conclusions and Recommendations

Bangladesh is suffering from the impacts of climate change in every aspect of the environment, which will transfer in the coming decades. The southeastern coastal belt, including the islands, is most affected. Their climate has caused fluctuations in temperature, enhancing the severity and frequency of different types of disasters and at the same time having impacts on the health of the inhabitants of the St. Martin's Island. Technology, lack of proper treatment facilities, weather-dependent and weak communication system, unavailability of doctors, deficiency of medical equipment, and absence of appropriate governance worsen the situation. Despite various limitations, the study tries to bring out the actual scenarios of climatic change-induced disasters and the human health impact on St. Martin's Island. Emphasis on the severity of climate change-influenced disasters and social health vulnerability should be a given. Priority should be given to those issues and adaptation, and mitigation measures should be taken immediately to controlling the situation as soon as possible. The information and findings from this study would be valuable for the policy- and decision-making process relating to climate change-related disaster vulnerability and adaptation regarding human health. However, the following specific measures may be taken to address the health impact due to climate change:

- Massive tree plantation in the beach area to prevent bay bank erosion and the natural embankment, protect from disaster, and limit the sea-level transgression.
- Ensure easy and smooth transport from St. Martin's to Taknaf the whole year round.
- Public awareness needs to be raised, vulnerable groups need to become involved in different awareness-building schemes, and there should be practical training for the vulnerable community on adaptation to the vulnerability.
- Temperature and waterborne diseases are the leading climate change impact on health problems on St. Martin's Island. Addressing this problem reduces the incidence of any climate-sensitive diseases, initiatives including policy decisions,

scientific research, monitoring, and broad research to identify the current condition; institutional capacity building to handle the consequences needs to be considered.

- Th number of health professionals needs to be increased, and they should be trained on the health impacts of climate change that they might have to deal with in the future.
- Treatment facilities should be increased, assigning more doctors and specialized doctors. Specialized health care facilities for women and children should be introduced, and the number of medical instruments should be increased.
- The availability of pure drinking water needs to be ensured in the shelter center during disasters and post-disaster on the Island.
- The government in connection with NGOs/research organizations should be working on reducing the impact of climate change disasters, rehabilitation, and relief distribution, and training programs on health issues for health professionals should be initiated.
- The availability of permanent doctors, nurses and health-care specialists should be arranged at all times.

References

Aalst MKV (2006) The impacts of climate change on the risk of natural disasters. Disasters 30(1):5–18. https://doi.org/10.1111/j.1467-9523.2006.00303.x

Abdul Ahad Biswas AKM, Mostafa Zaman AKM, Abdus Sattar Md, Shahidul Islam Md, Afjal Hossain Md, Faisal Md (2015) Assessment of disaster impact on the health of women and children. Journal of Health and Environmental Research 1(3):19–28. https://doi.org/10.11648/j.jher.20150103.11

Ahmed AU (2006) Bangladesh climate change impacts and vulnerability a synthesis Bangladesh climate change impacts and vulnerability. Climate Change Cell, Department of Environment, Comprehensive Disaster Management Programme, Bangladesh

Ahmad MSA, Ashraf M, Ali Q (2010) Soil salinity as a selection pressure is a key determinant for the evolution of salt tolerance in blue Panicgrass (*Panicum antidotale* Retz.). Flora (Jena) 205(1):37–45

Ali K, Barkat A, Molla M, Hossain N, Hoshen A, Rahman M, Billah M (2015) Climate change induced disasters in the southern coastal Belt of Bangladesh. Asian Journal of Water Environment 2(1):1–15

Asia Foundation (2012) A situation analysis of climate change adaptation initiatives in Bangladesh. The Asia Foundation Bangladesh, Dhaka

Bangladesh: Climate Change Strategy and Action Plan (BCCASP 2009) (2009) Climate change Laws of the world. The Department of Environment, Ministry of Environment and Government of Bangladesh, Dhaka

Banglapedia (2015) St. Martin Island. https://en.Banglapedia.org/index.php/St.martinIsland. Accessed 6 Dec 2019

Barkat KMA, Molla MH et al (2015) Climate change induced disasters in the Southeastern Coastal Belt of Bangladesh. Asian J Water Environ 2(1):1–15

Basu R (2009) High ambient temperature and mortality: a review of epidemiologic studies from 2001 to 2008. Environ Health 2009(8)

Basu R, Samet JM (2002) Relation between elevated ambient temperature and mortality: a review of the epidemiologic evidence. Epidemiol Rev 2002(24):190–202

BBS (2013) Bangladesh Bureau Statistics, Statistics and Informatics Division, Ministry of Planning, Dhaka, Bangladesh

BCCSAP (2009) Bangladesh Climate Change Strategy and Action Plan, 2009. Ministry of environment and forest. Government of Bangladesh

Brammer H (2002) Land use and land use planning in Bangladesh. University Press Ltd, Dhaka

Butler CD (ed) (2014) Climate change and global health. CABI, Wallongford

BWDB (Bangladesh Water Development Board) (2013) Coastal embankment improvement project, phase-I (CEIP-I). Government of the People's Republic of Bangladesh. Ministry of Water Resources, Bangladesh

CCC (Climate Change Cell) (2009) Climate change and health impacts in Bangladesh. Department of Environment, Ministry of Environment and Forests, Bangladesh

CCKNFP (Climate Change Knowledge Network Foundation Paper) (2001) Vulnerability and adaptation to climate change: concepts, issues. Assessment Methods, Santiago Olmos

CCPB (Climate Change Profile: Bangladesh) (2018) Ministry of Foreign Affairs Netherlands

Cruz RV et al. (2007) Asia. M. L. Parry ML et al.(eds.) (2007) Climate change 2007: impacts, adaptation and vulnerability. Contribution of Working Group II to the Fourth Assessment Report of the Intergovernmental Panel on Climate Change. Cambridge University Press, Cambridge

Disaster Management Bureau (DMB) (2008) Policy, plans & statements Bangladesh: National disaster management plan 2008–2015 draft (2008), Disaster Management Bureau (DMB); Ministry of Food and Disaster Management (MOFDM), Dhaka, Bangladesh

Disaster Management Bureau (DMB) (2012) Policy, plans & statements Bangladesh: National disaster management plan 2012–2015 draft (2012), Disaster Management Bureau (DMB); Ministry of Food and Disaster Management (MOFDM), Dhaka, Bangladesh

Elahi KM (2016) Climate change and health impacts in Bangladesh. Springer, Cham

Feeroz MM (2009) Effects of environmental degradation on food security in the St. Martin's island of Bangladesh. The National Food Policy Capacity Strengthening Programme (NFPCSP), Dhaka, Bangladesh

Germanwatch (2013) Global climate risk index 2013. Who suffers most from extreme weather events? Weather-related loss events in 2011 and 1992 to 2011. Harmeling S, Eckstein, D. Briefing paper. Germanwatch, Bonn. Available at: http://germanwatch.org/en/download/7170.pdf

Haines A, Kovats RS, Campbell-Lendrum D, Corvalan C (2006) Climate change and human health: impacts, vulnerability and mitigation. The Lancet Series 367:2101–2109

Hasan MK, Ahmed MM, Miah MG (2008) Agro-economic performance of jackfruit-pineapple agroforestry system in Madhupur tract. J Agric Rural Dev 6(1&2):147–156

Hossain D, Musa MH, Talib J, Jol H (2010) Effects of nitrogen, phosphorus and potassium levels on Kenaf (*Hibiscus cannabinus* L.) growth and photosynthesis under nutrient solution. J Agric Sci 2(2):49–57

Hughes L, McMichael T (2011) The critical decade: climate change and health. The Climate Commission Secretariat, Australia

Huq S, Alam M (2003) Flood management and vulnerability of Dhaka City. Bangladesh Center of Advance Studies (BCAS), Dhaka

Huq S, Karim Z, Asaduzzaman M, Mahtab F (2003) Vulnerability and adaptation to climate change for Bangladesh. Kluwer, London

ICZMP (2001) Delineation of the coastal zone. In: Project development Office for Integrated Coastal Zone Management Plan. December 2003, Dhaka, Bangladesh

IPCC (2001a) Intergovernmental Panel on Climate Change. Climate Change 2001: third assessment report, impacts, adaptations and vulnerability of climate change. Cambridge University Press Cambridge

IPCC (2001b) Synthesis report, third assessment report. Cambridge University Press, Cambridge

IPCC (2007) Climate change impacts, adaptation and vulnerability, contribution of working group II to the fourth assessment report of the Intergovernmental Panel on Climate Change. Cambridge University Press, Cambridge

IPCC (2014) Climate change 2014: impacts, adaptation, and vulnerability, part a: global and sectorial aspects, contribution of working group II the fifth assessment report of the Intergovernmental Panel on Climate Change. Cambridge University Press, Cambridge

Islam MR (ed) (2004) Where land meets the sea: a profile of the coastal zone of Bangladesh. The University Press Limited, Dhaka

Islam MS (2001) Sea-level changes in Bangladesh: the last ten thousand years. Asiatic Society of Bangladesh, Dhaka

IUCN (2011) Climate change and health in Bangladesh, information brief. Government of the People's Republic of Bangladesh, Ministry of Environment and Forests

Kabir R, Khan HTA, Ball E, Caldwell K (2016) Climate change impact: the experience of the coastal areas of Bangladesh affected by cyclones Sidr and Aila. J Environ Public Health 2016:1–9

Laila F (2013) Assessment on social vulnerabilities to climate change – a study on South-Western coastal region of Bangladesh. Published MS thesis, Department of Earth Science, Uppsala University, Uppsala

Luber G, Prudent N (2009) Climate change and human health. Trans Am Clin Climatol Assoc 120:113–117

Luber G, Knowlton K, Balbus J, Frumkin H, Hayden M, Hess J, McGeehin M, Sheats N, Backer L, Beard CB et al. (2014) Human health. In Melillo JM, Richmond TTC, Yohe GW (eds) Climate change impacts in the United States: the third national climate assessment. U.S. Global Change Research Program, Washington, DC, Chapter 9: 220–256

Minar MH, Hossain MB, Shamsuddin MD (2013) Climate change and coastal zone of Bangladesh: vulnerability, resilience and adaptability. Middle East J Sci Res 13(1):114–120

Ministry of Water Resources (MoWR) (1999) Integrated coastal zone management: concepts and issues; a government of Bangladesh (GoB) policy note. Ministry of Water Resources, Bangladesh, Dhaka

MoEF (Ministry of Environment and Forests) (2001a) Survey of Flora, National Conservation Strategy Implementation Project-1. Ministry of Environment and Forests, Dhaka

MoEF (Ministry of Environment and Forests) (2001b) Survey of Fauna, National Conservation Strategy Implementation Project-1. Ministry of Environment and Forests, Dhaka

MoEF (Ministry of Environment and Forests) (2005) National Adaptation Programmes of Action (NAPA)

MoEF (Ministry of Environment and Forests) (2009) Bangladesh climate change strategy and action plan. Government of the People's Republic of Bangladesh, Ministry of Environment and Forests

Nazem NI, Mahboob AQM (1992) Development of coastal region of Bangladesh. In: Elahi KM, Sharif AHMR, Kalam AKMA (ed.) Bangladesh: geography; environment and development. Bangladesh National Geographical Association (BNGA), Dhaka

Oxfam America (2009) Exposed: social vulnerability and climate change in the US southeast. Oxfam America Inc, Boston, pp 1–24

Patz JA, Mcgeehin MA, Bernard SM, Ebi KL, Epstein PR, Gubler DJ, Reiter P, Romieu I, Rose JB, Samet JM (2000) The potential health impacts of climate variability and change for the United States: executive summary of the report of the health sector of the U.S. National Assessment. Environ Health Perspect 108(4):367–376

Paul G (2010) The science of climate change. Das Gupta & Company Private Limited, Kolkata

PDO-ICZMP (2003) Local level institutional arrangement in ECFC project. In: Program development Office for Integrated Coastal Zone Management Plan Project; working paper WP023. Water Resources Planning Organization, Ministry of Water Resources, Bangladesh, Dhaka

Petkova EP, Ebi KL, Culp D, Redlener I (2015) Climate change and health on the U.S. Gulf Coast: public health adaptation is needed to address future risks. Int J Environ Res Public Health 2015(12):9342–9356

Rabbani G, Huq S (2010) Impacts of climate change on water resources and human health: empirical evidences from a coastal district (Satkhira) in Bangladesh. Impact Clim Change Water Health 20(8):272–285

Rahman MM, Kumar S, Hori S (2010) Assessment of vulnerability and adaptation options for climate change in coastal region of Bangladesh. J Geo-Environ 8(2008):42–60

Rasheed KBS (2016) Bangladesh resource and environmental profile. Development Publishing House, Dhaka

Rashid AKMM, Singha D, Imam H (2009) Climate change vulnerability in Bangladesh: strategic position of DSK/DCA in the field of climate change adaptation initiatives in Bangladesh. The Dustha Shastha Kendra

Rodó XM, Pascual GF, Faruque ASG (2002) ENSO and cholera: a nonstationary link related to climate change? Proc Natl Acad Sci 99:12901–12906

Smit B (1993) Adaptation to climate variability and change. Report of the task force on climate adaptation. Canadian Climate Program Board, Department of Geography, University of Guelph, occasional paper no. 19. Environment Canada. Downsview, Ontario, Canada

Thomas TS, Mainuddin K, Chiang C, Rahman A, Haque A, Islam N, Quasem S, Sun Y (2013) Agriculture and adaptation in Bangladesh: current and projected impacts of climate change. IFPRI discussion paper 01281

Thomson PA, Islam MA (2008) Environmental profile of St. Martin's Island, United Nations Development Programme (UNDP), Bangladesh

Thompson P, Islam M, Khan M, Islam A, Rahman M, Hebara M, Denyshchyk O, Islam R, Saif S, Haque Q, Mohit M, Sharif A, Haq M (2010) Environmental profiles of St. Martin's island. Coastal and Wetlands Biodiversity Management Project, United Nations Development Programme (UNDP), Bangladesh, ISBN: 978-984-33-0779

Thompson P, Islam M, Khan M, Islam A, Rahman M, Hebara M, Denyshchyk O, Islam R, Saif S, Haque Q, Mohit M, Sharif A, Haq M (2008) Environmental profiles of St. Martin's island. Coastal and Wetlands Biodiversity Management Project, United Nations Development Programme (UNDP), Bangladesh, ISBN: 978-984-33-0779

Tomascik T (Tom) (1997) Management plan for coral resources of Narikel Jinjira (St. Martin's island). https://doi.org/10.13140/RG.2.1.1269.7200

Trenberth K (2005) Climate: uncertainty in hurricanes and global warming. Science (New York, N.Y.) 308:1753–1754. https://doi.org/10.1126/science.1112551

Tsurita I, Prabhakar SVRK, Sano D (2013) Approaches to climate change adaptation: a case study of agricultural initiatives in Japan. In: Ha H, Dhaka TN (eds) Enhanced governance approaches for mitigation and adaptation to climate change – an Asia focus. Palgrave Macmillan Publishers, UK

UNDP (2010) Gender, climate change and community-based adaptation. UNDP, New York

UNICEF (2016) Learning to live in a changing climate: the impact of climate change on children in Bangladesh. United Nations Children's Fund, Bangladesh

United Nations Framework Convention on Climate change (UNFCCC) (2007) Climate change: impacts, vulnerabilities and adaptation in developing countries. UNFCCCSecretariat, Bonn. http://unfccc.int/resource/docs/publications/impacts.pdf

USGCRP (2016) The impacts of climate change on human health in the United States: a scientific assessment. Crimmins A, Balbus J, Gamble JL, Beard CB, Bell JE, Dodgen D, Eisen RJ, Fann N, Hawkins MD, Herring SC, Jantarasami L, Mills DM, Saha S, Sarofim MC, Trtanj J, Ziska L (eds) U.S. Global Change Research Program. Washington, DC, 312pp. https://doi.org/10.7930/J0R49NQX

Warrick RA, Ahmed QK (1996) The implementation of climate and sea level change for Bangladesh. Kluwer, Dordrecht

Water Resources Planning Organization (2006) Coastal development strategy. Ministry of Water Resources, Bangladesh

Watson RT, Zinyoera MC, Moss RH. (1996) Climate change 1995: impacts, adaptations and mitigation of climate change: scientific-technical analysis. Contribution of Working Group II to the Second Assessment Report of the Intergovernmental Panel on Climate Change. Cambridge University Press, Cambridge

Watson RT, Zinyoera MC, Moss RH (1998) The regional impacts of climate change: an assessment of vulnerability. A special report of IPCC Working Group II. Cambridge University Press, Cambridge

Wikipedia (2012) St. Martin's Island. https://en.wikipedia.org/wiki/St._Martin's_Island. Accessed on 6 December 2019

WHO (1948) WHO remains firmly committed to the principles set out in the preamble to the Constitution, the World Health Organization as a specialized agency within the terms of Article 57 of the Charter of the United Nations

Chapter 4
Measuring Resilience of Coastal Fishing Communities of Bangladesh to Climatic Impacts

Md. Monirul Islam, Md. Manjurul Islam, Md. Kawser Ahmed, Makidul Islam Khan, Angkur Mohammad Imteazzaman, and Zakir Hossain

Abstract Measuring resilience to the impacts of climate change is essential to finding ways to help communities to adapt to climate change and to improve their livelihoods. This study has measured the resilience to the impacts of climate change of two coastal fishing communities of Bangladesh – at the village of Bolgram Sura in Bhola Sadar and the village of Chita Kundu at Manpura Upazila of Bhola district, using 52 attributes under six specific types of capital – natural, human, physical, social, financial, and institutional – where a lower attribute score represents a higher resilience. This study has found significant variation among the resilience scores under each of the six specific types of capital of Chita Kundu (p value = 0.037), but not among the types of capital of Bolgram Sura (p value = 0.312). Some specific attributes have shown a highly significant difference between the two communities, such as fish landing center (p value < 0.001), power supply (p value = 0.004) and number of children (p value = 0.005), and a significant difference in family thriftiness (p value = 0.024). Overall, this study has reported that the Bolgram Sura fishing community is a little bit more resilient (though without statistical significance) in four types of capital – natural, physical, financial and institutional – compared with the Chita Kundu counterparts. Two other types of capital – human and social –

M. M. Islam · M. I. Khan (✉)
Department of Fisheries, University of Dhaka, Dhaka, Bangladesh

M. M. Islam · M. K. Ahmed
Department of Oceanography, University of Dhaka, Dhaka, Bangladesh

A. M. Imteazzaman
ECOFISH-Bangladesh Project, WorldFish, Bangladesh and South Asia Office, Bhola, Bangladesh

Z. Hossain
Department of Statistics, University of Dhaka, Dhaka, Bangladesh

M. N. Islam, A. van Amstel, *Bangladesh II: Climate Change Impacts, Mitigation and Adaptation in Developing Countries*, Springer Climate,
https://doi.org/10.1007/978-3-030-71950-0_4

have shown the opposite results. This study has identified specific attributes and types of capital for policy intervention to reduce vulnerability and increase resilience of coastal fishing communities. In order to scale up, more such studies need to be conducted from different contexts.

Keywords Bangladesh · Resilience · Climate change · Coastal zone · Fishing community

4.1 Introduction

Globally coastal communities are mostly vulnerable to the impacts of climate change, especially those in developing countries (Chinowsky et al. 2011; Rahman et al. 2020). About 38% of the world's population lives in coastal areas, of which over three-quarters are in developing countries (Barbier 2015) with a relatively poorer capacity to adapt to the impacts of climate change (Chinowsky et al. 2011). Bangladesh, particularly known for its vulnerability to climate change, is ranked the 7th most affected country globally according to the Climate Risk Index (CRI) during a period of 1999 to 2018 owing to the climatic hazards (Eckstein et al. 2019). The coastal communities of Bangladesh are more vulnerable to climatic hazards such as cyclones, floods, tidal surges, sea-level rise, riverbank erosion, and salinity intrusion (Islam et al. 2014a, 2019; Ahmed et al. 2016).

Climate change has affected ecosystems, societies, and economies across the world (IPCC 2014). Fisheries and fisherfolk are affected by the impacts of climate change in a wide range of ways including from cyclones, floods, and other extreme weather events (Nicholls et al. 2007). These hazards ultimately impose negative impacts on the fishing activities, yields, and livelihoods of the fishers and fisheries management (Cochrane et al. 2009; Belhabib et al. 2016; Ding et al. 2017). The direct impacts can be damage to fishers' physical assets (e.g., houses and fishing crafts) and missed fishing trips owing to cyclones and other extreme climatic events.

In Bangladesh, the fisheries sector plays an important role in nutrient supply, national GDP (3.50%), foreign export earnings (1.50%), livelihood security (11% of the total population) and employment (DoF 2019). But there is an increasing concern about the impact of climate change on this sector. In particular, coastal fishing communities of Bangladesh are more vulnerable to climatic hazards than inland fishing communities (Islam et al. 2019) as they are directly exposed to the Bay of Bengal. Consequently, climate change has added an extra risk factor to fishing communities in coastal areas as they are fragile and sensitive to these adverse climatic conditions.

Resilience has become a popular conceptual tool in research and policy in fostering the understanding within climate change adaptation and development contexts (Adger et al. 2011). Emerging from a wide range of disciplines (Alexander 2013; Bahadur et al. 2013), resilience is referred to as the ability of a system, community, or society to bounce back to normality to pursue its social, ecological, and economic development and growth objectives to adapt to climate change (Keating et al. 2014). This implies the return of the functions of an individual, household, community, or ecosystem to previous conditions, with as little damage and disruption as possible following climatic impacts (Tanner et al. 2015). The ability to prepare for, recover from, and adapt to the aforementioned climatic impacts is called resilience. Resilience is the capacity to withstand climatic impacts and rebuild when indispensable (Hoque et al. 2019). The concept of renewal is a key aspect of resilience in the context of climate change rather than adjusting to keep things the way they are. Resilience is the product of multiple activities, interactions, and relationships, and is frequently considered an attribute of a system (IFRC 2011). These attributes jointly or separately help to adapt to the impacts of climate change. For example, some resilience attributes may reduce the initial damage from an extreme climatic event such as a cyclone, whereas other attributes may improve recovery time, and others may help to overcome the future uncertainty of climatic hazards to reduce damage and improve resilience over time (Ewing 2014).

Climate change may push coastal communities across thresholds (Blythe 2015). For example, cyclones, sea-level rise, floods, and tidal surges may be the biggest concern for those communities who live in a coastal area or on a low-lying island. When this happens, climate-resilient communities and nations will have the capacity for transformation that will help them to rebuild in ways that work under the new conditions. Climate change resilience will demand that social, economic, and ecological systems become capable of reorganizing so as to maintain their key functions, identity, and structure, while also maintaining their capacity for adaptation, learning, and transformation. This will pave the way toward sustainable livelihood development as long as the structural inequalities that drive poverty, social exclusion, and vulnerability are addressed.

The concepts of resilience are inter-linked to a sustainable livelihood framework in terms of climate vulnerability (Sok and Yu 2015). Employing a sustainable livelihood framework in the context of climate change may help to extend the concept of access to five types of livelihood assets as indicators of resilience over time, with a focus on long-term flexibility (De Haan and Zoomers 2005). The sustainable livelihood framework encompasses five types of capital assets: natural, human, physical, social, and financial (DFID 1999). This framework characterizes the wellbeing of a community holistically, while at the same time suggesting ways to enhance its resilience. This study has included institutional indicators with the aforementioned types of livelihood capital followed by Stanford et al. (2017) for building resilience in a community. Institutional capital helps to measure the ability of the community to influence decisions to enhance community resilience. It is important to note that there are interdependencies across the types of livelihood capital to evaluate the resilience of a community (Twigg and Calderone 2019).

Community resilience measurement is crucial for helping coastal communities to adapt to climate change and improve their livelihoods. Most of the studies on coastal communities are focused on vulnerability to climate change, impacts, and adaptation strategies (Islam et al. 2014a, b, 2019, 2020; Khan et al. 2018), which are different than community resilience to climate change. Some studies have been carried out in Bangladesh in the context of coastal communities' resilience to climate change (Ahmed et al. 2016; Sharifuzzaman et al. 2018; Hoque et al. 2019). However, limited attention has been given to examining the coastal fishing community resilience to the impacts of climate change. The available study investigates resilience mostly in narrative forms, with very little effort made to measure it. Measurement of resilience is often necessary to know the extent of the components or variables of resilience, which in turn helps to make robust policy. The objective of this study is therefore to measure the resilience of two coastal fishing communities of Bangladesh to the impacts of climate change.

4.2 Methodology

In this study, an initial scoping study was conducted in the coastal regions of Bangladesh to finalize the study sites to assess the resilience to the impacts of climate change of the coastal fishing communities. Based on the findings of a scoping study, such as increasing frequency and magnitudes of climatic hazards, this study selected Bhola district for the assessment of resilience.

4.2.1 Study Area

In this study, two coastal fishery-dependent communities – Bolgram Sura at Dhania Union in Bhola Sadar Upazila and Chita Kundu at Manpura Union in Manpura Upazila, Bhola district, Bangladesh (Figs. 4.1 and 4.2) – were selected to evaluate their resilience to the impacts of climate change. These two communities are frequently affected by climatic hazards such as cyclones, floods, tidal surges, riverbank erosion, and salinity intrusion.

4.2.1.1 Bolgram Sura

Bolgram Sura is situated on the banks of the Meghna River. The total number of households in Bolgram Sura is 280 with a total population of 1484. Most of the people in this community are involved in fishery-related activities directly or indirectly. Hilsa (*Tenualosa ilisha*) fishing is the major income source (70% of the total households) of the local community. Besides these, they catch other finfishes such as Pungas catfish (*Pangasius pangasius*), Gangetic goonch (*Bagarius bagarius*), Lanceolate goby (*Pseudapocryptes elongatus*), Rubicundus eel goby (*Odontamblyopus rubicundus*), Tiger goby (*Eugnathogobius oligactis*), Gangetic whiting (*Sillaginopsis panijus*), Rita (*Rita rita*), and Paradise threadfin (*Polynemus paradiseus*). The rest of the people of this community are involved with

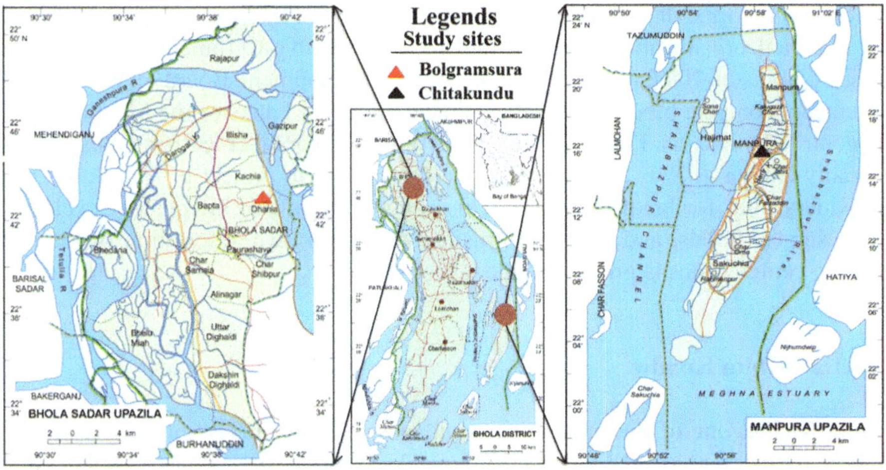

Fig. 4.1 Map of the study sites – Bolgram Sura at Dhania union in Bhola Sadar Upazila and Chita Kundu at Manpura union in Manpur Upazila, Bhola district, Bangladesh

Fig. 4.2 Typical settings of the coastal fishing communities of Bolgram Sura and Chita Kundu, Bhola district (fishers' house affected by cyclones and riverbank erosion on the left-hand side and fishing boats equipped with fishing gear and solar panel in the Meghna River on the right-hand side)

agro-farming, net mending, boat making, day labor, etc. Some people rear livestock such as goats and poultry. Until the 1990s, most of the fishers used large nonmechanized boats to catch fish, but later they started to use mechanized boats. Some fishers take informal credits or advance money *(dadon)* from the money lender (*Mahajan*) with some informal rules and conditions. The fishers who take the advance money are bound to sell their catch to the money lender with commission.

In this community, around 70% of fishers have earthen houses, i.e., houses made of low-quality materials such as split bamboo, soil, straw, and corrugated tin (Fig. 4.2) and the rest of the fishers have semi-concrete houses, i.e., the roof or the wall is made of concrete (bricks joined together with cement, iron rods, or steel). Bolgram Sura is dominated by Muslim fishers. The community is 6 km away from Upazila town and the District Headquarters. It has been affected by severe riverbank erosion over the last couple of years. During the scoping study, one key informant reported that nearly 50% of the households of the community are affected by riverbank erosion and as a result, they have moved to government-owned land or other places. In 2007, the Bangladesh Water Development Board (BWDB) constructed a concrete block-made embankment along the banks of the Meghna River to control the erosion. Later, in 2014, the BWDB built another earthen embankment (locally known as *Beri-bandh*) inside the community to protect it from floods and provide shelter for landless fishers. This community is also affected by cyclones, tidal surges, floods, and salinity intrusion. A comparative description of the two fishing communities is given in Table 4.1.

4.2.1.2 Chita Kundu

Chita Kundu, one of the most ancient villages in Manpura, is surrounded by the Meghna River and located in Manpura Upazila, Bhola district. It has a total population of 1400 in 244 households. Like Bolgram Sura, hilsa fishing is also the chief income source for the people of Chita Kundu. In this community, nearly two-thirds of the total households rely on hilsa fishery for their livelihoods. Fishers of this community use mechanized boats for fishing and they use solar panels in their boats (Fig. 4.2), which is an exceptional example from most of the fishing communities in the region. They also take informal credit from money lenders like the fishers of Bolgram Sura. In this community, money lenders hire fishers from outside of Manpura to catch hilsa on a daily basis on their boats. As a result, many local fishers who do not have their own fishing boat cannot go fishing and their income from fishing has reduced.

In this community, around 90% of fishers have earthen houses and the rest have semi-concrete houses. About 70% of the households here are Muslim and the rest are Hindu. The nearest town (Hazirhat Upazila) is 8 km from this community. Ramnewaz Ghat, a key entrance point in Manpura Upazila, is 2.5 km away. Waterway transportation is the only option for communicating with other important places from Ramnewaz Ghat. Owing to its geographical location, this community is also frequently affected by riverbank erosion. Over the last couple of years, nearly 1.5 km of the land was engulfed in the Meghna River and as a result, two-thirds of the total population lost their land and moved to government-owned land (source: key informant interviews). Like Bolgram Sura, this community is also affected by cyclones, floods, tidal surges, salinity intrusion, and riverbank erosion.

4.2.2 *Data Collection*

In this study, data were collected using a mixed method approach (Fig. 4.3). Data collection methods are described below.

Table 4.1 Socio-demographic features of the two fishing communities in Bolgram Sura and Chita Kundu, Bhola district

Variables	Bolgram Sura	Chita Kundu
Number of total population	1665	1460
Total number of households	325	244
Number of fishery-dependent households	255	216
Livelihood dependency in terms of income (%)		
Fisheries	65	60
Day -labor (agro)	11	8
Day labor (non-agro)	17	11
Others (e.g., skilled worker, agro-farming, driving, business and service holders)	7	21
Family size (mean)	5.3	4.1
Education level (at least can sign only) (%)	45	34
Religion (%)		
Muslim	100	0
Hindu	65	35
Housing condition (%)		
Earthen	70	90
Semi-concrete house-building	30	10
Access to training facilities (%)	40	30
Access to drinking water facilities (tube well) (%)	100	99.8
Transportation system	Medium	Poor
Distance to nearby medical facilities	Community clinic (0.5 km) and government hospital (5 km)	Community clinic (0.7 km) and government hospital (8 km)
Access to sanitary toilet (%)	70	82
Access to electricity (%)	100	–
Access to solar energy (%)	–	40
School within the community	Absent	Absent
Cyclone shelter within the community	Absent (nearest cyclone shelter 2 km away)	Present
Fishing boats (%)		
Mechanized	40	75
Nonmechanized	60	25
Type of fishing gear	Gill net, drag net, set bag net, cast net, long line	Gill net, drag net, cast net
Women's involvement in shrimp post-larvae collection (%)	30	70
Mean monthly income (US$[a])	89	115

Source: Household surveys, focus group discussions, key informant interviews, and BBS 2011
[a]US$ 1 = BDT84.45 (Date: 14 April 2020)

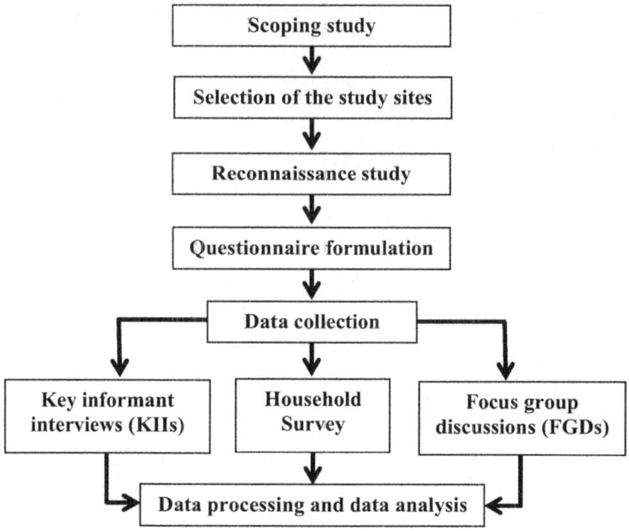

Fig. 4.3 Diagram of the data collection steps and data collection tools followed in this study

4.2.2.1 Household Survey

In this study, the sampling frame, collected from fisheries offices in Bhola Sadar Upazila and Manpura Upazila, was used to select the respondents for household interviews using a simple random sampling technique. In this study, 20 respondents were selected for semi-structured interviews to collect data on the resilience of fisheries to climatic hazards. Questionnaire formulation was one of the most important parts of this study as it helped to identify attributes to evaluate resilience to the impacts of climate change. A total of 52 attributes were used to characterize fishers' resilience in six specific types of capital – natural, human, physical, social, financial, and institutional (Table 4.2). In addition, the socio-economic factors of fishers were also taken into account to justify their resilience. Only the heads of household were interviewed as they had detailed knowledge about the attributes used in this study to evaluate the resilience of the fishing communities. Face-to-face interviews were conducted and data were handwritten after taking the respondents' consent. During household interviews, each question was asked in a similar way, which ensured the accumulation of equivalent answers as much as possible on which factual investigation could be performed.

4.2.2.2 Focus Group Discussions

Focus group discussion (FGD) is an important participatory and effective method of collecting data through community discussion. FGDs were conducted at two phases: during the scoping study and during the main data collection period. The objectives of FGDs were to gather data on fishers' resilience to climatic hazards in the fishing

Table 4.2 List of attributes used to collect data to evaluate the resilience of the fishing communities of Bolgram Sura and Chita Kundu, Bhola district

Specific types of capital[a]	Attribute name
Natural (7)	Geographical position, fish landing center, access to coastal resources, biodiversity, coastal resource management, land use plan, and natural hazards (e.g., cyclones, floods, tidal surges, salinity intrusion, and riverbank erosion)
Human (14)	Readiness to save, market awareness, work ethic, occupational multiplicity, entrepreneurial spirit, wives' involvement in fishery, number of children, education for children, family ability to provide education, retirement planning, family thriftiness, main earner thriftiness, risk knowledge, and goals
Physical (9)	Ownership of fishing boat, adequate fishing gear, ownership of physical assets, processing/added value, ice availability, power supply, housing/sanitation, structural design, and market
Social (8)	Community co-operation, trust/honesty, leadership, social security, equity – right to speak, fairness, emergency responses, and cultural activities
Financial (7)	Ability to save, collateral for credit, access to credit, ability to repay, current savings, remittances, and supplementary income
Institutional (7)	Extension, community livelihoods program, household livelihood program, advocacy, training and capacity building, warning and response, and disaster recovery

Adapted from Stanford et al. (2017)
[a]Number of attributes in parenthesis

communities. Four FGDs were conducted in each of the two fishing communities, with homogenous household heads and all responses were voice-recorded after taking consent from the respondents. Around 7–8 household heads were involved in each FGD and the session continued for about 2.5 h.

4.2.2.3 Key Informant Interviews

Key informant interviews (KIIs) can provide in-depth knowledge on the resilience of fishers to climatic hazards. Like FGDs, the KIIs were also conducted at two phases and voice-recorded. For KIIs, experts who had adequate knowledge about the fishing communities and the impacts of climatic hazards on the communities were selected from inside (e.g., community leader) and outside (e.g., Upazila fisheries officer, representatives of government and nongovernment officials) of the communities. Experts were interviewed to evaluate the resilience of fishers to climatic hazards. In this study, six KIIs were carried out in two fishing communities and each informant was interviewed for around 40 min to 1.5 h.

4.2.3 Data Processing and Data Analysis

All voice-recordings from FGDs and KIIs were listened to carefully before transcribing. To reduce confusion and increase the reliability of the data, voice recordings were translated into the local language. After that, coding of qualitative data was performed and data were analyzed using grounded theory (Strauss and Corbin 1990). In addition, key findings were revised and re-checked with the original transcripts to increase further reliability of the data. Major findings were also triangulated through FGDs and KIIs. The attributes were evaluated on a scale from 1 to 4 (1 = resilient, 2 = nearly resilient, 3 = moderately vulnerable, and 4 = vulnerable) followed by RAPFISH (a rapid assessment of fisheries sustainability) multidimensional scaling (Stanford et al. 2017). In this scoring system, a lower score represents greater resilience to the impact of climate change and vice versa. Using the software SPSS (version 23) the quantitative data were first analyzed by descriptive statistics. Difference in resilience scores in both fishing communities were analyzed using independent sample *t* test. Analysis of variance (ANOVA) was carried out to test the variation among six specific types of capital in each community. Results were represented in graphical and tabular forms for better understanding of the data.

4.3 Results

This study has measured the resilience of two coastal fishing communities – Bolgram Sura and Chita Kundu – based on the 52 livelihood attributes. These are described below.

4.3.1 Resilience of Bolgram Sura Fishing Communities

This study has found that there is no capital-wise difference in resilience to the impacts of climate change between the Bolgram Sura fishing communities (ANOVA: $F_{5,46} = 1.226$, p-value = 0.312). However, among the six specific types of capital of the Bolgram Sura fishing community, social capital is identified as the most resilient type of capital (mean score 18.7) (Fig. 4.4). In contrast, financial capital is identified as the least resilient type of capital (mean score 27.0) in the fishing community.

In Bolgram Sura, eight livelihood attributes (geographical position, access to coastal resources, market awareness, coastal resource management, wives' involvement in fishery, community cooperation, readiness to save, and fairness) are identified as being more resilient (mean score < 1.5; Fig. 4.5). On the other hand, fish landing center is considered to be the least resilient (vulnerable) attribute (mean score 4) in Bolgram Sura.

4.3.2 Resilience of Chita Kundu Fishing Community

This study has shown a significant difference among the six specific types of capital of the Chita Kundu fishing community (ANOVA: $F_{5,46} = 2.614$, p value = 0.037). Like Bolgram Sura, in this community, social capital is identified to be the most

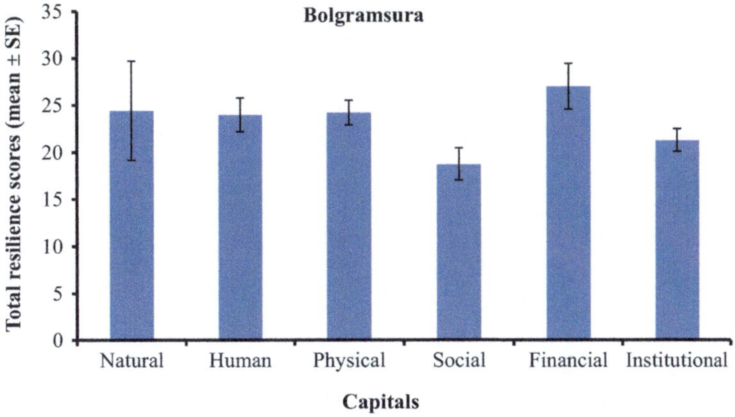

Fig. 4.4 Resilience scores (mean ± SE) of six specific types of capital of the coastal fishing community of Bolgram Sura, Bhola district. Attributes were scored on a scale from 1 to 4 (1 = resilient, 2 = nearly resilient, 3 = moderately vulnerable and 4 = vulnerable). (Source: Household surveys and focus group discussions)

resilient type of capital (mean score 18.1) and financial capital is identified to be the least resilient type of capital (mean score 28.0) among the six specific types of capital (Fig. 4.6).

In Chita Kundu, only six livelihood attributes (access to coastal resources, readiness to save, fairness, market awareness, trust/honesty, and community cooperation) are identified as being more resilient (resilience scores < 1.5) than the other 46 attributes in Chita Kundu (Fig. 4.7). Both natural hazards and lower ability to save attributes are considered to be the least resilient (vulnerable) attributes (resilience score 3.9) in Chita Kundu.

4.3.3 Comparison of Resilience Between Bolgram Sura and Chita Kundu Fishing Communities

Results have shown that there is no significant difference or variation in resilience to the impact of climate change between the Bolgram Sura and Chita Kundu fishing communities (p value = 0.946; Table 4.3). Based on all attributes, it is observed that the overall mean scores for Bolgram Sura (121.3) and Chita Kundu (121.1) are similar. However, the Bolgram Sura fishing community is slightly more resilient (with no statistical significance) in four types of capital – natural, physical, financial, and institutional – compared with their Chita Kundu counterparts. In contrast, the other two types of capital – human and social – have shown opposing results (Table 4.3).

This study has found a highly significant difference among three livelihood attributes: fish landing center (p value< 0.001), power supply (p value = 0.004), and number of children (p value = 0.005). It also found a significant difference in family thriftiness (p value = 0.024) between the two coastal fishing communities (Table 4.4).

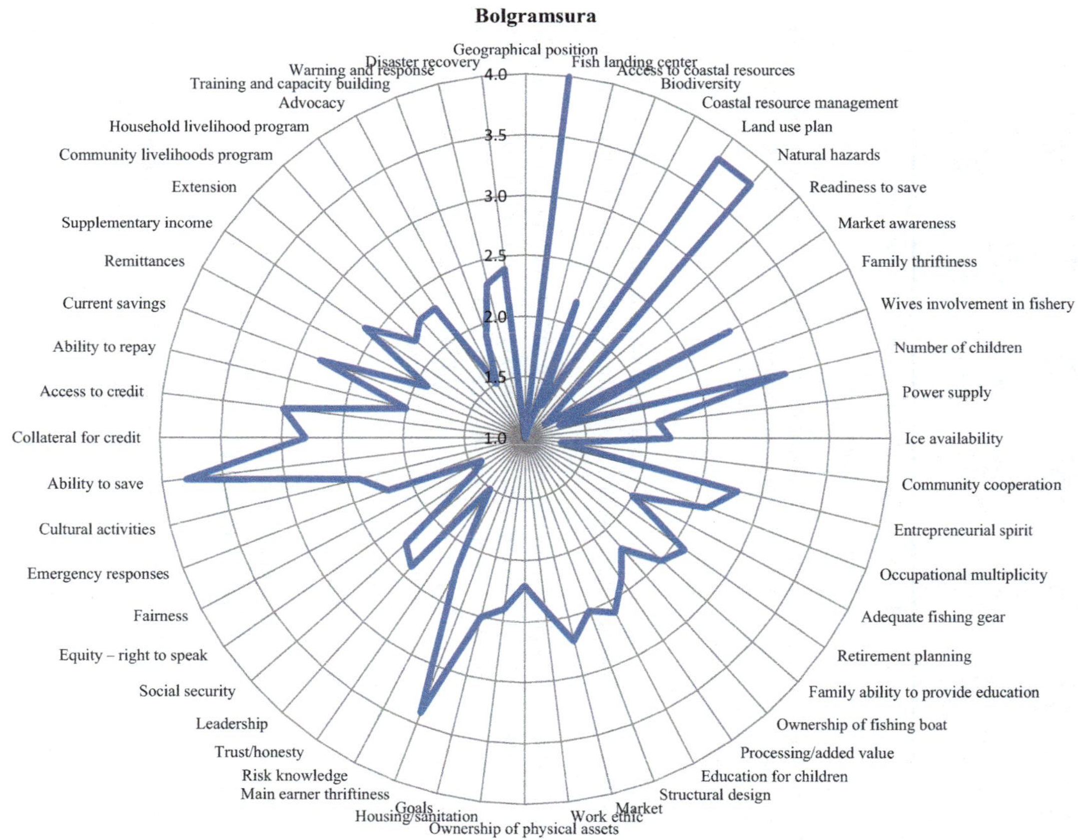

Fig. 4.5 Distribution of livelihood attributes in examining the resilience of the coastal fishing community to the impacts of climate change in Bolgram Sura, Bhola district. Attributes were scored on a scale from 1 to 4 (1 = resilient, 2 = nearly resilient, 3 = moderately vulnerable, and 4 = vulnerable). (Source: Household surveys)

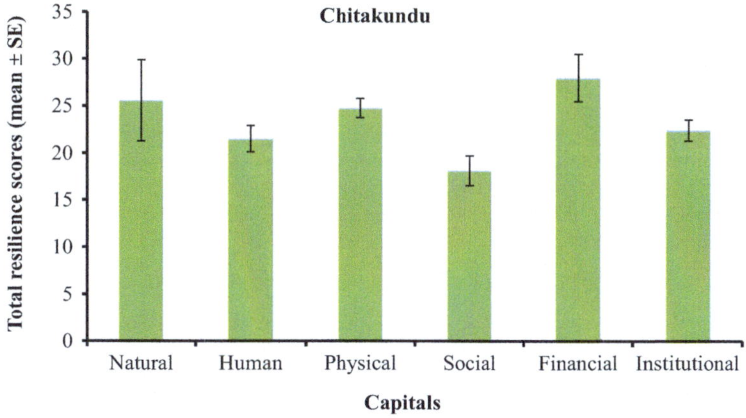

Fig. 4.6 Resilience scores (mean ± SE) in six specific types of capital of the coastal fishing community of Chita Kundu, Bhola district. Attributes were scored on a scale from 1 to 4 (1 = resilient, 2 = nearly resilient, 3 = moderately vulnerable, and 4 = vulnerable). (Source: Household surveys and focus group discussions)

The rest of the attributes of six specific types of capital in both fishing communities are not statistically significant (p value > 0.05).

4.4 Discussion

This study has found that there are a few attributes in each type of livelihood capital that are responsible for the resilience of fishing communities at both study sites. For example, the better geographical position and communication system in Bolgram Sura have influenced its resilience scores, making it a little more resilient than Chita Kundu. According to FGDs and KIIs, Manpura Island (where Chita Kundu is situated) lies in a disaster-prone zone. One key informant in Chita Kundu reported that most of time they are affected by various coastal disasters such as cyclones and tidal surges. In addition, poor communication systems (e.g., waterway transportation) obstructs fishing communities from moving quickly to safe places during any climatic hazard in Chita Kundu. In addition, there are not enough cyclone shelters in either of the fishing communities. Another key informant in Bolgram Sura reported that the cyclone shelters are inadequate and cannot accommodate the large number of people needing them during any natural disaster. This study agrees with McDaniels et al. (2008), who also reported that resilient infrastructural systems, particularly electricity, water, and other public services, are crucial in minimizing the adverse impacts of climatic hazards. Integration of local transport infrastructure may help people to move swiftly from an at-risk place to a safe place during any climatic hazard.

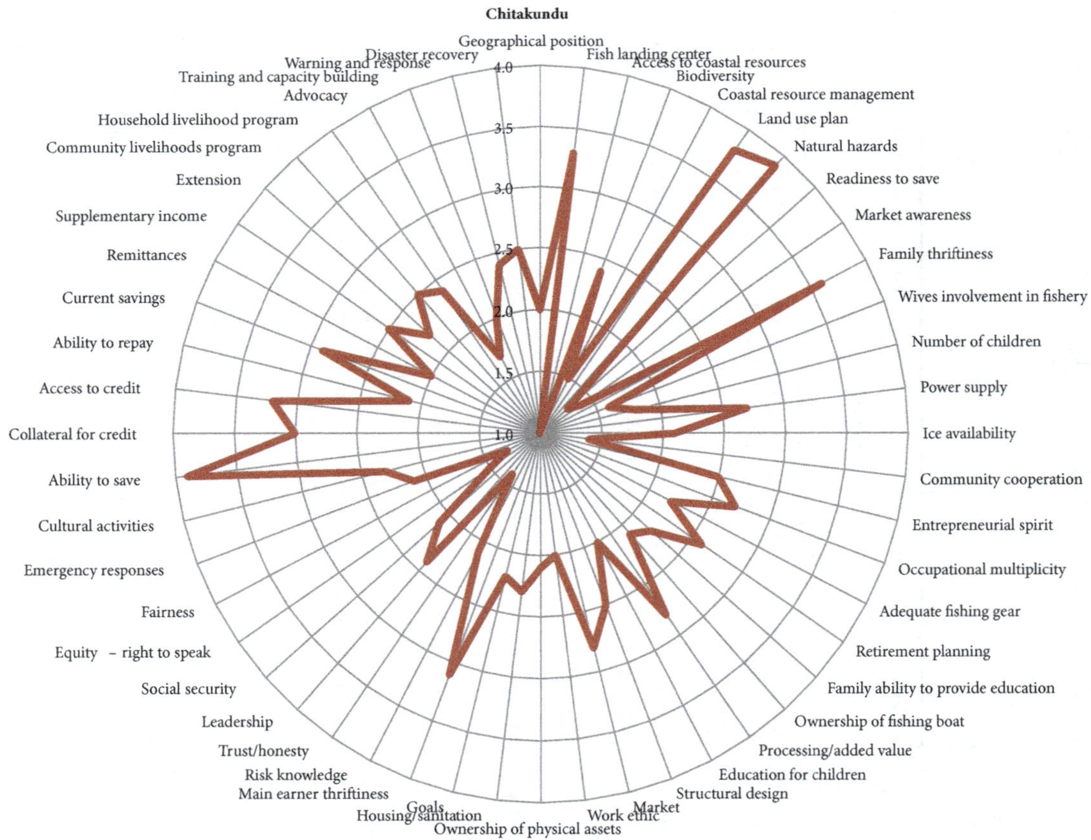

Fig. 4.7 Distribution of livelihood attributes in examining the resilience of the coastal fishing community to the impacts of climate change in Chita Kundu, Bhola district. Attributes were scored on a scale from 1 to 4 (1 = resilient, 2 = nearly resilient, 3 = moderately vulnerable, and 4 = vulnerable). (Source: Household surveys)

Table 4.3 Mean resilience scores, standard errors (SE), 95% confidence intervals (CI) along with p value of t test for the Bolgram Sura and Chita Kundu fishing communities, Bhola district, by different types of capital

Specific type of capital	Fishing community	Mean	SE	95% CI	p value
Natural	Bolgram Sura	17.1	0.42	(−1.88, 0.28)	0.137
	Chita Kundu	17.9	0.31	(−1.88, 0.28)	
Human	Bolgram Sura	33.6	1.27	(−0.58, 5.58)	0.105
	Chita Kundu	31.1	0.74	(−0.63, 5.63)	
Physical	Bolgram Sura	21.8	1.04	(−3.37, 2.37)	0.718
	Chita Kundu	22.3	0.88	(−3.37, 2.37)	
Financial	Bolgram Sura	18.9	1.03	(−3.01, 1.61)	0.533
	Chita Kundu	19.6	0.40	(−3.11, 1.71)	
Social	Bolgram Sura	15.0	0.49	(−1.13, 2.13)	0.529
	Chita Kundu	14.5	0.60	(−1.14, 2.14)	
Institutional	Bolgram Sura	14.9	0.89	(−3.33, 1.73)	0.516
	Chita Kundu	15.7	0.82	(−3.34, 1.74)	
Overall	Bolgram Sura	121.3	2.43	(−5.89, 6.29)	0.946
	Chita Kundu	121.1	1.59	(−5.96, 6.36)	

Geographical position and limited access to livelihood assets in Chita Kundu have compromised the fishers' capacity for resilience. Because of a direct land connection with Bhola district, the Bolgram Sura fishing community has an adequate power supply (e.g., grid electricity) so that fish can be preserved. In contrast, there is no electricity facility in Chita Kundu. Poor coastal resource management reduces the capacity of the fishing community in Chita Kundu for resilience to climate change compared with Bolgram Sura. Fishers of Bolgram Sura have greater access to credit because of their land-based collateral than their Chita Kundu counterparts. Hossain et al. (2013) reported that available natural assets help to enhance the resilience of coastal fishing communities to withstand the impacts of climatic hazards. In addition, Bolgram Sura fishers have an adequate supply of fishing gear and their fishing gear is also less affected by climatic hazards than that of their Chita Kundu counterparts. Consequently, their capacity for resilience has increased because of their greater ability to save. Moreover, better institutional management such as higher rates of advocacy, disaster recovery, training, and capacity building influence their capacity for resilience to climate change. For example, fishers who diversified their livelihoods are more resilient than traditional fishers.

This study reported that Bolgram Sura showed lower resilience scores than Chita Kundu in terms of human and social livelihood capital. This is because there is a higher number of family members in Bolgram Sura (more than five per family) than in Chita Kundu (four per family). A lower rate of social security and cultural activities because of less emergency support compromised their capacity for resilience compared with their Chita Kundu counterparts. But in a vulnerable fishing community, interfaith bonding and social harmony may increase the capacity for resilience to climatic disasters (Perry and Sumaila 2007).

Table 4.4 Resilience scores (mean ± SD) of 52 attributes under six specific types of capital along with p value of t test for the fishing communities of Bolgram Sura and Chita Kundu, Bhola district. Attributes were scored on a scale from 1 to 4 (1 = resilient, 2 = nearly resilient, 3 = moderately vulnerable, and 4 = vulnerable)

Specific types of capitals	Name of attribute	Bolgram Sura		Chita Kundu		p value
		Mean	(±) SD	Mean	(±) SD	
Natural						
	Geographical position	1.0	0.00	2.0	0.00	NA
	Fish landing center	4.0	0.00	3.3	0.48	<0.001[a]
	Access to coastal resources	1.0	0.00	1.0	0.00	NA
	Biodiversity	2.2	0.63	2.4	0.52	0.449
	Coastal resource management	1.3	0.48	1.5	0.53	0.388
	Land use plan	3.8	0.42	3.8	0.42	1.000
	Natural hazards	3.8	0.42	3.9	0.32	0.556
Human						
	Readiness to save	1.4	0.52	1.3	0.48	0.660
	Market awareness	1.2	0.42	1.4	0.52	0.355
	Work ethic	2.4	0.70	2.0	0.47	0.151
	Occupational multiplicity	2.6	0.52	2.7	0.48	0.660
	Entrepreneurial spirit	2.8	0.63	2.5	0.71	0.331
	Wives' involvement in fishery	1.3	0.67	1.6	0.52	0.279
	Number of children	3.2	0.92	1.8	1.03	0.005[a]
	Education for children	2.6	0.84	2.0	0.47	0.065
	Family ability to provide education	2.5	0.71	2.2	0.63	0.331
	Retirement planning	2.6	0.70	2.6	0.52	1.000
	Family thriftiness	2.9	0.74	3.6	0.52	0.024[b]
	Main earner thriftiness	3.4	0.84	3.1	0.74	0.408
	Risk knowledge	2.2	0.42	2.1	0.32	0.556
	Goals	2.5	0.53	2.2	0.79	0.331
Physical						
	Ownership of fishing boat	2.2	0.63	2.1	0.74	0.749
	Adequate fishing gear	2.0	0.67	2.2	0.63	0.500
	Ownership of physical asset	3.3	0.67	2.8	0.79	0.145
	Processing/added value	2.4	0.70	2.8	0.42	0.139
	Ice availability	2.2	0.63	2.1	0.32	0.660
	Power supply	2.1	0.32	2.7	0.48	0.004[a]
	Housing/sanitation	2.4	0.52	2.3	0.48	0.660
	Structural design	2.5	0.53	2.5	0.53	1.000
	Market	2.7	0.48	2.8	0.42	0.628
Social						
	Community cooperation	1.3	0.48	1.4	0.52	0.660
	Trust/honesty	1.5	0.53	1.4	0.52	0.673
	Leadership	2.4	0.52	2.4	0.52	1.000

(continued)

Table 4.4 (continued)

Specific types of capitals	Name of attribute	Bolgram Sura		Chita Kundu		
		Mean	(±) SD	Mean	(±) SD	*p* value
	Social security	2.3	0.48	2.1	0.32	0.288
	Equity – right to speak	1.5	0.53	1.5	0.53	1.000
	Fairness	1.4	0.52	1.3	0.48	0.660
	Emergency responses	2.2	0.42	2.1	0.32	0.556
	Cultural activities	2.4	0.52	2.3	0.48	0.660
Financial						
	Ability to save	3.8	0.63	3.9	0.32	0.660
	Collateral for credit	2.8	0.63	3.0	0.00	0.331
	Access to credit	3.0	0.82	3.2	0.42	0.500
	Ability to repay	2.0	0.47	2.1	0.32	0.584
	Current savings	2.8	0.63	2.9	0.32	0.660
	Remittances	1.9	0.32	2.0	0.00	0.331
	Supplementary income	2.6	0.52	2.5	0.53	0.673
Institutional						
	Extension	2.2	0.42	2.2	0.42	1.000
	Community livelihoods program	2.3	0.48	2.5	0.53	0.388
	Household livelihood program	2.3	0.50	2.4	0.53	0.660
	Advocacy	1.5	0.53	1.7	0.48	0.388
	Training and capacity building	1.9	0.74	2.0	0.82	0.777
	Warning and response	2.3	0.48	2.4	0.52	0.660
	Disaster recovery	2.4	0.52	2.5	0.53	0.673

Source: Household surveys
NA not applicable
[a]Highly significant
[b]Significant

Although the resilience scores varied across the communities, the level of resilience in tackling the climatic impacts successfully was mostly nonsignificant ($p > 0.05$) in both fishing communities. A community is resilient when it can absorb or withstand the shock of climatic impacts, when it can bounce back to its original condition, and when it can maintain or improve the current status after any climatic shock. In this study, both communities failed to absorb the impacts of climatic shocks owing to their limited resources or capacity, such as inadequate physical assets or lack of collateral for credit. In a few cases, both communities bounced back to their original condition because of their better coastal resource management, readiness to save, and community cooperation. However, this study could not measure the difference in their livelihood status before and after any climatic impact. Owing to the remoteness of the study sites, this study could not focus on these issues thoroughly enough to measure the resilience of the coastal fishing communities to the impacts of climate change. Future research should focus on these topics robustly when measuring the resilience of coastal fishing communities.

Assessment of resilience in coastal fishing communities identifies the entry points for livelihood upgrading in the contests between different types of climatic stresses. The impacts of climate change reduce the capacity for resilience of these coastal island fishing communities compared with other communities on the mainland because of the combined risk from multiple interacting weather and climate extremes; higher dependency on coastal resources, particularly fisheries; limited physical properties; unfavorable financial resources; and their socio-cultural sensitivity.

Better management of coastal resources may sustain livelihood services and reduce their vulnerability to coastal hazards. For example, the livelihood of multispecies fisheries can make the fishers more resilient to environmental change and future uncertainty than the livelihood of single-species fisheries (Johnson and Welch 2009). Effective land use and structural design of a community may help to adapt to climatic hazards. Increasing risk knowledge and emergency responses are needed for community members to take precautionary actions before any disaster and address emergency needs at the community level. Accurate early warning is obligatory to alert the coastal fishing communities to the onset of any climatic hazards. Disaster recovery plans help to accelerate disaster recovery activities, reduce vulnerability, and increase resilience capacity.

4.5 Conclusions

This study has measured the resilience of two coastal fishing communities in Bangladesh using 52 attributes under six specific types of capital. It found that one fishing community is slightly more resilient than the other with regard to four types of capital – natural, financial, physical, and institutional. The households of one fishing community show a significant difference in resilience under six specific types of capital. Some specific attributes (power supply, number of children, family thriftiness, and fish landing center) have shown significant variations between the two fishing communities. Some other attributes have also shown variations (with statistical significance) between these communities including geographical position, coastal resource management, wives' involvement in the fishery, ownership of physical assets, social security, access to credit, and advocacy. The findings of this study may be different from those in other coastal communities, as this study has particularly focused on two small fishing communities in Bhola district. However, measuring coastal community resilience is indispensable for developing the coastal area of Bangladesh. By planning and monitoring well, resilient livelihoods can be made for the coastal communities. This study may help to formulate policies to reduce vulnerability, and increase the resilience of the coastal fishing communities of Bangladesh. Special consideration should be made to scale up such studies to a national level, as well as in an international context across the world.

Acknowledgements We are thankful to all the participants for their voluntary participation.

Conflicts of Interest The authors declare that there are no conflicts of interest.

References

Adger WN, Brown K, Nelson DR, Berkes F, Eakin H, Folke C, Galvin K, Gunderson L, Goulden M, O'Brien K, Ruitenbeek J (2011) Resilience implications of policy responses to climate change. Wiley Interdiscip Rev Clim Chang 2(5):757–766. https://doi.org/10.1002/wcc.133

Ahmed B, Kelman I, Fehr HK, Saha M (2016) Community resilience to cyclone disasters in coastal Bangladesh. Sustainability 8(8):805. https://doi.org/10.3390/su8080805

Alexander DE (2013) Resilience and disaster risk reduction: an etymological journey. Nat Hazards Earth Syst Sci Discuss 1(2):1257–1284. https://doi.org/10.5194/nhess-13-2707-2013

Bahadur AV, Ibrahim M, Tanner T (2013) Characterising resilience: unpacking the concept for tackling climate change and development. Clim Dev 5(1):55–65. https://doi.org/10.1080/17565529.2012.762334

Barbier EB (2015) Climate change impacts on rural poverty in low-elevation coastal zones. Policy Research Working Paper No. WPS 7475. World Bank Group, Washington, DC, pp 1–43. http://documents.worldbank.org/curated/en/462581468197362959/pdf/WPS7475.pdf. Accessed 13 May 2020

BBS (2013) Bangladesh population and housing census 2011. Community report: Bhola. Statistics and Informatics Division, Ministry of Planning. Government of the People's Republic of Bangladesh. Bangladesh Bureau of Statistics, Dhaka, pp 1–287

Belhabib D, Lam VW, Cheung WW (2016) Overview of West African fisheries under climate change: impacts, vulnerabilities and adaptive responses of the artisanal and industrial sectors. Mar Policy 71:15–28. https://doi.org/10.1016/j.marpol.2016.05.009

Blythe JL (2015) Resilience and social thresholds in small-scale fishing communities. Sustain Sci 10(1):157–165. https://doi.org/10.1007/s11625-014-0253-9

Chinowsky P, Hayles C, Schweikert A, Strzepek N, Strzepek K, Schlosser CA (2011) Climate change: comparative impact on developing and developed countries. Eng Proj Organ J 1(1):67–80. https://doi.org/10.1080/21573727.2010.549608

Cochrane K, De Young C, Soto D, Bahri T (eds) (2009) Climate change implications for fisheries and aquaculture: overview of current scientific knowledge. FAO Fisheries and Aquaculture Technical Paper No. 530. Food and Agricultural Organization of the United Nations, Rome, pp 1–212

De Haan L, Zoomers A (2005) Exploring the frontier of livelihoods research. Dev Chang 36(1):27–47. https://doi.org/10.1111/j.0012-155X.2005.00401.x

DFID (1999) Sustainable livelihoods guidance sheets. Department for International Development, London. https://www.livelihoodscentre.org/documents/114097690/114438878/Sustainable+livelihoods+guidance+sheets.pdf/594e5ea6-99a9-2a4e-f288-cbb4ae4bea8b?t=1569512091877. Accessed 13 May 2020

Ding Q, Chen X, Hilborn R, Chen Y (2017) Vulnerability to impacts of climate change on marine fisheries and food security. Mar Policy 83:55–61. https://doi.org/10.1016/j.marpol.2017.05.011

DoF (2019) National fish week 2019 compendium (in Bengali). Ministry of Fisheries and Livestock, Government of the People's Republic of Bangladesh. Department of Fisheries, Dhaka, pp 1–160

Eckstein D, Künzel V, Schäfer L, Winges M (2019) Global climate risk index 2020. Who suffers most from extreme weather events? Weather-related loss events in 2018 and 1999 to 2018. Germanwatch, Bonn, pp 1–42. https://germanwatch.org/sites/germanwatch.org/files/20-2-01e%20Global%20Climate%20Risk%20Index%202020_14.pdf. Accessed 13 May 2020

Ewing LC (2014) Community resilience to coastal disasters. PhD dissertation, University of Southern California, pp 1–186. http://digitallibrary.usc.edu/cdm/compoundobject/collection/p15799coll3/id/454096/rec/1. Accessed 13 May 2020

Hoque MZ, Cui S, Lilai X, Islam I, Ali G, Tang J (2019) Resilience of coastal communities to climate change in Bangladesh: research gaps and future directions. Watershed Ecol Environ 1:42–56. https://doi.org/10.1016/j.wsee.2019.10.001

Hossain MS, Rahman MF, Thompson S, Nabi MR, Kibria MM (2013) Climate change resilience assessment using livelihood assets of coastal fishing community in Nijhum Dwip, Bangladesh. Pertanika J Sci Technol 21(2):397–422

IFRC (2011) Characteristics of a safe and resilient community: community based disaster risk reduction study. Arup International Development. International Federation for Red Cross and Red Crescent Societies, Geneva, pp 1–62

IPCC (2014) Summary for policymakers. In: Field CB, Barros VR, Dokken DJ, Mach KJ, Mastrandrea MD, Bilir TE, Chatterjee M, Ebi KL, Estrada YO, Genova RC, Girma B, Kissel ES, Levy AN, MacCracken S, Mastrangrea PR, White LL (eds) Climate change 2014: impacts, adaptation and vulnerability (Part A: global and sectoral aspects. Contribution of Working Group II to the fifth assessment report of the Intergovernmental Panel on Climate Change). Cambridge University Press, Cambridge/New York, pp 1–32

Islam MM, Sallu S, Hubacek K, Paavola J (2014a) Vulnerability of fishery-based livelihoods to the impacts of climate variability and change: insights from coastal Bangladesh. Reg Environ Chang 14(1):281–294. https://doi.org/10.1007/s10113-013-0487-6

Islam MM, Sallu S, Hubacek K, Paavola J (2014b) Migrating to tackle climate variability and change? Insights from coastal fishing communities of Bangladesh. Clim Chang 124:733–746. https://doi.org/10.1007/s10584-014-1135-y

Islam MM, Barman A, Kundu GK, Kabir MA, Paul B (2019) Vulnerability of inland and coastal aquaculture to climate change: evidence from a developing country. Aquac Fish 4(5):183–189. https://doi.org/10.1016/j.aaf.2019.02.007

Islam MM, Islam F, Akter MS, Kundu GK, Aparna B, Khan MI (2020) Transformative adaptations to climate change: cases from the Jamuna river fishing communities of Bangladesh. J Fish Environ. (in press)

Johnson JE, Welch DJ (2009) Marine fisheries management in a changing climate: a review of vulnerability and future options. Rev Fish Sci 18(1):106–124. https://doi.org/10.1080/10641260903434557

Keating A, Mechler R, Mochizuki J, Kunreuther H, Bayer J, Hanger S, McCallum I, See L, Williges K, Hochrainer-Stigler S, Egan C (2014) Operationalizing resilience against natural disaster risk: opportunities, barriers, and a way forward. Zurich Flood Resilience Alliance, pp 1–43. https://riskcenter.wharton.upenn.edu/wp-content/uploads/2020/03/ZAlliance-Operationalizing-Reslience.pdf. Accessed 13 May 2020

Khan MI, Kundu GK, Akter MS, Mallick B, Islam M (2018) Climatic impacts and responses of migratory and non-migratory fishers of the Padma River, Bangladesh. Soc Sci 7(12):254. https://doi.org/10.3390/socsci7120254

McDaniels T, Chang S, Cole D, Mikawoz J, Longstaff H (2008) Fostering resilience to extreme events within infrastructure systems: characterizing decision contexts for mitigation and adaptation. Glob Environ Chang 18(2):310–318. https://doi.org/10.1016/j.gloenvcha.2008.03.001

Nicholls RJ, Wong PP, Burkett VR, Codignotto JO, Hay JE, McLean RF, Ragoonaden S, Woodroffe CD (2007) Coastal systems and low-lying areas. In: Parry ML, Canziani OF, Palutikof JP, van der Linden PJ, Hanson CE (eds) Climate change 2007: Impacts, adaptation and vulnerability. Contribution of Working Group II to the fourth assessment report of the Intergovernmental

Panel on Climate Change. Cambridge University Press, Cambridge, pp 315–356. https://ro.uow.edu.au/cgi/viewcontent.cgi?article=1192&context=scipapers. Accessed 13 May 2020

Perry RI, Sumaila UR (2007) Marine ecosystem variability and human community responses: the example of Ghana, West Africa. Mar Policy 31(2):125–134. https://doi.org/10.1016/j.marpol.2006.05.011

Rahman MM, Haque SM, Galib SM, Islam MA, Parvez MT, Hoque MN, Wahab MA, Egna H, Brown C (2020) Mud crab fishery in climate vulnerable coastal Bangladesh: an analysis towards sustainable development. Aquac Int 28:1243–1268. https://doi.org/10.1007/s10499-020-00523-2

Sharifuzzaman SM, Hossain MS, Chowdhury SR, Sarker S, Chowdhury MSN, Chowdhury MZR (2018) Elements of fishing community resilience to climate change in the coastal zone of Bangladesh. J Coast Conserv 22(6):1167–1176. https://doi.org/10.1007/s11852-018-0626-9

Sok S, Yu X (2015) Adaptation, resilience and sustainable livelihoods in the communities of the Lower Mekong Basin, Cambodia. Int J Water Resour Dev 31(4):575–588. https://doi.org/10.1080/07900627.2015.1012659

Stanford RJ, Wiryawan B, Bengen DG, Febriamansyah R, Haluan J (2017) The fisheries livelihoods resilience check (FLIRES check): a tool for evaluating resilience in fisher communities. Fish Fish 18(6):1011–1025. https://doi.org/10.1111/faf.12220

Strauss A, Corbin J (1990) Basics of qualitative research: grounded theory procedures and techniques. Sage, London

Tanner T, Lewis D, Wrathall D, Bronen R, Cradock-Henry N, Huq S, Lawless C, Nawrotzki R, Prasad V, Rahman MA, Alaniz R (2015) Livelihood resilience in the face of climate change. Nat Clim Chang 5(1):23–26. https://doi.org/10.1038/nclimate2431

Twigg J, Calderone M (2019) Building livelihood and community resilience: lessons from Somalia and Zimbabwe. Overseas Development Institute and CESVI Working Paper No. 545. London. https://www.odi.org/sites/odi.org.uk/files/resource-documents/12585.pdf. Accessed 27 May 2020

Chapter 5
Seaweed: A Powerful Tool for Climate Change Mitigation That Provides Various Ecological Services

Md. Simul Bhuyan, Md. Nazrul Islam, Abu Sayeed Muhammad Sharif, Md. Mohidul Islam, Md. Shafiqul Islam, and Md. Enamul Hoq

Abstract Seaweed production (both culture and natural) has increased compared with in the past. It occupies a strong position in the food supply and meets global food demand. Seaweed emerges as a powerful tool to mitigate and adapt to climate change. It acts as a carbon sink by sequestrating carbon from the atmosphere into the ocean. It can reduce the carbon emission from agricultural fields by improving the soil quality. It also minimizes the emissions of methane gas when mixed in cattle food. Seaweed increases the pH of water thus reducing the ocean acidification phenomena. As a result, aquatic organisms such as finfish, shellfish, corals, and invertebrates find a suitable place to live in. It produces trace gas (e.g., volatile brominated and iodinated halocarbons) that deplete the ozone. Seaweed dampens wave energy during storms and protects the coast as climate change adaptation. Seaweed provides oxygen to the ocean water, which minimizes the issue of de-oxygenation. It offers

M. S. Bhuyan (✉)
Faculty of Marine Sciences and Fisheries, University of Chittagong, Chittagong, Bangladesh

Bangladesh Research Consultancy Firm, Dhaka, Bangladesh

M. N. Islam
Department of Geography and Environment, Jahangirnagar University, Savar, Dhaka, Bangladesh
e-mail: nazrul_geo@juniv.edu

A. S. M. Sharif
Bangladesh Oceanographic Research Institute, Cox's Bazar, Bangladesh

M. M. Islam
Marine Fisheries and Technology Station, Bangladesh Fisheries Research Institute, Cox's Bazar, Bangladesh

M. S. Islam
Faculty of Marine Sciences and Fisheries, University of Chittagong, Chittagong, Bangladesh

M. E. Hoq
Bangladesh Fisheries Research Institute, Mymensing, Bangladesh

M. N. Islam, A. van Amstel, *Bangladesh II: Climate Change Impacts, Mitigation and Adaptation in Developing Countries*, Springer Climate,
https://doi.org/10.1007/978-3-030-71950-0_5

habitats and food for important components of the marine ecosystem that have a great impact on the climate. Seaweed provides biofuels, fertilizer, medicine, and food for human consumption. In this review, we emphasize the role of seaweed in climate change mitigation and adaptation. Seaweed cultivation can be optimized to get maximum climate benefits and increase the livelihood status of the seaweed farmer.

Keywords Seaweed · Climate change · Mitigation and adaptation · Ecological services · Emission

5.1 Introduction

According to FAO, about 131.4 million tonnes of fish, aquatic animals, and aquatic plants produced worldwide in 2014 (FAO 2016a). Seaweed, marine aquatic plants contribute over 20% of this total production, with a growth of 8% per year over the past decade (FAO 2016a). Seaweed is regarded as an important component of marine aquaculture, which will be the main weapon to meet global food security over the next 30 years (Langton et al. 2019). As the world population is increasing rapidly, it will be a challenge to feed this huge population (Hasselström et al. 2020). The cultivation of seaweed is dominated by Asian countries although European countries (Ireland, Spain, Scotland, Norway, and Denmark) have started seaweed culture over the last 15–20 years (Kraan et al. 2000; Kerrison et al. 2015; Peteiro et al. 2016).

Seaweeds or marine macroalgae are commonly known as a vital source of ocean primary productivity (Mann 1973; Dayton 1985; Okey et al. 2004; Ruiz and Wolff 2011), which comprises 8000–10,500 species. There are three main categories of seaweed (i.e., green, red, or brown algae) (Lüning 1990; Thomas 2002; Hurd et al. 2014). Seaweed provides various ecological services and is regarded as the most diverse and productive habitat on earth (Mann 1973; Dayton 1985; Boden et al. 2017). The ecological services include habitat (feeding, breeding, and nursery ground), biodiversity, food web subsidy, nutrient cycling, and removal of excess nutrients, carbon sequestration and shore protection, environmental restoration and nursery grounds, and protecting juvenile invertebrates and fish from predators (Smale et al. 2013; Langton et al. 2019).

The provision of habitat is a great ecological service of seaweed. It provides physical structure, habitat, shading, and acts as good a source of food (Arsenault 2018). Seaweeds are the primary producers of the ocean and support secondary productivity and three-dimensional habitat structure for many commercially important marine organisms (invertebrates, fish, and marine top-predators, such as seabirds and sea mammals) (Lorentsen et al. 2010; Arsenault 2018). Seaweed is a significant biological resource as their detritus is exported to other habitats; this process increases the productivity of that particular area (Arsenault 2018). Seaweed takes up necessary nitrogen, phosphorus, and carbon dioxide required for its growth and production of energy storage products (Kim et al. 2017).

Climate change mitigation is an important role of seaweed (Langton et al. 2019). The impact of climate change on seaweed abundance, distribution, and quality is a global concern (Straub et al. 2016). Seaweed has a certain degree of resilience to global climate change (Krumhansl et al. 2016), and its biomass availability can vary on a spatial basis (Bell et al. 2015; Boden et al. 2017). Seaweed acts as a sponge for carbon dioxide and reducing ocean acidification (Duarte et al. 2017). *Gracilaria tikvahiae* (red seaweed) and *Saccharina latissima* (brown seaweed) assimilate carbon rapidly in Long Island Sound and the Bronx River Estuary of New York (Kim et al. 2014, 2015a). Bjerregaard et al. (2016) reported that if 0.03% ocean surface area can be cultured then it will be able to remove about 135 million tons of carbon from the ocean water. That means it will remove approximately 3.2% of carbon annually inputted to ocean water from the atmosphere.

Uptake of excess nitrogen, phosphorous, and some toxic chemical by seaweed reduces coastal eutrophication and pollution (Kim et al. 2014; Marinho et al. 2015; Rose et al. 2015). That reduces the harmful algal blooms such as red tides (Imai et al. 2006). For example, it was reported that the richness index of the red tide species *Skeleton emacostatum* declined from 0.32 to 0.05 during the growing season of *Porphyra yezoensis* in the Jiangsu Province in China (Wu et al. 2015). Thirty percent of the introduced nitrogen can be removed if 0.03% of the ocean surface area can be brought under seaweed culture (Bjerregaard et al. 2016; Kim et al. 2017). This way, seaweed can remove inorganic nutrients from ocean water and have a great impact on the mitigation of adverse environmental impacts (Neori et al. 2004, 2007; Corey et al. 2012, 2014; Kim et al. 2013, 2014, 2015b; Rose et al. 2015; Wu et al. 2017).

Overharvesting or degradation of marine algae habitat can be detrimental to marine biodiversity (Arsenault 2018). It will bring important changes into the benthic community structure. This phenomenon will decrease the functional diversity and overall productivity of the ocean (Bodkin 1988; Graham 2004; Lilley and Schiel 2006). Moreover, it will cut the amount of "*blue carbon*" stored in submerged marine habitats. Consequently, it will change the global weather patterns that will have negative impacts on the coastal residents, their livelihoods, and food security (Nelleman et al. 2009; Byrnes et al. 2011). The losses of seaweed also affect marine biodiversity such as manatees, dugongs, and green turtles who are herbivores (West et al. 2017).

5.2 Methodology

Related articles were collected from different databases, including Scopus, Web of Knowledge, Google Scholar, Dimension, and PubMed, using the keywords "Climate change mitigation by seaweed" or "Role of seaweed in climate change mitigation and adaptation" or "Ecological services of seaweed" or "Ecosystem services of seaweed" or "Carbon sequestration by seaweed" or "Carbon absorption by seaweed" or "Role of seaweed in reducing ocean acidification" or "Nutrients removal by seaweed" or "Uptake of nutrients by seaweed" or "Role of seaweed in reducing

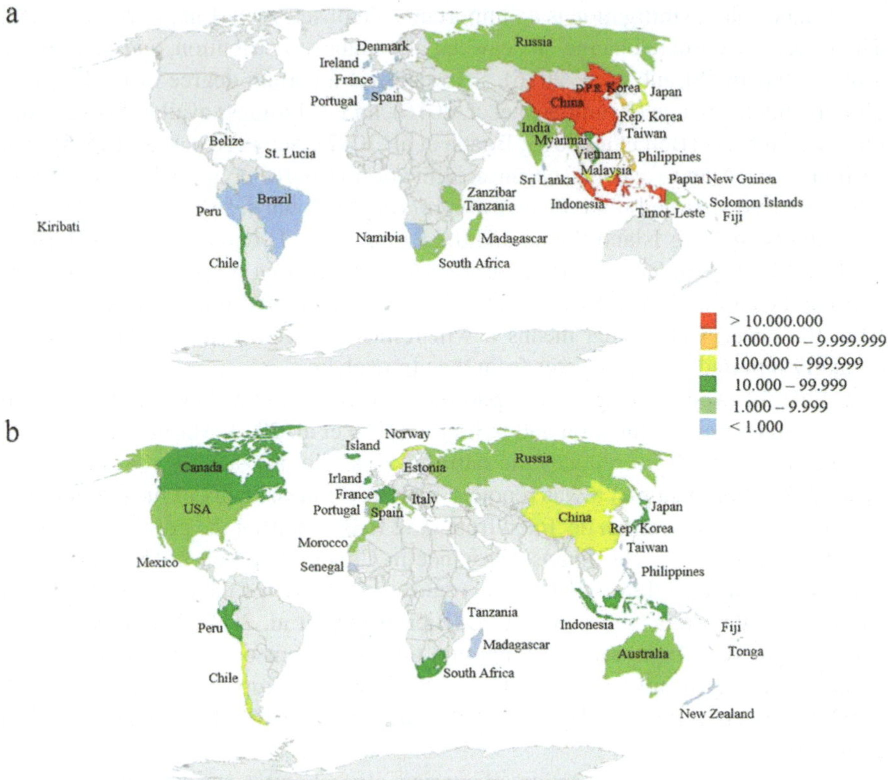

Fig. 5.1 Seaweed production in 2018 by different countries of the world. (**a**) Culture and (**b**) capture. Color scale in wet metric tonnes. (Source: FAO 2018)

eutrophication" or "Trace gases produced by seaweed" or "Shore protection by seaweed," "Dampening wave energy by seaweed" or "Absorption of heavy metals by seaweed" or "Bioabsorption of heavy metals by seaweed" or "Oxygen production by seaweed" or "Seaweed acts as best primary producer" or "Regulation of biogeochemical cycle by seaweed" (Fig. 5.1).

5.3 Worldwide Seaweed Production Status

In the past, seaweed production was higher from the wild than from culture. Production from culture increased in the 1960s (FAO 2018). Brown seaweed was the most abundant followed by red seaweed and green seaweed respectively (Fig. 5.2).

Now, seaweed contributes to 27% of the total marine aquaculture production (FAO 2016a). In 1984, income from brown seaweed was US$737,400.90 whereas it was US$5,944,093 in 2017 (FAO 2018). In the case of red seaweed, US$751,614.6

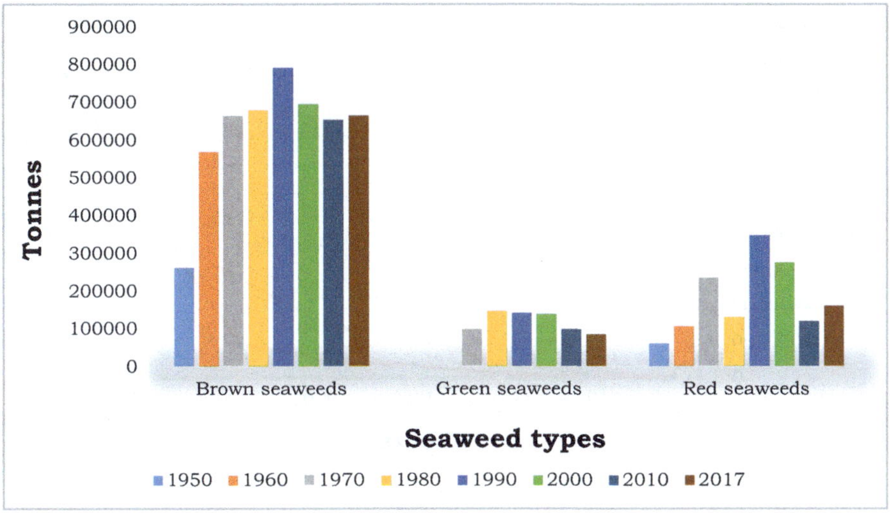

Fig. 5.2 Global capture of production of seaweed (tonnes). (Source: FAO 2018)

was made, while this figure converts into US$5,272,332 in 2017 (FAO 2018). Recently world, red seaweed has become the target species for the extraction of valuable chemicals (e.g., agar, carrageenan). Consequently, red seaweed production has increased and has surpassed the production of brown seaweed (Fig. 5.3).

5.4 Role of Seaweed in Climate Change Mitigation and Adaptation

Climate change mitigation is the process of cutting down or limiting greenhouse gas emissions to reduce future global warming. Mitigation can be done using new technologies, making available technologies more energy efficient, using clean energy sources, and changing people's behavior (IPCC 2014). The term climate change adaptation is different than the term climate change mitigation. According to IPCC (2014), climate change adaptation is the process of adjustment to the actual or expected climate and its effects. In human systems, adaptation seeks to moderate or avoid harm or exploit beneficial opportunities, whereas in natural systems it refers to human intervention to facilitate adjustment to the expected climate and its effects. Seaweed is the ideal candidate for climate change mitigation and adaptation. We emphasize seaweed as it has been providing a service for many years as a natural shield against violent storms. It protects coastal regions and provides human food. It also acts as a natural buffer (reducing ocean acidification and ocean deoxygenation) and restores the vulnerable ecosystems. The climate change benefits of cultivation are briefly described in Fig. 5.4.

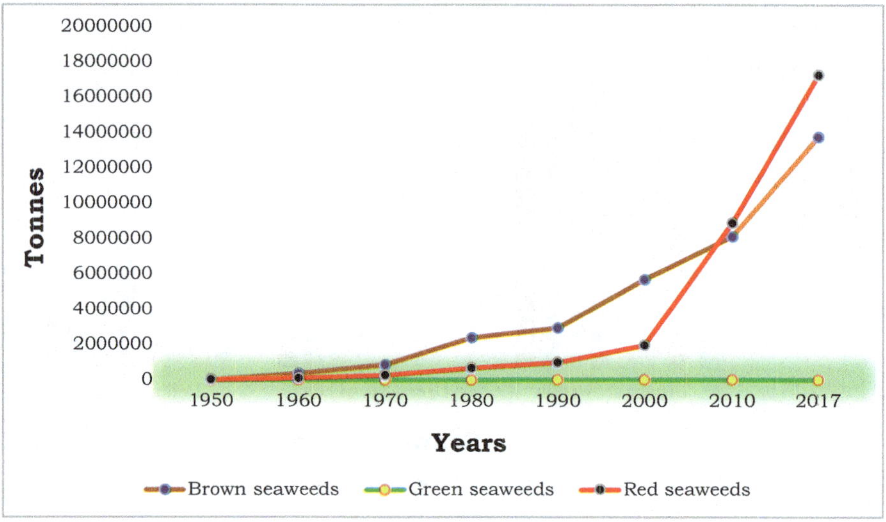

Fig. 5.3 Global culture production of seaweed (tonnes). (Source: FAO 2018)

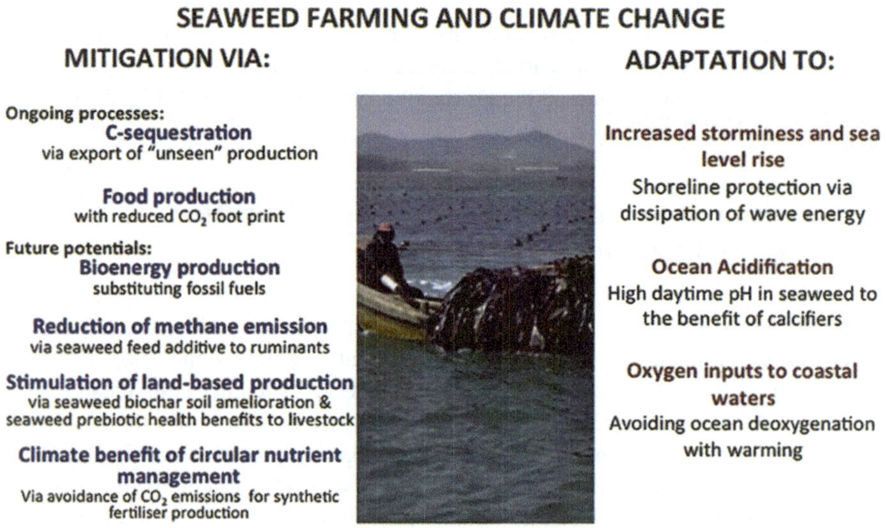

Fig. 5.4 Benefits of seaweed farming in climate change mitigation and adaptation (Source: Reproduced from Duarte et al. 2017)

5.5 Ecological Services of Seaweed

Seaweed provides various ecological services. Supporting and regulating services fall under the term ecological services (Table 5.1). Ecological services are crucial for climate change mitigation and adaptation. Although there are some

Table 5.1 Ecosystem services provided by seaweed

	Ecosystem services	Motivating factors for status classification
Supporting	S1. Biogeochemical cycling	Oxygen cycle, nutrient status, carbon cycle (low pH).
	S2. Primary production	Elevated phytoplankton concentrations, loss of eelgrass, and macroalgae.
	S3. Food web dynamics	Fish populations, bottom fauna, and habitats.
	S4. Biodiversity	Habitats, species abundance.
	S5. Habitat	Biological oxygen demand, bottom fauna, physical disturbance
	S6. Resilience	Observed regime shifts, loss of habitats, and biodiversity.
Regulating	R1. Climate and atmospheric regulation	Marine regulation of climate has good potential, but not sufficient given human greenhouse gas emissions.
	R2. Sediment retention	Pressures from bottom trawling and shipping, coastal zone vegetation.
	R3. Regulation of eutrophication	Coastal and pelagic nutrient concentration.
	R4. Biological regulation	Deterioration of top-down food web dynamics increased transport of parasitic microorganisms from agricultural land to marine systems due to climate change (precipitation patterns).
	R5. Regulation of toxic substances	Seafloor activities release embedded toxic substances, observed concentrations in commercial fish species, and sea birds.
Provisioning	P1. Food	Current status of commercial fish species abundance.
	P2. Raw material	Current status of commercial fish species abundance (e.g. for feed).
	P3. Genetic resources	Genetic material from within and between species biodiversity. Potential supply exceeds demand.
	P4. Chemical resources	Resources e.g. pharmaceuticals and food ingredients. Potential supply exceeds demand.
	P5. Ornamental resources	Current use is mainly sustainable. Potential supply exceeds demand.
	P6. Energy (from biomass only)	Current production is mainly sustainable. Potential supply exceeds demand.
	P7. Space and waterways	Space is currently abundant but increased competition expected.
Cultural	C1. Recreation	Eutrophication status, the abundance of recreational fish species, the satisfaction of recreationists (survey), bathing water quality.
	C2. Aesthetic values	Litter abundance, probability of oil spills.
	C3. Science & education	Increasing scientific interest in marine environments.
	C4. Cultural heritage	Loss of culturally important activities in coastal villages.
	C5. Inspiration	Inspiration to e.g. culture. Loose connection to water quality.
	C6. Natural heritage	Related to current water quality status.

Sources: Swedish EPA (2008), Bryhn et al. (2015), and Hasselström et al. (2018)

environmental risks associated with seaweed farming, these are much lower than ecological services it provides (Knox et al. 2015; Cabral et al. 2016; Kim et al. 2017; Walls 2017; Campbell et al. 2018; Lotze et al. 2019).

5.6 Supporting Services

Supporting services include biogeochemical cycling, primary production, food web dynamics, biodiversity, habitat, and resilience. All cycles are linked. For example, the photosynthetic conversion of CO_2 and other inorganic nutrients dissolved into organic material and oxygen by primary producers, such as algae, has a bearing on several of the cycles. A seaweed farm could influence the dynamic food web interactions that organisms have with the ecosystem. The long-term ability to cope with a changing environment is reflected in the resilience of an ecosystem. It is expected that resilience is affected by biodiversity in terms of, for example, species richness (Tilman et al. 1998).

5.7 Feeding, Breeding, Nursery Ground of Marine Organisms

Habitat-forming species such as seaweeds are popularly known as biological engineers (Jones et al. 1994). Seaweed modifies the existing ecological features (light, nutrients, sediments, physical scour, and water flow, etc.) and resources to make them favorable for other species (Jones et al. 1994; Bertness and Callaway 1994; Jones et al. 1997). Almost 8000 individual macroinvertebrates were found in a single kelp plant (Christie et al. 2003).

Holdfast, stipe, and lamina of seaweed provide a primary habitat (Rinde et al. 1992), whereas epiphytes (*Palmaria palmata*, *Phyllophora* spp., *Delesseria sanguinea*, *Polysiphonia* spp., *Ceramium* spp. *Lithothamnion* spp., etc.) provide secondary habitats for the colonization of organisms (Whittick 1983; Teagle et al. 2017). Holdfast traps sediment/detritus, is a good source of food, and provides a stable environment for the fish and invertebrates (Moore 1972; Schaal et al. 2009).

Holdfast is regarded as the most diverse species habitat, which supports 30–70 macrofaunal species per holdfast (Edwards 1980; Christie et al. 2003; Blight and Thompson 2008). Most of the organisms were found in the holdfasts than in other parts of the seaweed (Jones 1972; Moore 1972; Thiel and Vásquez 2000; Teagle et al. 2017). Epiphytes support highly diverse and abundant species that vary spatiotemporally (Christie et al. 2003).

Seaweed beds are the most productive habitats on Earth and provide three-dimensional habitats for many organisms in the coastal sea (Mann 1973, 2000; Graham 2004; Reed et al. 2008; Christie et al. 2009; Bustamante et al. 2014).

Seaweed habitat is vital for the promotion of species diversity. Macrocystis algae provide habitats in California that support genetic diversity. A 19-year observation study in the Channel Islands National Park showed that 90% of species were common in the giant kelp regions (Graham 2004).

Laminaria hyperborean is a canopy-forming species that supports huge species diversity in the northeast Atlantic (Smale et al. 2013). Approximately 130 species and 8000 individual species were recorded on a single *Laminaria hyperborea* sporophyte in Norway (Christie et al. 2003). Canopy plays an important role in the richness of species diversity. More than 40 species were regularly found under the kelp canopies (Maggs 1993). The elimination of canopy-forming *Cystoseira* species in the Mediterranean reduced the number of invertebrate species that relied on it (Benedetti-Cecchi et al. 2001; Bulleri et al. 2002; Mangialajo et al. 2008). Eriksson et al. (2006) reported that species diversity was higher beneath a canopy of *Fucus* in the Baltic Sea. Lilley and Schiel (2006) also showed that 36–44% of biological diversity declined because of the removal of the canopy of the *Hormosera banksii* species. Seventy-seven percent of the commercial species use seaweed beds as their habitat. Many commercial fish species rely on the seaweed beds as a nursery and feeding ground. These productive habitats increase the fish survival rates, hence increasing the yield of fish (Smale et al. 2013; Seitz et al. 2014).

Over the last 60+ years, a considerable amount of research has been conducted on the seaweed-associated biodiversity in the northeast Atlantic (Ebling et al. 1948; Sloane et al. 1957; Jones 1971; Moore 1971, 1973; Norton et al. 1977; Norderhaug et al. 2002; Christie et al. 2003; Blight and Thompson 2008; Walls et al. 2016, 2017; Teagle et al. 2017).

5.8 Habitat for Fish

Seaweed habitats are very favorable for the increase in diversity and abundance in fishes (Bodkin 1988). The complication of rocky substratum act as a suitable habitat for reef fishes to protect themselves from the predators (Quast 1968a,b; Miller and Geibel 1973; Russell 1977; Ebeling et al. 1980; Wheeler 1980; Bodkin 1988; Larson and DeMartini 1984; Stephens et al. 1984; Norderhaug et al. 2002). The structure of the substratum appears nearly flat, with little three-dimensional structure to large rocky outcrops. High vertical relief and complex structures are also available in the substratum (Bodkin 1988). Larson and DeMartini (1984) reported that low relief of seaweed beds is favorable for the increase in assemblage of fishes. The substratum structure plays a vital role in the increase in fish in the seaweed vegetated area (Stephens et al. 1984). Laur et al. (1988) reported that a huge amount of fishes found in the kelp-dominated regions of southern San Luis Obispo. Ebeling and Laur (1988) mentioned that fish diversity or species richness is high in the seaweed-dominated area of Santa Barbara, California. Murphy et al. (2000) found massive species richness in the filamentous algae-dominant regions of Alaska. *Sargassum* provides a vital habitat for many species and serves as a nursery ground for larvae and juveniles (Coston-Clements et al. 1991).

Seaweed beds or kelp forests are suitable spawning and reproduction grounds for many fishes (Gordon 1983; Schultze et al. 1990). Fishes use algae to make their habitats where they lay their eggs. Some fish species lay sticky eggs that stick to the seaweed or substratum. Gordon (1983) reported that spherical holdfasts of *Saccorhiza polyschides* are a favorite nesting place for headed clingfish (*Apletodon microcephalus*) and two-spotted gobies (*Gobiusculus flavescens*). The eggs of *Agonus cataphractus* are found in the rhizoid of *Laminaria* (Schultze et al. 1990). *Labrus bergylta* (Ballan wrasse) and *Ctenolabrus rupestris* (Goldsinny wrasse) feed on kelp-associated invertebrates (Norderhaug et al. 2005). Sardines, grunts, barracuda, and sharks were found in the seaweed bed of the Caribbean and Pacific coasts of Costa Rica (Langton et al. 2019).

Besides nesting and breeding grounds, seaweed beds are also used as a nursery ground for the growth of juvenile fishes (Carr 1983; Shaffer 2003; Lorentsen et al. 2004). Juvenile gadoids, cod (*Gadus morhua*), lumpsucker (*Cyclopterus lumpus*), striped sea snail (*Liparis liparis*), shore rockling (*Gaidropsarus mediterraneus*), Goldsinny wrasse (*Ctenolabrus rupestris*), and Montagu's sea snail (*Liparis montagui*) used seaweed beds as a nursery ground (Schultze et al. 1990; Fossa 1995; Borg et al. 1997; Sjøtun and Lorentsen 2003). Juvenile fishes have been found in the benthopelagic zone and canopy (*Sebastes* sp.) of seaweed (Carr 1983; Murphy et al. 2000). On the coast of Washington, juvenile salmon (i.e., *Oncorhynchus tshawytscha*) and forage fish (i.e., *Hypomesus pretiosus*) use kelp habitat (Shaffer 2003). *Macrocystis pyrifera* and *Nereocystis* spp. found on the western coast of the USA and Canada are suitable sites for fish to live in (Quast 1968a, b; Miller and Geibel 1973; Russell 1977; Leaman 1980; Ebeling et al. 1980; Ebling and Laur 1988; Laur et al. 1988). The compilation of Norwegian kelp forest species was conducted by Hoeisaeter and Fossa (1993).

Worldwide, a large number of studies have been carried out on the comparison between seaweed vegetated and non-vegetated fish assemblage and the effects of seaweed removal on the fish diversity (Limbaugh 1955; Moore 1972, 1973; Abbott and Perkins 1977; Perkins et al. 1978; Gordon 1983; Larson and DeMartini 1984; Stephens et al. 1984; Bodkin 1988; Schultze et al. 1990; Erwin et al. 1990; Fossa 1995; Murphy et al. 2000; Shears and Babcock 2003; Sjøtun and Lorentsen 2003; Burrows 2012).

5.9 Habitat for Invertebrates

Seaweed habitat is regarded as the most dynamic and biologically diverse habitat on the planet (Birkett et al. 1988). Seaweed slows down or prevents suspended particles from transportation from the overlying water column to the sea bed (Eckman et al. 1989). Seaweed beds are a hub/habitat for a large number of invertebrates (e.g., gastropod mollusks, crustaceans, and echinoderms), which are of great ecological and economic importance (Jones and Kain 1967; Kitching and Thain 1983; Christie et al. 2003). Seaweed/kelp creates microniches that support large decapods such as lobster and crayfish. Amphipods and gastropods are the most diverse and dominant invertebrate groups found on the seaweed bed (Christie et al. 2003; Wagge-Nielsen et al. 2003).

Polychaetes are also found in the kelp bed, as reported by Healy and McGrath (1998). Edwards (1980), Ball et al. (1995), and Healy and McGrath (1998) recorded various types of invertebrates from or within the holdfasts of seaweed off the coast of Ireland. Christie et al. (2003) and Wagge-Nielsen et al. (2003) made a checklist of invertebrates found in the Norwegian laminaria. Birkett et al. (1988) listed 1260 invertebrate species, of which 173 species belong to polychaetes. *Saccharina latissima* and other seaweed provide habitat where gastropods and crabs have been observed feeding on the seaweed. Hydrozoans (*Obelia* spp.) and harpacticoid copepods were recorded in farmed kelp in the spring (Peteiro and Freire 2013). Caribbean spiny lobster (*Panulirus argus*) pueruli post-larvae complete metamorphosis into the seaweed-associated substrate (Acosta and Butler 1999). Seaweed provides a surface for algicidal bacteria that can mitigate eutrophication (Imai et al. 2006).

Holdfast, stipe, and fronds of seaweed support different invertebrate organisms. Three-dimensional holdfast, with its internal spaces, provides a suitable habitat for moving species of polychaetes (e.g., *Anaitides*, *Eulalia*, *Harmothoe*, *Hediste*, *Kefersteinia*, *Lagisca*, *Lepidonotus*), crustaceans (e.g., *Bodotria*, *Idotea*, *Apherusa*, *Jassa*, *Melita*, *Porcellana*), and echinoderms (e.g., *Amphipholis*, *Asterina*, *Ophiothrix*, *Asterias*, *Psammechinus*, *Pawsonia*, and *Ocnus*) (Christie et al. 2003; Jørgensen and Christie 2003). The lower part of the stipe also supports polychaetes (e.g., *Amblyosyllis*, *Brania*, *Pionosyllis*, *Trypanosyllis*), crustaceans (*Caprella*, *Pariambus*, *Ammothelia*, *Anoplodactylus*), mollusks (*Onoba*, *Tricolia*, *Elysia*), and echinoderms (e.g., *Echinus*, *Psammechinus*, *Henricia*) (Kelly 2005).

5.10 Habitat for Birds

Seaweed provides foraging habitat for birds as the seaweed bed and its associated habitat are rich in diverse fishes and invertebrates. Furthermore, seaweed can dampen the wave energy (e.g., storms) and protect the shore, as well providing sheltered foraging habitat for the birds. Kelp Forests are underwater ecosystems formed in shallow water by the dense growth of several different species known as kelps. Though they look very much like plants, kelps are actually extremely large brown algae. Generally speaking, kelps live further from the tropics than coral reefs, mangrove forests, and warm-water seagrass beds, so kelp forests do not overlap with those systems. Like those systems, though, kelp forests provide important three-dimensional, underwater habitat that is home to hundreds or thousands of species of invertebrates, fishes, and other algae. Some species aggregate and spawn in kelp forests or utilize these areas as juvenile nursery habitat. Besides, a kelp forest acts as a natural barrier from the surge effects of waves, particularly in the case of storms, and therefore provides a more sheltered foraging environment for birds.

Seaweed provides three types of foraging habitats for birds (Foster and Schiel 1985):

- Living attached plants associated with rocky substrata (kelp forests).
- Drift kelp floating in the open sea.
- Wrack-detached kelp washed up on the shoreline

5.11 Food Provider/Primary Production

Seaweed is the best primary producer of marine ecosystems in the world with net production of 1521 Tg carbon/year. This amount of primary production by seaweed requires an area of over 3.5 million km^2 (Smith 1981; Steneck et al. 2002; Duarte et al. 2005; Krause-Jensen and Duarte 2016; Langton et al. 2019). Seaweed productivity largely depends on the availability of nutrients, temperature, wave exposure, light, and disturbance (Reed et al. 2008; Langton et al. 2019). Seaweed primary production is always greater than phytoplankton productivity. Seaweed primary production in the Atlantic regions is estimated to be over 1000 g C/m^2/year (Mann 1973; Smale et al. 2013), whereas phytoplankton production in the temperate areas is between 100 and 300 g C/m^2/year (Mann 2000). The primary production of cultivated seaweed is lower than that of the wild seaweed as cultivated seaweed grows only in summer and there is no further production once harvested (Yoshikawa et al. 2001). Estimated primary production by seaweed in different zones of Strangford Lough is summarized in Table 5.2.

Through the photosynthesis process, seaweed produces organic matter required for the growth and energy metabolism of higher trophic level organisms (Langton et al. 2019). Seaweed biomass is directly taken by herbivorous fish and invertebrates such as the blue-rayed limpet (*Patella pellucida*) (Langton et al. 2019). A very small amount is taken by herbivores and most of the seaweed biomass (>80%) enters the carbon cycle as detritus or dissolved organic matter (Gili and Comma 1998; Christie et al. 2009; Krumhansl and Scheibling 2012; Krumhansl and Scheibling 2012). This seaweed detritus settles locally or is transported to the adjacent or remote areas where detritus used as an ideal food source for benthic invertebrates and some other organisms (Duggins et al. 1989, 1990; Fredriksen 2003; Norderhaug et al. 2003; Norderhaug et al. 2003; Vanderklift and Wernberg 2008; Tallis 2009; Schaal et al.

Table 5.2 Primary production as tonnes of carbon in Strangford Lough

	Intertidal	Sub-tidal <10 m	>10 m	Total
Intertidal macroalgae	24,098			24,098
Subtidal macroalgae		68,582		68,582
Phytoplankton	812	5952	3394	10,158

Source: Kelly (2005)

2012; Leclerc et al. 2013a). Carbon derived from seaweed is used by suspension feeders, detrital grazers (i.e., limpets and *Littorina littorea*), and deposit feeders (Bustamante and Branch 1996; Leclerc et al. 2013b). Seaweed-derived carbon provides food for gastropod grazers, benthic suspension feeders, fish, and seabirds (Fredriksen 2003). Most of the seaweed biomass releases about 43% of its production in the water as particulate organic matter (POM; detritus) and dissolved organic matter (DOM) (Duarte and Cebrian, 1996; Krumhansl and Scheibling 2012; Filbee-Dexter and Scheibling 2014; Barron et al. 2014; Barrón and Duarte 2015; Hill et al. 2015). Organic matter derived from kelp provides more than 30% of the diet of kelp-associated organisms and is used as the ideal habitat for over half a million organisms/m^2 (Kaehler et al. 2000; Christie et al. 2009).

5.12 Food Provider of Fish

The seaweed bed is a hub of food for many fishes. Many moveable macrofauna (e.g., crustaceans and mollusks) are abundant in the kelp forest (*Laminaria hyperborea*). The macrofauna occupies an important place in the fish diet (Nelson 1979; Kennelly 1983, 1991; Holmlund et al. 1990; Nordeide and Fossa 1992; Hoeisaeter and Fossa 1993; Fossa 1995; Føsne and Gjøsaeter 1996; Jorgensen and Christie 2003; Christie et al. 2003). The abundance of the macrofauna in the kelp forest makes the habitat a vital source of prey for many top-down predatory consumers (Jorgensen and Christie 2003; Christie et al. 2003). Christie et al. (2003) reported that the average density of the macrofauna in the Norwegian *Laminaria hyperborea* kelp forest could be 100,000 ind/m^2.

Amphipods and gastropods are dominant in the Norwegian kelp forest and these are the favorite food of many fishes (Moore 1972, 1973; Gordon 1983; Schultze et al. 1990; Fossa 1995; Fossa et al. 1998; Christie et al. 1998, 2003; Norderhaug et al. 2002; Fredriksen 2003). Spatio-temporal variation in the prey species in the kelp bed also changes the availability of food and fish species dependent on them (Deady 1995; Deady and Fives 1995a, b; Varian 1998; Zemke-White and Clements 2004). Moreover, the occurrence and abundance of macroinvertebrates also rely on the age and size of the seaweed (Schultze et al. 1990). The number of invertebrates increased with the increase in seaweed age and size (Rinde et al. 1992). Consequently, the number of fishes in the kelp bed increases with the increase in seaweed age.

5.13 Food Provider of Invertebrates

Seaweed is directly used as a food source for invertebrates, gastropods (e.g., *Patella* and *Helicon*), and some echinoderms (e.g., *Echinus* and *Psammechinus*). Gastropods *Patella pellucida* and *Lacuna vincta* directly graze on seaweed for their food. Sea urchins *Strongylocentrotus droebachiensis* and *Paracentrotus lividus* also rely on seaweed for their food (Steneck et al. 2002; Molis et al. 2010; Leclerc et al. 2013b). In the northeast Atlantic, common limpet *Patella vulgata* feeds on drift kelp. An indirect form of seaweed (i.e., particulate organic matter) is used by the suspension and deposit feeders as their food (Dugan et al. 2003). Sponges, terebellids, sabellids, serpulids, spirorbids, bivalves, cirripeds, bryozoans, holothurians, crinoids, and tunicates used particulate organic matter for their growth and energy metabolism. Seaweed is also used as the food source for cnidarians, scale worms, syllids, hesionids, phyllodocids, nereids, isopods, lobster, and crab, etc. Along the coast, seaweed roots provide organic matter for the amphipods *Malacoceros* and *Capitella*.

5.14 Food Provider of Birds

Seaweed provides food for birds indirectly. In the seaweed bed food chain, kelp detritus inputs organic matter into the nutrients in poor coastal regions (particularly sandy beaches). Seaweed detritus provides nutrients that are a suitable habitat for many intertidal macroinvertebrate communities (secondary production) and fishes. These macroinvertebrates and fishes of the seaweed bed are regarded as prey/food for birds (Duggins et al. 1989).

5.15 Shore Protection

Seaweed acts as a buffer against various natural calamities (i.e., flood, storm surges, extreme wind, etc.) (Smale et al. 2013). It is a bioengineering structure in the nearshore or coastal areas such as salt marshes and mangroves. During flooding and storm events, the seaweed structure changes the water motion and dampens breaking wave velocity, protecting shore or coastal areas from possible damage (Lovas and Torum 2001). It protects the shore from erosion by sediment retention (Mork 1996; Lovas and Torum 2001). Seaweed beds are very important where climate change phenomena such as sea-level rise and storms are frequent. There is little information about the storm protection capability of seaweed beds. The magnitude of wave or storms is site specific and species dependent (Firth et al. 2016). The degree of water flow largely depends on the assemblage, density, and morphology of the seaweed (Eckman et al. 1989; Gaylord et al. 2007). *Laminaria hyperborea* beds reduced the height of the waves in Norway by 60% (Mork 1996). Similar findings were observed in the UK and Ireland in the case of shore protection. The importance of seaweed cultivation or naturally growing seaweed will increased soon as the climate is changing rapidly.

Besides, fronds and stipes are exposed to faster water currents and greater effects of wave action. Kelp stipe is often colonized by highly abundant and diverse flora and fauna, which varies considerably spatio-temporally (Christie et al. 2003).

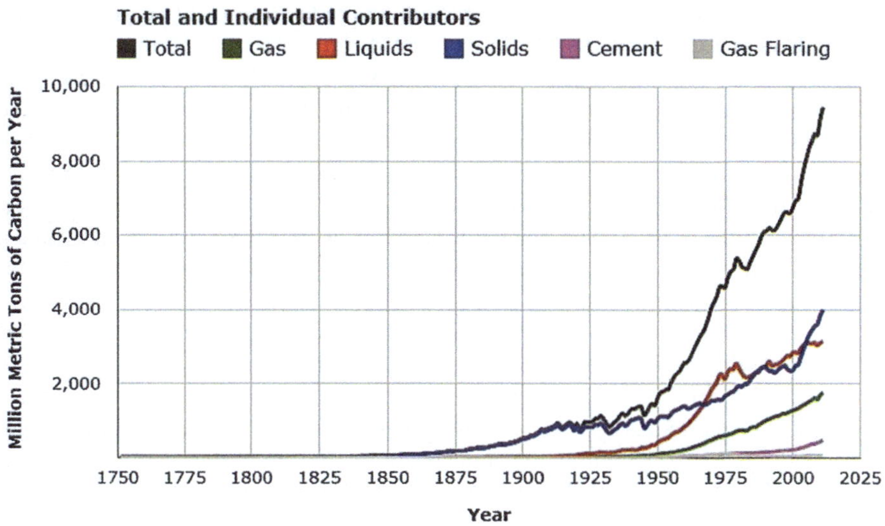

Fig. 5.5 Different contributing factors in the global carbon production trend (Source: Modified from Boden et al. 2017)

5.16 Carbon Sequestration and Climate Change Regulation

"Blue Carbon" is the carbon that is sequestered by both living and non-living biomass in the ocean and coastal habitats and provides many ecological services (Nellemann et al. 2009; Howard et al. 2014; Vierros 2017; Queirós et al. 2019). Worldwide carbon production is increasing at an alarming rate (Fig. 5.5). The average CO_2 concentration increased from 315 ppm to 380 ppm over 47 years (1960–2007). Worldwide, there has been an estimated 35% increase in CO_2 emission since 1990 (IPCC 2007). The ocean acts as a hub for the sink of carbon dioxide (Arsenault 2018; Froehlich et al. 2019; Ortega et al. 2019). Seaweed, phytoplankton, and seagrasses remove CO_2 from the atmosphere (Zou 2005; Kaladharan et al. 2009; Arsenault 2018). Seaweed is the permanent or long-term sequester of carbon dioxide (Nellemann et al. 2009; McLeod et al. 2011; Hughes et al. 2012a, b). By reducing CO_2 from the seawater it minimizes the issue of ocean acidification (Arsenault 2018). Seaweed stabilizes the pH concentration of the ocean water by taking CO_2 and by releasing oxygen during the photosynthesis process. The process that converts CO_2 into seaweed biomass and releases oxygen into the surrounding environment is light driven (Langton et al. 2019). Seaweed converts CO_2 into organic matter (N'Yeurt et al. 2012; Chung et al. 2013; Duarte et al. 2017) and this organic carbon cannot go back into the atmosphere (Hill et al. 2015; Trevathan-Tackett et al. 2015). Seaweed respires at night, but the concentration of oxygen consumption and CO_2 production do not exceed the amount of daytime O_2 production and CO_2 absorption (Duarte and Cebrian, 1996; Langton et al. 2019).

Seaweed cultivation showed a net increase in pH and oxygen levels (Liu et al. 2009). Shellfish (e.g., mollusks and crustaceans) respire CO_2 while seaweed receives the CO_2. This is a mutual aspect of the benefit that reduces the acidification of water (Langton et al. 2019). Excess CO_2 in the water forms carbonic acid that dissociates into bicarbonate and hydrogen ions, which lowers the pH. This lower pH largely hampers the formation of the shell (Langton et al. 2019). Lower pH changes the availability of shell-forming minerals required by corals, mollusks, and myriad microorganisms (Gatusso and Hansson 2011). Consequently, the shell-forming animals are declining. Seaweed helps in the mitigation of CO_2 and regulates the environmental impacts of climate change (Duarte et al. 2017) such as risk to human health, loss of biodiversity, increased risk of extreme weather events, and loss of agricultural productivity (Isacs et al. 2016).

An excessive volume of atmospheric carbon dioxide (CO_2) creates a serious adverse situation for marine organisms. Ocean acidification phenomena and an increase in sea surface temperatures are alarming issues (Feely et al. 2004; Meehl et al. 2007; Ciais et al. 2013). According to the IPCC (2013), CO_2 concentrations are expected to reach 1000 ppm in the atmosphere by the end of this century. This will increase dissolved CO_2 by ~2.5 fold. As a result, a decrease in pH (~0.4 units) will increase bicarbonate concentrations (by ~10%) and carbonate levels (approximately halve) (Feely et al. 2004; Raven et al. 2005). More than 30 countries have decided to increase the production of renewable resources to meet carbon emission targets (Bjerregaard et al. 2016). Seaweed cultivation plays a significant role in marine carbon sequestration (Chung et al. 2011, 2013; Duarte et al. 2017), which reduces ocean acidification and also provides human food, animal feed, and bioenergy (Kraan 2013; Krause-Jensen et al. 2015; Chen et al. 2015; Bjerregaard et al. 2016). This sequestrated carbon can be buried in sediments (Zhang et al. 2012), particularly in continental shelf sediments or in the deep sea (Krause-Jensen and Duarte 2016). A large flux of macroalgal carbon was exported to the offshore i.e., about 16.5 g carbon/m^2/day of giant kelp was exported through the Carmel Canyon, California. Approximately 7×10^{10} g carbon seaweed carbon reached a depth of 1800 m from the Bahaman shelf during a storm, while 0.4 g carbon/m^2/year of *Sargassum* reached a depth of 3600 m in the Northwest Atlantic region (Rowe and Staresinic 1979; Harrold et al. 1998; Dierssen et al. 2009). *Grypania spiralis* (the oldest dating of a multicellular organism) proved that macroalgae have contributed to carbon sequestration for over 2.1 billion years and act as a source of oil deposits (Han and Runnegar 1992; Sun et al. 2013; Xie et al. 2014). Macroalgal carbon ultimately finds its way into anoxic basins, submarine canyons, rocky shores, and the

Table 5.3 Nitrogen, phosphorus removal, and carbon rate by 500 million tonnes of dry seaweeds

Nitrogen removal	10,000,000 tons	Assumes nitrogen content to be 2% of dry weight. Equals 18% of the nitrogen added to oceans through fertilizer
Phosphorous removal	1000,000 tons	Assumes phosphorous content to be 0.2% of dry weight. Represents 61% of the phosphorous input as fertilizer
Carbon assimilation	135,000,000 tons	Assumes carbon content to be 27% of dry weight. Equals 6% of the carbon added annually to oceans from greenhouse gas emissions

Bjerregaard et al. (2016)

deep sea where sedimentation occurs (Wolff 1962; Canals et al. 2006; De Leo et al. 2010; Filbee-Dexter and Scheibling 2014; Barron et al. 2014; Renaud et al. 2015).

In 2010, emissions of carbon were about 8182 Tg from anthropogenic sources (Boden et al. 2010). Large-scale seaweed culture can remove huge amounts of carbon from the coastal water (Tang et al. 2011; Hughes et al. 2012a, b). For example, 500 million tons of seaweed production would absorb 135 million tons of carbon (Table 5.3). Krause-Jensen and Duarte (2016) reported that globally, about 173 Tg carbon/year (with a range of 61–268 Tg carbon/year) fixed by seaweeds, which is a relatively small proportion of total oceanic primary production (54–59 Pg carbon/year) and the increase in atmospheric CO_2 of 4 Pg carbon/year (Denman et al. 2007). This absorption process can add carbon credit as about 3.2% of the carbon is added annually to seawater from greenhouse gas emissions (Bjerregaard et al. 2016). It is reported that the seaweed biomass of the Indian coast can utilize 9052 tCO_2/day against 365 tCO_2/day emissions. This is a clear indication of a net carbon credit of 8687 tCO_2/day (Kaladharan et al. 2009). As a result, India is the biggest beneficiary of the carbon trade, and claims about 31% of the total world carbon trade (The Economic Times 2005).

CO_2 sequestration by seaweed was not fully incorporated with the *"Blue Carbon"* concept owing to the decomposition nature of seaweed (Nellemann et al. 2009; McLeod et al. 2011; Duarte et al. 2013). However, the thinking changed after the evidence that seaweed is the contributor to the carbon sink in the ocean (Hill et al. 2015; Sondak and Chung 2015; van der Heijden and Kamenos 2015; Trevathan-Tackett et al. 2015; Moreira and Pires 2016; Krause-Jensen and Duarte 2016). The role of seaweed in the *"Blue Carbon"* service and mitigation of climate change is now well accepted. Using seaweed biomass as biofuel or a seaweed-based food system to replace fossil fuel or intense carbon production could reduce the CO_2 emission (Fry et al. 2012; Kraan 2013; Chen et al. 2015). In Korea, a *"Blue Carbon"* program has been developed, even though they contribute only 6% of global seaweed production (Chung et al. 2013; Sondak and Chung 2015; FAO 2016b). To make this *"Blue Carbon"* program successful in mitigating climate change, world-leading seaweed producers (e.g., China, Indonesia, Philippines) can come forward.

5.17 Nutrients Uptake/Mitigation of Eutrophication

Eutrophication has recently become the emerging environmental concern throughout the world (Jiang et al. 2019). Oceans, especially coastal areas, receive nutrients from both natural and atmospheric sources (Paerl 1995; Prospero et al. 1996; Jickells 1998; Baker 2003). Moreover, nutrients are added from anthropogenic sources (e.g., finfish aquaculture, agriculture, and urban wastewater) (Smith 2003; Boesch et al. 2006) through the bio-deposition of feces and pseudofeces and the release of excess feed into the coastal water (Crawford et al. 2003; Kalantzi and Karakassis 2006; Forde et al. 2015). These excess nutrients can cause harmful algal bloom or eutrophication, which exerts negative impacts on the surrounding water

Table 5.4 Total nutrient removal by seaweed aquaculture in China and the nutrient removal capacity of Chinese seaweed farms per km^2

Total for China (2014)		
Seaweed production	2.00	million tonnes Dry-Wet (DW)
Seaweed area	1250	km^2
Nitrogen concentration[a]	3.76 ± 0.92	% DW
Phosphorus concentration[a]	0.47 ± 0.19	% DW
Nitrogen removal	75,371 ± 18,423	tonnes nitrogen per year
Phosphorus removal	9496 ± 3875	tonnes phosphorus per year
Per km^2 of seaweed farm and year		
Seaweed production	1604	tonnes DW
Nitrogen concentration[a]	3.76 ± 0.92	% DW
Phosphorus concentration[a]	0.47 ± 0.19	% DW
Nitrogen removal	60.31	tonnes nitrogen per km^2 per year
Phosphorus removal	7.60	tonnes phosphorus per km^2 per year
Nitrogen input[b]	3.38	tonnes nitrogen per km^2 per year
Phosphorus input[b]	0.06	tonnes phosphorus per km^2 per year
Seaweed farm nitrogen footprint area	17.8	km^2 of coastal ocean removed of nitrogen inputs per km^2 of seaweed farm
Seaweed farm phosphorus footprint area	126.7	km^2 of coastal ocean removed of phosphorus inputs per km^2 of seaweed farm

The seaweed farm nitrogen and phosphorus footprint area refer to the km^2 of Chinese coastal waters receiving nutrient inputs equivalent to those removed by 1 km^2 of seaweed farms. [a]The average tissue nutrient concentrations of Chinese seaweed, as weighted per species. [b]Nutrient input from the inventory integrating the riverine and atmosphere resources, weighted by the area of the East China Sea and the Yellow Sea
Source: Xiao et al. (2017)

quality (Bricker et al. 2008; Jiang et al. 2014; Glibert et al. 2018; Paerl et al. 2018). This polluted water is detrimental to both pelagic and benthic marine organisms (Shumway 1990; Anderson et al. 2002; Heisler et al. 2008; Chopin et al. 2008). Nitrogen and phosphorus are the main contributing agents for coastal pollution. Removal of these nutrients can be a great approach to mitigating the eutrophication issue worldwide (Conley et al. 2009; Holdt and Edwards 2014; Kim et al. 2015a).

Seaweed is the main weapon for removing nutrients from the coastal water (Fei 2004; Kang and Sui 2010; Liu et al. 2016; Roleda and Hurd 2019; Jiang et al. 2019). Cultivation of seaweed is regarded as the most promising tool for restoring the ecological balance (Buschmann et al. 2001, 2017; Chung et al. 2002; Neori et al. 2004; Yang et al. 2015a,b; Seghetta et al. 2016; Kim et al. 2017; Xiao et al. 2017). In China, excess nutrients have been removed significantly by cultured seaweed and seaweed farms (Table 5.4). Seaweed cultivated in suspended conditions along the coast can absorb inorganic nutrients from the water and absorption increases with the growth of seaweed (Troell et al. 1999; Neori et al. 2004; Troell et al. 2009; Kerrison et al. 2015; Marinho et al. 2015). Annually, seaweed removes 297 tonnes of nitrogen and 42 tonnes of phosphorus from Xiangshan Bay of the East China Sea

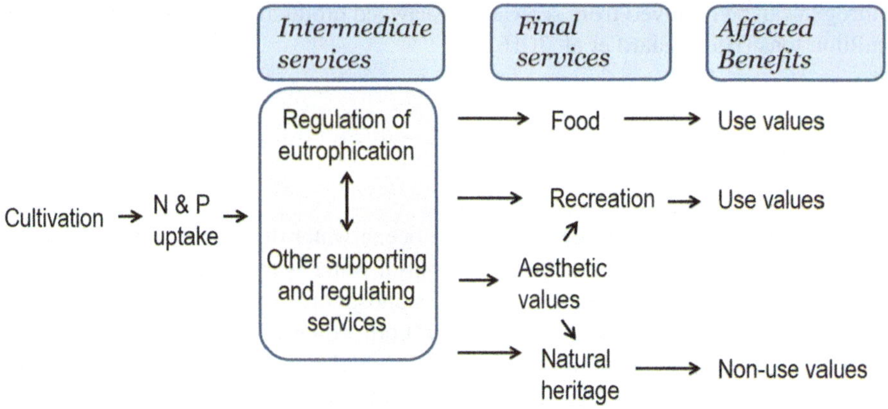

Fig. 5.6 Positive effects of removing excess nutrients from seawater through seaweed cultivation (Source: Modified from Hasselström et al. 2018)

(Jiang et al. 2019). Because of the decrease, less eutrophicated water or good-quality water has a positive impact on fish stocks and reproduction, habitat availability, and underwater vegetation (naturally growing kelp and bladderwrack density) (Kautsky et al. 1986; Paulsen 2007; Moy and Christie 2012). Furthermore, clear water is very important for the growth and succession of photosynthetic species and many other associated species (Kautsky et al. 1986; Svane and Gröndahl 1988; Jiang et al. 2019).

5.18 Nitrogen Removal

Globally, 124 million tons of nitrogen were used as fertilizer in 2014 for the growth of plants (Bjerregaard et al. 2016). But only half the amount was used by the plants and the remained unused. Finally, approximately 15–30% of the nitrogen was to find its way into the coastal water (Swaney et al. 2012; Lassaletta et al. 2014; FAO 2015). This excess nitrogen results in 245,000 km^2 of the polluted zone or biologically dead zone worldwide (Diaz and Rosenberg 2008). Removal of nitrogen is crucial because it is the main agent responsible for creating coastal eutrophication (Conley et al. 2009).

Seaweed cultivation can be a positive approach to removing this excess nitrogen from the coastal water (Bjerregaard et al. 2016). Cultivated seaweed removes nitrogen from seaweed, which exerts a positive impact on the environment (Fig. 5.6). Five hundred million tons of seaweed production would remove 10,000,000 tonnes of nitrogen (Table 5.3). Marine plants can produce 1000 tons dry weight per km^2 or 245 million tons dry weight, which can cover the dead zone area (Zhang et al. 2014; Kim et al. 2014, 2015a). In the case of dry seaweed, about 20 tons of nitrogen can be taken up per km^2 (Mišurcová 2012). It is postulated that 10 million tons of nitrogen can be removed from seawater if seaweed production could reach up to 500 million tons (Bjerregaard et al. 2016).

5.19 Phosphorus Removal

Phosphorus is not limiting nutrients in the ocean water or coastal water. Thus, the dead zone or eutrophication zone in the coastal water is less related to phosphorus (Bjerregaard et al. 2016). Removal of excess phosphorus from seawater by seaweed cultivation provides massive benefits for both aquatic organisms and humans (Fig. 5.6). The phosphorus reserve would be depleted in the next 50–100 years (Cordell et al. 2009). This nutrient reserve is declining owing to excessive use on the land and high energy production costs in manufacturing phosphate fertilizers. In 2014, 48 million tons of fertilizer was produced from phosphorus globally (FAO 2015). Phosphorus reserve in seaweed may be the best source of phosphorus for the future (Cordell et al. 2009). By-products of seaweed can be used as a potential source of phosphorus and as fertilizer or to replace the other forms of phosphorus use (Bjerregaard et al. 2016). Pechsiri et al. (2016) reported that 16 g nitrogen can be taken up by 1 kg of seaweed biomass in Sweden: 22.5–27.5 tons of seaweed (wet weight)/hectare/year can sequestrate 79.5–97 kg nitrogen/hectare/year. One million tons of phosphorus can be removed by cultivating 500 million tons of seaweed (Table 5.3).

5.20 Producer of Trace Gases

Seaweed acts as an important component that produces trace gases responsible for the depletion of ozone (Carpenter and Liss 2000). Macroalgae are global producers of trace gases that contain sulfur or halogens, such as volatile brominated and iodinated halocarbons. Various biotic and abiotic factors (i.e., physiological, mechanical, and oxidative stress) regulate the production of halogenated compounds (Mehrtens and Laturnus 1997; Manley 2002; Palmer et al. 2005; Leedham et al. 2015). Besides ozone layer depletion, trace gases are also responsible for playing an important role in global biochemical cycles, cloud formation, and the lifetime of other greenhouse gases (Laturnus 1996; Giese et al. 1999; Carpenter and Liss 2000; Leedham et al. 2013). Most of the studies are laboratory based with wild seaweed, which depends on the seasonal growth of seaweed (Zhou et al. 2005), whereas studies on farmed seaweed. Leedham et al. (2013) reported that cultured seaweed produced a low amount of halocarbons compared with naturally produced seaweed. Although Phang et al. (2015) mentioned that farmed seaweed can play a great role along with natural seaweed in terms of halocarbon production.

5.21 Nutrient Regulation and Biogeochemical Cycling

Seaweed plays a vital role in nutrient regulation and biogeochemical cycling in the coastal environment (Klinger 2015). This process is completed by nutrient assimilation, photosynthesis, and organic matter production, decomposition, and transportation (Klinger 2015). Seaweed is regarded as the best producer of the ocean. Seaweed productivity is greater than the productivity of phytoplankton, seagrasses, corals, and benthic microalgae (Mann 1973; Yokohama et al. 1987; Alongi 1998; Wada and Hama 2013). The productivity is the assimilation of carbon, nitrogen, phosphorous, silica, and other compounds from seawater. These compounds are then incorporated into organic matter. This organic matter is consumed by other organisms as POM or living tissue, used as food by herbivores and suspension feeders, and used as a substrate for microbial colonization and digestion. Finally, carbon and other nutrient content are transported into the system (Leclerc et al. 2013b; Yorke et al. 2013). Seaweed produces POM, which is higher than the production of phytoplankton (Bustamante and Branch 1996).

Seaweed also contributes to the organic matter in coastal water as DOM. Seaweed releases a considerable amount of DOM; up to 40% of seaweed production is released as DOM (Wada and Hama, 2013). DOM contributes to the organic carbon reservoir in the coastal water and fuels the microbial loop. POM and DOM export carbon in the offshore ecosystem. Seaweed plays an important role in the carbon cycle, which came into focus after the buzzword *"carbon sequestration"* was coined (Nellemann et al. 2009). Seaweed acts as a powerful tool in the context of mitigating climate change and ocean acidification.

5.22 Conclusion

From the above discussion, it can be concluded that seaweed plays a significant role in climate change mitigation and adaptation. If the culture of seaweed increased worldwide on a large scale it could be a powerful tool in controlling the climate. Ocean acidification and de-oxygenation are the effects of climate change. These issues can be minimized by seaweed cultivation. Moreover, it dampens the wave energy and protects coastal dwellers and their livelihood. It also provides food and habitat, and removes excess nutrients from the water, which is required for a healthy marine ecosystem. By reducing CO_2 from the atmosphere, people can benefit economically. Owing to the low investment required for the setup, seaweed culture gaining in popularity among the coastal people. Although some constraints are identified in the seaweed culture, climate change mitigation and adaptation features, as well as socio-economic benefits, make it a successful venture.

Acknowledgment Financial assistance from the Bangladesh Fisheries Research Institute, Cox's Bazar, is gratefully acknowledged.

References

Abbott OJ, Perkins EJ (1977) A survey of the littoral macrofauna at Siddick, Cumbria, 1971–1973; with particular reference to the fauna of *Laminaria* spp. holdfasts. Cumbria Sea Fish Committee Sci Rep 77

Acosta CA, Butler MJ (1999) Adaptive strategies that reduce predation on Caribbean spiny lobster postlarvae during onshore transport. Limnol Oceanogr 44:494–501

Alongi DM (1998) Coastal Ecosystem Processes. CRC Press, Boca Raton

Anderson DM, Glibert PM, Burkholder JM (2002) Harmful algal blooms and eutrophication: nutrient sources, composition, and consequences. Estuaries 25:704–726

Arsenault MM (2018) Marine biodiversity is important, and the roles played by marine algae (seaweed) are important to marine biodiversity and ecology. Beyond Pesticide. https://www.beyondpesticides.org/assets/media/documents/BP%20Comments%20on%20Marine%20Materials.final.pdf

Baker AR (2003) Atmospheric deposition of nutrients to the Atlantic Ocean. Geophys Res Lett 30:2296

Ball B, Costelloe J, Konnecker G, Keegan B (1995) The rocky subtidal assemblages of Kinsale Harbour (south coast of Ireland). In: Eleftheriou A, Ansell A, Smith C (eds) Biology and ecology of shallow coastal waters. Proceedings of the 28th European Marine Biological Symposium. Olsen and Olsen, pp 293–302

Barrón C, Duarte CM (2015) Dissolved organic carbon pools and export from the coastal ocean. Glob Biogeochem Cycles 29:1725–1738

Barron C, Apostolaki ET, Duarte CM (2014) Dissolved organic carbon fluxes by seagrass meadows and macroalgal beds. Front Mar Sci 1:42

Bell TW, Cavanaugh KC, Reed DC, Siegel DA (2015) Geographical variability in the controls of giant kelp biomass dynamics. J Biogeogr 42:2010–2021

Benedetti-Cecchi L, Pannacciulli P, Bulleri F, Moschella PS, Airoldi L, Relini G, Cinelli F (2001) Predicting the consequences of anthropogenic disturbance: large-scale effects of loss of canopy algae on rocky shores. Mar Ecol Prog Ser 214:137–150

Bertness MD, Callaway R (1994) Positive interactions in communities. Trends Ecol Evol 9:191–193

Birkett DA, Maggs CA, Dring MJ, Boaden PJS, Seed R (1988) Infralittoral reef biotopes with kelp species (volume VII). An overview of dynamic and sensitivity characteristics for conservation management of marine SACs. Scottish Association of Marine Science (UK Marine SACs Project). 174pp

Bjerregaard R, Valderrama D, Radulovich R, Diana J, Capron M, Mckinnie CA, Cedric M, Hopkins K, Yarish C, Goudey C, Forster J (2016) Seaweed aquaculture for food security, income generation and environmental health in tropical developing countries (English). World Bank Group, Washington, DC

Blight AJ, Thompson RC (2008) Epibiont species richness varies between holdfasts of a northern and southerly distributed kelp species. J Mar Biol Assoc 88:469–475

Boden TA, Marland G, Andres RJ (2010) Global, regional, and national fossil-fuel CO_2 emissions. Carbon Dioxide Information Analysis Center, Oak Ridge National Laboratory, U.S. Department of Energy, Oak Ridge

Boden TA, Marland G, Andres RJ (2017) Global, regional, and national fossil-fuel CO_2 emissions. Carbon Dioxide Information Analysis Center, Oak Ridge National Laboratory, U.S. Department of Energy, Oak Ridge

Bodkin J (1988) Effects of kelp forest removal on associated fish assemblages in central California. J Exp Mar Biol Ecol 117:227–238

Boesch D, Hecky R, O'Melia C, Schindler D, Seitzinger S (2006) Eutrophication of Swedish seas. In: Swedish Environmental Protection Agency, report 5509

Borg A, Pihl P, Wennhage H (1997) Habitat choice by juvenile cod (*Gadus morhua* L.) on sandy soft bottoms with different vegetation types. Helgol Meeresunt 51:197–212

Bricker SB, Longstaff B, Dennison WC, Jones AB, Woerner JL (2008) Effects of nutrient enrichment in the nation's estuaries: a decade of change. Harm Algae 8:21–32
Bryhn A, Lindegarth M, Bergström L, Bergström U (2015) Ekosystemtjänster från svenska hav. In: Status och påverkansfaktorer. Havs-och vattenmyndighetens rapport 2015:12.
Bulleri F, Benedetti-Cecchi L, Acunto S, Cinelli F, Hawkins SJ (2002) The influence of canopy algae on vertical patterns of distribution of low-shore assemblages on rocky coasts in the Northwest Mediterranean. J Exp Mar Biol Ecol 267:89–106
Burrows MT (2012) Influences of wave fetch, tidal flow and ocean colour on subtidal rocky communities. Mar Ecol Prog Ser 445:193–207
Buschmann AH, Correa JA, Westermeier R, Hernandez-Gonzalez MC, Norambuena R (2001) Red algal farming in Chile: a review. Aquaculture 194:203–220
Buschmann AH, Camus C, Infante J, Neori A, Israel Á, Hernández-González MC, Critchley AT (2017) Seaweed production: overview of the global state of exploitation, farming and emerging research activity. Eur J Phycol 52:391–406
Bustamante RH, Branch GM (1996) The dependence of intertidal consumers on kelp-derived organic matter on the west coast of South Africa. J Exp Mar Biol Ecol 196:1–28
Bustamante M, Tajadura J, Gorostiaga JM, Saiz-Salinas JI (2014) Response of rocky invertebrate diversity, structure and function to the vertical layering of vegetation. Estuar Coast Shelf Sci 147:148–155
Byrnes JE, Reed DC, Cardinale BJ, Cavanaugh KC, Holbrook SJ, Schmitt RJ (2011) Climate driven increases in storm frequency simplify kelp forest food webs. Glob Chang Biol 17:2513–2524
Cabral P, Levrel H, Viard F, Frangoudes K, Girard S, Scemama P (2016) Ecosystem services assessment and compensation costs for installing seaweed farms. Mar Pollut 71:157–165
Campbell I, Macleod A, Sahlmann C, Neves L, Funderud J, Overland M, Hughes A, Stanley M (2018) The environmental risks associated with the development of seaweed farming in Europe-prioritizing key knowledge gaps. Front Mar Sci 6:107
Canals M, Puig P, de Madron XD, Heussner S, Palanques A, Fabres J (2006) Flushing submarine canyons. Nature 444:354–357
Carpenter L, Liss P (2000) On temperate sources of bromoform and other reactive organic bromine gases. J Geophys Res Atmos 105:20539–20547
Carr M (1983) Spatial and temporal patterns of recruitment of young of the year rockfishes (genus *Sebastes*) into a central Californian kelp forest. MSc. Thesis, San Francisco State University, San Francisco. 104pp
Chen H, Zhou D, Luo G, Zhang S, Chen J (2015) Macroalgae for biofuels production: Progress and perspectives. Renew Sustain Energy Rev 47:427–437
Chopin T, Robinson SMC, Troell M, Neori A, Fang J (2008) Multitrophic integration for sustainable marine aquaculture. In: Jorgensen SE, Fath B (eds) Encyclopedia of ecology: ecological engineering. Elsevier, Oxford, pp 2463–2475
Christie H, Fredriksen S, Rinde E (1998) Regrowth of kelp and colonization of epiphyte and fauna community after kelp trawling at the coast of Norway. Hydrobiology 375(376):49–58
Christie H, Jorgensen NM, Norderhaug KM, Waage-Nielsen E (2003) Species distribution and habitat exploitation of fauna associated with kelp (*Laminaria hyperborea*) along the Norwegian coast. J Mar Biol Assoc 83:687–699
Christie H, Norderhaug KM, Fredriksen S (2009) Macrophytes as habitat for fauna. Mar Ecol Prog Ser 396:221–233
Chung IK, Kang YH, Yarish C, Kraemer G, Lee J (2002) Application of seaweed cultivation to the bioremediation of nutrient-rich effluent. Algae 17:187–194
Chung IK, Beardall J, Mehta S, Sahoo D, Stojkovic S (2011) Using marine macroalgae for carbon sequestration: a critical appraisal. J Appl Phycol 23:877–886
Chung IK, Oak JH, Lee JA, Shin JA, Kim JG, Park KS (2013) Installing kelp forest/seaweed beds for mitigation and adaptation against global warming: Korean Project overview. ICES J Mar Sci 70:1038–1044

Ciais P, Sabine C, Bala G, Bopp L, Brovkin V, Canadell J, Chhabra A, De Fries R, Galloway J, Heimann M, Jones C, Le Quéré C, Myneni RB, Piao S, Thornton P (2013) Carbon and other biogeochemical cycles. In: Stocker TF, Qin D, Plattner GK, Tignor M, Allen SK, Boschung J, Nauels A, Xia Y, Bex V, Midgley PM (eds) Climate change 2013: the physical science basis. Contribution of Working Group I to the Fifth Assessment Report of the Intergovernmental Panel on Climate Change. Cambridge University Press, Cambridge/New York

Conley DJ, Paerl HW, Howarth RW, Boesch DF, Seitzinger SP, Havens KE, Lancelot C, Likens GE (2009) Controlling eutrophication: nitrogen and phosphorus. Science 323:1014–1015

Cordell D, Drangert JO, White S (2009) The story of phosphorus: global food security and food for thought. Glob Environ Chang 19:292–305

Corey P, Kim JK, Garbary DJ, Prithiviraj B, Duston J (2012) Bioremediation potential of *Chondrus crispus* (basin head) and *Palmaria palmata*: effect of temperature and high nitrate on nutrient removal. J Appl Phycol 24:441–448

Corey P, Kim JK, Duston J, Garbary DJ (2014) Growth and nutrient uptake by *Palmaria palmata* integrated with Atlantic halibut in a land-based aquaculture system. Algae 29:35–45

Coston-Clements L, Settle LR, Hoss DE, Cross FA (1991) Utilization of the *Sargassum* habitat by marine invertebrates and vertebrates: a review. NOAA Technical Memorandum NMFS-SEFSC, 296, p 32

Crawford CM, Macleod CKA, Mitchell IM (2003) Effects of shellfish farming on the benthic environment. Aquaculture 224:117–140

Dayton PK (1985) Ecology of kelp communities. Annu Rev Ecol Evol Syst 16:215–245

De Leo FC, Smith CR, Rowden AA, Bowden DA, Clark MR (2010) Submarine canyons: hotspots of benthic biomass and productivity in the deep sea. Proc R Soc B 277:2783–2792

Deady S (1995) The biology and parasites of corkwing, *Crenilabrus melops* (L.) and ballan, *Labrus bergylta* Ascanius (Teleostei; Labridae) from Galway Bay and the Gulf of St. Malo. Ph.D. thesis, National University of Ireland, Galway. 189pp

Deady S, Fives JM (1995a) The diet of corkwing wrasse *Crenilabrus melops* in Galway Bay, Ireland in Dinard, France. J Mar Biol Assoc 75:635–649

Deady S, Fives JM (1995b) Diet of ballan wrasse *Labrus bergylta* and some comparisons with the diet of corkwing wrasse Crenilabrus melops. J Mar Biol Assoc 75:651–665

Denman KL, Brasseur GP, Chidthaisong A, Ciais P, Cox P, Dickinson RE, Hauglustaine DA, Heinze C, Holland EA, Jacob DJ, Lohmann U, Ramachandran S, Leite da silva Dias P, Wofsy SC, Zhang XY, Steffen W (2007) Coupling between changes in the climate system and biogeochemistry. ANU Research Publications. http://hdl.handle.net/1885/57818

Diaz RJ, Rosenberg R (2008) Spreading dead zones and consequences for marine ecosystems. Science 321:926–929

Dierssen HM, Zimmerman RC, Drake LA, Burdige DJ (2009) Potential export of unattached benthic macroalgae to the deep sea trough winddriven Langmuir circulation. Geophys Res Lett 36:L04602

Duarte CM, Cebrian J (1996) The fate of marine autotrophic production. Limnol Oceanogr 41:1758–1766

Duarte CM, Middelburg J, Caraco N (2005) Major role of marine vegetation on the oceanic carbon cycle. Biogeosciences 2:1–8

Duarte CM, Losada IJ, Hendriks IE, Mazarrasa I, Marbà N (2013) The role of coastal plant communities for climate change mitigation and adaptation. Nat Clim Chang 3:961–968

Duarte CM, Wu J, Xiao X, Bruhn A, Krause-Jensen D (2017) Can seaweed farming play a role in climate change mitigation and adaptation? Front Mar Sci 4:100

Dugan JE, Hubbard DM, McCrary MD, Pierson MO (2003) The response of macrofauna communities and shorebirds to macrophyte subsidies on exposed sandy beaches of southern California. Estuar Coast Shelf Sci 58S:25–40

Duggins DO, Simenstad CA, Estes JA (1989) Magnification of secondary production by kelp detritus in coastal marine ecosystems. Science 245:170–173

Duggins DO, Eckman JE, Sewell AT (1990) Ecology of understory kelp environments. II. Effects of kelps on recruitment of benthic invertebrates. J Mar Biol Assoc 143:27–45

Ebeling AW, Laur DR (1988) Fish populations in kelp forests without sea otters: effects of severe storm damage and destructive sea urchin grazing. In: The community ecology of sea otters. Van Blaricom GR, Estes JAA (eds) Springer: Berlin, pp 169–191

Ebeling AW, Larson RJ, Alevizon WS (1980) Habitat groups and island mainland distribution of kelp bed fishes off Santa Barbara, California. In: Power DM (ed) The California Islands: proceedings of a multidisciplinary symposium. Santa Barbara Museum of Natural History, Santa Barbara, pp 403–431

Ebling FJ, Kitching JA, Purchon RD, Bassindaleitish R (1948) The ecology of the lough Ine rapids with special reference to water currents. J Anim Ecol 17:223–244

Eckman JE, Duggins DO, Sewell AT (1989) Ecology of under story kelp environments. I Effects of kelps on flow and particle transport near the bottom. J Exp Mar Biol Ecol 129:173–187

Edwards A (1980) Ecological studies of the kelp, *Laminaria hyperborea*, and its associated fauna in South-West Ireland. Ophelia 19:47–60

Eriksson BK, Rubach A, Hillebrand H (2006) Biotic habitat complexity controls species diversity and nutrient effects on net biomass production. Ecology 87:246–254

Erwin DG, Picton BE, Connor DW, Howson CM, Gilleece P, Bogues MJ (1990) Inshore marine life of Northern Ireland. The report of a survey carried out by the diving team of the Botany and Zoology Department of the Ulster Museum in fulfilment of a contract with Conservation Branch of the Department of the Environment (N.I.). Ulster Museum, HMSO, Belfast

FAO (2015) World fertilizer trends and outlook to 2018. Food and Agriculture Organization of the United Nations, Rome

FAO (2016a) The state of world fisheries and aquaculture: contributing to food security and nutrition for all. Food and Agriculture Organization of the United Nations (FAO), Fisheries and Aquaculture Department, Rome

FAO (2016b) Fisheries and aquaculture software. FishStat Plus-Universal software for fishery statistical time series, in: FAO Fisheries and Aquaculture Department Rome. Updated 28 November 2013. Available online at: www.fao.org/fishery/statistics/software/fishstat/en. Accessed 22 Nov 2016

FAO (2018) Fishery statistical collections. Food and Agriculture Organization of the United Nations, Rome. http://www.fao.org/fishery/statistics/

Feely RA, Sabine CL, Lee K, Berelson W, Kleypas JA, Fabry VJ, Millero FJ (2004) Impact of anthropogenic CO2 on the CaCO3 system in the oceans. Science 305:362–366

Fei X (2004) Solving the coastal eutrophication problem by large scale seaweed cultivation. Hydrobiology 512:145–151

Filbee-Dexter K, Scheibling RE (2014) Detrital kelp subsidy supports high reproductive condition of deep living sea urchins in a sedimentary basin. Aquat Biol 23:71–86

Firth LB, Browne KA, Knights AM, Hawkins SJ, Nash R (2016) Eco-engineered rock pools: a concrete solution to biodiversity loss and urban sprawl in the marine environment. Environ Res Lett 11:94015

Forde J, O'Beirn FX, O'Carroll JPJ, Patterson A, Kennedy R (2015) Impact of intertidal oyster trestle cultivation on the ecological status of benthic habitats. Mar Pollut Bull 95:223–233

Føsne K, Gjøsaeter J (1996) Dietary composition and the potential food competition between O-group cod (*Gadus morhua* L.) and some other fish species in the littoral zone. J Mar Sci 53:757–770

Fosså JH (1995) Management of kelp. Prioritised research tasks. Institute of Marine Research (In Norwegian)

Fosså JH, Christie H, Sjøtun K (1998) *Laminaria hyperborea* beds as feeding chamber for fish. In: Abstracts of the 16th international seaweed symposium, Philippines

Foster MS, Schiel DS (1985) The ecology of Giant kelp forests in California: a community profile. US Fish Wildlife Serv Biol Rep 85:1–152

Fredriksen S (2003) Food web studies in a Norwegian kelp forest based on stable isotope (delta C-13 and delta N-15) analysis. Mar Ecol Prog Ser 260:71–81. https://doi.org/10.3354/meps260071

Froehlich HE, Afflerbach JC, Frazier M, Halpern BS (2019) Blue growth potential to mitigate climate change through seaweed offsetting. Curr Biol 29:3087–3093

Fry JM, Joyce PJ, Aumonier S (2012) Carbon footprint of seaweed as a biofuel. Prepared by Environmental Resources Management Limited (ERM), London for the crown Estate, p 64

Gattuso JP, Hansson L (2011) Ocean acidification: background and history. Ocean Acid:1–20

Gaylord BP, Rosman JH, Reed DC, Koseff JR, Fram J, MacIntyre S et al (2007) Spatial patterns of flow and their modification within and around a giant kelp forest. Limnol Oceanogr 52:1838–1852

Giese B, Laturnus F, Adams FC, Wiencke C (1999) Release of volatile iodinated c1-c4 hydrocarbons by marine macroalgae from various climate zones. Environ Sci Technol 33:2432–2439

Gili JM, Coma R (1998) Benthic suspension feeders: their paramount role in littoral marine food web. Trends Ecol Evol 13:316–321

Glibert PM, Berdalet E, Burford MA, Pitcher GC, Zhou M (2018) Global ecology and oceanography of harmful algal blooms, vol 232. Springer, Cham, pp 1–461

Gordon JCD (1983) Some notes on small kelp forest fish collected from *Saccorhiza polyschides* bulbs on the Isle of Cumbrae Scotland. Ophelia 22:173–183

Graham MH (2004) Effects of local deforestation on the diversity and structure of southern California giant kelp forest food webs. Ecosystems 7:341–357

Han T, Runnegar B (1992) Megascopic eukaryotic algae from the 2.1 billion year old Negaunee Iron Formation, Michigan. Science 257:232–235

Harrold C, Light K, Lisin S (1998) Organic enrichment of submarine canyon and continental shelf benthic communities by macroalgal drift imported from nearshore kelp forests. Limnol Oceanogr 43:669–678

Hasselström L, Visch W, Gröndahl F, Nylund GM, Pavia H (2018) The impact of seaweed cultivation on ecosystem services-a case study from the west coast of Sweden. Mar Pollut Bull 133:53–64

Hasselström L, Thomas JB, Nordström J, Cervin G, Nylund GM, Pavia H, Gröndahl F (2020) Socioeconomic prospects of a seaweed bioeconomy in Sweden. Sci Rep 10:1–7

Healy B, McGrath D (1998) Marine fauna of county Wexford, Ireland: the fauna of rocky shores and sandy beaches. Irish fisheries investigations new series, no. 2, pp 1–71

Heisler J, Glibert PM, Burkholder JM, Anderson DM, Cochlan W, Dennison WC et al (2008) Eutrophication and harmful algal blooms: a scientific consensus. Harmful Algae 8:3–13

Hill R, Bellgrove A, Macreadie PI, Petrou K, Beardall J, Steven A, Ralph PJ (2015) Can macroalgae contribute to blue carbon? An Australian perspective. Limnol Oceanogr 60:1689–1706

Hoeisaeter T, Fosså JH (1993) The kelp forest and its resident fish fauna. Institutt for Fiskeri og Marinbiologi, University of Bergen, Bergen

Holdt SL, Edwards MD (2014) Cost-effective IMTA: a comparison of the production efficiencies of mussels and seaweed. J Appl Phycol 26:933–945

Holmlund MB, Peterson CH, Hay ME (1990) Does algal morphology affect amphipod susceptibility to fish predation? J Exp Mar Biol Ecol 139:65–83

Howard J, Hoyt J, Isensee K, Pidgeon E, Telszewski M (2014) Coastal blue carbon: methods for conservation. International, Intergovernmental Oceanographic Commission of UNESCO, International Union for Conservation of Nature Factors in Mangroves, Tidal Salt Marshes, and Seagrasses Meadows. Conservation International, Inter-governmental Oceanographic Commission of UNESCO, International Union for Conservation of Nature, Arlington

Hughes AD, Black KD, Campbell I, Davidson K, Kelly MS, Stanley MS (2012a) Does seaweed offer a solution for bioenergy with biological carbon capture and storage? Greenhouse Gases Sci Technol 2:1–6

Hughes AD, Black KD, Campbell I, Davidson K, Kelly MS, Stanley MS (2012b) Does seaweed offer a solution for bioenergy with biological carbon capture and storage? Greenhouse Gases Sci Technol 2:402–407

Hurd CL, Harrison PJ, Bischof K, Lobban CS (2014) Seaweed ecology and physiology, 2nd edn. Cambridge University Press, Cambridge, 551pp

Imai I, Yamaguchi M, Hori Y (2006) Eutrophication and occurrences of harmful algal blooms in the Seto Inland Sea, Japan. Plank Benth Res 1:71–84

IPCC (2007) Synthesis report. Contribution of Working Groups of the Fourth Assessment Report of the Inter-governmental Panel on Climate Change. WMD and UNEP, 397 pp

IPCC (2013) In: Stocker TF, Qin D, Plattner GK, Tignor M, Allen SK, Boschung J, Nauels A, Xia Y, Bex V, Midgley PM (eds) Climate change 2013: the physical science basis. Contribution of Working Group I to the Fifth Assessment Report of the Intergovernmental Panel on Climate Change. Cambridge University Press, Cambridge, 1535 pp.

IPCC (2014) Summary for policymakers. In: Field CB, Barros VR, Dokken DJ, Mach KJ, Mastrandrea MD, Bilir TE, Chatterjee M, Ebi KL, Estrada YO, Genova RC, Girma B, Kissel ES, Levy AN, MacCracken S, Mastrandrea PR, White LL (eds) Climate change 2014: impacts, adaptation, and vulnerability. Part A: global and sectoral aspects. Contribution of Working Group II to the Fifth Assessment Report of the Intergovernmental Panel on Climate Change. Cambridge University Press, Cambridge/New York, pp 1–32

Isacs L, Finnveden G, Dahllöf L, Håkansson C, Petersson L, Steen B, Swanström L, Wikström A (2016) Choosing a monetary value of greenhouse gases in assessment tools. J Clean Prod. https://doi.org/10.1016/j.jclepro.2016.03.163

Jiang Z, Liu J, Chen J, Chen Q, Yan X, Xuan J, Zeng J (2014) Responses of summer phytoplankton community to drastic environmental changes in the Changjiang (Yangtze River) Estuary during the past 50 years. Water Res 54:1–11

Jiang Z, Liu J, Li S, Chen Y, Du P, Zhu Y, Chen J (2019) Kelp cultivation effectively improves water quality and regulates phytoplankton community in a turbid, highly eutrophic bay. Sci Total Environ 707:135561

Jickells TD (1998) Nutrient biogeochemistry of the coastal zone. Science 281:217–223

Jones DJ (1971) Ecological studies on macroinvertebrate populations associated with polluted kelp forests in the North Sea. Helgol Wissensch Meeresunt 22:417–441

Jones DJ (1972) Changes in the ecological balance of invertebrate communities in kelp holdfast habitats of some polluted North Sea waters. Helgoland Wissensch Meeresunters 23:248–260

Jones NS, Kain JM (1967) Subtidal algal colonisation following the removal of *Echinus*. Helgoland Wissensch Meeresunters 15:460–466

Jones CG, Lawton JH, Shachak M (1994) Organisms as ecosystem engineers. Oikos 69:373–386

Jones CG, Lawton JH, Shachak M (1997) Positive and negative effects of organisms as physical ecosystem engineers. Ecology 78:1946–1957

Jørgensen N, Christie H (2003) Diurnal, horizontal and vertical dispersal of kelp-associated fauna. Hydrobiologia 503:69–76. https://doi.org/10.1023/B:HYDR.0000008491.89382.e5

Kaehler S, Pakhomov EA, McQuaid CD (2000) Trophic structure of the marine food web at the Prince Edward Islands (Southern Ocean) determined by δ13C and δ15N analysis. Mar Ecol Prog Ser 208:13–20

Kaladharan P, Veena S, Vivekanandan E (2009) Carbon sequestration by a few marine algae: observation and projection. J Mar Biol Assoc India 51:107–110

Kalantzi I, Karakassis I (2006) Benthic impacts of fish farming: meta-analysis of community and geochemical data. Mar Pollut Bull 52:484–493

Kang KH, Sui Z (2010) Removal of eutrophication factors and heavy metal from a closed cultivation system using the macroalgae, *Gracilaria* sp. (Rhodophyta). Chin J Oceanol Limnol 28:1127–1130

Kautsky N, Kautsky H, Kautsky U, Waern M (1986) Decreased depth penetration of *Fucus vesiculosus* (L.) since 1940s indicates eutrophication of the Baltic Sea. Mar Ecol 28:1–8

Kelly E (2005) The role of kelp in the marine environment. Irish wildlife manuals, no. 17. National Parks and Wildlife Service, Department of Environment, Heritage and Local Government, Dublin

Kennelly SJ (1983) An experimental approach to the study of factors affecting algal colonization in a sublittoral kelp forest. J Exp Mar Biol Ecol 68:257–276
Kennelly SJ (1991) Caging experiments to examine the effects of fishes on understorey species in a sublittoral kelp community. J Exp Mar Biol Ecol 147:207–230
Kerrison PD, Stanley MS, Edwards MD, Black KD, Hughes AD (2015) The cultivation of European kelp for bioenergy: site and species selection. Biomass Bioenergy 80:229–242
Kim JK, Duston J, Corey P, Garbary DJ (2013) Marine finfish effluent bioremediation: effects of stocking density and temperature on nitrogen removal capacity of *Chondrus crispus* and *Palmaria palmata* (Rhodophyta). Aquaculture 414–415:210–216
Kim JK, Kraemer GP, Yarish C (2014) Field scale evaluation of seaweed aquaculture as a nutrient bioextraction strategy in Long Island Sound and the Bronx River Estuary. Aquaculture 433:148–156
Kim JK, Kottuparambilm S, Moh SH, Lee TK, Kim Y, Rhee J, Choi E, Kim BH, Yu YJ, Yarish C, Han T (2015a) Potential applications of nuisance microalgae blooms. J Appl Phycol 27:1223–1234
Kim JK, Kraemer GP, Yarish C (2015b) Use of sugar kelp aquaculture in Long Island Sound and the Bronx River Estuary for nutrient extraction. Mar Ecol Prog Ser 531:155–166
Kim JK, Yarish CY, Hwang EK, Park M, Kim Y (2017) Seaweed aquaculture: cultivation technologies, challenges and its ecosystem services. Algae 32:1–13
Kitching JA, Thain VM (1983) The ecological impact of the sea urchin *Paracentrotus lividus* (Lamarck) in Lough Ine, Ireland. Philos Trans R Soc Lond B Biol Sci 300:513–552
Klinger T (2015) The role of seaweeds in the modern ocean. Persp Phycol 2:31–40
Knox OGG, Marsden TJ, Warnick S, Birch G, Scherbatskoy MN, Wilson DB, Harvie BA (2015) Improved sustainability and ecosystem services from seaweed additions to an old agricultural production system. J Ecol Environ Sci 3:28–37
Kraan S (2013) Mass-cultivation of carbohydrate rich macroalgae, a possible solution for sustainable biofuel production. Mitig Adapt Strateg Glob Chang 18:27–46
Kraan S, Verges Tramullas A, Guiry MD (2000) The edible brown seaweed *Alaria esculenta* (Phaeophyceae, Laminariales): hybridization, growth and genetic comparisons of six Irish populations. J Appl Phycol 12:577–583
Krause-Jensen D, Duarte CM (2016) Substantial role of macroalgae in marine carbon sequestration. Nat Geosci 9:737–742
Krause-Jensen D, Duarte CM, Hendriks IE, Meire L, Blicher ME, Marbà N, Sejr MK (2015) Macroalgae contribute to nested mosaics of pH variability in a sub-Arctic fjord. Biogeosciences 12:4895–4911
Krumhansl KA, Scheibling RE (2012) Production and fate of kelp detritus. Mar Ecol Prog Ser 467:281–302
Krumhansl KAK, Okamoto DDKD, Rassweiler A, Novak M, Bolton JJ, Cavanaugh KC, Connell SD, Johnson CR, Konar B, Ling SD, Micheli F, Norderhaug KM, Pérez-Matus A, Sousa-Pinto I, Reed DC, Salomon AK, Shears NT, Wernberg T, Anderson RJ, Barrett NS, Buschmann AH, Carr MH, Caselle JE, Derrien-Courtel S, Edgar GJ, Edwards M, Estes JA, Goodwin C, Kenner MC, Kushner DJ, Moy FE, Nunn J, Steneck RS, Vásquez J, Watson J, Witman JD, Byrnes JEK (2016) Global patterns of kelp forest change over the past half-century. Proc Natl Acad Sci USA 113:13785–13790
Langton R, Augyte S, Price N, Forster J, Noji T, Grebe G, Gelais AS, Byron CJ (2019) An ecosystem approach to the culture of seaweed. NOAA technical memoirs. NMFS-F/SPO-195, 24 pp
Larson RJ, DeMartini EE (1984) Abundance and vertical distribution of fishes in a cobble-bottom kelp forest off San Onofre, California. U S Natl Mar Fish Serv Fish Bull 82:37–53
Lassaletta L, Billen G, Grizzetti B, Anglade J, Garnier J (2014) 50 year trends in nitrogen use efficiency of world cropping systems: the relationship between yield and nitrogen input to cropland. Environ Res Lett 9:105011
Laturnus F (1996) Volatile halocarbons released from Arctic macroalgae. Mar Chem 55:359–366

Laur DR, Ebeling AW, Coon DA (1988) Effects of sea otter foraging in subtidal reef communities off Central California. In: Van Blaricom GR, Estes JA (eds) The community ecology of sea otters. Springer, Berlin, pp 151–167

Leamon BM (1980) The ecology of fishes in British Columbia kelp beds. I. Barkley Sound Nereocystis beds. Fisheries development report 22. British Colombia Ministry of the Environment, Nanaimo, 100 pp

Leclerc JC, Riera P, Leroux C, Lévêque L, Laurans M, Schaal G, Davoult D (2013a) Trophic significance of kelps in kelp communities in Brittany (France) inferred from isotopic comparisons. Mar Biol 160:3249–3258

Leclerc J, Riera P, Leroux C, Lévêque L, Davoult D (2013b) Temporal variation in organic matter supply in kelp forests: linking structure to trophic functioning. Mar Ecol Prog Ser 494:87–105

Leedham EC, Hughes C, Keng F, Sturges WT (2013) Emission of atmospherically significant halocarbons by naturally occurring and farmed tropical macroalgae. Biogeosciences 10:3615–3633

Leedham EC, Phang SM, Sturges WT, Malin G (2015) Effect of desiccation on the emission of volatile bromocarbons from two common temperate macroalgae. Biogeosciences 12:387–398

Lilley S, Schiel D (2006) Community effects following the deletion of a habitat-forming alga from rocky marine shores. Oecologia 148:672–681

Limbaugh C (1955) Fish life in the kelp beds and the effects of kelp harvesting. University of California Institute of Marine Research. IMR Reference 55–59, pp 1–158

Liu D, Keesing JK, Xing Q, Shi P (2009) World's largest macroalgal bloom caused by expansion of seaweed aquaculture in China. Mar Pollut Bull 58:888–895

Liu H, Wang F, Wang Q, Dong S, Tian X (2016) A comparative study of the nutrient uptake and growth capacities of seaweeds *Caulerpa lentillifera* and *Gracilaria lichenoides*. J Appl Phycol 28:3083–3089

Lorentsen SH, Gremillet D, Nymoen GH (2004) Annual variation in diet of breeding Great Cormorants: does it reflect varying recruitment of gadoids? Waterbirds 27:161–169

Lorentsen SH, Sjøtun K, Grémillet D (2010) Multi-trophic consequences of kelp harvest. Biol Conserv 143:2054–2062

Lotze HK, Milewski I, Fast J, Kay L, Worm B (2019) Ecosystem-based management of seaweed harvesting. Bot Mar 62:395–409

Lovas SM, Torum A (2001) Effect of the kelp *Laminaria hyperborea* upon sand dune erosion and water particle velocities. Coastal Eng 44:37–63

Lüning K (1990) Seaweeds: their environment, biogeography, and ecophysiology. Wiley, New York, p 527

Maggs T (1993) Three decades of Iron age research in South Africa: some personal reflections. The South African Archaeological Bulletin 48(158):70–76. https://doi.org/10.2307/3888944

Mangialajo L, Chiantore M, Cattaneo-Vitti R (2008) Loss of fucoid algae along a gradient of urbanisation and relationships with the structure of benthic assemblages. Mar Ecol Prog Ser 358:63–74

Manley SL (2002) Phytogenesis of halomethanes: a product of selection or a metabolic accident? Biogeochemistry 60:163–180

Mann KH (1973) Seaweeds: their productivity and strategy for growth. Science 182:975–981

Mann KH (2000) Ecology of coastal waters. Blackwell, Malden

Marinho GS, Holdt SL, Birkeland MJ, Angelidaki I (2015) Commercial cultivation and bioremediation potential of sugar kelp, *Saccharina latissima*, in Danish waters. J Appl Phycol 27:1963–1973

Mcleod E, Chmura GL, Bouillon S, Salm R, Björk M, Duarte CM, Lovelock CE, Schlesinger WH, Silliman BT (2011) A blueprint for blue carbon: toward an improved understanding of the role of vegetated coastal habitats in sequestering CO_2. Front Ecol Environ 9:552–560

Meehl GA, Stocker TF, Collins WD, Friedlingstein P, Gaye AT, Gregory JM, Kitoh A, Knutti R, Murphy JM, Noda A et al (2007) Global climate projections. In: Solomon S, Qin D, Manning M, Chen Z, Marquis M, Averyt KB, Tignor M, Miller HL (eds) Climate change 2007: the physical science basis. Contribution of Working Group I to the Fourth Assessment Report of the Intergovernmental Panel on Climate Change. Cambridge University Press, Cambridge, pp 749–845

Mehrtens G, Laturnus F (1997) Halogenating activity in an arctic population of brown macroalga *Laminaria saccharina* (L.) Lamour. Polar Res 16:19–25

Miller DD, Geibel JJ (1973) Summary of blue rockfish and lingcod life histories: a reef ecology study; and giant kelp *Macrocystis pyrifera*, experiments in Monterey Bay, California. Califor Dept Fish Game Fish Bull 158:1–137

Mišurcová L (2012) Chemical composition of seaweeds. In: Kim SK (ed) Handbook of marine macroalgae. Wiley Blackwell, West Sussex, pp 173–192

Molis M, Enge A, Karsten U (2010) Grazing impact of, and indirect interactions between mesograzers associated with kelp (*Laminaria digitata*). J Appl Phycol 46:76–84

Moore PG (1971) The nematode fauna associated with holdfasts of kelp (*Laminaria hyperborea*) in Northeast Britain. J Mar Biol Assoc 51:589–604

Moore PG (1972) The kelp fauna of Northeast Britain. 1. Introduction and the physical environment. J Exp Biol Ecol 13:97–125

Moore PG (1973) The kelp fauna of Northeast Britain. 1. Multivariate classification: turbidity as an ecological factor. J Exp Biol Eco 13:127–163

Moreira D, Pires JC (2016) Atmospheric CO_2 capture by algae: negative carbon dioxide emission path. Bioresour Technol 215:371–379

Mork M (1996) Wave attenuation due to bottom vegetation. In: Grue J, Gjevik B, Weber JE (eds) Waves and nonlinear processes in hydrodynamics. Kluwer Academic Publishing, Oslo, pp 371–382

Moy FE, Christie H (2012) Large-scale shift from sugar kelp (*Saccharina latissima*) to ephemeral algae along the south and west coast of Norway. Mar Biol Res 8:309–321

Murphy ML, Johnson SW, Csepp DJ (2000) A comparison of fish assemblages in eelgrass and adjacent subtidal habitats near Craig, Alaska. Alaska Fish Res Bull 7:11–21

N'Yeurt A, Chynoweth D, Capron ME, Stewart J, Hasan M (2012) Negative carbon via ocean afforestation. Process Safe Environ Protect 90:467–474

Nellemann C, Corcoran E, Duarte CM, Valdés L, De Young C, Fonseca L, Grimsditch G (2009) Blue carbon. The role of healthy oceans in binding carbon. A rapid response assessment. In: United Nations Environment Programme, GRIDArendal.

Nelson WG (1979) Experimental studies of selective predation on amphipods: consequences for amphipod distribution and abundance. J Exp Biol Ecol 38:225–245

Neori A, Chopin T, Troell M, Buschmann AH, Kraemer GP, Halling C, Shpigel M, Yarish C (2004) Integrated aquaculture: rationale, evolution and state of the art emphasizing seaweed biofiltration in modern aquaculture. Aquaculture 231:361–391

Neori A, Troell M, Chopin T, Yarish C, Critchley A, Buschmann AH (2007) The need for a balanced ecosystem approach to blue revolution aquaculture. Environment 49:36–43

Nordeide JT, Fosså JH (1992) Diet overlap between two subsequent year classes of juvenile coastal cod (*Gadus morhua* L.) and wild and reared cod. Sarsia 77:111–117

Norderhaug KM, Christie H, Rinde E (2002) Colonization of kelp limitations by epiphyte and holdfast fauna; a study of mobility patterns. Mar Biol 141:965–973

Norderhaug KM, Fredriksen S, Nygaard K (2003) Trophic importance of *Laminaria hyperborea* to kelp forest consumers and the importance of bacterial degradation to food quality. Mar Ecol Prog Ser 255:135–144

Norderhaug KM, Christensen JD, Fossa JH, Fredriksen S (2005) Fish-macrofauna interactions in a kelp (*Laminaria hyperborea*) forest. J Mar Biol Assoc 85:1279–1286

Norton TA, Hoscock K, Kitching JA (1977) The ecology of Lough Ine. XX The *Laminaria* forest at Carrigathorna. J Ecol 65:919–941

Okey TA, Banks S, Born AF, Bustamante RH, Calvopiña M, Edgar GJ, Espinoza E, Farina JM, Garske L, Reck G, Salazar S, Shepard S, ToralGranda V, Wallem P (2004) A trophic model of a Galápagos subtidal rocky reef for evaluating fisheries and conservation strategies. Ecol Model 172:383–401

Ortega A, Geraldi NR, Alam I, Kamau AA, Acinas SG, Logares R, Duarte CM (2019) Important contribution of macroalgae to oceanic carbon sequestration. Nat Geosci 12:748–754
Paerl HW (1995) Coastal eutrophication in relation to atmospheric nitrogen deposition: current perspectives. Ophelia 41:237–259
Paerl HW, Otten TG, Kudela R (2018) Mitigating the expansion of harmful algal blooms across the freshwater-to-marine continuum. Environ Sci Technol 52:5519–5529
Palmer CJ, Anders TL, Carpenter LJ, Küpper FC, McFiggans GB (2005) Iodine and halocarbon response of *Laminaria digitata* to oxidative stress and links to atmospheric new particle production. Environ Chem 2:282–290
Paulsen S (2007) Topics on the ecological economics of coastal zones; linking land uses, marine eutrophication, and fisheries. Doctoral thesis. Swedish University of Agricultural Sciences (SLU), Uppsala
Pechsiri JS, Thomas JBE, Risén E, Ribeiro MS, Malmström M, Nylund G, Jansson A, Welander U, Pavia H, Gröndahl F (2016) Energy performance and greenhouse gas emissions of kelp cultivation for biogas and fertilizer recovery in Sweden. Sci Total Environ 573:347–355
Perkins EJ, Abbott OJ, Martin DJ (1978) A further study of the littoral macrofauna at Siddick, Cumbria 1974–1976, with particular reference to that associated with holdfasts of *Laminaria* spp. Cumbria Sea Fish Committee Sci Rep 78:1–6
Peteiro C, Freire Ó (2013) Epiphytism on blades of the edible kelps *Undaria pinnatifida* and *Saccharina latissima* farmed under different abiotic conditions. J World Aqua Soc 44:706–715
Peteiro C, Sánchez N, Martínez B (2016) Mariculture of the Asian kelp *Undaria pinnatifida* and the native kelp *Saccharina latissima* along the Atlantic coast of southern Europe: an overview. Algal Res 15:9–23
Phang SM, Keng FSL, Paramjeet-Kaur MS (2015) Can seaweed farming in the tropics contribute to climate change through emission of short-lived halocarbons. Malays J Sci 34:8–19
Prospero JM, Barrett K, Church T, Dentener F, Duce RA, Galloway JN et al (1996) Atmospheric deposition of nutrients to the North Atlantic Basin. Biogeochemistry 35:27–73
Quast JC (1968a) Fish fauna of the rocky inshore zone. In: North WJ, Hubbs CL (eds) Utilisation of kelp bed resources in southern California. California department of Fisheries Game Fisheries Bulletin, vol 139, pp 35–55
Quast JC (1968b) Estimates of the population and standing crop of fishes. In: North WJ, Hubbs CL (eds) Utilisation of kelp bed resources in southern California. California department of Fisheries Game Fisheries Bulletin, vol 139, pp 57–79
Queirós AM, Stephens N, Widdicombe S, Tait K, McCoy SJ, Ingels J, Somerfield PJ (2019) Connected macroalgal-sediment systems: blue carbon and food webs in the deep coastal ocean. Ecol Monogr 89:e01366
Raven JA, Caldeira K, Elderfield H, Hoegh-Guldberg O, Liss PS, Riebesell U, Shepherd J, Turley C, Watson A (2005) Ocean acidification due to increasing atmospheric carbon dioxide. Royal Society policy document 12/05. The Royal Society, London
Reed DC, Rassweiler A, Arkema KK (2008) Biomass rather than growth rate determines variation in net primary production by giant kelp. Ecology 89:2493–2505
Renaud PE, Løkken TS, Jørgensen LL, Berge J, Johnson BJ (2015) Macroalgal detritus and food-web subsidies along an Arctic fjord depth gradient. Front Mar Sci 2:31
Rinde EH, Christie H, Fredriksen S, Sivertsen A (1992) Ecological consequences of kelp trawling: importance of the structure of the kelp forest for abundance of fauna in the kelp holdfasts, benthic fauna and epiphytes. NINA Oppdragsme 127:1–37
Roleda MY, Hurd CL (2019) Seaweed nutrient physiology: application of concepts to aquaculture and bioremediation. Phycologia 58:552–562
Rose JM, Bricker SB, Deonarine S, Ferreira JG, Getchis T, Grant J, Kim JK, Krumholz JS, Kraemer GP, Stephenson K (2015) Nutrient bioextraction. Encycl Sustain Sci Technol:1–33
Rowe GT, Staresinic N (1979) Sources of organic matter to the deep sea benthos. Ambio Special Rep 1:19–23

Ruiz D, Wolff M (2011) The Bolivar Channel ecosystem of the Galápagos Marine Reserve: energy flow structure and role of keystone groups. J Sea Res 66:123–134
Russell BC (1977) Population and standing crop estimates for rocky reef fishes of north eastern New Zealand. N Z J Mar Fresh Res 11:23–36
Schaal G, Riera P, Leroux C (2009) Trophic significance of the kelp *Laminaria digitata*, (Lamour.) for the associated food web: a between-sites comparison. Estuar Coast Shelf Sci 85:565–572
Schaal G, Riera P, Leroux C (2012) Food web structure within kelp holdfasts (*Laminaria*): a stable isotope study. J Mar Ecol 33:370–376
Schultze K, Janke K, Kruess A, Weidermann W (1990) The macrofauna and macroflora associated with *Laminaria digitata* and *L. hyperborea* at the island of Helgoland (German Bight, North Sea). Helgol Meeresunter 44:39–51
Seghetta M, TøRring D, Bruhn A, Thomsen M (2016) Bioextraction potential of seaweed in Denmark-an instrument for circular nutrient management. Sci Total Environ 563–564:513–529
Seitz RD, Wennhage H, Bergstrom U, Lipcius RN, Ysebaert T (2014) Ecological value of coastal habitats for commercially and ecologically important species. ICES J Mar Sci 71:648–665
Shaffer S (2003) Preferential use of nearshore kelp habitats by juvenile salmon and forage fish. In: Proceedings of the Georgia Basin/Puget Sound Research Conference, 11 pp
Shears NT, Babcock RC (2003) Continuing trophic cascade effects after 25 years of no-take marine reserve protection. Mar Ecol Prog Ser 246:1–16
Shumway SE (1990) A review of the effects of algal blooms on shellfish and aquaculture. J World Aquac Soc 21:65–104
Sjøtun K, Lorentsen SH (2003) Kelp forest (*Laminaria hyperborea*) as habitat for juvenile gadoids. Poster presented at the 3rd European Phycological Congress, Belfast, North-Ireland, 21–26 July 2003
Sloane JF, Ebling FJ, Kitching JA, Lilly SJ (1957) The ecology of the Lough Ine rapids with special reference to water currents. V The sedentary fauna of the laminarian algae in the Lough Ine area. J Anim Ecol 26:197–211
Smale DA, Burrows MT, Moore P, O'Connor N, Hawkins SJ (2013) Threats and knowledge gaps for ecosystem services provided by kelp forests: a Northeast Atlantic perspective. Ecol Evol 3:4016–4038
Smith SV (1981) Marine macrophytes as a global carbon sink. Science 211:838–840
Smith VH (2003) Eutrophication of freshwater and coastal marine ecosystems – a global problem. Environ Sci Pollut Res 10:126–139
Sondak CF, Chung IK (2015) Potential blue carbon from coastal ecosystems in the Republic of Korea. Ocean Sci J 50:1–8
Steneck RS, Graham MH, Bourque BJ, Corbett D, Erlandson JM, Estes JA, Tegner MJ (2002) Kelp forest ecosystems: biodiversity, stability, resilience and future. Environ Conserv 29:436–459
Stephens JS Jr, Morris PA, Zerba K, Love M (1984) Factors affecting fish diversity on a temperate reef: the fish assemblage of Palos Verdes Point 1974–1981. Environ Biol Fish 11:259–275
Straub S, Thomsen M, Wernberg T (2016) The dynamic biogeography of the Anthropocene: the speed of recent range shifts in seaweeds. In: Hu ZM, Fraser C (eds) Seaweed phylogeography. Springer, Amsterdam, pp 63–93
Sun Y, Mao S, Wang F, Peng P, Chai P (2013) Identification of the Kukersite type source rocks in the Ordovician stratigraphy from the Tarim Basin, NW China. Chin Sci Bull 58:4450–4458
Svane I, Gröndahl F (1988) Epibioses of Gullmarfjorden: an underwater stereophotographical transect analysis in comparison with the investigations of Gislén 1926–29. Ophelia 28:95–110
Swaney DP, Hong B, Ti C, Howarth RW, Humborg C (2012) Net anthropogenic nitrogen inputs to watersheds and riverine N export to coastal waters: a brief overview. Curr Opin Environ Sustain 4:203–211
Swedish Environmental Protection Agency (2008) Ecosystem services provided by the Baltic Sea and Skagerrak. Report 5873. Swedish Environmental Protection Agency, Stockholm
Tallis H (2009) Kelp and rivers subsidize rocky intertidal communities in the Pacific Northwest (USA). Mar Ecol Prog Ser 389:85–96

Tang Q, Zhang J, Fang J (2011) Shellfish and seaweed mariculture increase atmospheric CO_2 absorption by coastal ecosystems. Mar Ecol Prog Ser 424:97–104
Teagle H, Hawkins SJ, Moore PJ, Smale DA (2017) The role of kelp species as biogenic habitat formers in coastal marine ecosystems. J Exp Mar Biol Ecol 492:81–98
The Economic Times (2005) Hey, what exactly are carbon credits? https://economictimes.indiatimes.com/hey-what-exactly-are-carbon-credits/articleshow/1212812.cms?from=mdr
Thiel M, Vásquez J (2000) Are kelp holdfasts islands on the ocean floor? Indication for temporarily closed aggregations of peracarid crustaceans. Hydrobiology 440:45–54
Thomas D (2002) Seaweeds. Life series. Natural History Museum, London, p 96
Tilman D, Lehman CL, Bristow CE (1998) Diversity-stability relationships: statistical inevitability or ecological consequence. Am Nat 151:277–282
Trevathan-Tackett SM, Kelleway J, Macreadie PI, Beardall J, Ralph P, Bellgrove A (2015) The capacity of marine macrophytes to contribute to blue carbon sequestration. Ecology 96:3043–3057
Troell M, Rönnbäck P, Halling C, Kautsky N, Buschmann A (1999) Ecological engineering in aquaculture: use of seaweeds for removing nutrients from intensive mariculture. In: Sixteenth international seaweed symposium, pp 603–611
Troell M, Joyce A, Chopin T, Neori A, Buschmann AH, Fang JG (2009) Ecological engineering in aquaculture-potential for integrated multi-trophic aquaculture (IMTA) in marine offshore systems. Aquaculture 297:1–9
van der Heijden LH, Kamenos NA (2015) Reviews and syntheses: calculating the global contribution of coralline algae to carbon burial. Biogeosciences 12:6429–6441
Vanderklift MA, Wernberg T (2008) Detached kelps from distant sources are a food subsidy for sea urchins. Oecologia 157:327–335
Varian SJA (1998) The biology and behaviour of Goldsinny wrasse, *Ctenolabrus rupestris* (L.) (Teleostei; Labridae), and their use as cleaner-fish on salmon farms in the west of Ireland. PhD thesis, National University of Ireland, Galway. 183pp
Vierros M (2017) Communities and blue carbon: the role of traditional management systems in providing benefits for carbon storage, biodiversity conservation and livelihoods. Climate Change 140:89–100
Waage-Nielsen E, Christie H, Rinde E (2003) Short-term dispersal of kelp fauna to cleared (kelp-harvested) areas. Hydrobiology 503:77–91
Wada S, Hama T (2013) The contribution of macroalgae to the coastal dissolved organic matter pool. Estuar Coast Shelf Sci 129:77–85
Walls AM (2017) Ecosystem services and environmental impacts associated with commercial kelp aquaculture, NUI Galway. PhD theses, 270 pp
Walls AM, Kennedy R, Fitzgerald RD, Blight AJ, Johnson MP, Edwards MD (2016) Potential novel habitat created by holdfasts from cultivated *Laminaria digitata*: assessing the macroinvertebrate assemblages. Aquac Environ Interact 8:157–169
Walls AM, Edwards MD, Firth LB, Johnson MP (2017) Successional changes of epibiont fouling communities of the cultivated kelp *Alaria esculenta*: predictability and influences. Aquac Environ Interact 9:55–69
West JA, Calumpong HP, Martin G, Calumpong HP, van Gaever S (2017) Kelp forests and Seagrass meadows. In: United Nations (ed) The first global integrated marine assessment: World ocean assessment I. Cambridge University Press, Cambridge, pp 869–876
Wheeler A (1980) Fish-algal relations in temperate waters. In: Price JH et al (eds) The shore environment. Academic, London, pp 667–698
Whittick A (1983) Spatial and temporal distribution of dominant epiphytes on the stipes of *Laminaria hyperborea* (Gunn.) Fosl. (Phaeophyta: Laminariales) in S.E. Scotland. J Exp Mar Biol Ecol 73:1–10
Wolff T (1962) The systematics and biology of bathyal and abyssal Isopoda Aselotta. Galathea Rep 6:1–320

Wu H, Huo Y, Zhang J, Liu Y, Zhao Y, He P (2015) Bioremediation efficiency of the largest scale artificial *Porphyra yezoensis* cultivation in the open sea in China. Mar Pollut Bull 95:289–296
Wu H, Kim JK, Huo Y, Zhang J, He P (2017) Nutrient removal ability of seaweeds on *Pyropia yezoensis* aquaculture rafts in China's radial sandbanks. Aquat Bot 137:72–79
Xiao X, Agusti S, Lin F, Li K, Pan Y, Yu Y, Duarte CM (2017) Nutrient removal from Chinese coastal waters by large-scale seaweed aquaculture. Sci Rep 7(1):46613
Xie X, Volkman JK, Qin J, Borjigin T, Bian L, Zhen L (2014) Petrology and hydrocarbon potential of microalgal and macroalgal dominated oil shales from the Eocene Huadian formation, NE China. Int J Coal Geol 124:36–47
Yang Y, Chai Z, Wang Q, Chen W, He Z, Jiang S (2015a) Cultivation of seaweed *Gracilaria* in Chinese coastal waters and its contribution to environmental improvements. Algal Res 9:236–244
Yang Y, Liu Q, Chai Z, Tang Y (2015b) Inhibition of marine coastal bloom-forming phytoplankton by commercially cultivated *Gracilaria lemaneiformis* (Rhodophyta). J Appl Phycol 27:2341–2352
Yokohama Y, Tanaka J, Chihara M (1987) Productivity of the Ecklonia cava community in a bay of Izu peninsula on the Pacific coast of Japan. Bot Magaz Tokyo 100:129–141
Yorke CE, Miller RJ, Page HM, Reed DC (2013) Importance of kelp detritus as a component of suspended particulate organic matter in giant kelp *Macrocystis pyrifera* forests. Mar Ecol Prog Ser 493:113–125
Yoshikawa T, Takeuchi I, Furuya K (2001) Active erosion of *Undaria pinnatifida* Suringar (Laminariales, Phaeophyceae) mass-cultured in Otsuchi Bay in northeastern Japan. J Exp Mar Biol Ecol 266:51–65
Zemke-White WL, Clements KD (2004) The relationship between long-term changes in algal community structure and herbivore diet at the Three Kings Islands, New Zealand. N Z J Mar Fresh Res 38:837–844
Zhang J, Fang J, Wang W, Du M, Gao Y, Zhang M (2012) Growth and loss of mariculture kelp *Saccharina japonica* in Sungo Bay, China. J Appl Phycol 24:1209–1216
Zhang J, Huo Y, Wu H, Yu K, Kim JK, Yarish C, Qin Y, Liu C, Xu R, He P (2014) The origin of the *Ulva* macroalgal blooms in the Yellow Sea in 2013. Mar Pollut Bull 89:276–283
Zhou Y, Varner RK, Russo RS, Wingenter OW, Haase KB, Talbot R, Sive BC (2005) Coastal water source of short-lived halocarbons in New England. J Geophys Res Atmos 110:D21302
Zou D (2005) Effects of elevated atmospheric CO_2 on growth, photosynthesis and nitrogen metabolism in the economic brown seaweed, *Hizikia fusiforme* (Sargassaceae, Phaeophyta). Aquaculture 250:726–735

Chapter 6
Climate Change-Induced Natural Hazard: Population Displacement, Settlement Relocation, and Livelihood Change Due to Riverbank Erosion in Bangladesh

Md. Abdul Malak, Nahrin Jannat Hossain, Mohammad Abdul Quader, Tania Akter, and Md. Nazrul Islam

Abstract Human migration is one of the worst possible effects triggered by riverbank erosion. Riverbank erosion is a frequent event in Bangladesh, with severe consequences such as land loss and forced migration. This research is intended to understand the process and pattern of human migration due to riverbank erosion at a small administrative unit of a coastal District Bhola located in the Meghna estuary in Bangladesh. This study was conducted on the basis of primary data following a mixed-methods approach. Data were collected through a questionnaire survey and focus group discussion (FGD). In this paper, we argue that riverbank erosion has a long-term impact on livelihood security of displaced people. A significant portion of migrants have changed their livelihood occupation. The displaced people have shifted their shelter to live in road-come-embankment (A road-come-embankment is a thick wall of earth that is built to carry a road or railway over an area of low ground, or to prevent water from a river or the sea from flooding the area. They climbed a steep embankment), annually leased land, and "housing without rent for humanity" provided by the local government. Displaced people have scattered all over the study area as well as nearby and adjoining higher administrative units as in-migrants. People have also migrated to large cities such as Dhaka and nearby

M. A. Malak (✉)
Department of Geography and Environment, Jagannath University, Dhaka, Bangladesh

School of Geography and Sustainable Communities, University of Wollongong, Wollongong, NSW, Australia
e-mail: amalak@geography.jnu.ac.bd

N. J. Hossain · M. A. Quader · T. Akter
Department of Geography and Environment, Jagannath University, Dhaka, Bangladesh

M. N. Islam
Department of Geography and Environment, Jahangirnagar University, Savar, Dhaka, Bangladesh
e-mail: nazrul_geo@juniv.edu

M. N. Islam, A. van Amstel, *Bangladesh II: Climate Change Impacts, Mitigation and Adaptation in Developing Countries*, Springer Climate,
https://doi.org/10.1007/978-3-030-71950-0_6

towns to seek employment. The research would be beneficial for conducting further study on forced migration on a larger scale in more dynamic riparian areas.

Keywords Riverbank erosion · Displacement · Force migration · Livelihood · Coastal Bangladesh

6.1 Introduction

Global environmental change and its associated environmental stressors are the part of human life; however, the frequency and impact are not the same for everyone. People who depend on natural resources for their livelihood are especially more vulnerable in the face of environmental changes (Adhikary et al. 2000; Shamsuddoha et al. 2011; Martin et al. 2014; IPCC 2014). Research suggests that people living on the banks of a mighty river or on the coast in many less developed and developing countries such as Bangladesh, India, Philippine, and Indonesia are significantly vulnerable owing to erosion (Hall and Bouapao 2010; Das et al. 2014; Alam 2016). In light of this concern, riverbank erosion is a frequent event in Bangladesh with severe consequences, which creates an enormous amount of land loss, population displacement, and migration every year (Islam et al. 2011; Tanjinul and Jannatul 2015). According to a recent estimation, around 9.6 million people from 29 districts of the country will be forced to migrate domestically and internationally owing to natural disasters by 2050 (The Daily Star, 30 May 2012 cited in Ishtiaq 2012). Among them, 1.9 million people will be displaced because of riverbank erosion (The Daily Star, 30 May 2012 cited in Ishtiaq 2012).

There have been numerous predictions regarding the number of people affected by displacement and migration. The countries with fast-growing populations are experiencing hardship to secure livelihoods in their homelands because of extreme natural calamities. On account of this, almost 40 million people had moved out by 2010, and this number will increase up to 200 million by the year 2050 (Perch-Nielsen et al. 2008). Of this predicted 200 million, about 26 million people of Bangladesh are at risk of increasing global warming effects (Perch-Nielsen et al. 2008). Climate variability change-induced environmental disasters have had both direct and indirect effects on human life, society, and the economy in Bangladesh (Mahbub 1997; BWDB 2005; Brown 2008; Bhuiyan et al. 2017). For instance, riverbank erosion causes huge socio-economic and environmental problems to the riparian people (Rafatullah and Hossain 2019). A significant number of riparian people lost their shelter and livelihood, which forced them to relocate to a new place and livelihood alternatives (Islam and Rashid 2011). As the coastal people of Bangladesh are dependent on farming and fishing (Rasheed 2008), erosion are particularly adversely affected in their food security as well as the livelihood of the region (Islam and Rashid 2011).

Hydraulic action in the form of the sheer physical impact of floodwaters in the river channels is a major cause of erosion of the banks (Hey 1979; Lawler et al.

1997). Riverbanks are susceptible to slumping or erosion by river currents and wave action. Riverbank erosion causes channel shifting, the creation of new channels during floods, bank slumping due to undercutting, and local scour from turbulence due to obstruction (Haque and Hossain 1988; Ahmed and Neelormi 2008; Alam et al. 2017). Bangladesh is one of the lowest riparian communities of the Ganges–Brahmaputra–Meghna (GBM) river systems of South Asia (Kartiki 2011). Every year the GBM basin carries around 1.1 billion tons of sediment (EGIS 2000; Sarker et al. 2003) because of flooding and riverbank erosion in Bangladesh (Elahi et al. 1991). Satellite images of the GBM Rivers demonstrate varying proportions of bank erosion and accretion in different years with an annual erosion rate of nearly 9000 hectares of land (Ahmed et al. 2001; Ahmed and Neelormi 2008; Alam et al. 2017). Another study by the Center for Environment and Geographic Information Services (CEGIS) based upon analysis of a 30-year long-term series of satellite images reveals that the Jamuna and Padma rivers have widened more than 3 km and destroyed about 130,000 hectares of floodplain land (Hasnat 2018).

Bangladesh is one of the largest geosynclines with a vast population and one of the most vulnerable countries to flooding and riverbank erosion (Kartiki 2011). Most of the country is comprises the combined floodplain and deltaic deposits of the Ganges, the Brahmaputra, and the Meghna Rivers system and their tributaries and distributaries (Rasheed 2008; Islam 2016). The relief of a significant part of the country is low, varying between 1 and 2 m and being fluvial-deltaic in origin. The rivers generally have shallow gradients: 4–5 cm/km for the Ganges, 6–10 cm/km for the Brahmaputra, and 3 cm/km for the Meghna (Rashid and Pramanik 1990; Islam 2016). Moreover, a huge discharge of water from the GBM river system is more significant than any other river system in the world except the Amazon and Congo (Whitehead et al. 2015). Therefore, the excessive water during monsoon season results in widespread riverbank erosion and destruction of resources, which cause enormous hardship and hinder overall development. Thus, Bangladesh is faced with natural disasters such as riverbank erosion that claim life and property in the usual manner (Ansary et al. 2001; BBS 2013).

In addition, Bangladesh is a land of more than 700 rivers, including their tributaries and distributaries (Islam and Rashid 2011). According to the Coast Trust (2007), also cited by Islam and Rashid (2011), owing to riverbank erosion around 2400 km of riverbank line and 283 locations along with the bank line, as well as 85 towns, are defenseless in Bangladesh. A study projected that every year, Bangladesh lost 50,000 to 200,000 hectares of land owing to riverbank erosion (Kniveton et al. 2012; Rafatullah and Hossain 2019).

It is visible that, every year, a large number of people affected by riverbank erosion as well as suffering financial problems. As a consequence, people are displaced or forced to leave their place of origin, losing their belongings (Rahman 2010). Besides, riverbank erosion places enormous stress on the people who reside along the riverbanks as they lose their homestead, agricultural lands, and overall agricultural production (Shamsuddoha et al. 2011; Martin et al. 2014; Mollah and Ferdaush 2015). The cumulative effects of such losses are income reduction, which forces people to relocate their settlement, and poverty (most often marginal) expenditure in food consumption, education, and health care sectors (Ellis 2003). Besides, riparian people maintain a poor livelihood status because of poor economic conditions. The situation even becomes worse and makes the community more vulnerable to riverbank erosion (Uddin and Basak 2012). Uddin and Basak 2012 studied the permanent and temporary effects on the rural economy and society of riverbank erosion on the banks of the river Arial Khan. The estimated total amount of land affected by riverbank erosion of the Arial Khan was approximately 3000 m^2, of which about 2400 m^2 was agricultural land and the rests were water bodies, homesteads, and social forest. The consequences of riverbank erosion are not only economic but also social, either directly or indirectly. For instance, around 17,911 people from 39 villages were directly affected by riverbank erosion (Islam et al. 2004). However, affected people can survive by practicing indigenous knowledge-based strategies (Abrar and Azad 2003).

A study revealed that the floodplain of Bangladesh is the most vulnerable for the riverbank erosion that causes migration (Ericksen et al. 1993). Every year, because of natural disasters, a massive number of the population migrate both permanently and temporarily (Mollah and Ferdaush 2015). Bhola Island was endured devastating flooding in 2005; consequently, roughly 0.5 million people were uprooted from their permanent settlements (Mahbub 1997; BWDB 2005; Saifuzzaman and Alam 2006; Islam 2016). Therefore, Bhola district has great significance in studying migration due to riverbank erosion.

The natural setting of Bangladesh in the South Asian subcontinent and the characteristics of tropical monsoon climate are greatly responsible for the riverbank erosion and flood hazards (Elahi et al. 1991). Recognizing that, the government of Bangladesh identified riverbank erosion as a natural disaster in 1993 (GOB 1998 cited in Islam et al. 2011). After that, a significant number of studies were conducted focusing on riverbank erosion by the academician and development practitioners. However, there exists a paucity of in-depth field study and research. This research thrives on the understanding that the riverbank erosion-induced migration of Borhanuddin Upazila in Bhola district adjoining the Meghna River. This Upazila is vulnerable to riverbank erosion for various reasons. Therefore, the main concern of this research is to study migrants, the migration process and pattern, as well as mobility behavior, livelihood changes, and management. In addition, there is also an attempt to find out the migrants' socio-economic backgrounds and their resettlement process and pattern, especially those related to riverbank erosion (Mahbub 1997; BWDB 2005; Brown 2008; Tanjinul and Jannatul 2015). Migration is not a simple phenomenon to understand and there are numerous social and physical factors working together behind it all. Moreover, the roots of migration and various other characteristics of migrants and migration have given us a strong concept about the migration behavior of the people living in those off-shore islands. As there are numerous other off-shore islands and the populations have a very high density, this study will help to policymakers to think about the future scenario for those islands too and to help the authority to plan accordingly.

6.2 Riverbank Erosion, Settlement Relocation and Livelihood Recovery

Erosion has long been a part of life in Bangladesh, which sits on a massive river delta. The Padma's rushing waters constantly shift and transform the shape of the river, eating away at its sandy banks. Deforestation, weather extremes, strong currents, and the accumulation of silt all contribute to erosion. But researchers say a warming climate is accelerating today's risks by intensifying rains and floods—sinking communities deeper into poverty. During the monsoon season, Bangladesh's approximately 24,000 km of waterways carry huge amounts of silt, sand, and murky water. According to an August 2018 report published by the NASA Earth Observatory, over 66,000 hectares (256 square miles) of land has been lost due to erosion caused by the Padma River since 1967 (Uddin and Basak 2012). Every year, at least 200,000 people become environmental refugees due to land lost from soil erosion. Such predictions should be a part of our river management strategy. Because these predictions and early warnings could minimize erosion or reduce the damage to properties. The government can take initiatives for river bank protection and also for the relocation of the local people who are supposed to be affected by river erosion. But it will only be helpful if the government makes a national strategy for river management and works based on the predictions. Moreover, river erosion is a natural disaster and the Ministry of Disaster Management and Relief should be handling the issue. It is no longer a problem for the developed countries of the world, because they had taken care of the issue long ago. They have managed their rivers in a way that the rivers do not change their course anymore. However, since it is still a major issue in Bangladesh, it needs urgent attention of the authorities concerned (CEGIS 2010).

6.3 Materials and Methods

6.3.1 Study Context

The study was conducted in Hassan Nagar Union of Borhanuddin Upazila in Bhola District. Hassan Nagar Union is a small union among nine unions under Borhanuddin Upazila in the Division of Barisal, Bangladesh. It is the southern part of Borhanuddin Upazila, with an area of 5293 acres. Its geographical location is between 22°21′ and 22°34′ north latitude and between 90°35′ and 90°51′ east longitudes. The study area is bounded by Bhola Sadar Upazila and Daulatkhan Upazilas on the north side and by Lalmohan Upazila, in the south. In addition, the eastern side is bordered by Tazumuddin Upazila, and the western side by Bauphal Upazila of Ptauakhali District (Fig. 6.1).

The total population is 20,597, 10,977 male and 9620 female. There are 3220 households in the study area. The literacy rate is 45.94% (BBS 2011). There are five primary schools, two high schools and one Madrasha. Fishing, agriculture, small

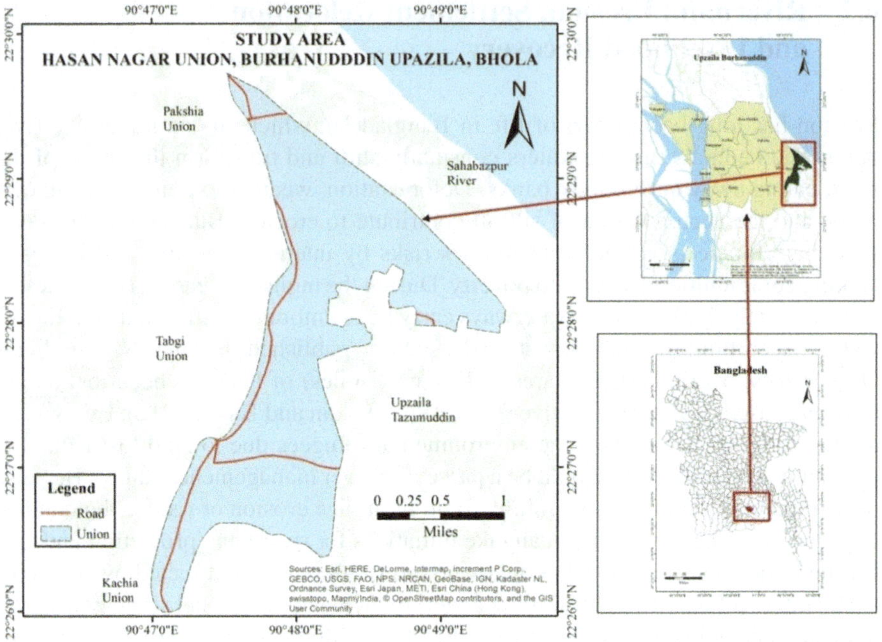

Fig. 6.1 Study area map, Hassan Nagar Union of Borhanuddin Upazila at Bhola District in Bangladesh

business, and seasonal work are the main occupations of this area. Women are mainly engaged in homestead gardening and poultry. Different types of rice are the main crops here. Coconut, betel nut, nut, palm, jackfruit, guava are the main fruits of this area. The communication system is not well developed. There are only 5 km of constructed roads; the rest of them are Kancha roads (Banglapedia 2014).

6.3.2 *Data Collection and Analysis*

In order to achieve the objectives, this research adopted a mixed research approach of geographical research for acquiring knowledge about experience, perception, and standpoint of the participant (Hammarberg et al. 2016). This descriptive type of research analyzes all the causal relationships of riparian people with riverbank erosion. In this study, the focus of the researcher was to find out about the riverbank erosion-induced displacers, migration process and pattern, new livelihood opportunity, and challenges in an offshore island in the Meghna Estuary region of Bangladesh. Both primary data and secondary sources were used in this study. Primary data were collected through semi-structured questionnaires from 100 migrated families, who were selected using the purposive sampling method. Face-to-face questionnaire surveys helped to gain both quantitative and qualitative responses from the participants. Two FGDs were conducted to gather an in-depth understanding of the pre-, during, and post-status of the affected people. One FGD was undertaken where erosion has taken place while another one was carried out in the place where the people migrated. Statistical yearbooks, journals, articles, published and unpublished theses, local administrative and various related sources were used as secondary data sources. The secondary data help to provide an understanding of the exact situation of Bangladesh, specifically the Meghna estuary, owing to riverbank erosion and migration. According to the objects and guidelines of the research, collected primary and secondary data were interpreted. At that time, all data were checked, verified, and edited to minimize errors. The data were analyzed using Statistical Package for Social Science (SPSS) and Microsoft Office Excel.

6.4 Results and Discussions

6.4.1 Socio-Economic Characteristics of the Participants

Riverbank erosion-induced migrants are helpless, and most of them are living a socio-economically marginal life compared with previous conditions. A number of problems such as loss of shelter, economic crises, loss of livelihood, and employment crises occur in their day-to-day life. As mentioned earlier, 100 migrated households were purposively selected from Hasan Nagar Union for this research. Table 6.1 depicts the demographic features of migrants in terms of age, educational, and occupational status. The average age of the participants is 40. The minimum and maximum ages of the respondents are 15 years old and 64 years old respectively. According to the 2011 Census, the number of males and females in Hasan Nagar Union is 10,977 and 9620 respectively. In this study, of the total participants, 62% are male and 38% are female. The average literacy rate of the study area was 40.5% and 34% in the census years of 2011 and 2001 respectively. From the survey, half (52%) of the respondents were literate, whereas 24% can only sign their name. In addition, 29% and 13% had completed their primary and secondary levels respectively. The main cause of this condition is poverty and unawareness concerning the importance of education. Like other parts of the country, the male literacy rate is higher than the female literacy rate in the research area. Bangladesh is an agricultural country, but in the coastal area, fishing is the main occupation. Accordingly, 26% and 12% were engaged in fishing and agriculture. Thirty-one percent of participants were living hand to mouth. Nine percent of respondents were involved in small businesses such as dispensaries and grocery shops, and 22% had other occupations such as student, service holder, primary school teacher, and housewife.

Agriculture, fishing, cattle, poultry farming, and homestead garden had been the predominant sources of income previously. Because of riverbank erosion cultivable and homestead lands were eroded, and the number of rearing cattle and poultry decreased because of the shortage of living place and the shortage of financial capital, which led to the participants having a poor income. Figures 6.2 and 6.3 show the

Table 6.1 Demographic background of the respondents

Age classification	Percentage	Educational status	Percentage	Occupational status	Percentage
				Agriculture	12
<24	13	Illiterate	28	Fishing	26
25–34	27	Literate	24	Day laborer	31
35–44	31	Primary	29	Business	9
45–54	18	Secondary	13	Others	22
55–64	11	Higher secondary	9		

Source: Field survey

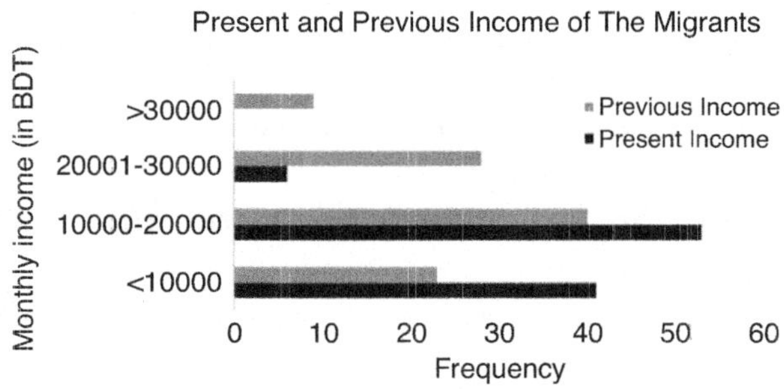

Fig. 6.2 The comparison of the proportions of present and previous monthly income

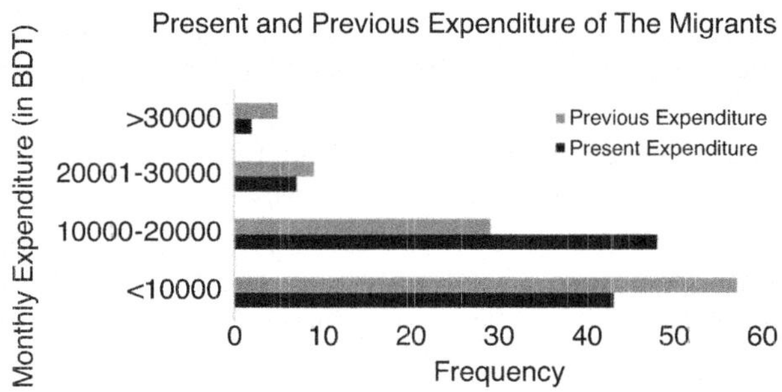

Fig. 6.3 The comparison of the proportions of present and previous monthly expenditure of the respondents

comparison of the proportions of present and previous monthly income and expenditure of the respondents. In a word, the previous condition of the respondent was better than the present economic condition. Regarding the present income, 53% of the respondents' present income is BDT10,000–20,000, whereas 40% of the respondents' previous income was within same range. Forty-one percent had a household income of less than BDT10,000 per month after migration. Only 6% had a good income ranging from BDT20,001 to 30,000 because they started a business after migration.

The majority of the respondents were unable to give information about the actual amount of income and expenditure; instead, they gave a rough estimation. Most of the respondents (48%) present monthly expenditure ranged from BDT10,000 to 20,000. About 43% of the respondents' present monthly expenditure was less than BDT10,000 BDT owing to a low income.

6.4.2 *The Banks of the Meghna: The Place of Origin*

Migration from the place of origin to the place of resettlement is one of the effective coping strategies to adopt with riverbank erosion. People do their level best to try to fight against the destructive riverbank erosion. After devastating destruction, 73% uprooted families lived temporarily either on the road or close to the bank, where some of their lands existed. The rest of the people did not stay in their place of origin, but rather moved to the study area immediately. The duration of living in their place of origin varied between 5 and 15 years. In some cases, households faced riverbank erosion more than once. In this study, over half of the migrants, around 54%, had been living at their previous place before migration for less than 5 years, whereas 29% and 17% of the total migrant respondents had been living in their place of origin for 6–10 years and 11–15 years respectively.

6.4.3 *Riverbank Erosion: Occurrence and Displacement*

Riverbank erosion is a natural process. It may occur at any time of the year, especially during rainy season. Typically, riverbank erosion is a common feature in the Meghna estuary region and has no fixed warning time. Sometimes, it takes few moments or few days to a few weeks. Figure 6.4 shows that more than half of the participants (53%) lost their shelter and land property because of riverbank erosion within a short time, which is less than 5 days (Fig. 6.5). The dwelling place and land of 27% of participants were eroded for 6–10 days. No systematic pattern for the erosion hazard has yet been observed, because of the involvement of a large number of variables in this natural process. In this regard, many hydrological studies have recorded both dramatic and gradual shifts of river channels (CEGIS 2010). The study found that 53% of the participants have been displaced owing to riverbank erosion at least three to four times, whereas only 15% and 13% had migrated 5 to 6 times and 6 to 7 times because of riverbank erosion respectively. For this reason,

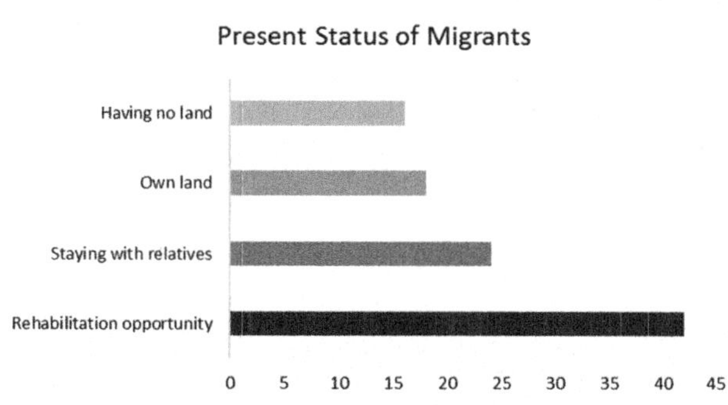

Fig. 6.4 Scenarios of occurrence and displacement by the riverbank erosion relatedmigration

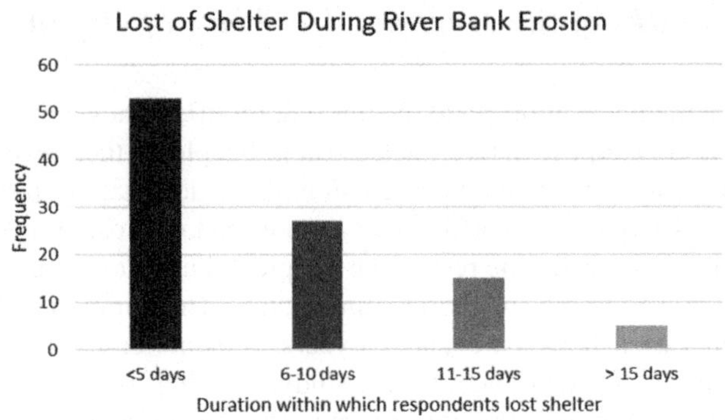

Fig. 6.5 Loss of shelter during riverbank erosion related migration in the study areas

people who are living on the cut bank[1] of the river are said to have temporary residence because of the dynamic nature of that bank.

6.4.4 Immediate Response to Riverbank Erosion

During bank erosion, the affected people tried to move their shelter elsewhere immediately. Preliminarily, their movement was around the place of origin where the bank erosion was occurring. The majority opined during the FGD that at that time they had taken shelter on the road, while a few of the participants got a rented house to live in temporarily. The second highest number of participants started living beside or near to the bank (Shamsuddoha et al. 2011; Martin et al. 2014). The rest moved to their relatives' houses. It is also revealed from the FGD that affected people could not even manage their livelihood. Most of the participants, about 62.82%, had to change their occupation after the event, whereas the rest were able to secure their former profession. The bulk of people who changed their profession became seasonal laborers and the rest chose either fishing or agricultural labor. People also identified some of the factors responsible for making them stay around the eroded riverbank. These include financial crisis, intending not to change their profession like fishing because of having a fishing net, a motor boat, and other equipment, not having any relatives to live with, not having any other land, not having any property, and not being willing to go far from their place of origin.

[1] A cut bank is a river cliff or river-cut cliff that is continually undergoing erosion.

6.4.5 *Riverbank Erosion and Migration*

Population mobility or movement includes all kind of spatial relocation, from routine daily commuting to permanent migration, which occurred over various distances, and in which the duration of moves varied from a few hours to many years (Mahubub 1997). People who are living along riverbank, are more vulnerable to riverbank erosion; therefore, they are displaced or force to migrate (Rashid 2013). In addition, CEGIS and Integrated Regional Information Networks (2008) found that, in around a year, almost 29,000 people lost their land and belongings owing to riverbank erosion. Resettlement is the process by which migrated households are to be rehabilitated because they are no longer able to stay in the area where they used to live. Every year, a large number of people in Bangladesh are made internally displaced owing to riverbank erosion. Displaced or forcefully migrated people try to manage their resettlement and livelihood, and this is their survival strategy.

6.4.6 *The Shelter for Migrants: The Place of Migrants*

Migration, either permanent or temporary, has always been one of the most important survival strategies adopted by people who face natural or man-made disasters (IOM 2008). Sometimes migration is a very common social problem and sometimes it is a blessing for the migrated or displaced family (Siddiqui 2010). After being affected by riverbank erosion, all families think about their safety, accommodation, and income sources. More than half of the respondents (56%) had been living in their present location, the study area, for less than 5 years. They migrated from their place of origin because of riverbank erosion and took shelter in the area for some reason. Among them, 42% of respondents has got the opportunity for rehabilitation (Fig. 6.4). They had been living in a colony, locally known as "Nodi Vanga Colony". This colony is authorized by the local government and is permitted only for people affected by the riverbank erosion. Around 24% of them had relatives to stay with. There are three categories of house identified where the respondents live. These are "Kancha" (made of mud and straw), "Pacca" (made out of brick, cement, and iron), and "Semi-Pacca" (made out of corrugated iron, wood, and brick). For housing purposes, people had to spend BDT5000 to BDT25,000 (Fig. 6.6), depending on their economic status. However, the money came from different sources, such as borrowing from relatives, a loan from a nongovernmental organization (NGO), and their personal savings.

People do not want to go far because of having limited accessibility in a new and unknown place. More than half of the respondents (53%) were able to manage their livelihood in their current living place. One participant stated during the FGDs: "(with a deep sigh) we lost our land and were forced to leave home because of riverbank erosion. We only know about this place; therefore, most of us tried to do something while living here to secure our livelihoods" (Kalam, 42, fisherman).

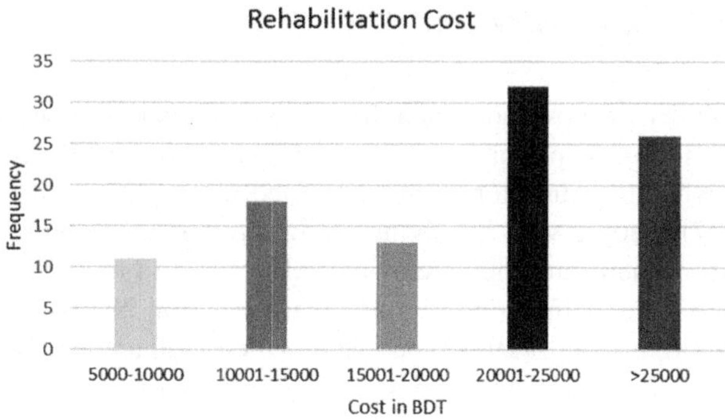

Fig. 6.6 Scenarios of rehabilitation cost depending on their economic status

Thirty-four percent of the total respondents worked adjacent to their living place. The rest 13% worked outside the administrative units, sometimes far from their living place, most preferably Dhaka city for a better livelihood.

Table 6.2 provides comparative information between the present and previous facilitated services of the study area. The facilitated services include educational services, health services, loan services, market facility, electricity services, and communication services of the present location in comparison with previously.

6.4.7 Challenges to Migration

Almost all the participants faced challenges during migration. They faced multiple challenges rather than single ones. The economic crisis was the most prominent for 81.6% of respondents during migration. The food crisis during migration was mentioned by 53.6% respondents. Besides, during migration, other problems such as moving house, transferring furniture, physical and mental stress, and transportation problems were faced by the participants and are shown in Fig. 6.7. However, 56% of the migrants obtained relief from NGOs as support. The relief provided rice, housing materials, tube wells, clothes, dry food, medicine, etc., although that was not sufficient to fulfil their demands at that time. A major portion of the migrants did not get any help from anywhere and remained unaided, supportless, shelterless, and above all helpless.

6.4.8 Livelihood Opportunity and Challenges

Migrants faced difficulties in every sphere of their survival, particularly in adopting a new livelihood strategy (Abrar and Azad 2003). Sometimes, people had to change their livelihood options to cope better with the situation. Changing livelihood options, whether positive or negative, may affect people in several ways. If it was

Table 6.2 Comparison of facilitated services between the present and the previous area

	Present locating area					Previous area				
Facilitated services	Very good	Good	Moderate	Bad	Very bad	Very good	Good	Moderate	Bad	Very bad
Education	✓					✓				
Health services			✓				✓			
Loan services (NGOs)				✓				✓		
Market facility		✓				✓				
Electricity services				✓				✓		
Communication services		✓				✓				

Source: Data compiled from FGD
NGO nongovernmental organization

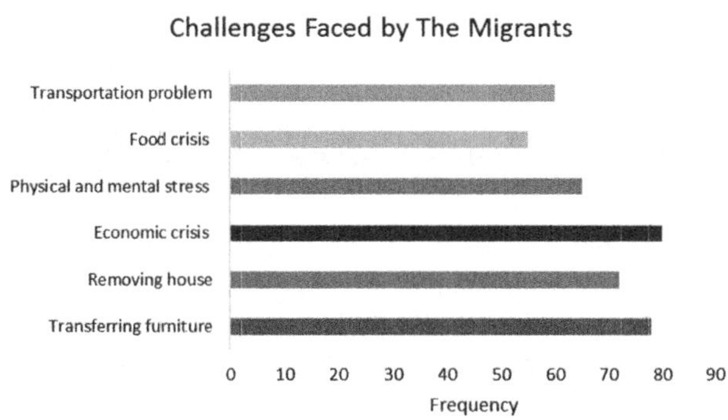

Fig. 6.7 Participants faced challenges during migration

caused by a sudden forceful event that consequently caused unwilling migration, the effect counted as severe (Islam and Rashid 2011). Loss of income compelled them to live a sub-standard life, and they could not continue their way of living the way they did before migration. Natural disaster-induced migration is different from other types of migration. They faced difficulties with finding new sources of income in a new settlement area. After riverbank erosion, the first priority is to have a safe shelter to live in, followed by a livelihood. Hasan Nagar (study area) is located beside the Meghna estuary; however, it is not safe from riverbank erosion, although people come and start living there because it is the nearest location to move to. Thirty-three percent of the respondents obtained shelter there with the help of relatives. Twenty-one percent of respondents bought land to stay on. The rest of them took shelter in the "Nodi Vanga Colony," which was provided by the local government. About 120 migrated families had been living here for 3 years.

It was found that 63% of respondents changed their occupation afterward, whereas one-third (37%) of respondents did not change their occupation because they had a fishing boat as well as their own agricultural land distant from the erosion

Table 6.3 Sources of income (pre- and post-riverbank erosion)

Main income sources of households	Post riverbank erosion (%)		Pre-riverbank erosion (%)	
	1st preference	2nd preference	1st preference	2nd preference
Agriculture	11	–	29	8
Homestead gardening	–	9	6	4
Business	17	–	18	13
Agricultural laborer	25	–	12	–
Non-agricultural laborer	28	–	9	5
Fishing	35	–	21	9
Working abroad	4	–	2	–

Note: Multiple responses considered

site. In addition, as they migrated within a short distance, near to their place of origin, the new place was more or less familiar to them. Table 6.3 illustrates the participants' livelihood status before and after migration.

6.4.9 Social and Economic Challenges Due to Migration

During displacement, participants moved from one place to another. At that time, they faced a few social as well as economic problems, which are cited in Figs. 6.8 and 6.9. Most of the participants (74.3%) claimed to have very few livelihood options. About 66.4% of respondents claimed to have a lack of social capital, i.e., having no relatives in the study area. In addition, there were other challenges including family crisis, crime intensity, and social clashes, which were mentioned by 59.5%, 48%, and 44.1% respectively (Fig. 6.8). Apart from the social challenges, a number of factors controlling economic challenges are noted. These are limited access to land for agriculture, less available land, low crop productivity, minimum profit, lack of employment, and low income opportunity. Almost one-tenth of the riverbank erosion-induced marginalized people migrated to an urban center in searching for livelihood options (Hossain 1984). A major proportion of the respondents (81.9%) claimed to have a crisis over agricultural land, whereas 39.5% declared a low profit in business. Most (78%) participants believed that limited income opportunities and limited income played a key role in their financial problems. Low crop production (52%) was another cause of their poor livelihood (Fig. 6.9). Owing to the problems mentioned above, 28% of the total participants had planned for future migration from their existing dwelling place, and 72% of the total participants had no future migration plane. They mainly choose the nearest place or where they can get land.

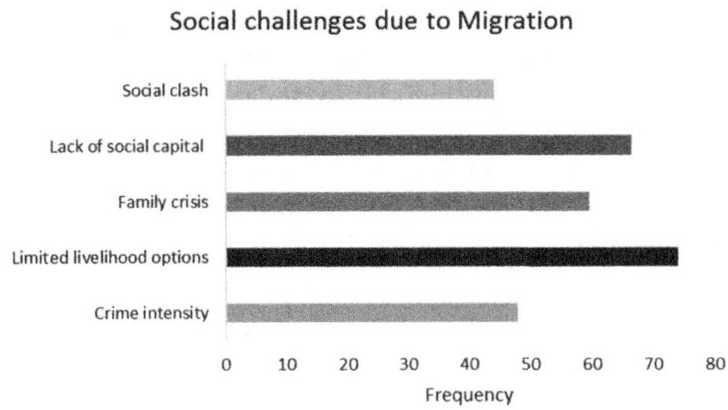

Fig. 6.8 Social challenges during displacement due to riverbank erosion and participants moved from one place to another

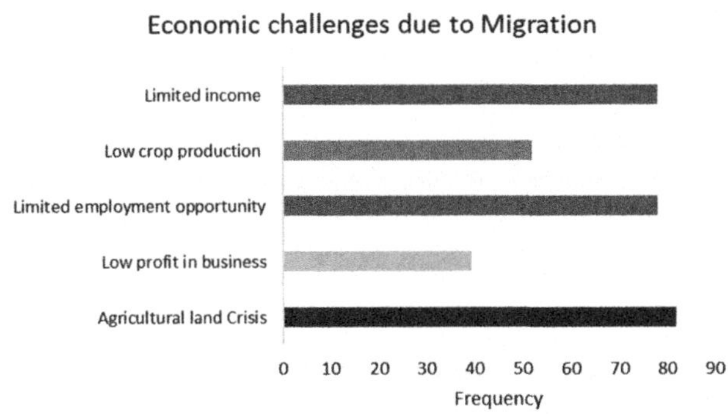

Fig. 6.9 Economic challenges during displacement due to riverbank erosion and participants moved from one place to another

6.5 Conclusion

Riverbank erosion is a successive and gradual process of loss. The consequences of riverbank erosion have long-term effects on the economic and social lives of humans. The devastating event takes place, washing away permanent settlements into the riverbed with the simultaneous loss of other sources of living such as cultivable land. The occurrence of migration due to riverbank erosion is evident in many ways, such as in material terms, economic terms, and social terms. In this study, an attempt was made to find out the migration pattern of riverbank erosion-induced migrants, their socio-economic background, their livelihood changes and management, and their challenges. Riverbank erosion is a push factor for migration; therefore, the migrants deserve special attention from the government as well as NGOs in arranging suitable and sustainable rehabilitation schemes at the places of destination. Rehabilitation schemes should be taken within rural regions as well as nearby urban centers. Despite the severe impact of erosion, a significant number of erosion victims are willing to stay in their village community and in most cases, they are reluctant to make a migratory move because they have employment opportunities in or near their villages. This is an opportunity for policy makers to create suitable livelihood schemes at the riverbank erosion sites so that those affected can stay with their families, kith and kin, and friends, because rehabilitating erosion-induced migrants away from their villages of origin requires a lot of money and resources to make proper resettlement schemes along with new livelihood provision.

References

Abrar CR, Azad SN (2003) Copping with displacement: riverbank erosion in Northwest Bangladesh. RMMRU, RDS and the North Bengal Institute, Dhaka/Rangpur

Adhikary KD, Ahmad QK, Malla SK, Pradhan BB, Rangachari R, Rasheed KBS, Verghese BG (eds) (2000) Corporation on eastern Himalayan rivers: opportunities and challenges. Konark Publishers Private Ltd/Unanayan Parishad, Delhi/Dhaka

Ahmed AU, Neelormi S (2008) Climate change, loss of livelihoods and forced displacements in Bangladesh: whither facilitated international migration. Campaign for Sustainable Rural Livelihoods and Centre for Global Change, Dhaka. Retrieved 12 Nov 2012, 12 p

Ahmed QK, Biswas AK, Rangchari RAND, Sainju MM (2001) Ganges-Brahmaputra-Meghna region: a framework for sustainable development. The University Press Ltd, Dhaka

Alam G (2016) An assessment of the livelihood vulnerability of the riverbank erosion hazard and its impact on food security for rural households in Bangladesh. University of Southern, Queensland

Alam GM, Alam K, Mushtaq S, Clarke ML (2017) Vulnerability to climatic change in riparian char and riverbank households in Bangladesh: implication for policy, livelihoods and social development. Ecol Indic 72:23–32

Ansary MNA, Rahman MM, Paul S (2001) Impact of riverbank erosion on human: case study of four unions in Faridpur District. Bhugol O Parivesh 1:113–125

Banglapedia (2014) Encyclopedia of Bangladesh

BBS (2011) Bangladesh Bureau of Statistics

BBS (2013) Population and housing census. Community Report, Bhola Zila, Statistics and Information Division, Bangladesh Bureau of Statistics. Ministry of Planning. Government of the Peoples Republic of Bangladesh

Bhuiyan MAH, Islam SDU, Azam G (2017) Exploring impacts and livelihood vulnerability of riverbank erosion hazard among rural household along the river Padma of Bangladesh. Environ Syst Res 6(1):25

Brown O (2008) Migration and climate change. IOM, Geneva

BWDB (2005) Rivers in Bangladesh (in Bangla). Bangladesh Water Development Board, Dhaka

Das TK, Haldar SK, Gupta ID, Sen S (2014) River bank erosion induced human displacement and its consequences. Living Rev Landsc Res 8(3):1–35

CEGIS-Centre for Environmental and Geographical Information Services (2010) Identification mission K2K, water mondiaal Delta Alliance reconnaissance visit to Dhaka

Elahi KM Ahmad QS, Mafizuddin M (1991) Riverbank erosion, flood and population displacement in Bangladesh. Riverbank erosion impact study. Jahangirnagar University, Savar, Dhaka

Ellis F (2003) A livelihoods approach to migration and poverty reduction. Department for International Development (DFID), Norwich

EGIS-Environmental Geographical Information Services (2000) Environmental baseline of Gorai river restoration project—environment and GIS support project for water sector planning

(EGIS-II). Ministry of Water Resources, Government of Bangladesh and Government, The Netherlands, pp 1–150

Ericksen NJ, Ahmad QK, Chowdhury AR (1993) Socio-economic implications of climate change for Bangladesh. Bangladesh Unnayan Parishad, Dhaka, p 1

Hammarberg K, Kirkman M, de Lacey S (2016) Qualitative research methods: when to use them and how to judge them, vol 31. Human Reproduction, Oxford, England. https://doi.org/10.1093/humrep/dev334

Hall D, Bouapao L (2010) Social impact monitoring and vulnerability assessment: regional report. Mekong River Commission, Vientiane, vol 7

Haque CE, Hossain MZ (1988) Riverbank erosion in Bangladesh. Geogr Rev 78(1):20–31

Hasnat MA (2018) Over 160,000 hectares lost to Padma, Jamuna in four decades. Dhaka Tribune, 12 October 2018, Nation

Hey RD (1979) Dynamic process-response model of river channel development. Earth Surf Process 4(1):59–72

Hossain MZ (1984) Riverbank erosion and population displacement: a case of Kazipur in Pabna. Unpublished thesis in Department of Geography and Environment, Jahangirnagar University, Savar, Dhaka

IOM (2008) Migration, development and environment. International Organization of Migration, Geneva

IPCC, CC (2014) Synthesis report. Contribution of Working Groups I, II and III to the fifth assessment report of the Intergovernmental Panel on Climate Change, vol 138

Islam SN (2016) Deltaic floodplains development and wetland ecosystems management in the Ganges–Brahmaputra–Meghna Rivers Delta in Bangladesh. Sustain Water Resour Manag 2:237–256. https://doi.org/10.1007/s40899-016-0047-6

Islam A et al (2011) Antagonistic Gcn5-Hda1 interactions revealed by mutations to the anaphase promoting complex in yeast. Cell Div 6(1):13

Islam MF, Rashid AB (2011) Riverbank erosion displaces in Bangladesh: need for institutional response and policy intervention. Bangladesh J Bioeth 2(2):4–19

Islam M, Lucky N, Islam MR, Ahad A, Das BR, Rahman MM, Siddiui MSI (2004) Haematological parameters of Fayoumi, Assil and local chickens reared in Sylhet region in Bangladesh. Int J Poult Sci 3(2):144–147

Kartiki K (2011) Climate change and migration: a case study from rural Bangladesh. Gend Dev 19(1):23–38

Kniveton DR, Smith CD, Black R (2012) Emerging migration flows in a changing climate in dryland Africa. Nat Clim Chang 2:444–447

Lawler DM, Couperthwaite J, Bull LJ, Harris NM (1997) Bank erosion events and processes in the Upper Severn basin. Hydrol Earth Syst Sci Discuss 1(3):523–534

Mahbub AQM (1997) Mobility behaviour of working people in Bangladesh, Dhaka. Urban Studies Forum (USP), Department of Geography and Environment, University of Dhaka

Martin M, Billah M, Siddiqui T, Abrar C, Black R, Kniveton D (2014) Climate-related migration in rural Bangladesh: a behavioural model. Popul Environ 36(1):85–110

Mollah TH, Ferdaush J (2015) Riverbank erosion, population migration and rural vulnerability in Bangladesh (a case study on Kazipur Upazila at Sirajgonj District). Environ Ecol Res 3(5):125–131

Perch-Nielsen SL, Bättig MB, Imboden D (2008) Exploring the link between climate change and migration. Clim Chang 91(3–4):375

Rafatullah NI, Hossain K (2019) Socio-economic vulnerability of riverbank erosion of displacees: case study of coastal villages in Bangladesh. Indian J Ecol 46(1):34–38

Rahman MR (2010) Impact of riverbank erosion hazard in the Jamuna floodplain areas in Bangladesh. J Sci Found 8(1–2):55–65

Rashid MM (2013) Lives and livelihoods of riverbank erosion displacees in Bangladesh: need for protection framework. J Intern Displacement 3(1):19–35

Rasheed KBS (2008) Bangladesh-resource and environmental profile. AHD Publishing House, Dhaka, Bangladesh

Saifuzzaman M, Alam S (2006) Erosion induced hazard assessment of the Brahmaputra (Jamuna) river floodplain using remote sensing & GIS data. In: 36th COSPAR Scientific Assembly, vol 36

Sarker MH, Hugue I, Alam M, Koudstaul R (2003) Rivers, chars and char dwellers of Bangladesh. International Journal of River Basin Management 1(1):61–80

Shamsuddoha M, Khan SMMH, Raihan S, Hossain T (2011) Displacement and migration from climate hot-spots in Bangladesh: causes and consequences. Center for Participatory Research and Development and Action Aid Bangladesh, Dhaka

Siddiqui T (2010) Climate change and human insecurity: migration as an adaptation strategy. Refugee and migratory movement research unit. Dhaka University, Bangladesh

Slingerland K, Bons K, Driel WV (Eds) (2010, June 20–24) Bangladesh, Mission report (draft). Wageningen University and Research Centre, The Netherlands, pp 1–22

Tanjinul HM, Jannatul F (2015) Riverbank Erosion, population migration and rural vulnerability in Bangladesh (a case study on Kazipur Upazila at Sirajgonj district). Environ Ecol Res 3(5):125–131

Uddin AFMA, Basak JK (2012) Effects of riverbank Erosion on livelihood. Unnayan Onneshan, Dhaka, Bangladesh, Dhaka

Whitehead PG, Barbour E, Futter MN, Sarkar S, Rodda H, Caesar J, Butterfield D, Jin L, Sinha R, Nicholls R, Salehin M (2015) Impacts of climate change and socio-economic scenarios on flow and water quality of the Ganges, Brahmaputra and Meghna (GBM) river systems: low flow and flood statistics. Environ Sci Process Impacts 17(6):1057–1069

Chapter 7
Impact of Climate-Induced Cyclonic Disaster on Sustainable Livelihood Assets and Its Relevance to Household Food Security

Tasnim Jahan, Nazmun Nahar, and Asib Ahmed

Abstract The food security of any household largely depends on the security or sustainability of livelihoods. The livelihood security of the households of the south-western coastal region of Bangladesh is immensely affected by various climatic conditions and disasters. The present research is intended to conduct a very in-depth study on the impact of cyclonic disaster on all types of livelihood capitals (i.e., natural, physical, human, social, and financial capital) and the relevance of such impacts to the food security conditions of the study area. The results of the present study show that the percentages of livelihood related to food production are decreasing because of a cyclonic disaster-induced unfavorable environment in the study area. The decreased employment opportunities reducing the income level also lowered the purchasing capacity for food and other daily basic necessities. Thus, switching occupation or migrating to other places seems to be the only option to them to survive. Such a change of occupation and land use that are related to food production would decrease the food supply of the area. In order to maintain sustainable livelihood and food security for households of the study area the long-term disaster resilience plan and policies should be taken and implemented for vulnerable coastal areas such as Shyamnagar Upazila.

Keywords Climate change · Coast · Disaster · Food security · Livelihood

T. Jahan · N. Nahar · A. Ahmed (✉)
Department of Geography and Environment, University of Dhaka, Dhaka, Bangladesh
e-mail: nazmun.geoenv@du.ac.bd; asib01geo@du.ac.bd

M. N. Islam, A. van Amstel, *Bangladesh II: Climate Change Impacts, Mitigation and Adaptation in Developing Countries*, Springer Climate,
https://doi.org/10.1007/978-3-030-71950-0_7

7.1 Introduction

Bangladesh is recognized as one of the most disaster-prone countries of the world owing to the situation arising from its distinct geographical setting, climatic conditions, and socio-economic circumstances of its people. Bangladesh is located in the humid tropics. In the north of the country are the Himalayas, and in the south, a funnel-shaped coast touching the Bay of Bengal (BoB). This unique geography helps to bring the vital monsoons that support the agriculture of the country, but unfortunately also makes the country vulnerable to devastating disasters, such as cyclones with concurrent storm surges, tornadoes, and floods, almost every year. Among these natural disasters, cyclones associated with storm surges are the most catastrophic for the people of Bangladesh, especially for those who live in the coastal zone of the country.

The frequent occurrences of tropical cyclones (TCs) are a notable climatic phenomenon in the BoB region (Das et al. 2016) (Fig. 7.1). Each year, about four TCs occur in the area, which shares about 5% of the total number of TCs around the world (Alam et al. 2003). Records show that a total number of 20 of the deadliest TCs took place around the world of which the BoB region experienced the highest number (i.e., 14) (Longshore 2008). However, theories proposed by Yu and McPhaden (2011) and Vissa et al. (2013) suggest that the role of warm oceans might be important in the process of forming and intensifying TCs in the region (Yesubabu

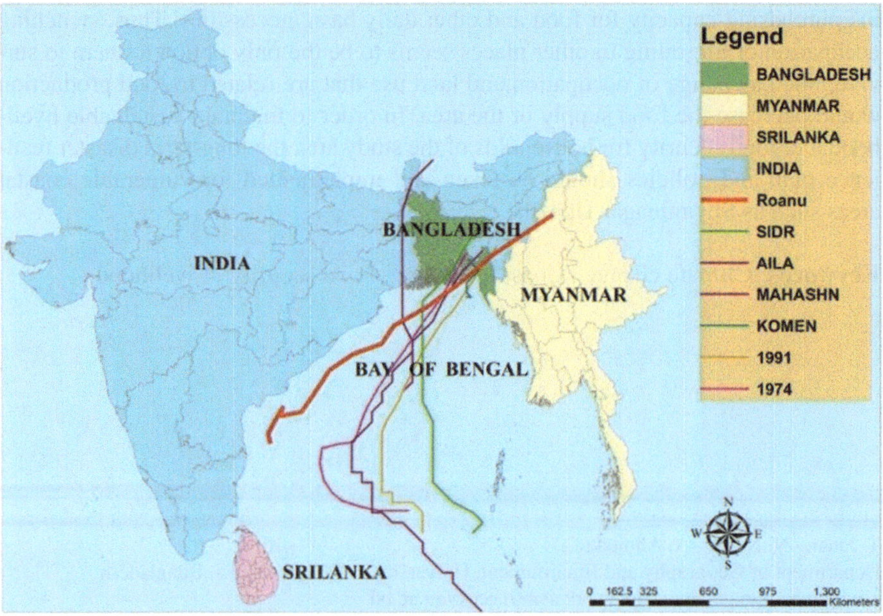

Fig. 7.1 The tracks of major cyclones (including Aila) in the BoB region. The landfall of Aila was very close to the Satkhira district of Bangladesh. (Adapted from Mita et al. 2018)

et al. 2019). The condition of high sea surface temperature (SST) was reported as an indicator of warm-ocean and consequent origin and strengthening of TCs in the region (Wada 2009). The stratified upper ocean is responsible for a shallow mixed layer (<30 m) and warmer (>28 °C) SST in the water regime of the region. In addition, the stronger winds of TCs make the mixed layer deeper in the BoB region (Bender et al., 1993). This condition is highly favorable for the formation of cyclones in the region (Prakash and Pant 2016).

Previous works identified the decreasing trend of the frequency of TCs worldwide. However, there is an increasing trend in the intensity of TCs that was identified for the BoB region during the pre- and post-monsoon seasons for the 1970s and 1980s (Gupta et al. 2019). This increasing trend of TCs in the BoB region could be attributed to the impacts of climate change (Gupta et al. 2019). The works of Emanuel et al. (2004), Persing and Montgomery (2003), and Montgomery et al. (2006) justified the increase in intensity and frequency of TCs under warmer climates (Sarthi et al. 2015). Furthermore, the outputs of climate models suggest an increase in the intensity of TCs in the BoB region in the future (Gupta et al. 2019). However, the coastal region of the BoB is highly vulnerable to cyclones and associated storm surges because of its dense population and the existence of a number of cities in the area (Yesubabu et al. 2019). The devastating effects of TCs in the coastal area of Bangladesh have already posed threats to the lives and livelihood of the coastal communities (Gupta et al. 2019), which ultimately lead to food insecurity (Habiba et al. 2015). The recurring cyclones in the BoB mostly affect the coastal areas of Bangladesh (Fig. 7.2) by taking away human lives and the lives of cattle, by damaging infrastructures (buildings, roads, and other communication facilities), washing away crops, damaging trees, destroying livelihood opportunities, etc. Cyclones make lives more vulnerable by giving rise to food insecurity, drinking water crises, lack of shelter, sanitation problems, waterborne diseases, and many other problems.

7.2 Background of the Study

7.2.1 Sustainable Livelihood

A person's livelihood simply refers to the means of income that enables a person to fulfill his basic needs and those of his family. According to Chamber and Conway (1991), "A livelihood comprises people, their capabilities and their means of living, including food, income and assets. Tangible assets are resources and stores, and intangible assets are claims and access. A livelihood is environmentally sustainable when it maintains or enhances the local and global assets on which livelihood depend, and has net beneficial effects on other livelihoods. A livelihood is socially sustainable which can cope with and recover from stress and shocks, and provide for future generations." Scoones (1998) stated that, "The term 'sustainable

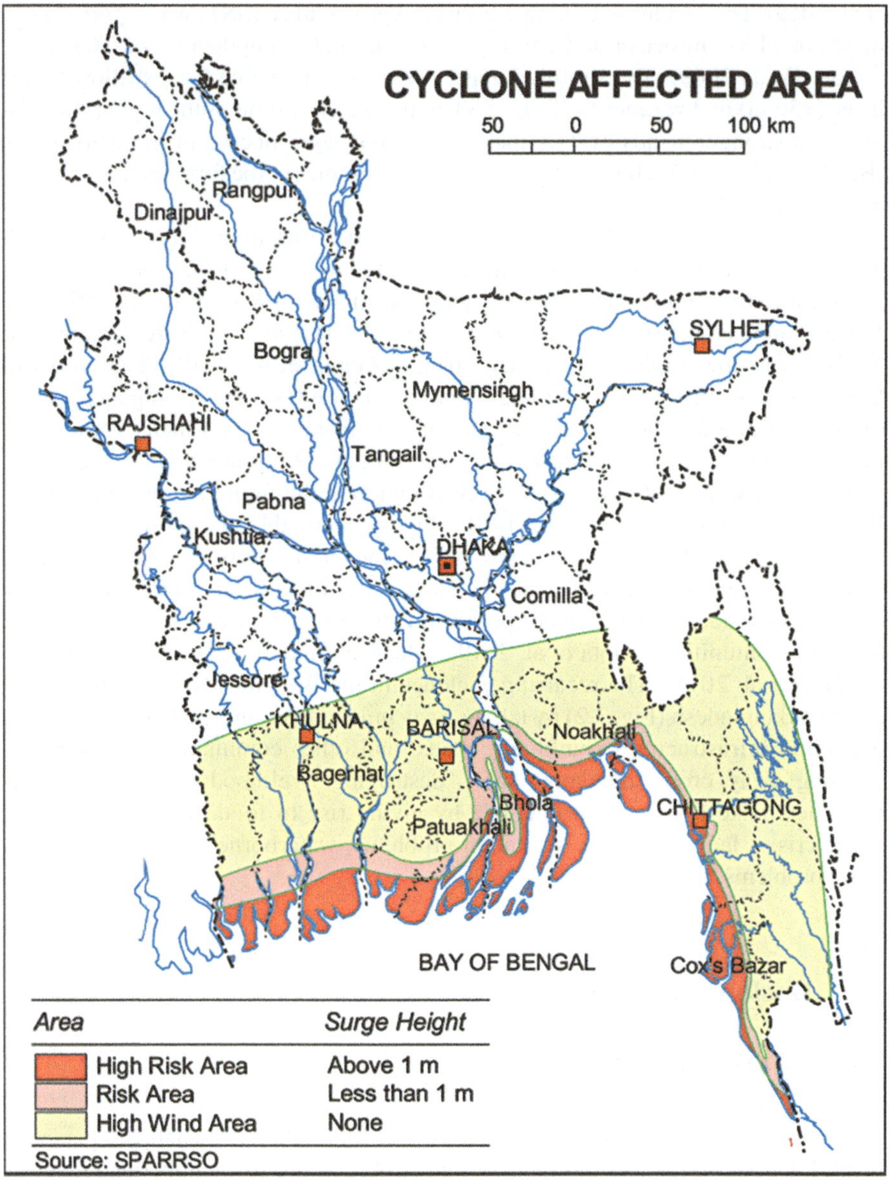

Fig. 7.2 The categories of cyclonic risks and associated surge heights in the coastal area of Bangladesh. (Adapted from CDMP 2006)

livelihoods' relates to a wide set of issues which encompass much of the broader debate about the relationships between poverty and environment." Scoones (1998) identified four types of capital that are necessary for pursuing the sustainable livelihood: natural, financial, human, and social capital. According to Scoones (1998), natural capital includes natural resource stocks (soil, water, air, genetic resources, etc.) and environmental services (hydrological cycle, pollution sinks, etc.); financial capital includes cash, credit/debit, savings, other economic assets, basic infrastructures, production equipment, and technologies; human capital includes skills, knowledge, ability to labor, good health and physical capability; social capital includes networks, social claims, social relations, affiliations, associations, etc.

According to the Department for International Development (DFID) (2000), "A livelihood comprises the capabilities, assets and activities required for a means of living. A livelihood is sustainable when it can cope with and recover from stresses and shocks and maintain or enhance its capabilities and assets both now and in the future, while not undermining the natural resource base."

The DFID (1999) also stated in its Sustainable Livelihood Guidance Sheets, "Sustainability has many dimensions, all of which are important to the sustainable livelihoods approach.

Livelihoods are sustainable when they:

- Are resilient in the face of external shocks and stresses
- Are not dependent upon external support (or if they are, this support itself should be economically and institutionally sustainable)
- Maintain the long-term productivity of natural resources
- Do not undermine the livelihoods of, or compromise the livelihood options open to, others.

The DFID (1999) includes five types of capital in their sustainable livelihood frameworks and these five capitals are human capital, natural capital, physical capital, social capital, and financial capital (Fig. 7.3).

Thus, a community's livelihood can be regarded as sustainable if it holds the long-term capacity of providing income or resources that help the community to

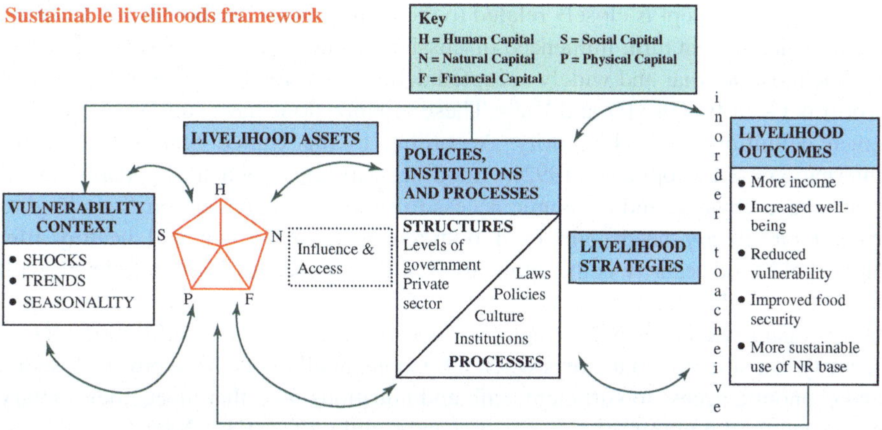

Fig. 7.3 Sustainable livelihoods framework according to the DFID (1999). (Adapted from Knutsson 2006)

meet their basic needs, even after shocks (disasters, economic ups and downs, etc.) and under stress. There are five basic types of capital for sustainable livelihood: natural, physical, human, financial, and social capital. One of these types of capital or a combination of two or more types of capital can provide a household with livelihood sustainability.

7.2.2 Understanding the Concept of Food Security

The concept of "food security" originated in the mid-1970s, when the World Food Conference (1974) defined food security in terms of food supply – assuring the availability and price stability of basic foodstuffs at the international and national level: "Availability at all times of adequate world food supplies of basic foodstuffs to sustain a steady expansion of food consumption and to offset fluctuations in production and prices." (FAO 2003a).

Concerns about food security up to the end of the 1970s were directed more at the national and international level, and concerned the ability of countries to secure adequate food supplies. Only later did the level of analysis shift to include a focus on food security at a local level, even down to households and individuals. In 1986, the World Bank report named "Poverty and Hunger" focused on the temporal dynamics of food insecurity. It introduced the distinction between "chronic food insecurity," which is associated with problems of continuing or structural poverty and low incomes, and "transitory food insecurity," which involves periods of intensified pressure caused by natural disasters, economic collapse, or conflict. As a result the concept of food security was further elaborated in terms of: "access of all people at all times to enough food for an active, healthy life." (FAO 2003b). The 1994 UNDP Human Development Report promoted the construct of human security, including a number of component aspects, and food security is one of those aspects. This concept is closely related to the human rights perspective on development, which has, in turn, influenced discussions about food security (FAO 2003b).

The most familiar and widely accepted definition of food security of the present world is given by USAID and FAO. These organizations have similarities in their given definition of food security. According to the United States Agency for International Development (1992), "food security exists when all people, at all times, have physical and economic access to sufficient, safe and nutritious food to meet their dietary needs and food preferences for an active and healthy life. Household food security is the application of this concept to the family level, with individuals within households as the focus of concern." (FAO 2003a, b). This definition is again redefined in The State of Food Insecurity in the World 2001, "Food security is a situation that exists when all people, at all times, have physical, social and economic access to sufficient, safe and nutritious food that meets their dietary needs and food preferences for an active and healthy life" (FAO 2002).

This widely accepted definition points to the following four dimensions of food security:

(a) Food availability: food availability refers to the availability of adequate amounts of food of the required quality, supplied through native or domestic production and imports (including food aid). It is defined by USAID (1992) as when "sufficient quantities of appropriate, necessary types of food from domestic production, commercial imports, commercial aid programs, or food stocks are consistently available to individuals or within their reach" (Woller et al. 2011).
(b) Food access: food access refers to the individual's access to resources to collect better quality food for a nutritious diet. "Individuals obtain food through: (i) own food production and consumption, (ii) purchases in the market place, or (iii) in-kind transfers or loans from relatives, members of the community, the government, or foreign donors" (USAID, 2010). It is defined by USAID (1992) as when "individuals have adequate assets or incomes to produce, purchase, or barter to obtain levels of appropriate foods needed to maintain consumption of an adequate diet or nutrition level" (Woller et al. 2011).
(c) Food utilization: food utilization refers to the individual's actual amount of food consumption, food preparation process, storage methods, nutritional knowledge and individual's health condition. Healthy diet, healthy body, and healthy physical environment, availability of safe drinking water, and hygienic sanitary conditions are all part of proper food utilization. It is defined by USAID (1992) as when "Food is properly used; proper food processing and storage techniques are used; adequate knowledge of nutrition and child care techniques exist and are applied; and adequate health and sanitation services exist" (Woller et al. 2011).
(d) Food stability: "Food stability is the fourth component of food security that cuts across the other three. Stability refers to the temporal dimension or time-frame of food security as implied by the wording 'at all times' in the USAID definition of food security" (USAID 2010). USAID defined food stability as "the ability to access and utilize appropriate levels of nutritious food over time" (Woller et al. 2011).

On the basis of the duration of food insecurity, two general types of food insecurity are found:

(i) Chronic food insecurity: chronic food insecurity is a long period of crisis situation where people become unable to obtain enough food to meet their food needs. The reasons behind this type of food insecurity could be long-term poverty, scarcity of assets, and insufficient ownership of productive or financial resources.
(ii) Transitory food insecurity: transitory food insecurity is a situation where a food crisis occurs for a short period of time owing to sudden shocks or stresses. Transitory food insecurity is divided into the following two types:

    (a) Cyclical or seasonal food insecurity: people can stay alert and easily survive against this type of food insecurity because it is predictable and occurs on a routine basis. For example, if in a region, there is a certain season or month (e.g., lean period) when the food production is at the lowest compared with other seasons and as a result food insecurity occurs in that particular season, then it is seasonal food insecurity. Many food security analysts consider seasonal food insecurity as chronic food insecurity as it is usually predictable and follows a sequence of known events, but is still ineradicable.

(b) Temporary food insecurity: temporary food insecurity occurs for a very brief or limited period of time and this is generally due to unpredictable phenomena or circumstances, such as natural disasters (e.g., cyclones).

Transitory and chronic food insecurity are closely linked. The recurrence of transitory food insecurity in a region can perish the assets or properties of the region and ultimately lead it to the edge of chronic food insecurity.

7.2.3 Nexus Between Sustainable Livelihood and Food Security

Sustainable livelihood can be seen as a precondition of food and nutrition security in a household. Food and nutritional security are subsets of livelihood security (FAO 2002, 2003a, b). "Food security will be achieved when equitable growth ensures that the poor and vulnerable have sustainable livelihoods" (Woller et al. 2011). Hence, households can be said to be food and nutritionally secure when their livelihoods are secure. But the recurrence of cyclonic disasters in Bangladesh destroys the livelihood sustainability of the inhabitants of the coastal region and creates food insecurity and a shortage of safe drinking water.

As the sources of livelihood of the cyclone-affected areas get damaged instantly by the sudden force of the cyclone and also gradually through saline water intrusion into the productive land and fresh water sources; thus, the natural resources-dependent people of cyclone areas lose a major proportion of their resources and become unemployed, which ultimately leads to food and fresh water shortages, homelessness, sanitation problems, health issues, etc. The sudden damage of natural capital caused by a cyclone gives rise to emergency or transitory food insecurity in a region, whereas the gradual loss of natural capital and recurrence of such disasters give rise to the possibility of chronic food insecurity in the future (Fig. 7.4).

After the occurrence of cyclones, the local markets usually become dysfunctional and inaccessible for a while and worsens the food crisis. Even when the water moves away from the road, the transport costs increase owing to damage to local roads and consequently, the price of food products increases. The loss of physical capital increases food insecurity and creates prolonged dependency on relief support. Social capital mostly includes help from neighbors, relatives, and association with various social organizations. But the neighbors and in most cases, the relatives, are also affected by cyclonic disasters and lose their resources. As a result, there is almost no chance or very little opportunity of getting help from neighbors and relatives.

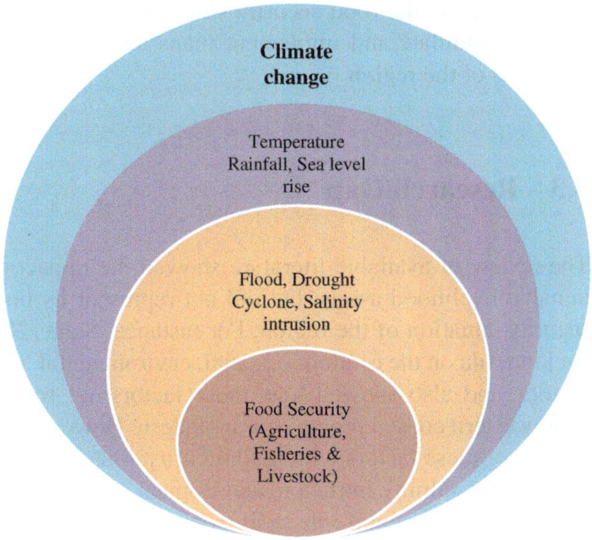

Fig. 7.4 The relationships between climate change due to cyclones and food security. (Adopted from Habiba et al. 2015)

Every cyclonic disaster poses a great threat to the livelihood opportunities of the inhabitants of coastal areas in Bangladesh. Even the previous two big cyclones that occurred in Bangladesh named Sidr (2007) and Aila (2009) inflicted great damage on coastal livelihoods. These cyclones also had a worst impact on local food security. Cyclone Aila was less intense than the super cyclone Sidr, but it inflicted a great threat to the livelihood opportunities and thus to the economy of the affected areas. The most significant losses that occurred because of cyclone Aila included damage to standing crops, damage to trees, loss of livestock and poultry, overflowing shrimp farms, loss of homestead vegetables, loss of ecosystem services from mangrove forests, loss of fresh water sources, etc. For instance, in Satkhira district alone, a total of 812 ha of croplands and 26,028 ha of shrimp farm were damaged (District Damage Assessment Report 2009) and a total of 15,004 livestock were killed because of cyclone Aila (Ministry of Food and Disaster Management 2009). Owing to the impact of devastating cyclone AILA, the inhabitants of the southwestern coastal region of Bangladesh lost their physical assets, which used to support their livelihood sustainability, such as loss of agricultural tools, loss of fishing boats and nets, and loss of nearby grocery shops. Most of the road network was disrupted and went under water because of the cyclone and its associated storm surge, and as a result the distribution of food and other necessary basic aid was hampered in the region.

Although the food insecurity derived from the loss of sustainable livelihood assets after cyclones seems to be transitory, the frequent occurrence of such disasters and the initiation of certain gradual processes after cyclones, such as salinity intrusion into the soil and fresh water, can lead to chronic food insecurity. Thus, it is important to find out the process and extent of damage to sustainable livelihood assets that happened because of cyclonic disasters and to understand how this damage affects the food security situation of the region. This study can afterward help to formulate and implement plans and policies to eradicate food insecurity problems of the region.

7.3 Research Gap

The review of available literature showed the impact of climatic disasters on sustainable livelihood assets, but did not represent its linkage to the vulnerable food security situation of the region. For instance, Saha (2015) explained the impact of cyclone Aila on the economic, social, environmental, institutional, and geographical factors, and also showed how these factors are framing the vulnerability in the southwestern coastal region of Bangladesh. However, the work did not establish the linkage of these factors with food security. Shameem et al. (2014) showed how land use change, salinity intrusion, and tropical cyclones are damaging the five types of capital (environmental, physical, social, human, and financial) of the sustainable livelihood approach. But the study was not concerned about the food security issue of the region. Mallick et al. (2011) tried to establish how the availability of sufficient and developed infrastructure can save lives and livelihoods, reduce food and water shortage, etc. However, this work does not focus on other assets of the sustainable livelihood approach, as the authors did not aim to study from that perspective. Hossain et al. (2012) explained how climate change is increasing the intensity of coastal flooding, drought, extreme weather events and other climatic disasters and how these disasters are affecting the natural resource base, assets owned by inhabitants, and their livelihoods. The aim of this paper was not to find the impact of damaged livelihood assets on food security situation. But the authors succeeded in attaining the goals that they were determined to achieve. Even the study by Garai (2014) focused on the impact of various climate change-induced disasters such as flooding, tidal surge, cyclones, and soil salinity on the livelihood opportunities, health problems, sanitation crisis, etc. He also explained the impact of such disasters on access to food, livestock, and natural resources, crop production, etc., but in a very brief context. Rahman et al. (2014) focused on how decreased income levels after natural disasters are reducing the food purchasing capacity of people and thus leading to food-insecure conditions. However, the authors did not take other assets of sustainable livelihoods into consideration that also affect the food security situation of the households. Some studies (Moslehuddin et al. 2015; Rabbani et al. 2015) only worked on the basis of secondary data. But they did not aim to explain the link between climate change-induced unsustainable livelihood situation and vulnerable food security conditions. Bala and Hossain (2010) used both primary and secondary databases in their paper and aimed to find out the food security status and ecological capacity of selected coastal areas, and succeeded in fulfilling that aim. Although this study gives an insight into the food security conditions of the study area, the study does not focus on the impacts of cyclones on sustainable livelihoods and the relationship between such unsustainable livelihoods and the food security situation.

However, in the present research we intend to conduct a very in-depth study on the impact of cyclonic disaster on all types of livelihood capital (i.e., natural, physical, human, social, and financial capital) and the relevance of such impacts to the food security conditions of the study area.

7.4 Objectives

In the present study, we aimed to find out whether the study area is food secure or insecure and if the area is food insecure, then what type(s) of food insecurity (transitory or chronic) occurs or exists in the study area and the causes behind it. The specific objectives of this study were:

- To identify the impact of climate-induced cyclone Aila on the sustainable livelihood assets of the Shyamnagar Upazila of Satkhira District
- To analyze the differences between pre-cyclone Aila and the present food security situation of the study area
- To explore the relationship between losses of sustainable livelihood assets caused by cyclone Ail and the food security situation of the study area.

7.5 Methods and Data

The current study is based upon both a quantitative and a qualitative approach and intended to utilize both primary and secondary databases. Primary data for this study were collected through questionnaire surveys at the household level and observation, whereas the secondary data were collected from various published and unpublished sources.

7.5.1 Selection of the Study Area

In the case of area sampling, the "Purposive or Judgmental Sampling" process was used. As the purpose was to conduct the study on the livelihood change and food security situation of the cyclone Aila-affected southwestern region, at first data were gathered about which area of the southwestern coastal region (Satkhira, Khulna, and Bagerhat) fall under which risk category (e.g., high-risk area, at-risk area, and wind-risk area). These data were collected from a map produced by the Vulnerability Analysis and Mapping (VAM) Unit of the World Food Programme Bangladesh and the unit has collected the data from SPARRSO. On the map, areas with a surge height of above 1 m are named high-risk areas, areas with less than 1 m surge height are named at-risk areas, and the remaining areas are named wind risk areas. The collected data show that Shyamnagar Upazila of Satkhira district, Dacope and Koyra Upazila of Khulna district, and Sarankhola Upazila of Bagerhat district have areas in all three risk categories. However, the high-risk and at-risk areas only cover the mangrove forest Sundarban and the areas inhabited by people fall under the wind-risk area. But, still being located adjacent to the at-risk area, these regions are considered to be more vulnerable to the cyclone impact than other areas. Then, further study on the impact of cyclone Aila revealed that Shyamnagar Upazila of Satkhira district experienced more damage than other upazilas in terms of affected population. A total of 160,432 people were affected in the upazila (Shyamnagar UNO 2009). Therefore, it was decided to conduct the survey in Shyamnagar Upazila of Satkhira district (Fig. 7.5).

Henceforth, the decision was made to select villages that could represent the whole of Shyamnagar Upazila. The loss and damage statistics and the report by the United Nations entitled "Cyclone Aila Joint UN Multisector Assessment and Response Framework" helped to identify severely affected unions, moderately affected unions, and less affected unions of the upazila. Then, two villages from severely affected Gabura union (i.e., Chakbara, Central Khalishabunia), one village from moderately affected Buri Goalini union (i.e., Buri Goalini) and one village (i.e., Badghata) from the less affected Shyamnagar union were selected for the study. The reason behind selecting two villages from Gabura union was to obtain a more detailed scenario of this severely damaged island union. About 19% of people in Gabura union and Buri Goalini union depend on the forest resources for their livelihood, 42% depend on crab fattening, 16% depend on shrimp farming, 23% people depend on agricultural day-laboring and other occupations such as fishing, van or auto bike (Wrench Electric Tricycle) puller, and small trades. The main crop being produced in Gabura union is paddy and the main crops of Buri Goalini union are paddy and vegetables (National Web Portal of Bangladesh, 2016). More than 75% people of the Gabura union and 64% people of the Buri Goalini union are living below the poverty line (National Web Portal of Bangladesh 2016).

7.5.2 Population Sampling

There were a total of 2528 households from the four selected villages (BBS 2011). In order to determine the sample, the Sample Size Calculator was used, which is presented as a public service of creative research systems. With the confidence level of 95%, a confidence interval of 7.63, and a total population of 2528 (total household number), the calculated sample size is 155. Among these 155 sample households, 60 households were surveyed in Chakbara village of Gabura union, 30 in Central Khalishabunia village of the same union, 40 in Buri Goalini village of Buri Goalini union, and 25 households in Badghata village of Shyamnagar union. These 155 households were surveyed according to the simple random sampling method.

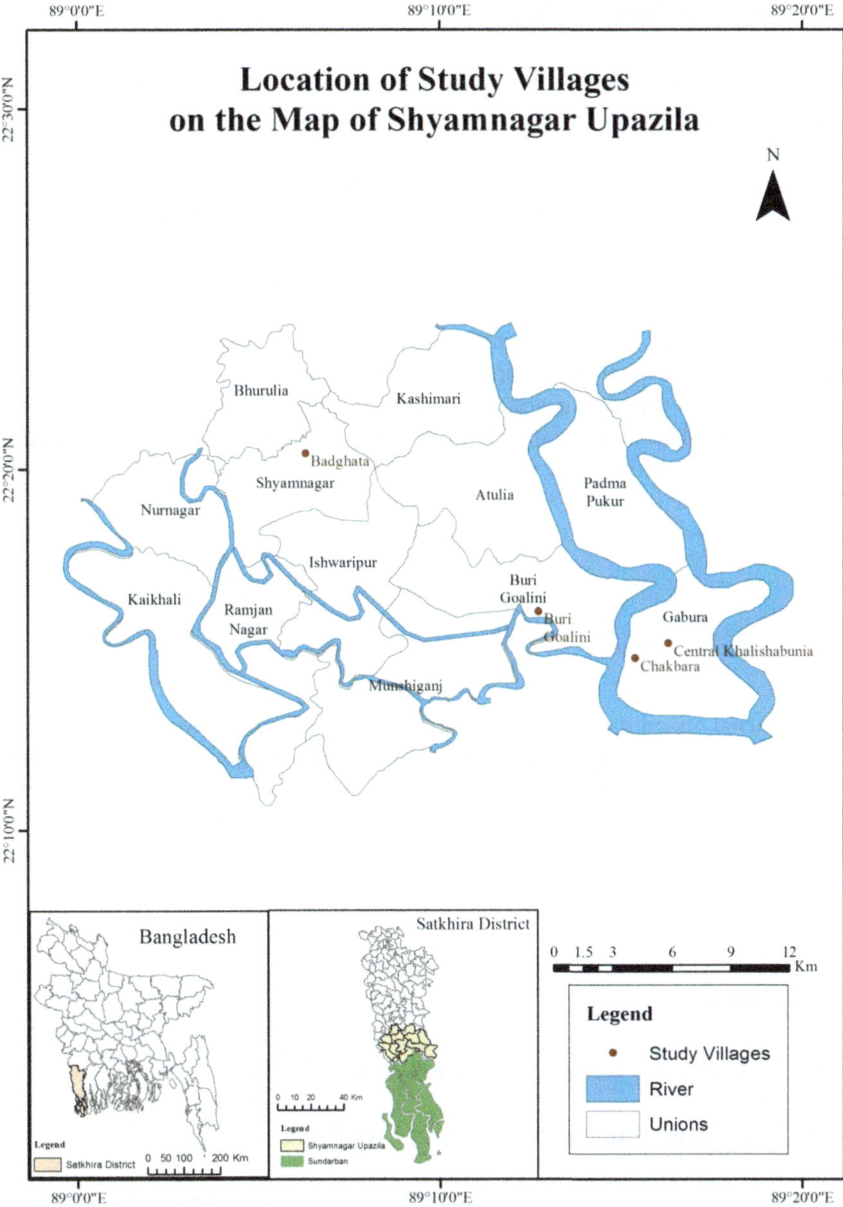

Fig. 7.5 The study area, which is located in the southwestern region of Bangladesh under the Shyamnagar Upazila of Satkhira district. Shyamnagar Upazila, the largest upazila of Satkhira district in terms of its area (1968.24 km^2). There are 12 unions under the upazila

7.5.3 *Questionnaire Design*

In order to acquire the necessary primary data, a questionnaire was prepared to survey the households. This questionnaire was made with the purpose of collecting data on change of occupation, loss of other livelihood assets (agricultural land, shrimp farm, livestock, forest resources, fuel supply, etc.) because of cyclone Aila, the process and extent of livelihood assets loss and data on change in income and expenditure conditions before and after the cyclone hit. The questionnaire also includes questions about sources of food (before, immediately after the cyclone, and at the present time), times/numbers of daily meal intake (before, immediately after the cyclone, and at present times), dietary diversity, and about problems they faced during food collection and preparation. This questionnaire also included the questions to collect the coping strategies that were applied by the sample population to reduce the food crisis immediately after the cyclone hit (relief, loan, help from relatives, etc.) and at the present time. Ultimately, the data collected through this questionnaire helped in analyzing the relationship between sustainable livelihood loss caused by cyclonic disaster and food insecurity.

7.5.4 *Data Analysis*

This study used mixed methods, which means that both qualitative and quantitative methods were used to analyze the collected data. The quantitative method was used to show the percentage of change or loss of livelihood assets or opportunities and to show change in income and expenditure in households pre- and post-cyclone. IBM SPSS Statistics and Microsoft Excel, etc., software were used to analyze and represent the collected data.

7.6 Results

7.6.1 *Impact of Cyclonic Disaster on Livelihood*

7.6.1.1 Immediate Impact on Sources of Livelihood

The strike of cyclone Aila on 25 May 2009 extensively damaged the livelihood options of the people of Shyamnagar Upazila. The analysis of primary data that were collected from four study villages of the Shyamnagar Upazila show that among total respondents, 91.67% people from Chakbara village of Gabura union, 70% people from Central Khalishabunia village of the same union and 77.50% people

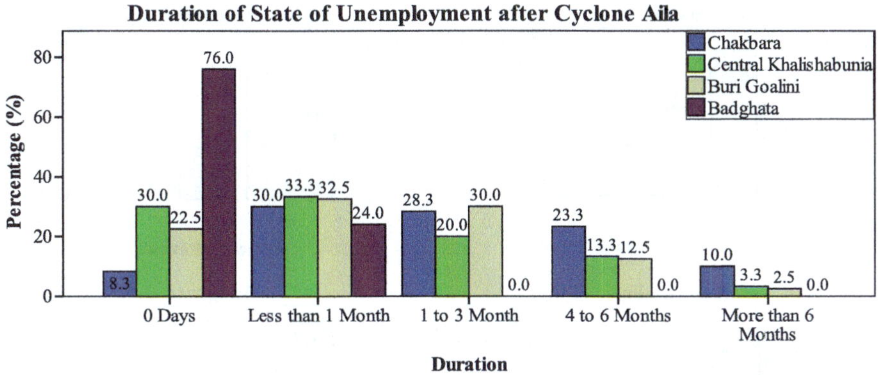

Fig. 7.6 Percentage of unemployed people after cyclone Aila on the basis of duration of their unemployment. (Source: Field Survey, 2015 and 2016)

from Buri Goalini village of Buri Goalini union and 24% people from Badghata village of Shyamnagar union lost their livelihood immediately after Cyclone Aila and the associated storm surge hit. The inhabitants of the study villages who lost their livelihood options remained unemployed for varying timespans. About 30% people from Chakbara, 33.33% from Central Khalishabunia, 32.5% from Buri Goalini, and 24% from Badghata village were unemployed for less than 1 month after cyclone Aila (Fig. 7.6). A few people also remained unemployed for even more than 6 months: 10% people from Chakbara, 3.3% from Central Khalishabunia, and 2.5% from Buri Goalini.

Such a sudden loss of income-generating options lowered the purchasing capacity and made it difficult for the inhabitants to meet their basic daily needs. These destroyed livelihood options also included agricultural work, livestock rearing, fish cultivation, etc. which are also the major sources of food supply of the area and such destruction of sources of food supply created an instant food insecurity situation in the study area.

7.6.1.2 Long-Term Impact on Sources of Livelihood

Owing to the occurrence of cyclone Aila, the percentage of households with more than one source of earnings decreased from 61.67% to 30.33% in Chakbara village of Gabura union and from 40% to 16.67% in Central Khalishabunia village from the same union (Fig. 7.7). The percentage of households with multiple sources of earnings also decreased for Buri Goalini union of Buri Goalini village from 35% to 27.5%. But the percentage of households with more than one source of earnings increased from 24% to 44% for Badghata village of Shyamnagar union.

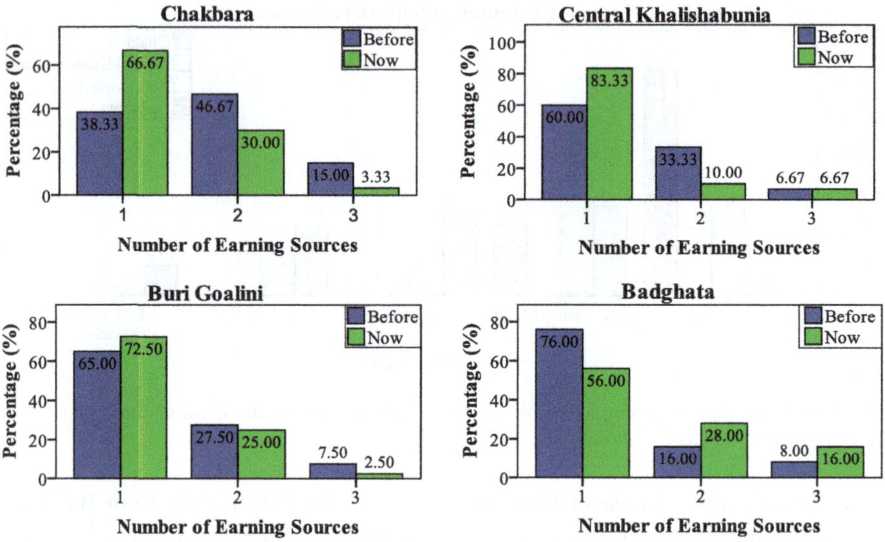

Fig. 7.7 Comparison of the number of sources of earnings of households between pre-cyclone Aila (2008) and the present time (2015 and 2016). (Source: Field Survey 2015 and 2016)

7.6.1.3 Change in Occupational Status

The sources of earnings for the study villages are a combination of primary, secondary, and tertiary economic activities. Before cyclone Aila, more respondents were dependent on primary economic activities such as agricultural work, livestock rearing, forest resources gathering, and fish cultivation. However, as the cyclonic disaster destroyed natural assets such as cultivable land, fresh water sources, grazing lands for livestock, and forest resources, people dependent on primary economic activities were compelled to change their occupation (Fig. 7.8).

The percentage of people involved with agricultural work decreased from 25% to 10% in Chakbara, from 13.3% to 6.67% in Central Khalishabunia, and from 15% to 10% in Buri Goalini village. The percentage of people engaged with forest resource collection as their means of livelihood decreased from 41.7% to 16.7% in Chakbara, from 30% to 16.7% in Central Khalishabunia, and from 30% to 12.5% in Buri Goalini village. The percentage of people engaged in fish cultivation and livestock rearing also decreased in Chakbara, Central Khalishabunia, and Buri Goalini village. However, the percentage of primary economic activities remained almost the same as before, even after the cyclone hit in Badghata village. About 63.33% of people from Chakbara village, 40% from Central Khalishabunia, 35% from Buri Goalini, and only 4% from Badghata village either changed the occupation they used to do before cyclone Aila or just left their alternative source of livelihood because of unfavorable conditions induced by cyclone Aila and the associated storm surge. The unfavorable conditions that forced them to leave their previous occupation were fewer employment opportunities in the previous occupation, better

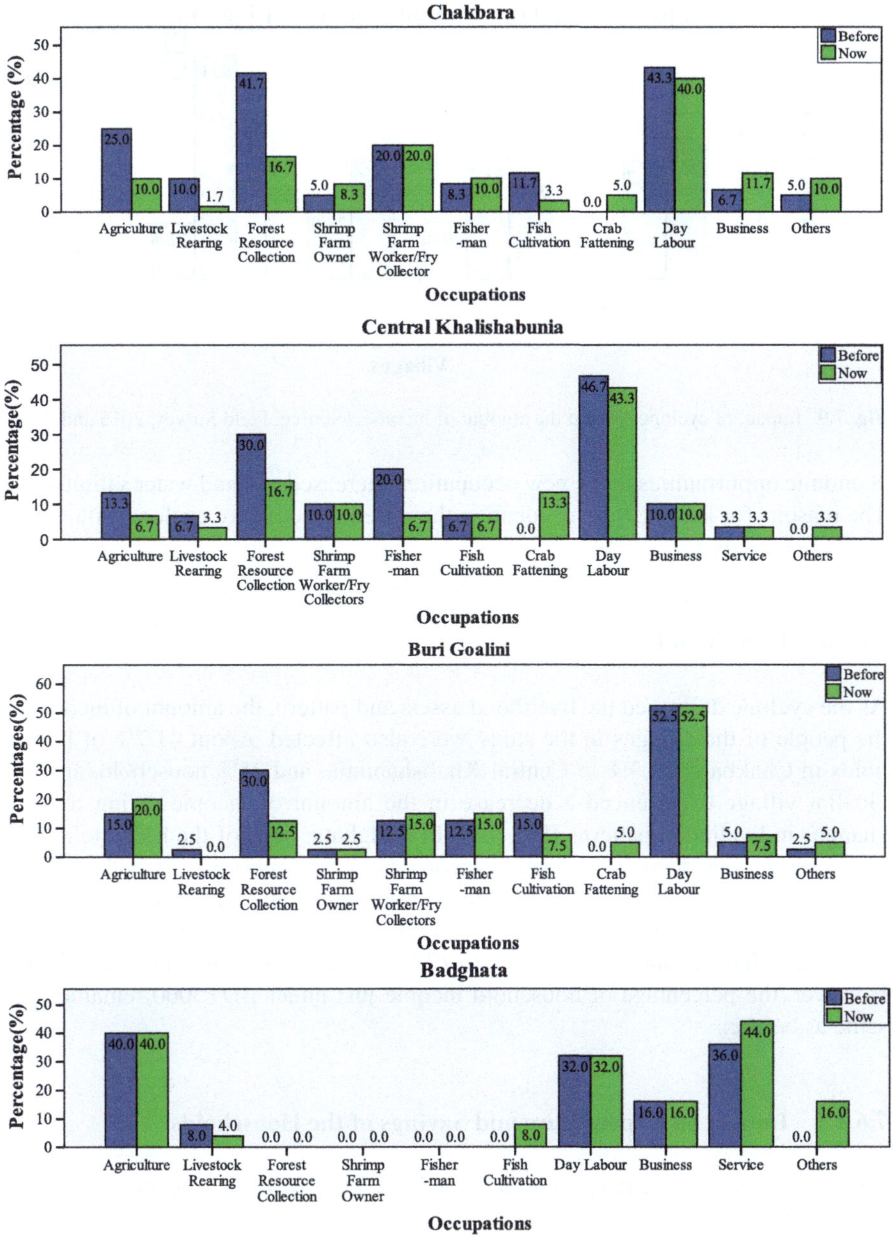

Fig. 7.8 Comparison of percentages of each occupation between pre-cyclone Aila (2008) and the present time (2015 and 2016). (Source: Field Survey 2015 and 2016)

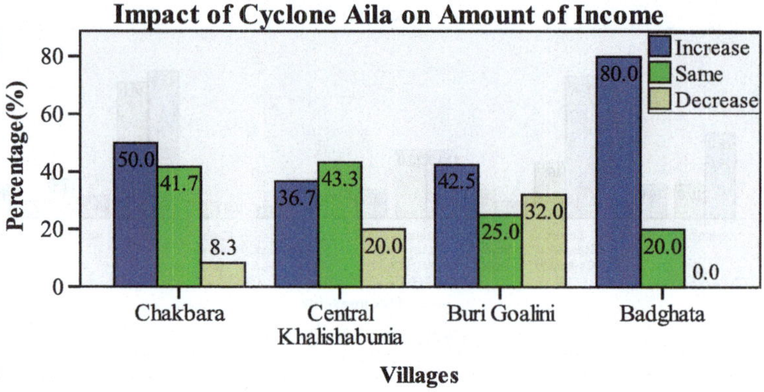

Fig. 7.9 Impact of cyclone Aila on the amount of income. (Source: Field Survey, 2015 and 2016)

economic opportunities in the new occupation, increased soil and water salinity, etc. The reasons for leaving their occupation they used to do before cyclone Aila varied by occupation type.

7.6.1.4 Impact on Income

As the cyclone destroyed the livelihood assets and pattern, the amount of income of the people of the villages in the study were also affected. About 41.7% of households in Chakbara, 43.3% in Central Khalishabunia, and 25% households in Buri Goalini village experienced a decrease in the amount of income owing to such changes in livelihood patterns (Fig. 7.9). In Badghata, none of the households had to experience any decrease in level of income after the occurrence of cyclone Aila.

The percentage of households with a monthly income just under BDT5000 increased after cyclone Aila from 46.7% to 56.6% in Chakbara, from 40% to 50% in Central Khalishabunia, and 67.50% to 72.50% in Buri Goalini village (Fig. 7.10). However, the percentage of household income just under BDT5000 remained the same as before.

7.6.1.5 Impact on Expenditure and Savings of the Households

The lack of employment opportunities and decrease in income level as a consequence of cyclone Aila as well as the rise in commodity prices have affected the expenditure and savings percentages of the households of study villages. The percentages of households who have to spend all of their monthly income to sustain life increased from 40% to 55% in Chakbara, from 30% to 60% in Central Khalishabunia, from 55% to 65% in Buri Goalini, and from 24% to 32% in Badghata village. Such an increase in the percentage of expenditure decreases the capacity to save a share

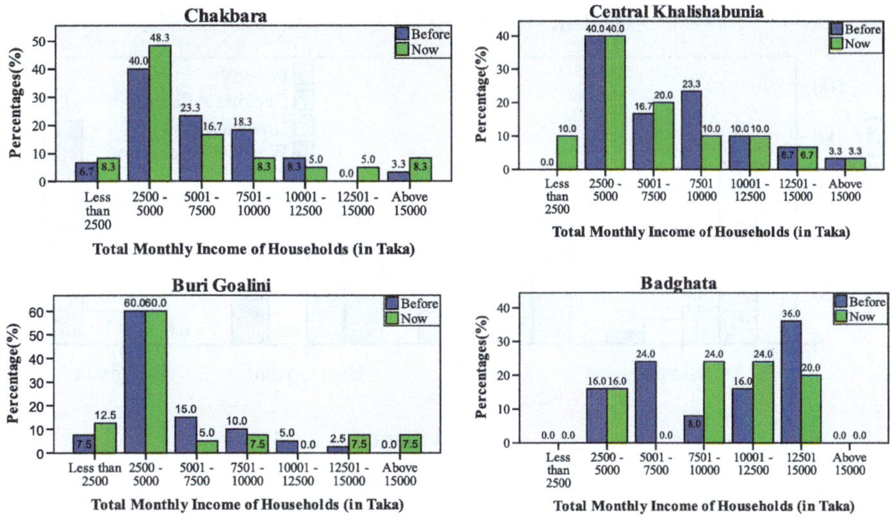

Fig. 7.10 Comparison of monthly income between pre-cyclone Aila (2008) and present times (2015 and 2016). (Source: Field Survey 2015 and 2016)

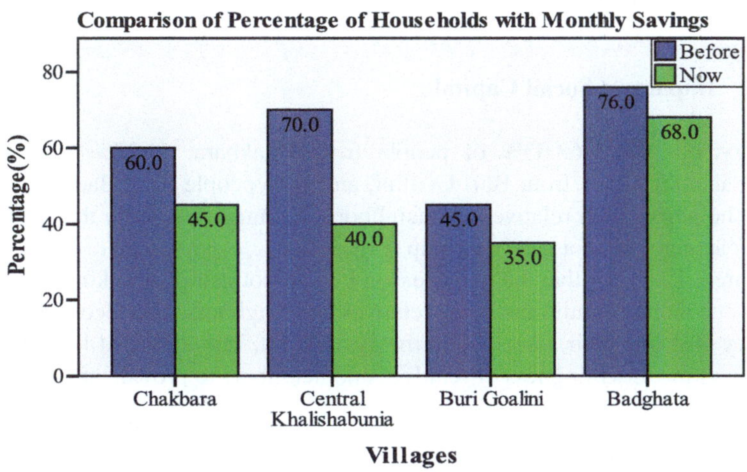

Fig. 7.11 Percentage of households with savings during the pre-cyclone Aila situation (2008) and the present time (2015 and 2016). (Source: Field Survey 2015 and 2016)

of their income for future necessities, especially the necessity of basic amenities during devastating disasters such as cyclones, storm surges, and waterlogging.

The percentage of households with monthly savings decreased from 60% to 45% in Chakbara, from 70% to 40% in Central Khalishabunia, from 45% to 35% in Buri Goalini, and from 76% to 68% in Badghata village (Fig. 7.11). This lack of savings makes the households more vulnerable to food, clothing, shelter insecurities, etc.,

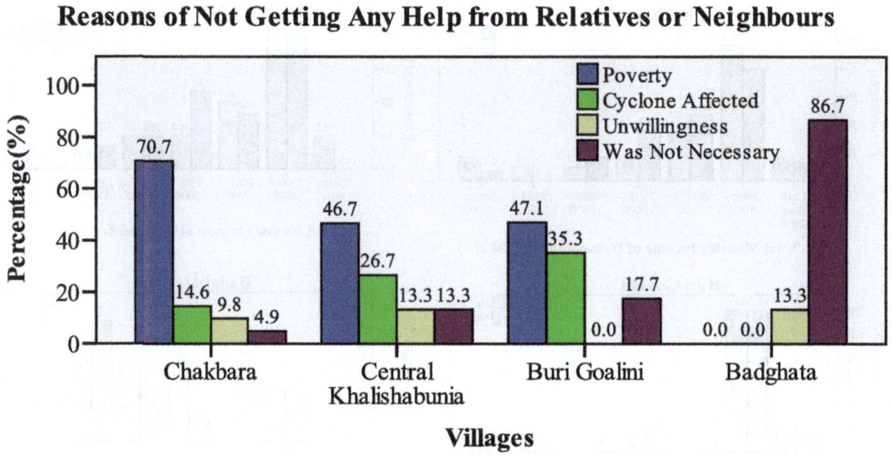

Fig. 7.12 Percentages of reasons for not getting any help from relatives and neighbors. (Source: Field Survey 2015 and 2016).

during the disaster and makes them more dependent on relief for basic needs, as well as increasing their debt.

7.6.1.6 Impact on Social Capital

After cyclone Aila, 68.33% of people from Chakbara, 53.33% from Central Khalishabunia, 42.5% from Buri Goalini, and 60% people from Badghata did not get any help from their relatives and neighbors. The main reason for this is that most of them actually did not need any help (Fig. 7.12).

Figure 7.12 shows that the main reason behind not getting any kind of help was the poor economic conditions of the relatives and neighbors. The second reason was that they also lost their assets when the cyclone hit. The third and less prominent reason was the unwillingness of relatives and neighbors to provide any help.

7.6.2 Relevance to Food Security

7.6.2.1 Sources of Food

The major source of food immediately after cyclone Aila was food from relief organizations (Fig. 7.13). Only a few of them would purchase food from markets as the nearby food markets were demolished by the cyclone and the markets that were open were located far away, lacked supplies, and had high prices. In Chakbara, the nearby food market was closed for 1 month, in Central Khalishabunia for 20 days, and in Buri Goalini for 14 days. The food markets were open in Badghata even immediately after the cyclone. About 41.7% of people from Chakbara, 40% from

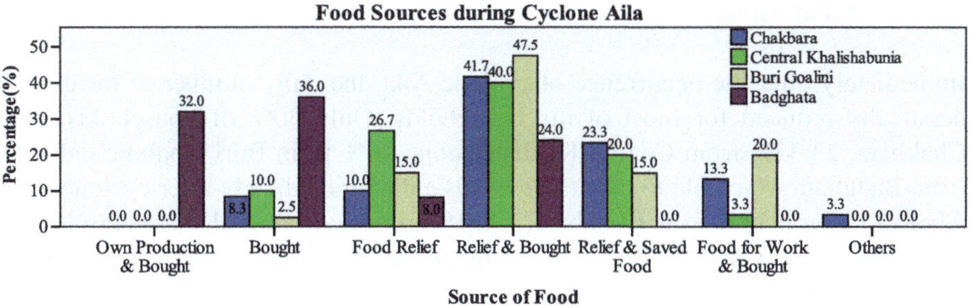

Fig. 7.13 Sources of food immediately after cyclone Aila. (Source: Field Survey 2015 and 2016)

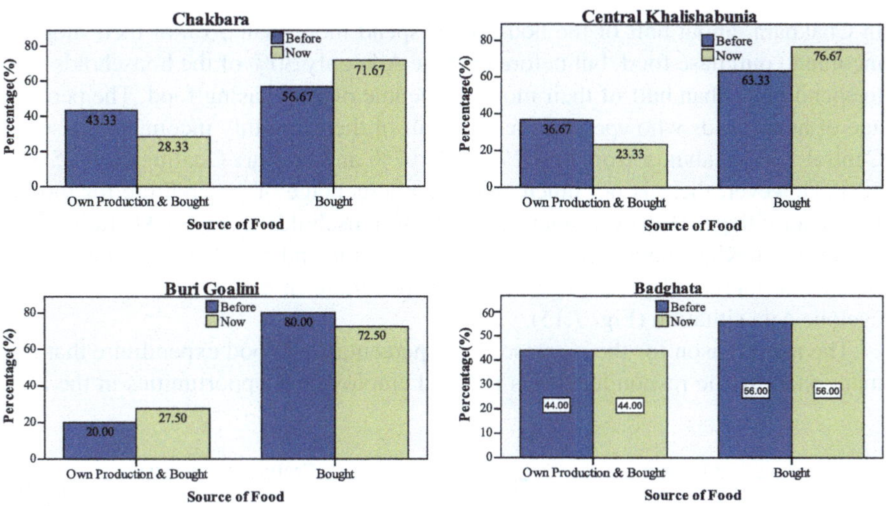

Fig. 7.14 Percentage of sources of foods of households pre-cyclone Aila (2008) and at the present time (2015 and 2016). (Source: Field Survey 2015 and 2016)

Central Khalishabunia, 47.5% from Buri Goalini, and 24% from Badghata used to depend on both relief and bought food.

The sources of food collection for the inhabitants of study villages were affected by cyclone Aila (Fig. 7.14). The majority of the households purchase their required food from the market and a few of them produce a small share of their required food through agricultural production, homestead gardening, fish cultivation, etc. But, after cyclone Aila the percentages of households who produced a share of their own food reduced from 43.33% to 28.33% in Chakbara and from 36.67% to 23.33% in Central Khalishabunia. However, the scenario in Buri Goalini is different than that of other villages, as more households are becoming engaged in homestead gardening by using seeds and techniques to manage soil salinity provided by nongovernmental organizations (NGOs). There was no change in the sources of food for Badghata village.

7.6.2.2 Meal Intake

Immediately after the occurrence of cyclone Aila, the daily number of meals was drastically reduced for most of the households. Only 30% of households from Chakbara, 23.33% from Central Khalishabunia, 20% from Buri Goalini, and 88% from Badghata were able to take three meals a day immediately after cyclone Aila. About 45.83% people from Chakbara, 73.91% from Central Khalishabunia, 60.61% from Buri Goalini, and 77.78% from Badghata had to take fewer meals per day, for at least 1 to 10 days.

7.6.2.3 Monthly Expenditure for Food

In Chakbara, about half of the households spend more than 50% of their monthly income to purchase food, but before cyclone Aila only 30% of the households used to spend more than half of their monthly income on purchasing food. The percentage of households who spend more than half of their monthly income increased in Central Khalishabunia from 16.67% to 26.67% and in Buri Goalini from 25% to 50%. However, there is not much change in percentage of expenditure for food in Badghata village. However, about 78.33% of households from Chakbara, 63.33% from Central Khalishabunia, 65% from Buri Goalini, and 56% from Badghata experienced an increased percentage of food expenditure in comparison with the pre-cyclone Aila situation (Fig. 7.15).

The main reason for the increase in the percentage of food expenditure that was mentioned by the respondents was reduced employment opportunities in the area.

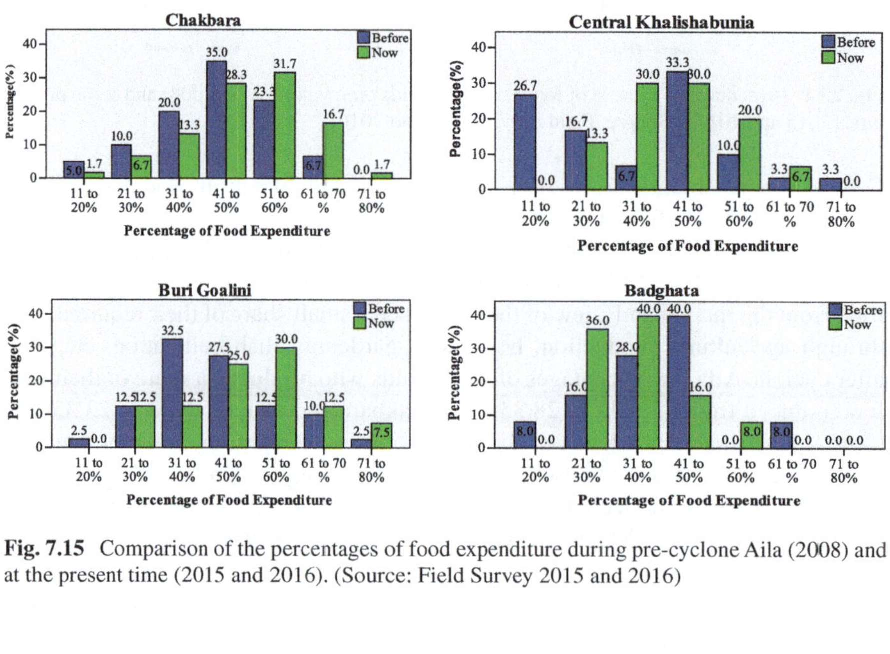

Fig. 7.15 Comparison of the percentages of food expenditure during pre-cyclone Aila (2008) and at the present time (2015 and 2016). (Source: Field Survey 2015 and 2016)

About 65.96% from Chakbara, 52.63% from Central Khalishabunia, and 53.84% from Buri Goalini mentioned reduced employment opportunities as the reason behind the increase in percentage of household food expenditure. On the other hand, the majority of the households of Badghata (78.57%) mentioned price hikes as the main reason behind increases in the percentage of food expenditure. The percentage of households who could have met their monthly food needs within BDT3000 decreased from 73.33% to 61.66% in Chakbara, from 80% to 70% in central Khalishabunia, from 85% to 77.67% in Buri Goalini, and from 44% to 10% in Badghata village.

7.6.2.4 Dietary Diversity

The situation of dietary diversity of the study villages was analyzed by finding the weekly consumption of food of the six major food groups by household members. The major six food groups include grain, vegetable, protein, fruit, dairy, and fat. The analysis shows that dietary diversity is very poor in Chakbara, Central Khalishabunia, and Buri Goalini villages. In Chakbara village, only 10% of households are able to consume all six major groups food every week and the percentage of households who consume food from all six food groups in a week is 17.5%, 16% in Buri Goalini and in Badghata only 28% (Table 7.1).

7.6.2.5 Cooking Fuel

The major fuel type of all study villages is wood, but a few of the households also use leaves, cow dung, and agricultural residue along with wood to cook food. About 41.67% of people from Chakbara, 60% from Central Khalishabunia, and 35% from Buri Goalini mentioned that they are facing problems in collecting cooking fuel after the occurrence of cyclone Aila. The respondents also mentioned that lack of money, strict security in Sundarban after cyclone Sidr, and the extinction of tree species are the behind the problems collecting fuel.

Table 7.1 Percentages of households by food variety consumed in a week

Food variety during a week	Villages			
	Chakbara (%)	Central Khalishabunia (%)	Buri Goalini (%)	Badghata (%)
2	5	6.67	–	–
3	11.67	23.33	20	
4	58.33	36.67	40	24
5	15	20	22.5	48
6	10	13.33	17.5	28

Source: Field Survey (2015 and 2016)

7.6.2.6 Sources of Water

One of the most devastated resources due to cyclone Aila in the study area is water. The storm surge associated with cyclone Aila inundated the surface drinking water sources with saline water and immediately made drinking water scarce in the area. The main sources of drinking and cooking water immediately after cyclone Aila hit were relief water, bought water, and rain water. In Chakbara, 31.67% of households used to depend on bought water, 33.33% on both rainwater and bought water and 18.33% on both relief and rainwater for drinking and cooking purposes. 43.33% of households in Central Khalishabunia used to depend both on relief and rainwater, 20% on rainwater and bought water, and 16.67% used to depend on bought water for drinking and cooking purposes. In Buri Goalini, 30% of households used to depend on bought water, 30% on both relief and rainwater, and 22.5% on both rainwater and bought water for drinking and cooking purposes. As the sources of drinking and cooking water did not get damaged in Badghata village, the sources of water stayed the same. For bathing and other uses, most of the households of the study villages used water logged from storms, ponds, and rivers.

Before the strike of cyclone Aila, the major source of drinking water for the study villages was pond water, and other sources included rainwater and water brought from the authorities. Before cyclone Aila hit, about 78.33% of households in Chakbara, 53.33% in Central Khalishabunia, 10% in Buri Goalini, and 40% in Badghata were dependent on pond water to meet their drinking water needs. However, now only 25% of households from Chakbara, 46.67% from Central Khalishabunia, and 7.5% from Buri Goalini depend on pond water as a drinking source. But, in Badghata, none of the households now uses pond water for drinking purposes.

At present, the major source of drinking water for Chakbara is water bought from the authorities. About, 51.67% of households depend on bought water and 23.33% depend on both rainwater and pond water. For Central Khalishabunia, the major source of drinking water is now rainwater and water from the authorities. About 50% of the respondents from Central Khalishabunia depend on rainwater and bought water combined. Seventy-five percent of households in Buri Goalini and 60% from Badghata take water from a pond sand filter for drinking purposes and that is why pond sand filters are the main source of drinking water in these two villages. In Badghata, 40% of people also drink water from tube wells. About 35% people from Chakbara, 46.67% from Central Khalishabunia, 22.5% from Buri Goalini, and 12% from Badghata village mentioned problems of pollution and salinity in their drinking water.

7.6.2.7 People's Perception

Most of the respondents from study villages are not satisfied with their food habits and think that they are suffering from food insecurity. About 68.3% of households from Chakbara, 40% from Central Khalishabunia, and 67.5% from Buri Goalini think that they are not food secure and need to change their food habits. Among the households who think they are food insecure, 36.6% of them are from Chakbara, 66.7% from Central Khalishabunia, and 18.5% from Buri Goalini village mentioned that they are consuming insufficient amounts of food. Among food-insecure households, 53.7% from Chakbara, 25% from Central Khalishabunia, and 48.2% from Buri Goalini mentioned that they are not economically capable of consuming good-quality food.

7.6.3 Coping Strategies

The most common coping strategy applied by respondents in surviving against vulnerable livelihood condition is migration to other places and the second most common coping strategy is taking loans from various sources (Table 7.2).

7.6.3.1 Temporary Migration

Owing to the destruction of livelihood options and prolonged water logging caused by associated storm surges, many people from the study area had to migrate temporarily to resume their livelihood and to avoid the suffering of living in such a damaged environment. About 28.33% people from Chakbara village, 30% from Central Khalishabunia, and 27.5% from Buri Goalini village had to migrate to other places for a short period of time in search of work and better living places where basic requirements for leading a life were available. There was no outward temporary migration from Badghata village after cyclone Aila hit.

The collected data show that the majority of the people who migrated outward from the study villages after cyclone Aila went to other districts and a few of them went to other sub-districts. About 76.47% of the temporarily migrated people from Chakbara, 66.6% from Central Khalishabunia, and 63.63% from Buri Goalini village migrated to other districts for a short period of time.

Table 7.2 Percentages of coping strategies applied by the respondents to survive against vulnerable livelihood situations

Coping strategies	Chakbara (%)	Central Khalishabunia (%)	Buri Goalini (%)	Badghata (%)
Migrated	51.67	50	40	–
Took loans	56.67	30	42.5	4
Spent savings	23.33	10	12.5	4
Others	5	–	–	–

Source: Field Survey (2015 and 2016)

7.6.3.2 Permanent Migration

The analysis of permanent occupational migration data show that 23.33% of people from Chakbara, 20% from Central Khalishabunia, and 12.5% from Buri Goalini village migrated permanently in different years after the occurrence of cyclone Aila. There was no outward permanent migration from Badghata village after cyclone Aila. Most of the permanent occupational migration from study villages happened in 2009 and 2010. In the former, 35.71% people from Chakbara, 33.33% from Central Khalishabunia, and 40% from Buri Goalini village migrated out of their village in search of work. In the latter, 42.86% people from Chakbara, 33.33% from Central Khalishabunia, and 20% from Buri Goalini village migrated out of their village in search of work. From Chakbara and Central Khalishabunia village the majority of permanently migrated inhabitants of the study villages went to other unions and other sub-districts. The data reveal that 35.71% of permanently migrated people from Chakbara and 50% from Central Khalishabunia went to other unions and then 35.71% from Chakbara and 33.33% from Central Khalishabunia migrated to other sub-districts. The majority of the migrated people from Buri Goalini went to other sub-districts and other districts. About 40% of people from Buri Goalini migrated to other sub-districts and 40% to other districts. This extensive migration that occurred after cyclone Aila proves the vulnerability of the livelihood opportunities in the study area.

7.6.3.3 Status of Loan

After the occurrence of cyclone Aila, 56.67% people from Chakbara, 30% from Central Khalishabunia, 42.5% from Buri Goalini, and 24% people from Badghata had to take out loans to meet the daily basic needs, to resume their livelihood, and to recover from the losses. The major reason behind taking loans from various sources after cyclone Aila was to repair or reconstruct the dwelling place of the inhabitants. It was not easy for all of them to get a loan to recover from losses caused by the cyclonic disaster. About 17.65% of loan takers in Chakbara, 6.67% in Central Khalishabunia, and 42.5% in Buri Goalini had to face various types of constraints in getting loans. The majority of the inhabitants of the study area took out loans from NGOs. About 44.10% of people from Chakbara, 77.78% from Central Khalishabunia, 52.94% from Buri Goalini, and 66.67% from Badghata took loans from NGOs. Among the people who took out loans to cope after the cyclone, 61.76% in Chakbara, 66.67% in Central Khalishabunia, 52.94% in Buri Goalini, and 33.33% in Badghata still could not repay the debt. They mentioned that the inability to repay the loans they took out may affect the possibility of getting loans for future necessity.

7.7 Discussion

The analyses of the secondary and primary data show that the villages of the island union named Gabura union and the villages of the Buri Goalini Union are more vulnerable to cyclonic disasters owing to their close proximity to the rivers Kholpetua, Kapotakkho, and Chunar. The collected and analyzed primary and secondary data revealed that the most damage to natural, physical, financial, human, and social capital due to cyclone Aila occurred in the villages of Gabura union and Buri Goalini union. People from these villages are more prone to changing their occupation and to migrating outward where they can find better availability of livelihood options. About 63.33% of people from Chakbara village, 40% from Central Khalishabunia, 35% from Buri Goalini, and only 4% from Badghata village changed the occupation they used to have before cyclone Aila or just left their alternative source of livelihood because of unfavorable conditions induced by cyclone Aila and the associated storm surge. About 23.33% of people from Chakbara, 20% from Central Khalishabunia, 12.5% from Buri Goalini village migrated permanently after the occurrence of cyclone Aila. In particular, food production- and collection-related livelihood options, such as agricultural work, decreased from 25% to 10% in Chakbara, from 13.3% to 6.7% in Central Khalishabunia, and from 15% to 10% in Buri Goalini village. The percentage of people engaged with forest resource collection as their means of livelihood also decreased from 41.7% to 16.7% in Chakbara, from 30% to 16.7%% in Central Khalishabunia, and from 30% to 12.5% in Buri Goalini village.

The tendency toward land use change is also more evident among inhabitants of these villages. The most prominent land use change of these villages is conversion of agricultural land to shrimp farms. Although most of the land use changes are consequences of various frequently occurring disasters, especially the frequent occurrence of cyclones and associated storm surges, which also cause waterlogging in these river front and island union villages. The income level of the majority of the inhabitants of these villages is not stable. About 41.7% of households in Chakbara, 43.3% in Central Khalishabunia, and 25% of households in Buri Goalini village experienced a decrease in the amount of income due to such changes in livelihood patterns.

The combination of reduced employment opportunities, decreased income levels, and reduced household level of food production caused by cyclonic disasters and increased salinity have increased food insecurity for the households of the Chakbara and Central Khalishabunia villages of Gabura union. As Gabura union is an island union and hard to reach, the support from various governmental and nongovernmental organizations do not reach them properly, which is another reason for vulnerable livelihoods and food security in the union. As Buri Goalini union is attached to the main land, various governmental and nongovernmental support reaches the inhabitants with regard to the livelihood and food security situation. As a result, the situation is slightly better in Buri Goalini union than in Gabura union. On the other hand, the village of Shyamnagar union, which is not a river front village and is also situated relatively far away from major rivers of the area than other villages, experienced less damage to livelihood assets. In the village of Shyamnagar union, the tendency to change occupation, to migrate outward, and to change the land use are rarely visible. The income level of respondents of Badghata village also seems to be more stable than other study villages.

7.8 Conclusion

The study showed that the livelihood and food security situation immediately becomes devastated after the occurrence of a cyclonic disaster. However, there are also some long-term or gradual impacts of such disasters on livelihood assets and thus on the food security of households. The livelihood options that are related to the production of food become damaged and reduce the food supply. The damage to livelihood also decreases the income level and thus lowers the purchasing capacity of the people. On the basis of the results of the study and also on the basis of the opinions given by the respondents, some short-and long-term recommendations might be effective in reducing cyclonic disaster-related livelihood vulnerability and food insecurity. People who are more vulnerable to devastation as a consequence of cyclonic disasters must consider the cyclone warning of utmost importance and get prepared by saving dry foods, bottled water, emergency medicines, etc. and must evacuate the place and reach a cyclone shelter in good time. People should take their livestock to the shelter or high places where the livestock can be safe. The immediate support by governmental and nongovernment organizations regarding victim rescue mission; medical assistance for injured people; distribution of food, water, and water-purifying tablets; the arrangement of temporary shelter and sanitation facilities, etc., must be provided rapidly after the disaster. Saline water must be pumped out, especially from the sources of drinking water, so that rainwater can be conserved there for drinking and cooking purposes. Employment opportunities should be created for those poor and vulnerable people who lost their livelihood because of the devastation of cyclonic disasters through programs such as "Food-for-Work" or "Cash-for-Work." Moreover, the cultivation of salinity and of submergence-tolerant crop varieties should be increased to improve the livelihoods of farmers and the food security situation of the area. As waterlogging is quite common after the occurrence of cyclonic disasters, the floating agricultural system should be introduced to farmers. However, there are various salinity management techniques, improved technologies for increasing crop yield and cropping intensity, etc., in use around the world. Farmers should be provided with the knowledge of and training on such improved and more viable disaster-resilient adaptation techniques and technologies. Rainwater harvesting, pond sand filter, re-excavation of ponds, etc., can help to minimize the problem regarding fresh water for drinking and irrigation. Furthermore, access to micro-credit can help farmers, small businessmen, and people in other occupations to restart their livelihood after cyclonic disasters. The preparedness measures for survival against such cyclonic disasters should be proactive instead of merely reactive measures.

References

Alam MM, Hossain MA, Shafee S (2003) Frequency of Bay of Bengal cyclonic storms and depressions crossing different coastal zones. Int J Climatol 23:1119–1125

Bala BK, Hossain MA (2010) Food security and ecological footprint of coastal zone of Bangladesh. Environ Dev Sustain 12(4):531–545

BBS (Bangladesh Bureau of Statistics) (2011) Bangladesh population and housing census 2011. Bangladesh Bureau of Statistics (BBS). Community Report, Satkhira. Available from http://www.bbs.gov.bd. Accessed 14 Aug 2015

Bender MA, Ginis I, Kurihara Y (1993) Numerical simulations of tropical cyclone-ocean interaction with a high resolution coupled model. J Geophys Res 98:245–263

Chambers R, Conway GR (1991) Sustainable rural livelihoods: practical concepts for the 21st century. IDS discussion paper 296. Institute of Development Studies, Brighton. Available from: https://www.ids.ac.uk/files/Dp296.pdf. Accessed 9 Aug 2015

Comprehensive Disaster Management Programme (CDMP) (2006) Climate change cell component of comprehensive disaster management programme [report]. CDMP, Dhaka

Cyclone Aila Joint UN Multisector Assessment & Response Framework (2009) Available from http://www.lcgbangladesh.org. Accessed 2 Aug 2015

Das S, Giri S, Das I, Chanda A, Akhand A, Mukhopadhyay A, Maity S, Hazra S (2016) Tide induced annual variability of selected physico-chemical characteristics in the northern Bay of Bengal (nBoB) with a special emphasis on Tropical Cyclone-Phailin. Indian J Geo-Marine Sci 45(8):952–959

DFID (Department for International Development), 1999. Sustainable livelihood guidance sheets. Available from http://www.eldis.org/vfile/upload/1/document/0901/section2.pdf. Accessed 4 Nov 2015

DFID (Department for International Development) (2000) Sustainable livelihood guidance sheets. Available at www.eldis.org/vfile/upload/1/document/0901/section1.pdf. Accessed 4 Nov 2015

Emanuel K, DesAutels C, Holloway C, Korty R (2004) Environmental control of tropical cyclone intensity. J Atmos Sci 61:843–858

FAO (Food and Agricultural Organization) (2002) State of food insecurity in the world 2001. Rome. Available from http://www.fao.org. Accessed 6 Nov 2015

FAO (Food and Agricultural Organization) (2003a) Trade reforms and food security: conceptualizing the linkages. Commodity Policy and Projection Service Commodities and Trade Division. Rome. Available from http://www.fao.org. Accessed 6 Nov 2015

FAO (Food and Agricultural Organization) (2003b) Corporate Document Repository. Available at: http://www.fao.org. Accessed 18 Nov 2015

Garai J (2014) The impacts of climate change on the livelihoods of coastal people in Bangladesh: a sociological study. In: Filho WL, Alves F, Caeiro S, Azeiteiro UM (eds) International perspectives on climate change: Latin America and beyond. Climate change management. Springer, Cham, pp 151–163

Gupta S, Jain I, Johari P, Lal M (2019) Impact of climate change on tropical cyclones frequency and intensity on Indian coasts. In: Rao PJ et al (eds) Proceedings of international conference on remote sensing for disaster management, Springer Series in geomechanics and geoengineering. Available from: https://doi.org/10.1007/978-3-319-77276-9_32. Accessed 10 Jan 2020

Habiba U, Abedin MA, Shaw R (2015) Introduction and overview of food security and risk reduction issues. In: Habiba U, Abedin MA, Hassan AWR, Shaw R (eds) Food security and risk reduction in Bangladesh. Springer, Tokyo, p 8

Hossain MA, Reza MI, Rahman S, Kayes I (2012) Climate change and its impacts on the livelihoods of the vulnerable people in the southwestern coastal zone in Bangladesh. In: Filho WL (ed) Climate change and the sustainable use of water resources. Climate change management. Springer, Berlin/Heidelberg, pp 237–259

Knutsson P (2006) The sustainable livelihoods approach: a framework for knowledge integration assessment. Hum Ecol Rev, 13(1):90–99

Longshore D (2008) Encyclopedia of hurricanes, typhoons, and cyclones. Checkmark Books, New York, p 468

Mallick B, Rahaman KR, Vogt J (2011) Coastal livelihood and physical infrastructure in Bangladesh after cyclone Aila. Mitig Adapt Strateg Glob Chang 16(6):629–648

Mita KS, Azad AA, Zaman MW, Sakib M (2018) Effectiveness of adaptive measures against storm surge hazard based on field experience from a real time cyclone in Bangladesh coast. In: 2nd international conference on sustainable development (ICSD), Dhaka, Bangladesh

Montgomery MT, Bell MM, Aberson SD, Black ML (2006) Hurricane Isabel 2003, new insights into the physics of intense storms, part I: mean vortex structure and maximum intensity estimates. Bull Am Meteorol Soc 87:1335–1347

Moslehuddin AZM, Abedin MA, Hossain MAR, Habiba U (2015) Soil health and food security: perspective from southwestern coastal region of Bangladesh. In: Habiba U, Abedin MA, Hassan AWR, Shaw R (eds) Food security and risk reduction in Bangladesh. Disaster risk reduction: methods, approaches and practices. Springer, Tokyo, pp 187–212

National Web Portal of Bangladesh (2016) Government of Bangladesh. Available at: http://gaburaup.satkhira.gov.bd and http://burigoaliniup.satkhira.gov.bd. Accessed 3 Mar 2016

Persing J, Montgomery MT (2003) Hurricane super intensity. J Atmos Sci 60:2349–2371

Prakash R, Pant V (2016) Upper oceanic response to tropical cyclone Phailin in the Bay of Bengal using a coupled atmosphere-ocean model. Ocean Dyn 67(1):51–64. https://doi.org/10.1007/s10236-016-1020-5

Rabbani MG, Rahman AA, Shoef IJ Khan ZM (2015) Climate change and food security in vulnerable coastal zones of Bangladesh. In: Habiba U, Abedin MA, Hassan AWR, Shaw R (eds) Food security and risk reduction in Bangladesh. disaster risk reduction: methods, approaches and practices. Springer, Tokyo, pp. 173–185

Rahman MA, Hossain ML, Foysal MA, Akter A (2014) Impact of climate change on food security in south-west coastal region of Bangladesh. Am J Exp Agric 4(12):1916–1934

Saha CK (2015) Dynamics of disaster-induced risk in southwestern coastal Bangladesh: an analysis on Tropical Cyclone Aila 2009. Nat Hazards 75(1):727–754

Sarthi PP, Agrawal A, Rana A (2015) Possible future changes in cyclonic storms in the Bay of Bengal, India under warmer climate. Int J Climatol 35(7):1267–1277

Scoones I (1998) Sustainable rural livelihoods: a framework for analysis. IDS working paper 72. Institute of Development Studies, Brighton. Available from: https://energypedia.info/images/a/a5/Scoones_1998_Sustainable_Rural_Livelihoods.pdf. Accessed 19 Aug 2015

Shameem MIM, Momtaz S, Rauscher R (2014) Vulnerability of rural livelihoods to multiple stressors: a case study from the southwest coastal region of Bangladesh. Ocean Coast Manage 102:79–87

Upazila Nirbahi Office, Shyamnagar (2016) Government of the Peoples' Republic of Bangladesh

USAID (United States Agency for International Development) (1992). Definition of food security. USAID Policy Determination 19. Washington, DC

USAID (United States Agency for International Development) (2010) Feed the future guide. Presidential Initiative on Hunger and Food Security. Draft, Washington, DC

Vissa NK, Satyanarayana ANV, Prasad KB (2013) Response of upper ocean and impact of barrier layer on Sidr cyclone induced sea surface cooling. Ocean Sci J 48(3):279–288

Wada A (2009) Idealized numerical experiments associated with the intensity and rapid intensification of stationary tropical cyclone-like vortex and its relation to initial sea-surface temperature and vortex induced sea-surface cooling. J Geophys Res 114. https://doi.org/10.1029/2009JD011993

Woller G, Wolfe J, Brand M, Fowler B, Parrot L, Thompson J, Dempsey J, Berkowitz L, Haeften BV (2011) Livelihood and food security conceptual framework. FHI360. USAID. Available at: https://www.microlinks.org. Accessed 19 Nov 2015

Yesubabu V, Kattamanchi VK, Vissa NK, Dasari HP, Sarangam VBR (2019) Impact of ocean mixed-layer depth initialization on the simulation of tropical cyclones over the Bay of Bengal using the WRF-ARW model. Meteorol Appl:1–18. https://doi.org/10.1002/met.1862

Yu L, McPhaden MJ (2011) Ocean pre-conditioning of Cyclone Nargis in the Bay of Bengal: interaction between Rossby waves, surface fresh waters, and sea surface temperatures. J Phys Oceanogr 41:1741–1755

Chapter 8
Analysis of Spatio-temporal Variation of Lightning Casualties in Bangladesh

Ummey Kulsum, Hasan Mahmud, and Md. Mizanoor Rahman

Abstract Recently, lightning has drawn the attention of researchers, social activists, policy makers and various governmental and nongovernmental organizations of disaster management fields because of its increasing trend of casualties. The literature indicates that the growing trend of lightning casualties has become an issue of concern in developing countries, especially in Bangladesh. Owing to its gigantic number of casualties in recent years, the Bangladesh Government has considered and declared lightning to be a disaster. This chapter attempts to portray the changing scenario of casualties due to lightning with respect to spatial and temporal variation in Bangladesh. In the past 10 years, the number of deaths, injuries, and casualties due to lightning in Bangladesh have been accelerating with spatial and temporal variations; the maximum was in 2018. Subsequently, spatial dissemination portrays the rising trend all over the country, but distinct zones, the northwestern, northeastern, and southern parts, show the most significant increasing tendency during the study period. In contrast, very specifically, district-wise analysis shows that in the case of casualties and deaths Chapai Nawabganj attained the highest position, whereas Jhalokati secured the lowest. But in the context of injury, Thakurgaon and Rajbari were found to be at the top and bottom of the list respectively. Nonetheless, the findings established a strong reason to consider lightning as an emerging disaster, and will help to draw attention to and extend the views of researchers related to the climate change field to find out the relationship with climatic variability, which will contribute to formulating appropriate adaptation strategies to minimize the changing stipulation.

Keywords Lightning · Lightning disaster · Spatio-temporal variation · Bangladesh

U. Kulsum · M. M. Rahman (✉)
Department of Geography and Environmental Studies, University of Rajshahi, Rajshahi, Bangladesh
e-mail: mizangeo@gmail.com

H. Mahmud
Institute of Environmental Science, University of Rajshahi, Rajshahi, Bangladesh

M. N. Islam, A. van Amstel, *Bangladesh II: Climate Change Impacts, Mitigation and Adaptation in Developing Countries*, Springer Climate,
https://doi.org/10.1007/978-3-030-71950-0_8

8.1 Introduction

Lightning is a temporary atmospheric, transient, and high electric discharge of current (Blumenthal et al. 2012) that originates in the cloud and the route generally expands over a distances kilometers in length (Uman 1987). Naturally, it occurs within the clouds or from the cloud to the ground, but most lightning discharges are within clouds (Sumangala and Kumar 2015).

There are several paths connecting cloud-to-ground lightning with people. Five mechanisms have been mentioned by Cooper and Holle (2010); in order of importance the ground current is the largest (50–55%) followed by side splash (30–35%), upward streamer (10–15%), contact injury (3–5%, and direct strike (3–5%). Lightning is an essential natural phenomenon, to maintain balance electric charges between the atmosphere and earth surface (Ahmed 2000) and one of the primary sources of nitrogen oxides ($NOx = NO + NO_2$) in the troposphere (Banerjee et al. 2014). Worldwide, about 2000 thunderstorms are produced in each moment, which produce about 100 lightning strikes each second. During lightning air is heated to above 30,000 K (29,726 °C) when lightning passes through it, leading to a sudden pressure increase and production of a powerful shock wave that is heard as thunder (Rafferty 2011). Lightning can travel at speeds of 60,000 m/s and heats nearby air to about 10,000 °C, and strikes within only one-fifth of a second (Rahman 2011). As temperature remains at the highest level in the afternoon of the day, lightning mostly occurs at that time (Dewan et al. 2017) and most of the lightning flashes occur in tropical areas between ±23° latitude (Nath et al. 2009). According to Albretch et al. (2016), the highest flash rate density is found in Lake Maracaibo, Venezuela, which is situated in 9°.45′ north latitude. In most countries, lightning mostly occurs during pre-summer or the summer season owing to suitable climatic preference, as it is inextricably correlated with climatic factors, even for each degree of surface warming lightning can be increased (Price 2008). Intensification of water vapor into the upper troposphere enhances the possibility of lightning (Singh et al. 2011); similarly, lightning can be increased with an increase in rainfall (Singh and Singh 2015). Earth's average surface temperature has increased by 0.74 °C ± 0.18 °C during the twentieth century (IPCC 2001) and the rate of increase was 0.17 °C per decade in the last 30 years. If the greenhouse emissions continue, climate models project that the earth's temperature will increase by an amount of 1.4–5.8 °C by the year 2100 (Ahmed and Shamsuddin 2011), indicating that the lightning rate will increase globally.

Every year, the world loses 24,000 people to lightning strikes. The International Decade for Natural Disaster Reduction (IDNDR), professed lightning to be one of the most dangerous natural disasters (Ma et al. 2008). There is a variation in lightning casualties between developed and developing countries, and in developing countries this rate is nearly 2-/3-fold higher than in developed countries. Increasing awareness, modern technology, and preparedness, along with a proportion of the rural people shifting to urban areas as the population shifts away from labor-intensive agriculture are the main contributors to reduced lightning casualties in developed countries (Dewan et al. 2017). From 2006 to 2017, a total of 376 people were killed by lightning in the USA (NOAA 2017). In Canada, on average 9–10 people die each year because of lightning (Mills et al. 2006), China experienced 360 people lost for the same reason during 1997–2009 (Holle and Cooper 2019). Even in Japan, England, Singapore, and Malaysia the lightning casualty rates are lower (Holle and Cooper 2019). Although lightning casualties have been reduced in developed countries in the last century, they have not yet decreased in developing countries.

On the contrary, the developing countries are agriculture oriented, people have to work outside, especially in fields, the main cause of increasing casualties of lightning. Moreover, people do not consider it a disaster owing to lack of knowledge, awareness, and preparedness. That is why most of the south Asian, South American and African inhabitants are heavily affected by lightning strikes (Gomes et al. 2006). For example, in India, during 1967–2012, an average of 1755 people died by lightning each year (Illiyas et al. 2014).

As a whole, the total Indian subcontinent is extremely prone to lightning strike, where a few hotspots are demarcated based on strike intensity (indicated by rectangles in Fig. 8.1). These are comparatively less frequent over the broad northwest plains (A), the highest strike densities are in the northeaster subcontinent within which Bangladesh is situated (B), the southwest coast of India (C), Sri Lanka (D), and the northwestern base of the Himalayas (E). In addition, there is also extensive lightning over the northern Indian Ocean (Fig. 8.1).

Bangladesh is one of the largest deltas of South Asia and one of the most populated areas of the world (Ericson et al. 2005), with a density of 968 people per square kilometer (BBS 2011); it is known as the country of natural disasters (Rahman et al. 2019). Geographically, it is situated between 20°34′ and 26°38′ north latitude and 88°01′ to 92°41′ east longitude, and surrounded by India, Myanmar, and the Bay of Bengal. According to Rahman (2020), the climate of this country is basically sub-tropical with the special influence of the monsoon, classified into three dominant seasons: pre-monsoon hot summer (March to June), wet and humid monsoon (July to November), and post-monsoon (December to January). Pre-monsoon is basically dominated by high temperatures with moderate rainfall, whereas the monsoon rainy season is dominated by moderate temperatures and high rainfall. Geographical setting dynamics, especially location, climatic parameters such as high temperature, rainfall, and humidity mean that Bangladesh has evolved as a suitable ground for a lightning strike. Casualties of lightning mostly occur during the pre-monsoon hot summer and the monsoon, as warmer weather, temperatures, rainfall, and humidity increase help to produce huge quantities of water vapor as well as producing more cumulonimbus cloud and ultimately increasing the tendency toward lighting occurrence (Rahman et al. 2019). Like other south Asian countries, Bangladesh has also been suffering with a high mortality rate from lightning disaster. During 1990–2016, a total of 3086 people have been killed by lightning strikes in Bangladesh (Dewan et al. 2017), and 2016 was recorded as the most

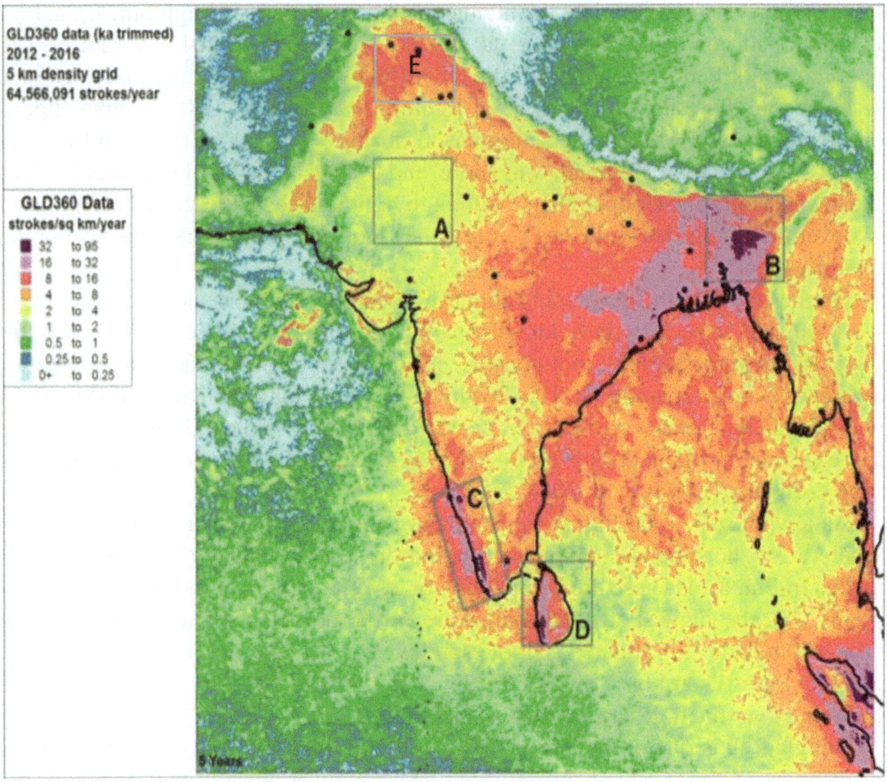

Fig. 8.1 Average annual lightning strike density over the Indian sub-continent from 2012 to 2016. (Modified by the authors by demarcating the area 'E'. Source: Nag et al. 2017)

affected year by lightning casualties. In response to a lightning event in 2016, when 81 lives were lost in just 2 days because of lightning, the government of Bangladesh declared lightning a natural disaster (Biswas et al. 2017).

8.2 Objectives

Specific objectives of this chapter are -

1. To investigate the dynamics of temporal change of lightning casualties in Bangladesh.
2. To discuss the spatial variation of lightning casualties over time in respect of regional analysis of the point of view of the study area.

8.3 Methodology

The spatio-temporal change of lightning was analyzed, basically emphasizing variation of lightning casualties (deaths and injuries) across the country in relation to time. Bangladesh consists of 64 districts, is considered as a spatial unit, and the last 10 years (2010–2019) were taken into account for temporal analysis. As both the pre-monsoon and monsoon seasons are highly prone to lightning strikes, analysis was performed highlighting the seasonal variation in relation to space. To fulfill the objectives of the study, the discussion was presented accordingly. First, to envisage changing patterns over time and space, trend analysis was conducted considering district level and also the whole country. Besides, to present seasonal variations, monthly distribution was also discussed. Second, the spatial pattern of lightning casualties for each year was analyzed to visualize the scenario over the country, especially identifying the highest and lowest prone districts. Third, from the regional point of view spatio-temporal change was presented emphasizing regional variation over time, and trend analysis over the years as the whole of Bangladesh with regard to district level.

Despite work on the topic being limited, a rigorous review of the literature was conducted to obtain a strong background for this study. To explore the variability of lightning events, data on casualties (deaths and injuries) were collected from intensive searching of various printed and online daily newspapers such as www.observerbd.com, m.bdnews24.com, www.thedailystar.net, en.prothomalo.com, www.kalerkantho.com, www.bd-pratidin.com, www.dhakatribune.com, www.daily-sun.com, www.bssnews.net, m.risingbd.com, www.bhorekagoj.com, m.u71news.com, Clickittefaq.com, m.theindependentbd.com for the period 2010 to 2019, because lightning casualty data have not yet been recorded properly in the country. To ensure the reliability of the collected data, they were compared with national disaster reports and published articles in various national and international journals. The collected data were sorted, categorized, and graphically presented using SPSS software. They were also used to detect changes in lightning casualties over time, both locally and nationally through regression analysis. Eventually, spatial variation of lightning casualties in all of the districts as well as the country during the study period was analyzed through mapping presentation using ArcGIS 10.3.1 software. Accordingly, regional analysis was conducted to picture regional variation of lightning casualties over time applying the digital elevation model (DEM) with the help of the same software. Aimed at a better understanding, five class intervals were applied to spatial and three class intervals to regional variation analysis using the equal class interval method.

8.4 Results

Anywhere there are thunderclouds, lightning can strike (Urashima 2007), and the trigger of lightning can be both living organisms and non-living substances, but draws more attention when any living organisms are injured or killed. During lightning strikes, a high voltage natural atmospheric electric charges exchange between earth and clouds (Kalair et al. 2013) through nonliving things such as air, rain, and also living things such as plant and animals' bodies. Its passage severely damages humans, trees, electrical structures, and other living and nonliving objects (IEEE 2005). Hence, any place can be smashed by a lightning strike, but outdoors is identified to be a riskier place than indoors, that is why the rate of casualties is higher among outdoor working group than among those who stay indoors. However, the geographical distribution pattern of economic activity has influenced the acceleration of the damage to properties and life in the recent past in Bangladesh. The following discussion is an effort to present the changing trend of lightning casualties (deaths and injuries) during the years 2010–2019 with regard to Bangladesh.

8.4.1 Temporal Variation Analysis

In Bangladesh, a significant increasing trend in the events (casualties, i.e., deaths and injuries) of lightning has been observed during the recent past, even higher than cyclones (42 in 2019, 18 in 2018, 26 in 2017), floods (114 in 2019, 42 in 2016), and landsides (152 in 2017). A similar result was observed by Gatewood and Zane (2004) in the USA that lightning was second only to flash floods and floods in weather-related deaths during the 30-year record compiled in storm data, ahead of earthquakes, tornadoes, and temperature-related deaths. A casualty is defined as when a person is killed or injured due to an accident or war or disasters. That means that casualties consist of total deaths and injuries. During the study period, the total number of casualties was 3277 persons, among which there were 2149 deaths and 1128 injuries.

From a year-wise distribution point of view, an increasing trend with a little fluctuation was perceived in all sorts of incidences such as casualties (deaths and injuries). In 2010, casualties scored 198 (6.04%), and reached 406 (12.38%) in 2019, with the highest score 442 (13.48%) in 2018. Likewise, a total of 122 persons (10.81%) expired in 2010, extending to 285 (25.26%) in 2019 with a peak of 305 (27%) in 2018. Subsequently, in the case of injury, the score was almost double from 76 (3.53%) in 2010 to 121 (5.63%) in 2019 with the highest score 160 (7.44%) in 2015. At this juncture, the regression coefficient value also justifies the significantly increasing trend in all events where total casualties, total deaths, and total injuries coefficient values were 25.727, 20.636 and 5.0909 respectively (Fig. 8.2). This trend is the opposite of that found in developed nations, which have a dramatically decreasing trend of lightning casualties (Ronald 2016).

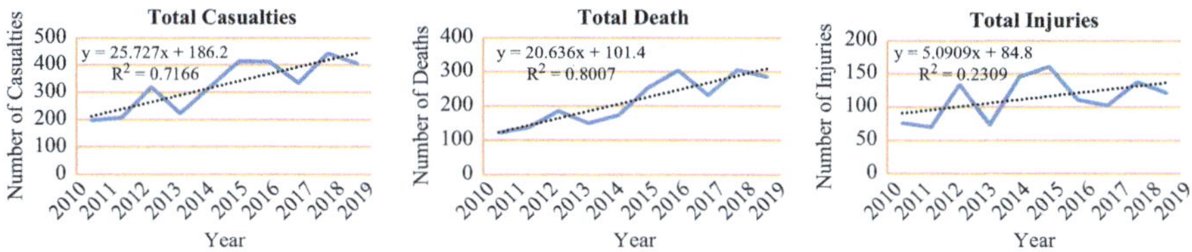

Fig. 8.2 Overall increasing trend of lightning casualties (deaths and injuries) during the years 2010–2019 in Bangladesh

Analysis of district-wise casualties (deaths and injuries) showed a mixed pattern of the trend over the study period. Significant upward trends in the case of casualties (deaths and injuries) are observed in respect to both the spatial and the temporal scale, and the comparison of average change of lightning events (casualties, including deaths and injuries) from 2010 to 2019 clearly indicates a variation among the lightning events, which are not uniform across the study area. Anyway, the individual entity analysis shows that in the case of casualties, out of 64 districts, an increasing trend was found in 52 districts, and a decreasing trend in 12 districts (Fig. 8.3). Similarly, in most of the districts an increasing trend in deaths was observed, totaling 56, but there was a decreasing trend in eight districts. Eventually, with regard to injuries, an upward trend was recorded in 41 districts, and a decreasing trend in 23 districts. However, comparison scenarios in lightning events indicate an overall continuous increase in lightning incidents in most of the areas (Fig. 8.3).

Analysis data suggest a varied changing trend of lightning casualties. At the top of the list of increasing trends were Sylhet, Netrokona, Sunamganj, Rajshahi, Moulvibazar, Magura, and Naogaon in descending order with a coefficient score from 2.1939 to 1.00. Contrariwise, a decreasing trend was found in nine districts, namely Chapai Nawabganj, Bhola, Manikganj, Gopalganj, Kurigram, and Gaibandha in a downward order containing coefficient values ranging from −1.66 to −0.2167 (Fig. 8.3).

Comparative scenarios of death cases caused by lightning, with both increasing and decreasing trends, were noticed. From the quantifying of changing the result during 2010–2019, among the 56 districts, Sylhet, Sunamganj, and Habiganj scored the highest in respect of lightning, with regression coefficient values of 1.3273, 1.105, and 1.00 respectively. Inversely, only eight districts achieved a merit in reducing the death rate from lightning; among them, Chapai Nawabganj and Lalmonirhat achieved top position with regression coefficient values of −0.9515 and −0.4121 respectively (Fig. 8.9, Appendix).

Likewise, from a comparative assessment of historical data of lightning injury, a mostly negative inclination was found, with some positive cases. Out of 41 districts, acceleration of the injury rate to the highest rate was observed in Moulvibazar, then Sylhet and Netrokona, with a regression coefficient value of 0.8727, 0.8667, and 0.8545 respectively. In contrast, the lowering trend in injury rate was observed in 23 districts, among them Chapai Nawabganj, Gaibandha, and Manikganj achieved the

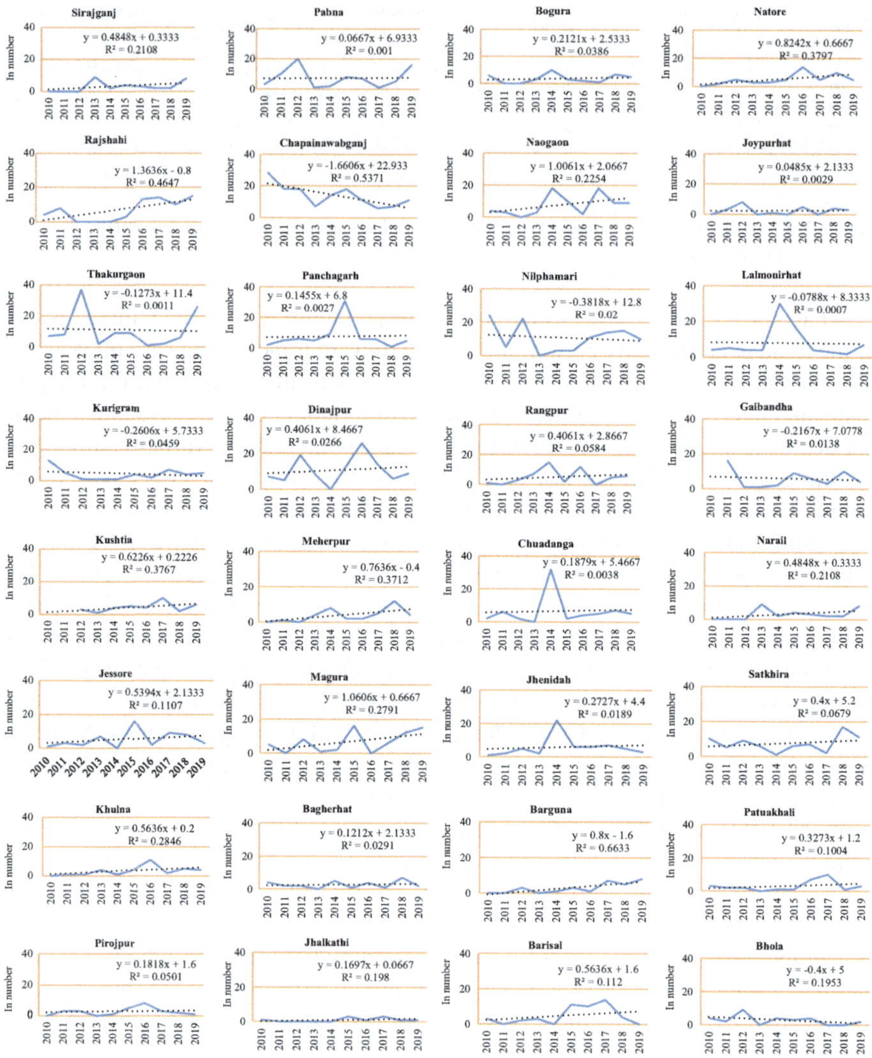

Fig. 8.3 Increasing trend of district-wise lightning casualties during the years 2010–2019 in Bangladesh

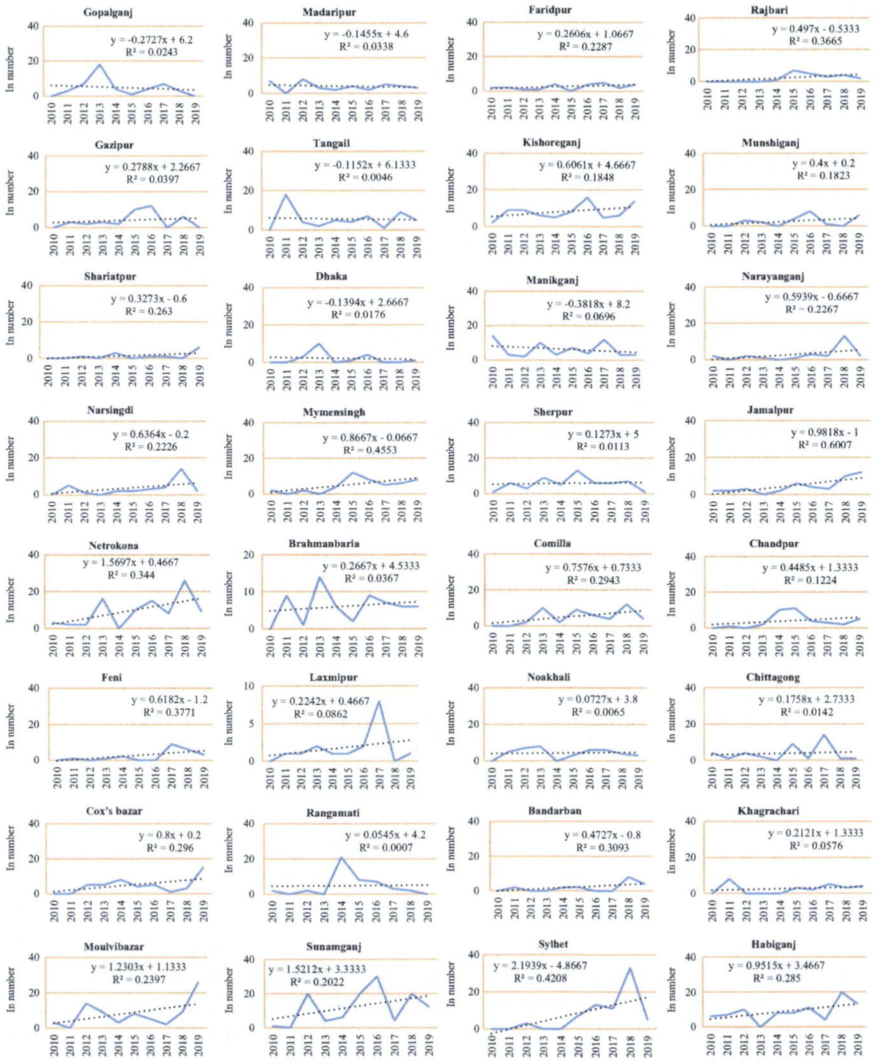

Fig. 8.3 (continued)

highest positions, with regression coefficient values ranging from −0.7091 to −0.3636 accordingly (Fig. 8.10, Appendix).

8.4.2 Spatial Variation Analysis

The spatial change of lighting events was analyzed in Bangladesh as a whole along with considering the district level during 2010–2019. The analysis result indicates that the whole of Bangladesh was affected by lightning events, but the intensity is not evenly distributed over the districts and years. During the study period, the average scenario of lightning casualties indicates that Chapai Nawabganj and Sunamganj were carrying the highest number, 138 (4.21%) and 113 (3.57%) respectively, whereas Jhalokati and Shariatpur had the lowest, 10 (0.30%) and 12 (0.36%) respectively. A similar distribution was observed in the case of deaths from lightning, Chapai Nawabganj and Sunamganj scored the peak with 99 (4.60%) and 90 (4.18%) deaths; in contrast, Jhalokati and Shariatpur had the lowest, with 7 (0.32%) and 8 (0.37%) deaths respectively. Subsequently, the maximum number of injuries took place in Thakurgaon and Nilphamari jointly, with 61 (5.40%), and the lowest numbers were 1 (0.08%) and 2 (0.17%) in Rajbari and Faridpur respectively (Fig. 8.4).

Again, temporal analysis visualizes the scenario of variation in district-wise spatial distribution from year to year (Fig. 8.5). A comparison result reveals that 2010 witnessed the lowest number of casualties, whereas in 2018 the highest number occurred. In 2010, the total number of casualties was 198; among these, Chapai Nawabganj and Nilphamari received the highest number, with 28 and 24 casualties respectively, whereas the lowest number of casualties was 1, received by Rangpur, Jessore, Jhenaidah, and Jhalokati. There was an evident increase in the following 2 years, changing the areas with the highest and lowest numbers of casualties. In 2011 and 2012, the areas with highest numbers of casualties were Chapai Nawabganj (18), Tangail (18), Gaibandha (16), Thakurgaon (37), and Nilphamari (22) respectively, whereas the lowest score was the same as 2010, including new districts.

Figure 8.5 further shows a slightly lowering trend in 2013, and then the casualties increased up to 2016. In 2013, new districts Gopalganj and Netrokona experienced the highest scores, 18 and 16 respectively. Eventually, as new areas Chuadanga and Lalmonirhat scored the highest number of casualties with 32 and 30 respectively in 2014 and both of the figures were higher than a recent previous year, 2012. Accordingly, 2015 included a new district as the area with the highest number of casualties, Panchagarh, with 31 casualties. Similarly, in 2016, Sunamganj and Dinajpur were found to be the area with the highest numbers of casualties, with values of 30 and 26 respectively. Once, in 2017, a fluctuation was observed; then, the trend was re-claimed until 2019, with the peak in 2018. Naogaon was secured as the new area with the highest number of casualties in 2017, with 18. In 2018, Sylhet had 33 casualties, which was the highest in that year. Finally, in 2019, both Moulvibazar and Thakurgaon scored 26, which was the highest of the year. From

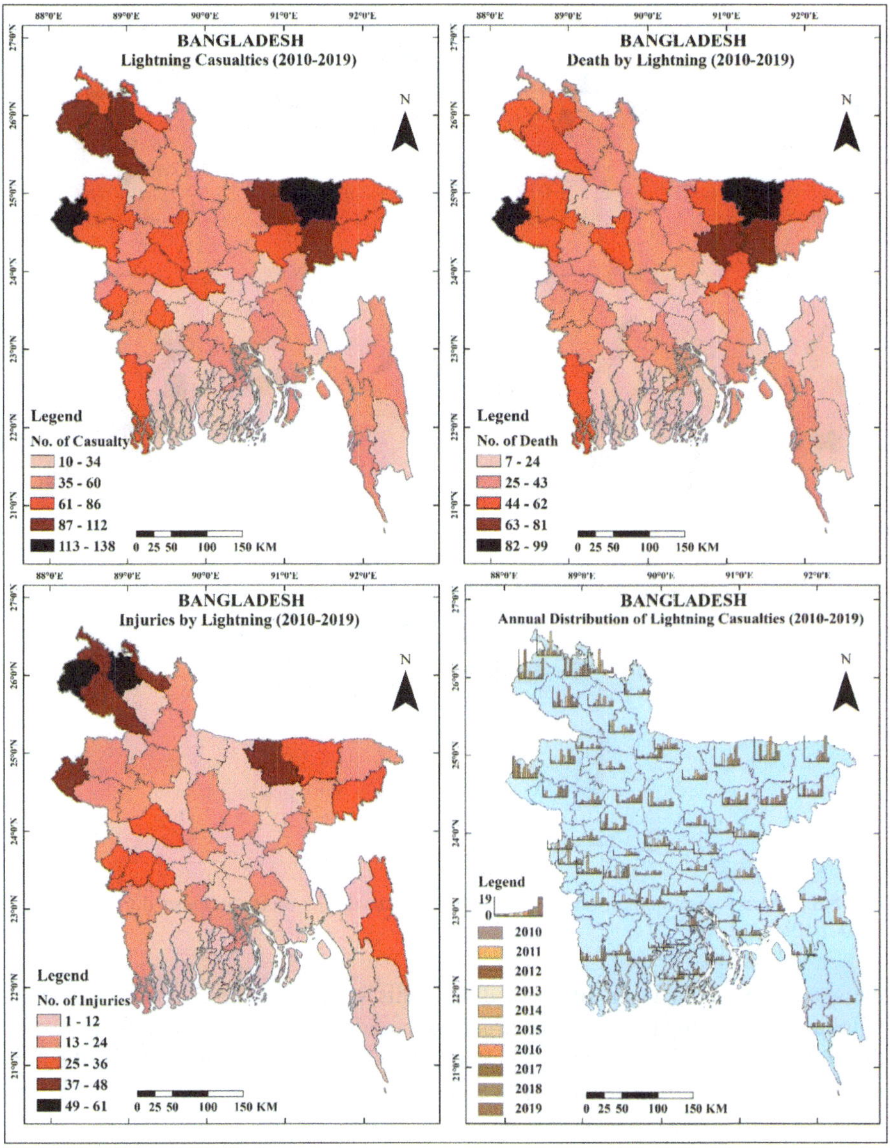

Fig. 8.4 Overall spatial variation of casualties (death and injuries) by lightning in Bangladesh during the years 2010–2019

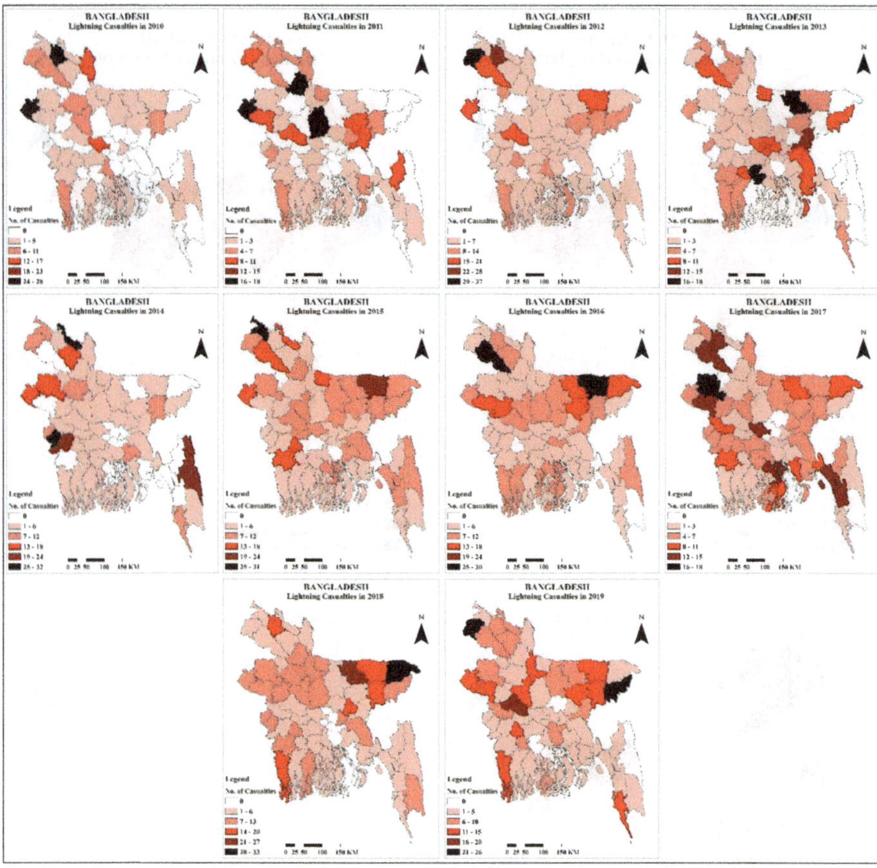

Fig. 8.5 District-wise spatial variation of lightning casualties in Bangladesh during the years 2010–2019

the spatial variation point of view, during this study period, Chapai Nawabganj, Nilphamari, Sunamganj, and Thakurgaon secured the peak of casualties twice, and the highest score was secured by Thakurgaon, 37 in 2012. On the contrary, the lowest score was 1, varying from four to eight districts over the years.

8.5 Discussion

Envisaging the spatially and temporally changing situation of lightning incidents over time and space was the primary aim of the study, and the following discussion is an attempt to present the scenario of the distribution of lightning events in Bangladesh on both a spatial and a temporal scale. The yearly casualty trend obtained from the analysis of collected data depicts the highly positive change with

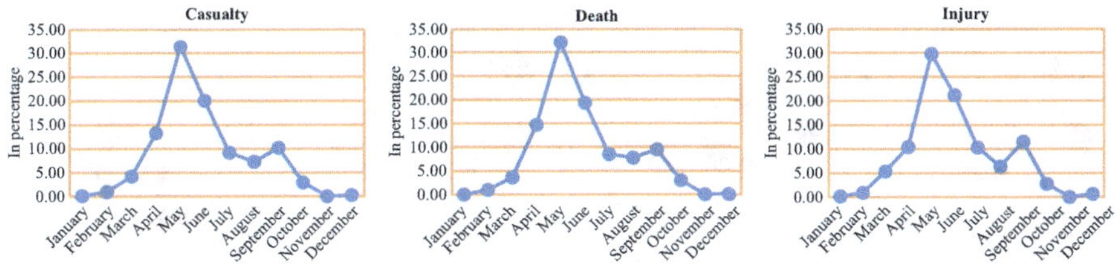

Fig. 8.6 Monthly distribution of casualties (deaths and injuries) during the years 2010–2019

a little fluctuation over the study period and encompasses new areas. Comparative assessment of Fig. 8.5 reveals that the spatial expansion of lightning casualty from 2010 onward is significant. In 2010, only 37 districts experienced of lightning casualties, which applied to 53 districts in both 2011 and 2012. Hence, a little fluctuation was found in 2013, shrinking to 44 districts. Apart from these, it was found that lightning casualties brought more districts gradually under its jurisdiction up to 2019, totaling 60 districts, with a pick of 61 districts in 2016.

Figure 8.6 shows the monthly variation of lightning events (casualties including deaths and injuries) received throughout the country during the study period. Analysis of the results reveals that over a 10-year period a wide variation of lightning events were experienced each month. As lightning events were significantly linked with year-round weather conditions, a wide winter season, during November to February, received the smallest number of casualties (deaths and injuries), which is very unusual, rather than it being continuous each year. In contrast, other months of the year, March to October, receive close to 100% of the casualties (death and injury). This period is established as the season of lightning events, which is most likely the findings of the research result in India (Singh and Singh 2015). The figure further indicates two prominent peak seasons, the first maximum during the pre-monsoon with the peak in May, and the second highest during the monsoon, with a maximum in September. The pre-monsoon is characterized by strong incoming solar radiation; thus, thunderstorms are very frequent during this time period (Chowdhury and De 1995). According to Dewan et al. (2017) the seasonal variation of lightning fatalities and injuries indicates the pre-monsoon season to have more casualties than others, which is similar to the findings of this study.

However, the findings from the monthly distribution of casualties show that the pre-monsoon season receives 69.08% of casualties, whereas May is perceived to be the highest individual month (31.37%); and the second highest season, the monsoon, manifesting 29.59% of lightning casualties, with September displaying the highest monthly score, 10.19% of total casualties. A similar distribution is found from death analysis, the pre-monsoon achieves 70.04% of cases of death whereas May secures the highest score, 32.20% individually; and the monsoon acquires 29.04% deaths, whereas September shows the top monthly score, 9.49% of total deaths. Accordingly, evaluation of the injury analyses results explores the very close outcome of casualties as a whole, and deaths in particular. Eventually, the pre-monsoon is exposed as the highest season, with 66.85% injuries, with May touching

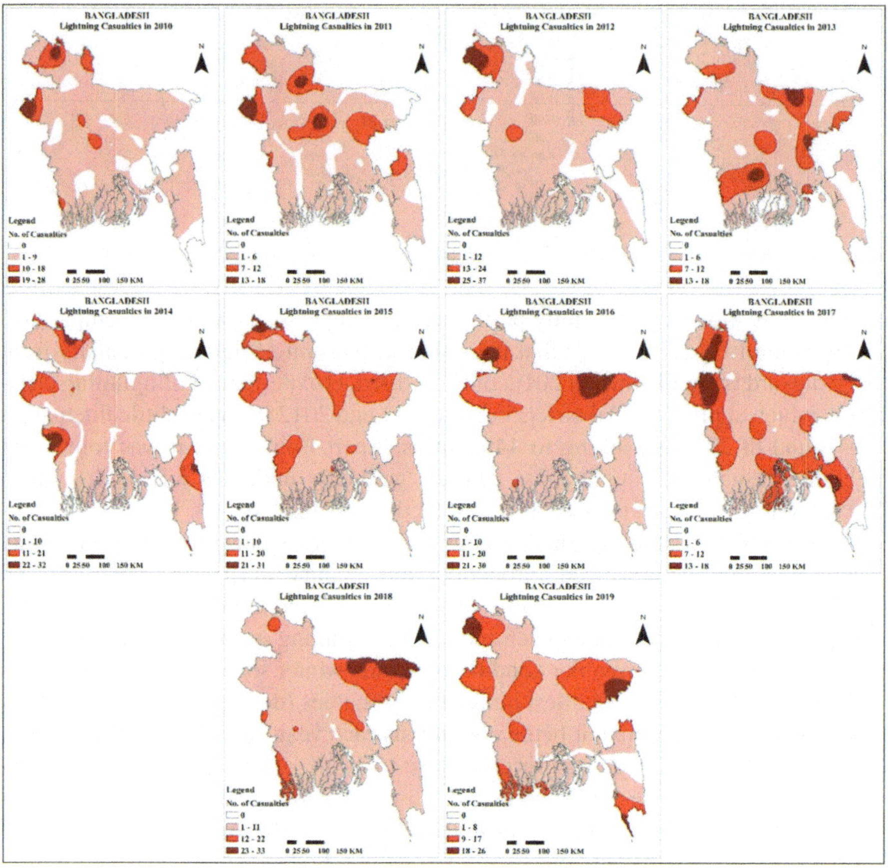

Fig. 8.7 Yearly regional variation of lightning casualties in Bangladesh during the years 2010–2019

the peak with a score of 29.79% injuries; and the monsoon caused 31.11% of injuries with the highest score of 11.52% in September.

Regional variation analysis of lightning casualties gives a better insight into different casualty-prone areas on a spatio-temporal scale over the country. From this point of view, a regional expansion trend of the highest number of casualties was found, either in scattered or in continuous form, along with higher contiguity in the greater northern region during the study period. The generalized distribution maps (Fig. 8.7) depict that northwestern (Chapai Nawabganj) and extreme northern parts (Nilphamari) of the country were exposed as the highest casualty areas, whereas other parts show moderate to low casualty rates, except in casualty-free districts of the country in 2010. The next year, 2011, indicates a wide-ranging area directed by the highest to moderate casualties from the northwestern part (Chapai Nawabganj, Thakurgaon) to the middle-eastern part (Kishoreganj, Brahmanbaria) of the country with few breaks. Hence, in both of the years, the greater Sylhet region was not within the outline of lightning casualty area. But in 2012, Netrokona was included in the moderate casualty region and held the highest casualty position with a remarkable extended area in a south direction in the next few years. Along with Netrokona region, the southwestern parts (Satkhira, Khulna, and Bagerhat) had the highest index of casualties in 2013. On the other hand, in 2014, lighting casualties basically dominated almost the whole western and at the same time southeastern coastal parts of the country. However, the northern region was found to be the area with the highest number of casualties in both 2015 and 2016. Surprisingly, massive spatial extension of the highest casualty areas was found alongside the board of the whole country in 2017. But this highest casualty area was limited to only the greater Sylhet region in 2018. Again, as the area with the highest number of casualties, the major portion of the whole northern region and coastal regions was exposed in 2019. Hence, during the study period, along with the area with the highest number of casualties, an expanding trend was found in the case of areas with moderate and low numbers of casualties, by occupying the casualty-free areas. However, the distribution of deaths and injuries depicted similar results to those of casualties as a whole.

Comparative assessment of overall lightning events of the entire study period shows a very comprehensive depiction of north to southward movement of a lowering tendency (Fig. 8.8). From the casualty distribution, it is perceived that there are two distinct hotspots, which are the zones with the most lightning (96–138 casualties) such as the northwestern and northeastern parts of the country. With regard to casualties, within the two highest zones, Chapai Nawabganj, Thakurgaon and Dinajpur in the northwest region, and Sunamganj and Netrokona in the northeastern zone were the districts with the highest number of casualties; those are surrounded by the moderate zone (53–95 casualties). Moreover, another moderate zone was found in the southwestern zone, which includes Satkhira, Jhenaidah, Chuadanga, Jessore, and Meherpur districts. Subsequently, a similar zonal pattern to that for casualties was observed in the case of deaths, with a change in the number of districts in the northwest region, where Thakurgaon and Dinajpur shifted position from the highest death zone to moderate. Additionally, a spatial extension was apparent in the moderate death area of the northeastern zone. Similarly, regional distribution of injury shows two highest zones (41–61 injuries), the northwestern and the northeastern part of the country. Eventually, these zones are surrounded by moderate zones (21–40 injuries). Moreover, the scenario of injury depicts two more moderate zones in the southwestern and southeastern parts of the country. Nonetheless, with respect to all events, considering the entire period, the whole country was found to be a low affected zone except for high and moderate zones (Fig. 8.8).

8.6 Conclusion

Like other hazards, lightning is a significant natural phenomenon in Bangladesh, considering the recent trends of casualties, especially deaths, it is placed at the top of the hazard list. Lightning trend analysis provides insights in real facts into changing patterns of casualty (deaths and injury) distribution over time and space. The

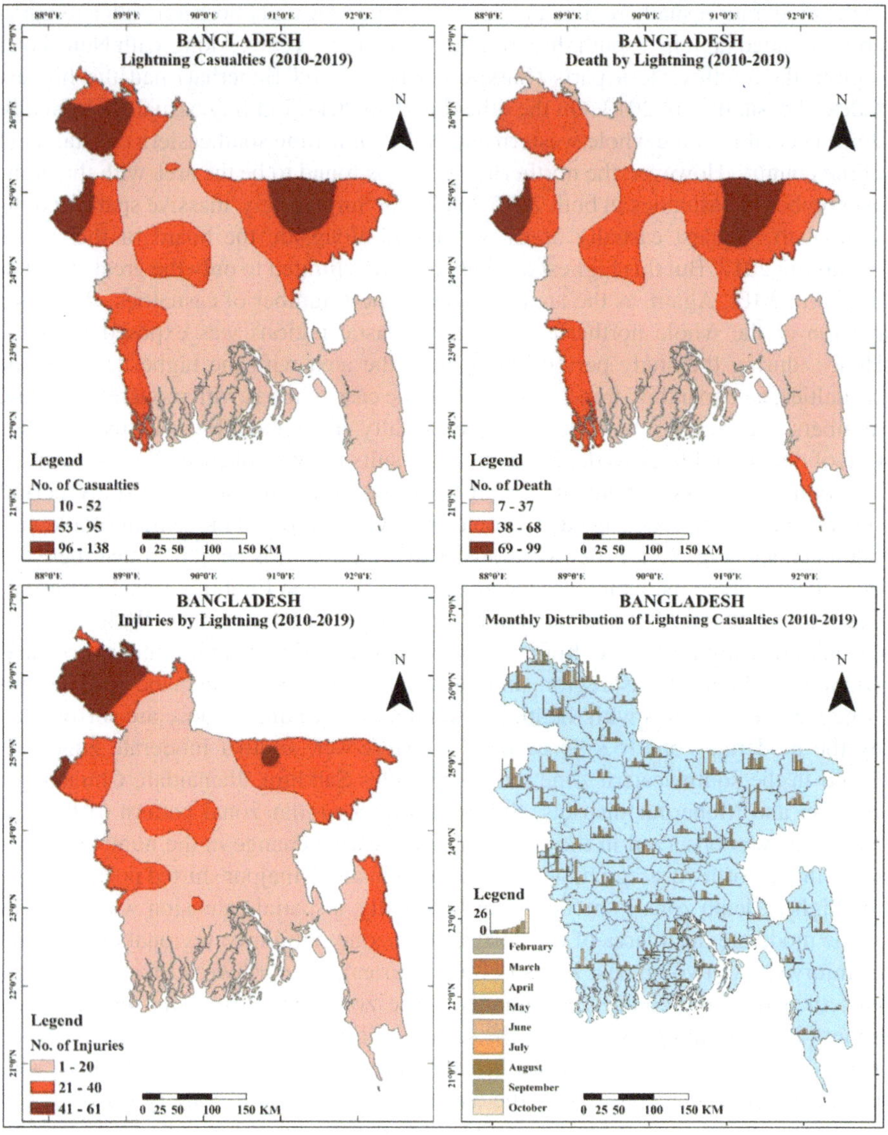

Fig. 8.8 Overall regional variation of lightning casualties (deaths and injuries), and monthly variation in Bangladesh during the years 2010–2019

result of variation in the yearly and seasonal lightning casualty trends, along with the spatial dimension over the 10-year period, is considered as the significant outcome of the study.

The present study found a significant upward trend in all sorts of lightning events (casualties as a whole, as well as deaths and injuries separately). Over the 10-year period, casualties as a whole increased by 6.34%, deaths by 14.45% and injuries by 2.10%. Although there were positive change approaches in all cases, differences were observed in the individual cases; casualties as a whole reached a peak in 2018, deaths in 2015 and injuries in 2018. Hence, compare with other hazards, damage to property from the lightning event is significantly lower, but casualties are substantially higher in recent times.

In contrast, the findings of spatial analysis show casualty to be unevenly distributed over the country; the districts with the highest numbers of casualties were Chapai Nawabganj (4.21%) and Sunamganj (3.57%), whereas the districts with the lowest numbers were Jhalokati (0.30%) and Shariatpur (0.36%) during the study period. Similar spatial distribution as for casualties was found with regard to death, showing Chapai Nawabganj (4.60%) and Sunamganj (4.18%) to be the most affected district, whereas Jhalokati (0.32%) and Shariatpur (0.37%) were the least affected. But a different spatial distribution the in case of injury was observed. Thakurgaon and Nilphamari jointly secured the highest position (5.40%), and the lowest was Rajbari (0.08%) and Faridpur (0.17%). Contrariwise, the yearly spatial pattern indicates inconstancy in the highest and the lowest district incorporated. Surprisingly, Chapai Nawabganj was credited with having the highest position with a decreasing casualty rate, but until now it had secured the top position in receiving total casualties.

From the regional point of view, initially, the northern part of the country was approaching being the area most prone to being affected by lightning casualties as well as death and injury but, in case of casualties and death an expansion trend was found toward the southwestern, and injury toward the southwestern and southeastern parts over the last 10-year.

During the collection of data, it was observed that most of the occurrences happened while people were working in agriculture-affiliated activities such as in crop fields, fishing, or any other activities that take place in the open air. That is why the pre-monsoon season accrues the highest number of casualties, the season of cultivating and collecting paddy crops, mango, and other fruits, as well as fishing. Besides, the observation also indicates that there is less awareness and seriousness among the people when considering that lightning can be the cause of casualties.

However, it is expected that the findings of the present study will help people in disaster management to think seriously about lightning as one of the prodigious hazards such as flood, drought, cyclone, etc., and contribute to formulating a suitable and sustainable development policy and program for the country. For instance, as for other hazards, forecasts and early warning systems can be developed so that people can take shelter in a safe place during lightning strikes. As most people care less about lighting strikes, the government should initiate the awareness-building program through publicity on different print and electronic media, along with topics on lightning in the school curriculum, so that people can take proper lessons right from the early stage of life. Moreover, medical care and emergency communications systems need to improve in order to provide quick treatment facilities to the affected people who may have made significant contributions to reducing the death rate of lightning strikes.

Appendix

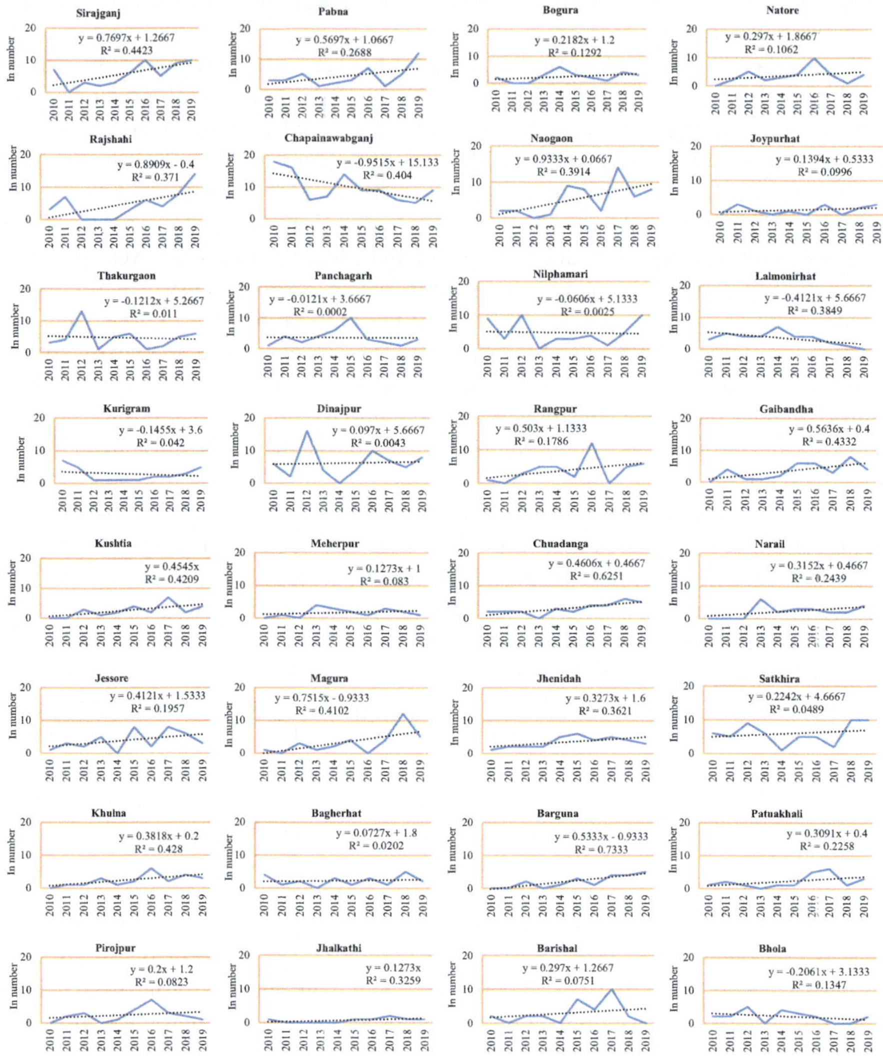

Fig. 8.9 District-wise lightning death: changing trends during the years 2010–2019 in Bangladesh

Fig. 8.9 (continued)

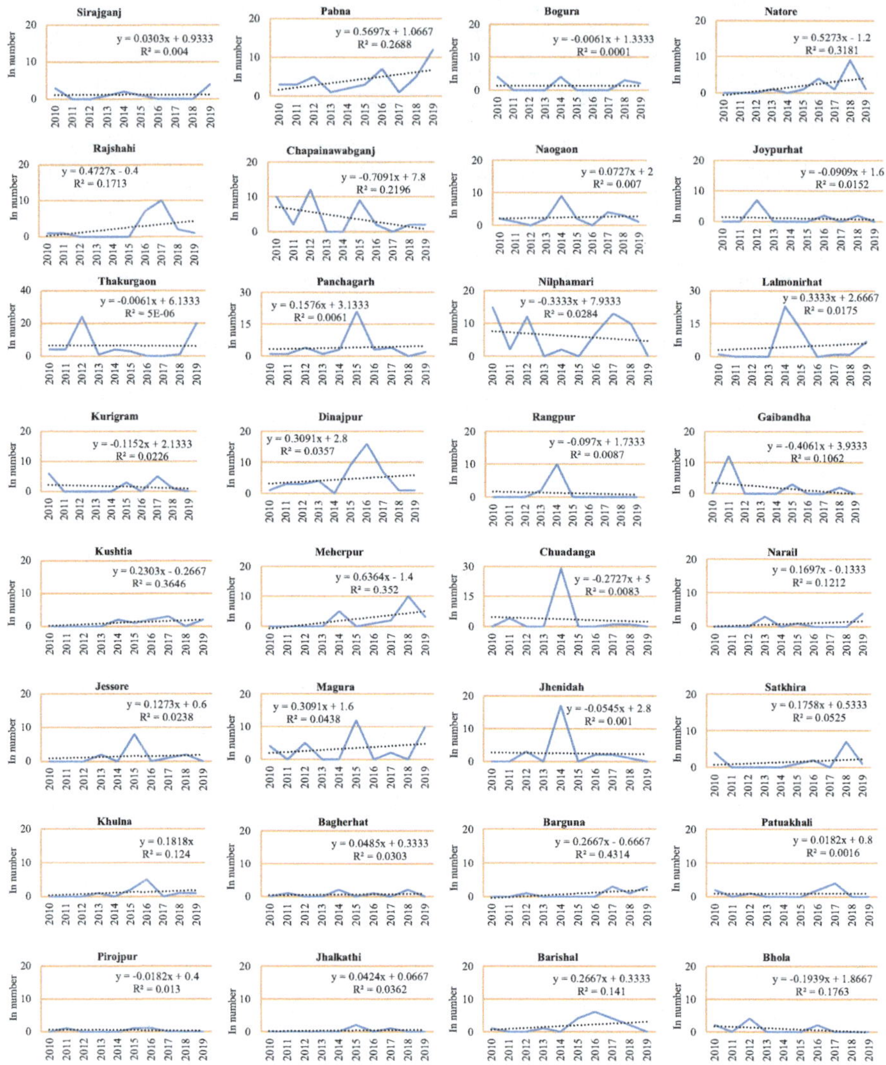

Fig. 8.10 District-wise lightning injury: changing trends during the years 2010–2019 in Bangladesh

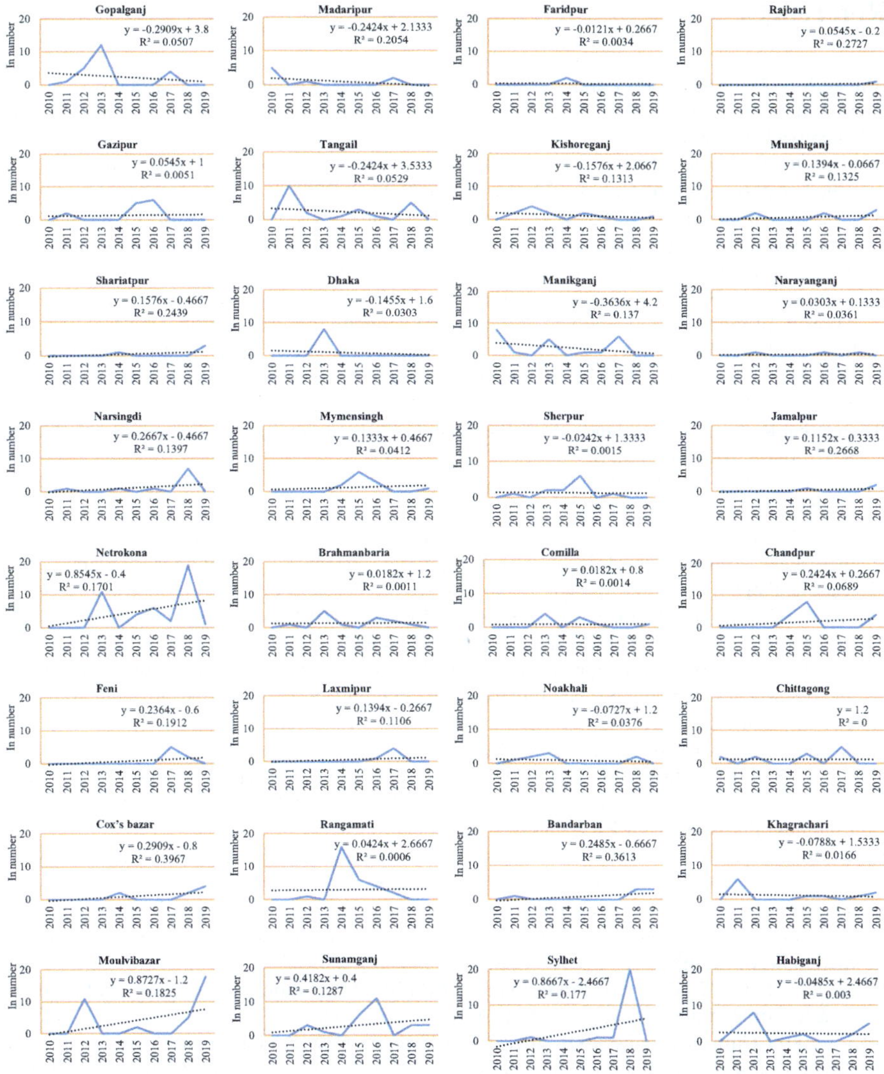

Fig. 8.10 (continued)

References

Ahmed R (2000) Meteorology and climatology. Uttaron Ofset Printing Press, Rajshahi

Ahmed R, Shamsuddin SD (2011) Climate change-issues and perspectives for Bangladesh. Shahitya Prakash, Dhaka

Albrecht RI, Goodman SJ, Buechler DE, Blakeslee RJ, Christian HJ (2016) Where are the lightning hotspots on Earth? Bull Am Meteorol Soc 97:2051–2068

Banerjee A, Archibald AT, Maycock AC, Telford P, Abraham NL, Pyle JA (2014) Lightning NO_x, a key chemistry-climate interaction: impacts of future climate change and consequences for tropospheric oxidising capacity. Atmos Chem Phys 14:9871–9881

BBS (Bangladesh Bureau of Statistics) (2011) Population and housing census: preliminary results. Ministry of Planning, Government of the People's Republic of Bangladesh, Dhaka

Bishwas A, Dalal K, Hossain U, Baset KU, Rahman F, Mashreky SR (2017) Lightning injury is a disaster in Bangladesh?-exploring its magnitude and public health needs. F1000 Res 2016 5:2931

Blumenthal R, Trengrove E, Jandrell IR, Saayman G (2012) Lightning medicine in South Africa. S Afr Med J 102(7):625–626

Chowdhury MAM, De UK (1995) Pre-monsoon thunderstorm activity over Bangladesh from 1983 to 1992. Terr Atmos Ocean Sci J 6:591–606. https://doi.org/10.3319/TAO.1995.6.4.591(A)

Cooper MA, Holle RL (2010) Mechanisms of lightning injury should affect lightning safety messages. Preprints, third international lightning meteorology conference Orlando, FL, Vaisala, 5 pp

Dewan A, Hossain MF, Rahman MM, Yamane Y, Holle RL (2017) Recent lightning-related fatalities and injuries in Bangladesh. Weather Clim Soc 9:575–589

Ericson JP, Vorosmarty CJ, Dingman SL, Ward LG, Meybeck M (2005) Effective sea-level rise and deltas: causes of change and human dimension implications. Glob Planet Chang 50:63–82

Gatewood MOK, Zane RD (2004) Lightning injuries. Emergency medicine clinic of North America. Elsevier Saunders 22(2004):369–403

Gomes C, Kithil R, Ahmed M (2006) Developing a lightning awareness program model third world based on American experience. Paper presented at 28th International Conference on Lightning Protection (ICLP), Kanasawa, Japan

Holle RL, Cooper MA (2019) Overview of lightning injuries around the world. The 11th Asia-Pacific international conference on lightning. Hong Kong, China

IEEE (2005) How to protect your house and its contents from lightning. IEEE guide for surge protection of equipment connected to AC power and communication circuits. Standards information network, IEEE

Illiyas FT, Mohan K, Mani SK, Pradeepkumar AP (2014) Lightning risk in India: challenges in disaster compensation. Econ Polit Wkly XLIX(23):23–27

IPCC (2001) Climate change 2001: the scientific basis, Contribution of Working Group 1 to the third assessment report of the Intergovernmental Panel on Climate Change. Cambridge University Press, Cambridge

Kalair A, Abas N, Khan N (2013) Lightning interactions with humans and lifelines. J Light Res 5:11–28

Ma M, Lyu WT, Zhang Y, Meng Q (2008) Analysis of lightning disasters in China and their correlative factors. Adv Earth Science 23(8):856–865

Mills B, Unrau D, Parkinson C, Jones B, Yessis J, Spring K (2006) Striking back: an assessment of lightning-related fatality and injury risk in Canada. Final technical report. Environment Canada, Waterloo, Canada

Nag A, Holle RL, Murphy MJ (2017) Cloud-to-ground lightning over the Indian subcontinent. Paper presented at the 8th conference on the meteorological applications of lightning data, Seattle, Washington

Nath A, Manohar GK, Dani KK, Devara PCS (2009) A study of lightning activity over land and oceanic regions of India. J Earth Syst Sci 118(5):467–481

NOAA (2017) Frequently asked questions about lightning. Available at https://www.nssl.noaa.gov/education/svrwx101/lightning/faq/
Price C (2008) Thunderstorms, lightning and climate change. Paper presented at 29th international conference on lightning protection, Uppsala, Sweden
Rafferty JP (2011) Dynamic earth: storms, violent winds and earth's atmosphere. Britannica Educational Publishing, New York
Rahman A (2011) Lightning protection practices in Bangladesh: an overview. Available at http://www.lightningsafety.com/nlsi_lhm/lightning-protection-bangladesh-2011.pdf
Rahman MA (2020) Analysis of spatio-temporal changes in temperature over Bangladesh using observed and CMIP5 HadGEM2-AO model data. In: Shamsuddin SD, Alam MS, Roni R (eds) Climate change-issues and perspective. Shahitya Prakash, Dhaka
Rahman SM, Hossain M, Jahan M (2019) Thunderstorms and lightning in Bangladesh. Bangladesh Med Res Counc Bull 45:1–2
Ronald LH (2016) A Summary of Recent National-Scale Lightning Fatality Studies. Weather Clim Soc 8(1):35–42
Siingh D, Singh RP, Singh AK, Kulkarni MN, Gautam AS, Singh AK (2011) Solar activity, lightning and climate. Surv Geophys 32:659–703
Singh O, Singh J (2015) Lightning fatalities over India: 1979-2011. Meteorol Appl 22(4):770–778
Sumangala CN, Kumar MP (2015) Lightning death: a case report. J Indian Acad Forensic Med 37(1):93–95
Uman MA (1987) The lightning discharge. Academic, Cambridge
Urashima K (2007) Considering of lightning damage protection and risk reduction for a safe and secure society. China Q 25:21–32

[illegible]

Chapter 9
Climate Change-Induced Livelihood Vulnerability and Adaptation of St. Martin's Island's Community, Bangladesh

Md. Kawser Ahmed, Farjana Yeasmin, Md. Monirul Islam, Aparna Barman, and Zakir Hossain

Abstract Bangladesh, amongst the countries most vulnerable to climatic impacts, is facing a significant impact on the livelihood of the community, especially in the coastal areas, where subsistence mostly depends on the natural resources. This study analyzes the livelihood vulnerability caused by climatic hazards in St. Martin's Island, Bangladesh, with reference to different well-being groups and their adaptation strategies. Both primary and secondary data were used. The primary data were collected from three well-being groups (well-off/rich, medium, and poor groups) of the community using key informant interviews, vulnerability matrices, and focus group discussions. The major climatic hazards identified include storms, cyclones, rainfall, salinity intrusion, and tidal water increase, which significantly affect the priority livelihood activities such as tourism business, agricultural practices, fishery practices, and other occupations. Cyclones momentously affect all three groups of people, despite their economic differences. Tidal water increase, on the other hand, affects mostly the poor and medium groups of people. Heavy rainfall causes damage mostly to the poor group. Interestingly, salinity intrusion has profound effects on the rich and medium groups of people compared with the poor group. Collectively, the rich (score: 8.25) and medium (score: 8.25) groups of people have a higher level of vulnerability due to major hazards than the poor group (score: 7.50), although the

M. K. Ahmed · F. Yeasmin
Department of Oceanography, University of Dhaka, Dhaka, Bangladesh

M. M. Islam (✉)
Department of Fisheries, University of Dhaka, Dhaka, Bangladesh

A. Barman
Sustainability Services Ltd., Dhaka, Bangladesh

Z. Hossain
Department of Statistics, University of Dhaka, Dhaka, Bangladesh

M. N. Islam, A. van Amstel, *Bangladesh II: Climate Change Impacts, Mitigation and Adaptation in Developing Countries*, Springer Climate,
https://doi.org/10.1007/978-3-030-71950-0_9

variation is not statistically significant (p value = 0.953). To cope with the impacts, the islanders are following short-term traditional adaptation strategies, i.e., changing crop variety, changing fishing patterns, selling more livestock, and selling corals, handmade ornaments, dry fish, and mollusks. However, to live with the climatic impacts in the longer-term and maintaining sustainable livelihoods, long-term, proactive and effective adaptation strategies need to be taken, e.g., enhancing cooperation among the community people, improving infrastructure, incorporating modern fishing technology, and producing new as well as diversified livelihood activities, amongst others.

Keywords Climate change · Livelihood · Vulnerability · Adaptation · St. Martin's Island · Bangladesh

9.1 Introduction

The subsistent development and future survival of mankind are being affected by global climate change (Tao et al. 2011), a broadly discussed issue that exacerbates the frequency and intensity of natural disaster (Kabir 2014) such as floods, tropical cyclones, salinity intrusion, tidal water surge, droughts, and heat-waves (IPCC 2014). These disasters bring about a devastating effect on a nation's agriculture, fisheries, water supply, food resources, health, and shelter (CFE-DM 2017). Bangladesh, a country with the seventh ranking in the Global Climate Risk Index among 170 countries, is among the most vulnerable to climate change (Rawlani and Sovacool 2011; Adem et al. 2017; Eckstein et al. 2019). The geographical location, low-lying topography and funnel-shaped coast expose the land to cyclones, tidal surges and seasonal flooding (Ali 1999; Rahman et al. 2011) and cause damage to lives and properties. In addition, a large population base, poor institutional development, and widespread poverty are making Bangladesh more vulnerable to climate variability and change (World Bank 2000). Coastal areas of the country are mostly exposed to these disasters. In Bangladesh, approximately 44 million people live in the coastal area, which made the situation worse, as some disasters such as cyclones are likely to hit in the coastal regions only (Mallick et al. 2017). About two-thirds of Bangladesh lies within 5 m above sea level with a high population density (Agrawal and Perrin 2009) whereby a 1-m increase in sea level will immerse 18% of the total land area, affecting 11% of the total population (Huq et al. 1995; Islam 2002; Alam 2003; Islam et al. 2014b; Ahammed et al. 2016). The predicted increase in sea level in Bangladesh is ten times more rapid than the global average, which is putting the country and its coastal areas at a four times higher risk (CFE-DM 2017) and can initiate permanent inundation, drainage congestion, salinity intrusion, and frequent storm surge inundation (Mohal et al. 2006).

From the last decade, vulnerability, adaptation, and adaptive capacity have emerged as the key concepts in illustrating the social implications of climate change (Füssel and Klein 2006). Since the publication of the IPCC fourth assessment report (2007), vulnerability and adaptation have gained the attention of numerous researchers throughout the world (Balasubramanian et al. 2007; Below et al. 2012; Gbetibouo et al. 2010; Iglesias et al. 2011; Malone and Brenkert 2008). This emphasizes the need to carry out research studies on climate change to understand the impacts, vulnerability, and adaptation. According to the Intergovernmental Panel on Climate Change (IPCC), vulnerability can be defined as the degree to which a system is susceptible to, or unable to cope with, adverse effects of climate change, including climate variability and extremes (IPCC 2001). It is a function of the character, magnitude, and rate of climate variation to which a system is exposed, its sensitivity, and its adaptive capacity (IPCC 2001), whereby adaptive capacity is the ability of a system to adjust to climate change (including climate variability and extremes), to moderate the potential damage from it, to take advantage of its opportunities, or to cope with its consequences (IPCC 2007).

Coastal districts of Bangladesh are facing an expanding range of stresses and traumas, the scale of which now poses a warning to the resilience of both humans and environmental coastal systems, and are possibly to be aggravated by climate change (IPCC 2007; Islam et al. 2014a). Major climatic hazards such as storms, cyclones, land erosion, salinity intrusion, tidal water increase, sea-level rise, and heavy rainfall invade the coastal regions of Bangladesh (Islam et al. 2014a, b). Cyclones and tidal water increases have frequently invaded these regions, annihilating lives and properties (Alam and Collins 2010). Approximately 5.5% of the world's tropical cyclones hit the Bay of Bengal (Ali 1996) and the frequency of severe cyclones is one in every 3 years (GoB 2009), which critically destroys lives and property. Salinity intrusion and tidal water increases have long-term social and economic impacts (IPCC 2007). The natural resource-dependent livelihood practices such as agriculture and fisheries are facing severe impacts due to these climatic hazards (Alam et al. 2013). The common perception is that poor income group communities are the most vulnerable to climate change-induced hazards, as they have less capacity to adapt to the adverse environmental conditions (IPCC 2001; Islam et al. 2014c). However, this phenomenon may not be applicable everywhere. It may change over time, place, and in a particular situation in any community considering their access to resources, knowledge, and information. Therefore, for better understanding and effective policy development, local-level study of vulnerability and adaptive strategies for all well-being groups is a prerequisite.

St. Martin's Island, the only coral island of Bangladesh, is situated in the Bay of Bengal. The community's dependence on natural resources expose them to climatic hazards (IPCC 2007; Paavola and Adger 2002; Mosse 1994). Besides, the panoramic landscapes, clear seawater, and natural treasures of coral colonies have made this island an enchanting tourist spot (Rani et al. 2020) with the convenience of tourism-related livelihoods. Historically, there have been various research studies on St. Martin's Island from food security (Feeroz 2009) to policy development (Islam 2001). Besides, the island's diversified biodiversity i.e., coral reef, seaweeds, fish, and shellfish (Alam et al. 2015; Haider 2008; Tomascik 1997; Mollah 1997) as well as their potential threats (Moudud 2010; Hasan 2009) have been a vast hotspot of research. But despite being the most exposed and most vulnerable coastal area of Bangladesh owing to climatic hazards, and despite the potential threat to the livelihood of the community due to the dependency on natural resources, research on the livelihood vulnerability of St. Martin's Island's community and how they adapt in such vulnerable circumstances has been disregarded among the researchers (Islam 2002; Alam 2003; Islam et al. 2014b; Ahammed et al. 2016). To understand the community's evidentiary condition and to implement sustainable adaptation practices we need to know the community level vulnerability and how they cope with the adverse situation. For this reason, this study was aimed at exploring how climate change is affecting the livelihoods of local communities, thus determining the vulnerability of different well-being groups and how they are trying to cope with and adapt to the climate hazards.

9.2 Materials and Methods

9.2.1 Study Area

St. Martin's Island is the southernmost union of Bangladesh with a total area of 835 acres, which lies 3.6 m above the mean sea level (Chowdhury 2012). The population of this island is only 6703, with a total of 1169 households (BBS 2011), where 90% of them consist of fishermen (Haider 2008). Myanmar's Arakan coast resides about 10 km away from the eastern side of the island; the southern and western sides are bounded by open sea and the only mainland edge, the Teknaf coast of Bangladesh, lies about 9.8 km north of the island (Hossain 2001) (Fig. 9.1).

9.2.2 Methodology

9.2.2.1 Data Collection

Both primary and secondary data were collected for this study during the period October 2017 to February 2018. To collect primary data, this study used key informant interviews (KIIs), climate vulnerability and capacity analysis (CVCA), and focus group discussions (FGDs).

A total of 12 KIIs were conducted to make up the three well-being groups of the sample people of St. Martin's Island, based on their income, food sufficiency, land ownership, and educational qualifications (Mosse 1994) (Table 9.1). According to the KII findings, the island contains approximately 30–50 rich households, 300–500 medium-rich households, and the remaining 800–1000 were poor households.

This study adopted CARE's CVCA method to explore the climatic consequences with the sense of well-being vulnerability and adaptation strategies of St. Martin's Island's community. This method provides a framework for analyzing vulnerability and adaptive capacity to climate change at the community level by combining local knowledge with scientific data to yield greater understanding about the local impacts of climate change (Daze et al. 2009). The CVCA process is different from other

Fig. 9.1 Map of St. Martin's Island on the map of Bangladesh. (Source: Banglapedia 2014)

forms of participatory methods because of its focus on climate change (how climate change will affect the lives and livelihoods of target populations) and its emphasis on multi-stakeholder analysis, collaborative learning, and dialog. It focuses on communities as well as examining enabling environments.

Among several participatory tools of CVCA, there are, for example, hazard mapping, seasonal calendars, historical timeline of livelihood activities, vulnerability matrix, and stakeholder analysis; vulnerability matrix was used in this study to

Table 9.1 Criteria for the categorization of the three well-being groups of St. Martin's Island's people

Criteria applied	Well-off/rich group	Medium group	Poor group
Food sufficiency and storage	Food sufficiency throughout the year and food storage for 1 year or more	Food sufficiency throughout the year and food storage for 6 months or less	Food sufficiency for 6 months or less
Land/boat/trawler	Have large area of irrigated and non-irrigated land, also boat and trawler	Limited boat and land	No irrigated land or no developed boats
Education and employment	Children admitted to schools and family has a large business	Children admitted to schools and at least one member per family has a job/has small business	Children admitted to schools but no job/business, do different forms of labor
Income	High income, around BDT15,000–20,000 per month and no loan	Medium income, around BDT5000–10,000 per month	Very low income, around BDT1000 per month; taken out a loan

Table 9.2 Impact scoring criteria for determining vulnerability

Impact scores	Criteria
3 (significant impact)	People's houses are destroyed. They have to live under the open sky until their house is repaired. Their coconut tree garden is also destroyed. The educational institution remains shut down for a long time. People have to starve because of lack of food
2 (medium impact)	People's houses are damaged. They have to suffer a lot until their house is repaired. Their coconut tree garden is also damaged. The educational institution remains shut down for a certain period. People suffer from lack of food
1 (low impact)	People's houses and coconut tree gardens are damaged. It takes time to repair them. Education system is hampered for some time. People cannot concentrate on their business because they are busy repairing their houses
0 (no impact)	No impact at all

determine the hazards with the most serious impacts on important livelihood resources and their vulnerability as well as to identify adaptation strategies used by the communities to address the identified hazards (Rawlani and Sovacool 2011; Adem et al. 2017). A total of three CVCAs were conducted each for a well-being group of 8–10 participants. During a CVCA, at first, the most important livelihood resources were identified and the priority resources were listed vertically. Then, the greatest hazards to the community's livelihood were perceived and the significant hazards were set horizontally. The final step was to put the score within a range of 0–3 to determine vulnerability (Table 9.2). During the process of building consensus for each score, the participants were encouraged to discuss the reasons for each score, related vulnerability, and adaptation strategies. The climate change-related impacts, adaptation, and vulnerability were prompted.

In addition, three FGDs, each containing 8–10 preselected household heads of different categories (well-being groups) were carried out to discuss the important issues in detail and triangulate the CVCA and KII data. For the FGDs a checklist was prepared beforehand seeking information about the perception of communities regarding climate change, impacts of climatic hazards, extreme climatic events, and response strategies. In addition, to cross-check the primary data, participant observation and secondary data were collected from various books, journals, and other published literature. Meteorological data of the last 20 years (1997–2007) from Teknaf station were collected from Bangladesh Meteorological Department (BMD) to compare and underpin the spatial and temporal variations in temperature and rainfall with the observed impacts.

9.2.3 Data Analysis

Content analysis was used to analyze the field data collected for FGDs, KIIs, and the discussion part of the CVCA matrices. A list of climate hazards and livelihood resources was prepared and then significant hazard and livelihood resources were itemized in the vulnerability matrix table to be scored from 0 to 3 (3 = significant impact, 2 = medium impact, 1 = low impact, 0 = no impact) by the respondents and the total scores were calculated. Scores obtained from the vulnerability matrix table were delineated in charts and graphs using MS Excel to better understand the results. Finally, analysis of variance (ANOVA) F test was executed to compare how the vulnerability varied among the three well-being groups owing to the impact of major hazards.

9.3 Results and Discussion

9.3.1 Impacts of Climate Change on Livelihood

St. Martin's Island has tropical sunny weather with an average dry bulb temperature fluctuating from almost 25.5 to 26.45 °C. The highest annual maximum temperature ever, 38.1 °C, was recorded here in the year 2000. Average precipitation in this island ranges from 3200 to 5500 mm. The climatic variables significantly affect the occurrence of natural hazards such as storms, cyclones, weather fluctuations, and seasonal environmental stresses such as rainfall, salinity intrusion, and tidal water increase that affect the livelihood resources of the community. The livelihood of the community of this island is predominantly dependent on the natural resources i.e., fish, coconut, coral, and other natural amenities (Rawlani and Sovacool 2011; Adem et al. 2017). Fishing is the main source of income for most of the people on the island, which covers 90% of the total community (Haider 2008). In addition, hotel/

cottage and resorts, restaurant businesses, coral and shell collection, fish drying, small shop, tourism business, algae collection, coconut selling and agricultural practices, are also important livelihood resources. Among these, fishing, fish drying, and agricultural practices related to coconut selling are the most common livelihood options throughout the year, although coconut and dry fish selling, shell extraction, tourist van-pulling, shop-keeping, and hotel and motel services only run during the peak tourist season from November to April (Nafi and Ahmed 2017).

The major climatic hazards, i.e., storms, cyclones, heavy rainfall, salinity intrusion, and tidal water increase the impact on the principal livelihood resources in various ways. Cyclones and tidal surges, which are considered the world's foremost climatic hazards, have frequently destroyed lives and properties (Alam and Collins 2010). As the island is only accessible by local mechanized wooden boat (tourist vessels are only seasonally available), the usual communication with the mainland is cut off under disastrous circumstances; thus, the conventional food supply becomes entangled and shatters the supply–demand chain of the island, leaving a crisis of basic needs, namely, food, water, health, sanitation, and transportation. Lack of environmental health knowledge among the affected people creates the most important problem of water and sanitation in the post-disaster period (Alam and Collins 2010), mostly among the poor and middle-income group people. The CVCA discussion reveals that agricultural practices on the St. Martin's Island are reducing rapidly because of salinity intrusion, excessive population pressure, and increasing construction activities. Salinity intrusion also affects livestock rearing by reducing local vegetation and grass production (Rawlani and Sovacool 2011; Adem et al. 2017). Fishing activity also stops under hazardous environmental conditions such as storms, cyclones, and heavy rainfall. Hence, only a few months during monsoon and the tourist season are the local poor community able to feed themselves by securing substantiated income, mainly by selling fish, coconuts, dry fishes, and pulling vans.

9.3.2 *Differential Vulnerability of Well-Being Groups*

To identify the vulnerability of three well-being groups of communities of St. Martin's Island this study has used cyclone, tidal water increase, heavy rainfall, and salinity intrusion as major climatic hazards. Prioritized livelihood resources and activities were tourism business (e.g., hotels, cottages or resorts, small or big shops), fishing by using boats/trawlers, agricultural practices, i.e., crops and coconut gardens, and other jobs. Each hazard was scored according to the significant impacts on the livelihood resources and activities. The people of the community have given their vulnerability score on the basis of their destruction and losses due to the hazards. Tables 9.3, 9.4, and 9.5 represent the vulnerability matrices of well-off, medium, and poor well-being groups respectively.

The results from Tables 9.3, 9.4, and 9.5 show that cyclones have a significant impact on all three well-being groups despite their economic distinction, as they severely damage and destroy the homesteads and livelihoods of the people (Mallick

Table 9.3 Vulnerability matrix for the well-off/rich well-being group of people of St. Martin's Island

	Major hazards				
Major livelihood resources and activities	Cyclone	Heavy rainfall	Tidal water increase	Salinity intrusion	Total vulnerability score
Tourism business/shop	3	3	3	0	9
Agricultural practices	3	3	3	3	12
Fishing (boat/trawler)	2	2	2	0	6
Job/education	3	2	1	0	6
Total score	11	10	9	3	

Table 9.4 Vulnerability matrix for the medium well-being group of people of St. Martin's Island

	Major hazards				
Major livelihood resources and activities	Cyclone	Heavy rainfall	Tidal water increase	Salinity intrusion	Total vulnerability score
Tourism business/shop	3	2	3	0	8
Agricultural practices	2	3	3	3	11
Fishing (boat/trawler)	3	2	2	0	7
Job/education	3	2	2	0	7
Total score	11	9	10	3	

Table 9.5 Vulnerability matrix for the poor well-being group of people of St. Martin's Island

	Major hazards				
Major livelihood resources and activities	Cyclone	Heavy rainfall	Tidal water increase	Salinity intrusion	Total vulnerability score
Tourism business/shop	2	2	3	0	7
Agricultural practices	3	2	3	0	11
Fishing (boat/trawler)	3	2	2	1	8
Job/education	3	2	2	0	7
Total score	11	8	10	1	

et al. 2017). Heavy rainfall, at the same time, has a profound impact on the well-off group (as it mainly affects the agricultural land and tourism business) followed by the medium and poor groups of people. On the other hand, tidal water increases usually affect the medium and poor groups of people, mostly because of the structure and position of their household, which become inundated during flooding situations and they need to take shelter in a safe place. Sometimes, poor people build their temporary house by using polythene and plastic, whereas the middle group of people use bamboo and wood. They eat less than they require and some have no food to eat. The study says that approximately 54% of people of this island do not get sufficient food at the market owing to disruption of the food supply from the mainland after the disaster (Feeroz 2009). Interestingly, salinity intrusion has a significant impact on the well-off and medium-income groups of people as salinity mainly affects the agricultural lands, which are mainly owned by these two groups of people. Thus, in most climatic hazards, both well-off and medium groups of people are more affected than the poor group of people.

The priority livelihood resources of the people of this island are tourism business (e.g., hotels/resorts, food hotels, and small shops), agricultural practices (i.e., cultivation of lands and coconut gardens), fishery activities (fishing in the sea using boats or trawlers, fish drying) and other jobs or educational services. All these livelihood options are mostly dependent on the natural resources of the island; therefore, climatic hazards can easily affect the livelihood of all income groups of people. The tourism business, for example, is mainly dependent on the aesthetic beauty of this island. In any natural disturbance such as a storm or a cyclone, the island becomes separated from the mainland, interrupting the tourism business and affecting dependent people as no tourists can come onto the island. This study has found that, in the case of the tourism business and agricultural practices, climatic hazards mostly affect the rich group of people followed by the medium and poor well-being groups (see Tables 9.3, 9.4, and 9.5) as the rich group of people have most of the tourism-related business and most own agricultural land and coconut gardens. Disturbance in fishery activities, e.g., fishing in the sea using a small boat or working on a trawler as a wage earner mostly affect the poor well-being group, as their livelihood is often dependent on these, followed by the medium and rich groups of people. At the same time, inconvenience in the job and other sectors due to climatic hazards also simultaneously affects the medium and poor groups of people. Thus, from the CVCA score it is evident that the rich group of people has the greatest vulnerability in the case of the tourism business and agricultural practices; the poor group of people has the highest vulnerability in fishery activities and the job sector, whereas the medium group are only highly vulnerable in the job sector.

Following the ANOVA *F* test (Table 9.6) it is perceptible that the combined impact of major hazards has the highest impact on the rich (score: 8.25, standard

Table 9.6 Comparison of the vulnerability scores of the three well-being groups of people on St. Martin's Island due to major climatic hazards along with standard error (SE), 95% confidence interval (CI), and *p* value of the ANOVA *F* test

		Vulnerability scores for well-being groups			*p* value (ANOVA *F* test)
		Rich/well-off	Medium	Poor	
Major hazards	Cyclone	11	11	11	0.953
	Heavy rainfall	10	9	8	
	Tidal water surge	9	10	10	
	Salinity intrusion	3	3	1	
	Total score	33	33	30	
	Mean	8.25	8.25	7.50	
	SE	1.80	1.80	2.26	
	95% CI for mean	(2.53, 13.97)	(2.53, 13.97)	(0.32, 14.68)	

Table 9.7 Some studied proactive and reactive adaptation practices adopted by the three well-being groups of people of St. Martin's Island

Well-being status	Proactive adaptations	Reactive adaptations
Well-off	Change fishing and cropping pattern, purchase land, lend money, access new business options	Keep storage of fish and crops, cash savings
Medium	Change business patterns, find different skill-based work, livelihood diversification	Keep storage of fish and crops, cash savings, selling products on the market
Poor	Change work pattern, sale of property	Sell corals, mollusks, echinoderms, handmade ornaments, and dry fish to the tourists

error [SE] = 1.80, 95% confidence interval [CI]: 2.53 to 13.97) and medium (score: 8.25, SE = 1.80, 95% CI: 2.53 to 13.97) well-being groups than on the poor (score: 7.50, SE = 2.26, 95% CI: 0.32 to 14.68) group of people, although this variation is not statistically significant (p value = 0.953). Although it is conceivable that climate change-related natural disasters damage the livelihood of poor people by making them more vulnerable (Kabir et al. 2016), in this study we found that the rich and middle-income groups of people are also correspondingly vulnerable. Even the well-off group of people faces difficulties in resolving potential vulnerability.

9.3.3 Adaptation Strategies

Despite the impacts posed by different climatic hazards, communities of St. Martin's Island try to adapt by diversifying their livelihood resources. Communities adopt both proactive and reactive adaptation practices based on their capacity (Table 9.7). But in most of the cases, these adaptation practices are short term and traditional, such as diversification of livelihoods, changing techniques and patterns in farming and fishing, adopting supplementary livelihood options.

Table 9.7 clearly illustrates that poor households are unable to adopt some adaptation options such as changing crop variety, changing fishing pattern, and saving more livestock owing to their lack of resources and cash. In the tourist season, they change their work options such as selling corals, handmade ornaments, dry fish, and mollusks, but at other times they have fewer options and most of them are reactive rather than proactive. On the other hand, the well-off households save cash, purchase land, store food, and lend money to the poor with high interest. Also, most of the well-off households and some medium households explore new fishing and agriculture technology, new crop varieties, new business paths, and diversify their livelihood options. Adaptation strategies by rural communities are mostly autonomous and reactive rather than strategic (Smit et al. 2000), which are in most cases inefficient and could be unsuccessful (IPCC 2007). Evaluation of adaptation options can be based on criteria such as costs, benefits, urgency, efficiency, and implementation ability (Smit and Wandel 2006); thus, the poor households have limited access to long-term effective proactive adaptation measures owing to lack of resources and low-income strategy. They have the greatest vulnerability in contrast to the well-off households, with enough resource ownership and better financial access. This statement is also supported by another statement by Reardon and Taylor (1996): rich households are expected to have more flexibility in adapting to climate change because they can afford more expensive strategies. The study has revealed that all the adaptation practices identified by the households are mainly traditional practices based on their local knowledge and experience in response to climate hazards. A similar result was also found in a study by Alam (2017) where respondents applied their perception and long-term knowledge in adapting to climate change. In general, adaptation practices adopted in the countryside are mostly related to crop diversification, irrigation, water management, and lending money (IPCC 2007; Gentle and Maraseni 2012). These short-term traditional adaptation strategies might be helpful in the current situation, but with the additional burden of climate change, they might not be able to cope in the future.

9.3.4 Reducing Impacts and Increasing Adaptations

Increasing the competency of individuals or groups to adapt to changes by building adaptive capacity and transforming that competency into practice are the two factors in implementing adaptation (Adger et al. 2005). Improved risk management, increasing knowledge, and development of technology can be involved in an adaptation strategy (West and Gawith 2005). But it should be continuous and reflect social norms and processes (Adger et al. 2005). On small islands, supplying incessant power facilities, increasing tourism facilities and activity, and enhancing services related to the coastal resources can increase adaptation (Nurse et al. 2014). As most of the adaptation practices adopted by the St. Martin's Island community are short term, for effective and sustainable adaptation, long-term, proactive, and integrated adaptation measures on social, economic, technological, institutional, and infrastructural dimensions are truly needed.

9.4 Conclusions

As one of the most vulnerable countries in the world, the livelihoods of the coastal areas of Bangladesh are being heavily impacted because of major natural and climatic hazards. With the study it was intended to explore the vulnerability of different well-being groups of St. Martin's Island community in Bangladesh to climatic hazards by using vulnerability matrices, FGDs, and KIIs. It identified four major hazards, namely, cyclones, heavy rainfall, salinity intrusion, and tidal water increase, which affect nature-dependent major livelihood resources and activities, i.e., tourism businesses, agricultural practices, fishery practices, and other occupations. In spite of having economic disparity, cyclones affect all three well-being groups, whereas poor and medium-income groups of people are mostly affected by tidal water increases. Interestingly, although heavy rainfall affects the poor group most, salinity intrusion affects rich and medium-income groups of people more. It was found that the rich and medium groups of people have a higher level of vulnerability to major hazards than the poor group, although the variation is not statistically significant (p value = 0.953). All groups struggle to adapt under changing environmental conditions by adopting short-term traditional adaptation strategies such as the diversification of livelihoods, changing crop varieties, selling more livestock, change fishing patterns, and storing foods. However, to ensure effective adaptation and sustainable livelihoods of the community, long-term, proactive adaptations by the governmental and nongovernmental bodies are needed, for example, by:

(a) Enhancing co-operation: as most of the people on the island are fishermen, to build cooperation among them (despite their economic differences), the community's association with microfinancing facilities can be built to incorporate savings accounts and increase access to credit by providing emergency loan schemes.
(b) Ensuring basic needs: as there is insufficient fulfilment of basic needs during and after disasters, ensuring a continuous supply of food, water, and health services from the mainland during these periods will increase the adaptation capacity of the community.
(c) Improving infrastructure and incorporating modern technology: improving infrastructural facilities such as solar power supply for everyone and improving communication with the mainland all the year round will augment the tourism business; increase the livelihood capacity of all income groups of people, thereby increasing their adaptation. In addition, technological advancement of fishing boats and trawlers can increase the catch by facilitating deep-sea fishing.
(d) Producing new livelihood opportunities: as the poor well-being group of people have fewer opportunities to diversify their livelihoods, some additional income-generating opportunities, namely, producing value-added fish products, and hand crafting (by both women and men), can be created outside of the tourist season. In addition, introducing seaweed culture and crab culture may have a profound impact on the livelihoods of the island's people.

References

Adem A, Deering K, Molla S (2017) Building adaptive communities through integrated programming: CARE Ethiopia's experience with Climate Vulnerability and Capacity Analysis (CVCA). In: Climate change adaptation in Africa. Springer, Cham, pp 217–234

Adger WN, Arnell NW, Tompkins EL (2005) Successful adaptation to climate change across scales. Glob Environ Chang 15(2):77–86. https://doi.org/10.1016/j.gloenvcha.2004.12.005

Agrawal A, Perrin N (2009) Climate adaptation, local institutions and rural livelihoods. In: Adapting to climate change: thresholds, values, governance. Cambridge University Press, Cambridge, pp 350–367

Ahammed SS, Hossain MA, Abedin MZ, Khaleque MA (2016) A study of environmental impacts on the coral resources in the vicinity of the Saint Martin Island, Bangladesh. Int J Sci Technol Res 5(1):37–39

Alam M (2003) Bangladesh country case study. National Adaptation Programme of Action (NAPA) workshop, 9–11 September 2003, Bhutan

Alam GM (2017) Livelihood cycle and vulnerability of rural households to climate change and hazards in Bangladesh. Environ Manag 59(5):777–791

Alam E, Collins AE (2010) Cyclone disaster vulnerability and response experiences in coastal Bangladesh. Disasters 34(4):931–954. https://doi.org/10.1111/j.1467-7717.2010.01176.x

Alam M, Ahammad R, Nandy P, Rahman S (2013) Coastal livelihood adaptation in changing climate: Bangladesh experience of NAPA priority project implementation. In: Climate change adaptation actions in Bangladesh. Springer, Tokyo, pp 253–276

Alam O, Deng TL, Uddin MN, Alamgir M (2015) Application of environmental ethics for sustainable development and conservation of Saint Martin s Island in Bangladesh. J Environ Sci Nat Resourc 8(1):19–27. https://doi.org/10.3329/jesnr.v8i1.24628

Ali A (1996) Vulnerability of Bangladesh to climate change and sea level rise through tropical cyclones and storm surges. In: Climate change vulnerability and adaptation in Asia and the Pacific. Springer, Dordrecht, pp 171–179

Ali A (1999) Climate change impacts and adaptation assessment in Bangladesh. Clim Res 12(2–3):109–116. https://doi.org/10.3354/cr012109

Balasubramanian TN, Nambi AA, Paul DIYA (2007) A suggested index for assessment of vulnerability of a location to climate change. J Agrometeorol 9(2):129–137

Banglapedia (2014) Banglapedia: national encyclopedia of Bangladesh. Available online http://en.banglapedia.org/index.php?title=File:StMartinsIsland.jpg. Accessed 2 June 2020

BBS (2011) Bangladesh population and housing census. National report 2: union statistics. Bangladesh Bureau of Statistics, Statistics and Information Division, Ministry of Planning. Govt. of the People's Republic of Bangladesh, Dhaka

Below TB, Mutabazi KD, Kirschke D, Franke C, Sieber S, Siebert R, Tscherning K (2012) Can farmers' adaptation to climate change be explained by socio-economic household-level variables? Glob Environ Chang 22(1):223–235

CFE-DM (2017) Bangladesh disaster management reference handbook. https://www.preventionweb.net/go/54099. Accessed 5 May 2020

Chowdhury SQ (2012) St Martin's Island. In: Islam S, Jamal AA (eds) Banglapedia: national encyclopedia of Bangladesh, 2nd edn. Asiatic Society of Bangladesh, Dhaka

Daze A, Ambrose K, Ehrhart C (2009) Climate vulnerability and capacity analysis handbook. Care International, London

Eckstein D, Künzel V, Schäfer L, Winges M (2019) Global climate risk index 2020. Germanwatch, Bonn

Feeroz M (2009) Effects of environmental degradation on food security in the St. Martin's Island of Bangladesh. National Food Policy Capacity Strengthening Program, Dhaka

Füssel HM, Klein RJ (2006) Climate change vulnerability assessments: an evolution of conceptual thinking. Clim Chang 75(3):301–329

Gbetibouo GA, Hassan RM, Ringler C (2010) Modelling farmers' adaptation strategies for climate change and variability: the case of the Limpopo Basin, South Africa. Agrekon 49(2):217–234

Gentle P, Maraseni TN (2012) Climate change, poverty and livelihoods: adaptation practices by rural mountain communities in Nepal. Environ Sci Pol 21:24–34. https://doi.org/10.1016/j.envsci.2012.03.007

GoB (2009) Bangladesh climate change and strategic action plan 2009. https://www.iucn.org/content/bangladesh-climate-change-strategy-and-action-plan-2009. Accessed 5 May 2020

Haider T (2008) Saving the coral biodiversity of St. Martin's Island. Ground report. http://www.groundreport.com/World/Saving-the-coral-biodiversity-of-St-Martins-island. Accessed 5 May 2020

Hasan MM (2009) Tourism and conservation of biodiversity: a case study of St. Martins Island, Bangladesh. Law Soc Justice Glob Dev J 1(1):45–55
Hossain MS (2001) Biological aspects of the coastal and marine environment of Bangladesh. Ocean Coast Manag 44(3–4):261–282. https://doi.org/10.1016/S0964-5691(01)00049-7
Huq S, Ali SI, Rahman AA (1995) Sea-level rise and Bangladesh: a preliminary analysis. J Coast Res Spec Issue 14:44–53
Iglesias A, Quiroga S, Diz A (2011) Looking into the future of agriculture in a changing climate. Eur Rev Agric Econ 38(3):427–447
IPCC (2001) Climate change 2001: impacts, adaptation and vulnerability. In: McCarthy JJ, Canziani OF, Leary NA, Dokken DJ, White KS (eds) Contribution of Working Group II to the third assessment report of the Intergovernmental Panel on Climate Change. Cambridge University Press, Cambridge
IPCC (2007) Climate change 2007: impacts, adaptation and vulnerability. In: Parry ML, Canziani OF, Palutikof JP, Van Der Linden PJ, Hanson CE (eds) Contribution of Working Group II to the fourth assessment report of the Intergovernmental Panel on Climate Change. Cambridge University Press, Cambridge
IPCC (2014) Summary for policymakers. In: Edenhofer O, Pichs-Madruga R, Sokona Y, Farahani E, Kadner S, Seyboth K (eds) Climate change 2014: mitigation of climate change. Contribution of Working Group III to the fifth assessment report of the Intergovernmental Panel on Climate Change. Cambridge University Press, Cambridge/New York
Islam MZ (2001) St Martin pilot project, national conservation strategy implementation project-1. Draft final report. Ministry of Environment and Forests, Dhaka
Islam MZ (2002) Threats to sea turtles in St. Martin's island, Bangladesh. Kachhapa 6:6–10
Islam MM, Sallu S, Hubacek K, et al. (2014a) Vulnerability of fishery-based livelihoods to the impacts of climate variability and change: insights from coastal Bangladesh. Reg Environ Chang 14:281–294. https://doi.org/10.1007/s10113-013-0487-6
Islam MM, Sallu S, Hubacek K, Paavola J (2014b) Limits and barriers to adaptation to climate variability and change in Bangladeshi coastal fishing communities. Mar Policy 43:208–216. ISSN 0308-597X. https://doi.org/10.1016/j.marpol.2013.06.007
Islam MM, Sallu S, Hubacek K, Paavola J (2014c) Migrating to tackle climate variability and change? Insights from coastal fishing communities of Bangladesh. Clim Chang 124:733–746. https://doi.org/10.1007/s10584-014-1135-y
Kabir R (2014) The impacts of cyclones Sidr and Aila on the health of the coastal people of Bangladesh. Doctoral dissertation, Middlesex University
Kabir R, Khan HT, Ball E, Caldwell K (2016) Climate change impact: the experience of the coastal areas of Bangladesh affected by cyclones Sidr and Aila. J Environ Public Health 2016. https://doi.org/10.1155/2016/9654753
Mallick B, Ahmed B, Vogt J (2017) Living with the risks of cyclone disasters in the south-western coastal region of Bangladesh. Environment 4(1):13. https://doi.org/10.3390/environments4010013
Malone EL, Brenkert AL (2008) Uncertainty in resilience to climate change in India and Indian states. Clim Chang 91(3–4):451–476
Mohal N, Khan ZH, Rahman N (2006) Impact of sea level rise on coastal rivers of Bangladesh. Institute of Water Modelling (IWM). Assessment conducted for WARPO, an organization under Ministry of Water Resources, Dhaka
Mollah AR (1997) Status of coral and associated resources in Bangladesh. In: Proceedings of the regional workshop on the conservation and sustainable management of coral reefs, MSSRF and BOBP, FAO/UN, December, pp B37–B51
Mosse D (1994) Authority, gender and knowledge: theoretical reflections on the practice of participatory rural appraisal. Dev Chang 25(3):497–526. https://doi.org/10.1111/j.1467-7660.1994.tb00524.x
Moudud HJ (2010) St. Martin's Island and its unique biodiversity face serious threats. IUCN, International news release

Nafi SM, Ahmed T (2017) Sustainable tourism in Saint Martin Island: an observation on young tourist perception and awareness level. IOSR J Humanit Soc Sci (IOSR-JHSS) 22(10):73–80. https://doi.org/10.9790/0837-2210117380

Nurse LA, McLean RF, Agard J, Briguglio LP, Duvat-Magnan V, Pelesikoti N, Tompkins E, Webb A (2014) Small islands. In: Barros VR, Field CB, Dokken DJ, Mastrandrea MD, Mach KJ, Bilir TE, Chatterjee M, Ebi KL, Estrada YO, Genova RC, Girma B, Kissel ES, Levy AN, MacCracken S, Mastrandrea PR, White LL (eds) Climate change 2014: impacts, adaptation, and vulnerability. Part B: regional aspects. Contribution of Working Group II to the fifth assessment report of the Intergovernmental Panel on Climate Change. Cambridge University Press, Cambridge/New York, pp 1613–1654

Paavola J, Adger WN (2002) Justice and adaptation to climate change. Tyndall Centre Working Paper, 23

Rahman MS, Haque MM, Khan MBK (2011) Perception on climate change: an exploratory study on urban citizens. Eur J Econ Finance Adm Sci 31:17–30

Rani S, Ahmed MK, Xiongzhi X, Yuhuan J, Keliang C, Islam MM (2020) Economic valuation and conservation, restoration & management strategies of Saint Martin's coral island, Bangladesh. Ocean Coast Manag 183:105024

Rawlani AK, Sovacool BK (2011) Building responsiveness to climate change through community based adaptation in Bangladesh. Mitig Adapt Strateg Glob Chang 16(8):845–863

Reardon T, Taylor JE (1996) Agroclimatic shock, income inequality, and poverty: evidence from Burkina Faso. World Dev 24(5):901–914. https://doi.org/10.1016/0305-750X(96)00009-5

Smit B, Wandel J (2006) Adaptation, adaptive capacity and vulnerability. Glob Environ Chang 16(3):282–292. https://doi.org/10.1016/j.gloenvcha.2006.03.008

Smit B, Burton I, Klein RJ, Wandel J (2000) An anatomy of adaptation to climate change and variability. In: Societal adaptation to climate variability and change. Springer, Dordrecht, pp 223–251

Tao S, Xu Y, Liu K, Pan J, Gou S (2011) Research progress in agricultural vulnerability to climate change. Adv Clim Chang Res 2(4):203–210. https://doi.org/10.3724/SP.J.1248.2011.00203

Tomascik T (1997) Management plan for coral resources of NarikelJinjira (St. Martin's Island). Final report. National Conservation Strategy Implementation Project-1, Ministry of Environment and Forest, Government of Bangladesh, pp 125

West C, Gawith M (2005) Measuring progress: preparing for climate change through the UK climate impacts programme, Technical report. UKCIP, Oxford

Chapter 10
Migration Patterns and Adaptation Strategies to Natural Hazards Outside the Coastal Embankment of Bangladesh

Md. Bahuddin Sikder, A. H. M. Abdul Baqee, Sameena Begum, and Md. Najmul Kabir

Abstract The coastal area of Bangladesh is frequently affected by natural hazards. For that reason, the people living in this zone regularly need to migrate and adapt to an unfavorable environment. The main objectives of this study are to observe the migration scenario and adaptation strategies outside the coastal embankment of Barguna Sadar Upazila. To deal with these objectives, a field survey was conducted based on the socio-economic conditions as well as the places of origin and destination, causes, and types of migration. Along with that, we aimed to observe what strategies are followed in the study area to protect it from catastrophic disasters. All these necessary data were collected through a questionnaire survey, in-depth interviews, focus group discussion, and field observation. In addition, secondary data were collected from the Naltona and Baliatali Union Parishad offices. The descriptive and inferential statistical techniques were used to analyze the data. It was found from the collected data that 504.28 people live in every square kilometer of the study area. About 90.3% of respondents of the study area are the owners of a one-storied house, which are mainly built of Nypa Palm, tall reed, bamboo, wood, and tin. Almost 96.87% of respondents are engaged with the professions directly related to natural resources such as fisherman or boatman. The main source of drinking water is a deep-tube well, but there are not enough of them in the study

M. B. Sikder · M. N. Kabir (✉)
Department of Geography and Environment, Shahjalal University of Science and Technology, Sylhet, Bangladesh
e-mail: n.kabir-gee@sust.edu

A. H. M. Abdul Baqee
Department of Geography and Environment, University of Dhaka, Dhaka, Bangladesh

S. Begum
Department of Geography and Environment, Jagannath University, Dhaka, Bangladesh

M. N. Islam, A. van Amstel, *Bangladesh II: Climate Change Impacts, Mitigation and Adaptation in Developing Countries*, Springer Climate,
https://doi.org/10.1007/978-3-030-71950-0_10

area. There are no paved roads and the condition of the roads is rough. There are 35% of respondents identified as migrants in the study area. Shifting of the riverbank, environmental degradation after catastrophic disasters, available land near the seashore, plenty of job opportunities, and the chance to live with the same class of people – all of these act as the key factors for migration. There is only in-migration seen rather than out-migration. The inhabitants who migrate from other areas, permanently live here to carry out their daily livelihoods. It is spotted that respondents of the study area held native adaptation techniques to deal with the adverse environment. It is expected from this study that researchers and policymakers understand the probable causes of migration, its pattern, and become familiar with adaptation strategies that will help them in the near future to develop a regional plan outside the coastal embankment in different places.

Keywords Coastal embankment · Hazard · Migration · Adaptation

Abbreviation

DDM	Disaster Management Bureau
IOM	International Organization for Migration
IPCC	Intergovernmental Panel on Climate Change
JPL	Jet Propulsion Laboratory
MoDMR	Ministry of Disaster Management and Relief
MoEF	Ministry of Environment and Forests
MoFDM	Ministry of Food and Disaster Management
MoWR	Ministry of Water Resources
NASA	National Aeronautics and Space Administration
NGO	Nongovernmental organization
OECD	Organization for Economic Cooperation and Development
OSM	Open Street Map
WB	World Bank

10.1 Background of the Study

The coastal area of Bangladesh regularly faces the fury of natural hazards, namely tropical cyclones and storm surges, riverbank erosion, floods, as well as landslide and drought owing to its unique position. The Bay of Bengal is the perfect breeding ground for tropical cyclones. Low altitude with gentle slopes and funnel-shape characteristics near the adjacent seashore gives way to creating a bigger storm surge in the coastal inland area. These hazards make the life of people miserable, along with damaging the infrastructure and breaking down economic progress.

According to the assessment of the Intergovernmental Panel on Climate Change (IPCC) (2007), small island states as well as coastal regions that have low altitude are in a vulnerable position because of the rise of sea level, which poses a direct threat to their existence. Bangladesh is the most affected country amongst them.

Human activities have accelerated to the sea-level rise. Experts are alarmed that the sea level may rise 11–38 inches (27.94 to 96.52 cm) (DDM 2010). Along with that, if there are any major cracks occurring in the Greenland and Antarctic region, then the situation will get worse. It is suggested from an estimation that if the sea level rises 1 m from the current level, then more than 3 crore people will become refugees in Bangladesh (Uzzaman 2014). A NASA report based on the trend of sea-level change indicated that the port city of Chittagong is situated in a vulnerable position and it has a chance of becoming completely inundated by seawater in the next 100 years (Larour et al. 2017).

Because of the rising sea level, the daily livelihood pattern of the coastal area will be adversely affected. Environmentalists assume that vast stretches of coastal arable land may turn into barren areas as the salinity level is increasing rapidly. Researchers have said that some parts of the coastal area are already facing this situation. According to the study by the Soil Resource Development Institute (SRDI), 8.33 lakh hectares area of land in 19 coastal districts were affected by salinity in 1973, but now this situation stands at 10.2 lakh. Experts have assessed that rice production would be decreased at a rate of 3.9% by 2050 (MoFDM 2010). Thus, people who depend on agricultural practices need to be shifted from their jobs or culture farming activities in a new area. In addition, saltwater is assimilated with freshwater, putting water safety at risk. Owing to the scarcity of freshwater, man's dependency on groundwater will be increased, which raises the risk of arsenic contamination, which has a bad effect on the human body.

Bangladesh has 12 out of 19 coastal districts directly exposed to the sea. One-fourth of the total population of Bangladesh live in the coastal area. Among them, more than 55% live within 100 km to 710 km from the coastal belt (MoDMR 2014). Most of the people are always fighting poverty as their livelihood generally depends on the weather and natural resources. This type of people are specifically exposed to the threat of cyclones and storm surges. For that reason, large-scale displacement has occurred during periods of natural disasters.

In recent years, Bangladesh has experienced the wrath of nature. More than 6,500,000 people were displaced and 3447 killed by cyclone Sidr in 2007 and around 4.82 million were affected by the cyclone Aila. Cyclone Roanu, which hit the coastal belt in 2009, forced 2 million people to flee from their houses (Akter 2009). Cyclone Mora, which occurred in 2017, displaced more than 5 million people from their homeland (Al Jazeera and News Agencies 2017). Along with displacement, the causal factor was also large. Owing to the aggression of nature, people were forced to migrate from their homeland and became environmental migrants (IOM). It is considered a key tool to be able to cope with rapid-onset natural hazards, as well as adaptation techniques to deal with slow onset processes. Migrants who are displaced by cyclones generally move for a short period and in particular choose the shortest distance. They usually find a temporary job. The influx of those populations is known as the swelling population. Climatologists expect that the intensity of cyclones will soon be increased because of climate change; thus, these short time spans will continue. Therefore, it is high time for research into migration patterns and adaptation techniques of the coast area.

10.1.1 Objectives of the Study

The present study is primarily focused on how the poor denizens make a fresh start to make a living in a fragile environment. Specifically, the study focuses on:

1. Carrying out an in-depth study on the life and livelihood of people outside the embankment.
2. Explaining why people migrate from a more stable site to outside the embankment.
3. Making an elaborate attempt of the migration field.
4. Investigating and evaluating the adaptation techniques to different natural hazards undertaken by the people outside the embankment.

10.1.2 Rationale of the Study

Barguna region is geographically one of the most volatile areas. Rivers and flood waters carry sediments, as well as the cyclonic sea storms, for the formation of new lands on the one hand and activation of riverbank erosion on the other. During the period between 2012 and 2017, Barguna Sadar Upazila lost about 124.07 ha of land, whereas there was 336.89 ha of land rise. In this place, accretion and erosion are both active because of the presence of Payra (Burishwar), Bishkhali, and Baleshwar rivers, which formed many estuaries, as well as this area directly exposed to the sea. This is the area that constantly faces storm surges, and, owing to the low elevation line (less than 1 m), it is regularly the subject of sea-level change.

These two factors increase the salinity level, which caused the loss of 38,000 ha of arable land. Because of these adverse effects of climate, people who live outside the coastal embankment area are more vulnerable populations. They need to regularly migrate or adapt to this scenario. But there was no exact database on the outside of the coastal embankment linked with lifestyle, migration pattern, and adaptation technology. The topmost priority may be given to finding out the various risk factors in a better way for adaptation and to reduce the rate of migration. This study is aimed at observing migration patterns and adaptation techniques based on the real situation, which may be worked on as a perspective for regional planning.

10.1.3 Expected Outcomes and Its Relevance

From this study, we expected to:

- Find out about the lifestyle of the people who live outside the coastal embankment area
- Find out the reason why people migrate, who are migrants, their destination, and finally the pattern of migration
- Determine various adaptation strategies that they follow and assess the viability of those strategies
- Bring out the people who stayed outside the coastal embankment, in the spotlight, which is badly needed for them. Without in front page, the government and various aid organizations will not know the extremes condition these people are living in.

10.2 Literature Review

This section includes the available literature on natural disasters, level of vulnerability, migration status, as well as adaptation strategies of residents, government, and nongovernmental organizations (NGOs).

10.2.1 Search for Research Trends on Natural Hazards

Bangladesh experienced one of the top ten disasters in its history between 1988 and 2013 (EM-DAT 2018). This country frequently faces floods and cyclones every year because of its geographical position. The rivers, which originate from the Ganges-Brahmaputra basin, penetrate this region and eventually fall into the Bay of Bengal, which is usually responsible for flooding every year. In addition, because of the strong cyclones, unexpected floods occur. (Mirza and Monirul 2011).

The worst floods happened in 1988 and 2007 and were responsible for the deaths of 1517 and 1110 people respectively (EM-DAT 2018). Harris (1999) estimated that in 1988, almost 84% of the land area was inundated, and 45 million people were directly affected because of this catastrophic incident. This flood lasted for 65 days and caused the deaths of 1050 people. The amount of economic loss was three billion US dollars. Khatun (2013) identified that about 68% of the area was inundated in 2004 and almost 137 million people were affected in the 2007 floods. In the previous record, it was found that 69 floods (1988–2013) occurred in this country on a large or a small scale and the frequency rate was decreased (Auerbach et al. 2015).

Islam (2004) calculated that cyclones, which regularly occur in the Bangladesh coastal region during early summer or the late monsoon season, covered 47,211 km^2.

From the period1877–1995, Bangladesh faced more than 154 cyclones, five of them severe. As the climate changed, more cyclones were expected to hit the coastal belt of Bangladesh. Sidr and Aila, two of the most severe cyclones, which occurred in the years 2007 and 2009, were responsible for killing 4234 and 330 people respectively. The 1991 cyclone killed 0.14 million people and there were 154 million victims. Economically, the cost was two billion US dollars. In 2013, cyclone Mahasen hit the coastal area of Bangladesh. Because of the improved early warning system, people were already aware of this fact and were able to evacuate from risky places. Only 17 people were killed in this incident (EM-DAT 2018).

Cyclone Sidr had a wind speed of more than 220 km/h and covered 26% of the vulnerable zones. If the same cyclone were to occur in 2050, then it would covered almost 46% of the area. In addition, the projection reveals that sea level rise and high intense wind speed would be responsible for increasing 14% vulnerable zone with 1-meter inundation depth and 69% vulnerable zone with more than 3 meter inundation depth during the period of storm surge. From this projection study, it was found out that almost 9.12 million people would be directly exposed to storm surges. The financial loss from Sidr was estimated to be US$1.67 billion. If this type of cyclone were to occur in 2050, then the loss would be increased. The predicted amount would be US$4.560 billion at that time. If human deaths and injuries were to be considered, then US$1.03 billion would need to be added. (Dasgupta et al. 2010).

Landslide, one of the common phenomena in the southeastern part of Bangladesh, was responsible for the deaths of 122 people as well as the direct suffering of 0.35 million people in 2012. In 1994, Bangladesh faced a severe drought, which affected 53% of the total population. (EM-DAT 2018).

10.2.2 Search for Research Trends on the Vulnerability Level of the Coastal Area

Sarwar (2011) published a journal entitled "Vulnerability Assessment of the Coast of Bangladesh Using Geographic Information System," which stated that the Coastal Vulnerability Index (CVI) showed a common pattern for the Bangladesh coast. From this index base study, it was found out that the Meghna river estuary area was the most vulnerable area, and included Manpura and Hatiya Islands, the banks of the Meghna river, and the shoreline of the Bhola coastal zone. Amongst these areas, Hatiya was less vulnerable because of the high accretion rate in the southern part. The western coast, which included the Sundarban area, and parts of Barguna, Patuakhali, and the Mongla coastal zone, was highly vulnerable. But the presence of mangrove forests in this part reduced the vulnerability level to less than the Meghna River estuary area. It was revealed from this study that the vulnerability level was moderate to low in the Noakhali Feni coastal zone, the Chittagong coastal zone, and the Cox's Bazar coastal zone, which also added Kutubdia and Maheskhali islands. This result was found because the sea level rise and tidal range variables were low.

A recent study revealed that a 57.9 km coastline of the Ganges delta in Bangladesh faced a vulnerable situation, of which 50 km were highly vulnerable. This scenario was found because of the presence of intertidal and supratidal flats in which elevation was very low from the sea level as well as a moderate erosion rate at the coastline and a highly susceptible tendency during the period of storm surge. All of those characteristics made the Sundarban, Kuakata Upazila, and isolated islands of Patuakhali districts highly vulnerable. On the other hand, almost 61 km coastline was moderately vulnerable as the slope was moderate along with supratidal flats in those areas and the erosion rate was moderate. Amongst 286.2 km of shoreline, only 117 km had fallen into a lower spectrum with regard to the vulnerability scale. These areas were found on the southeast coast of Bhola Island, the west coast of Barabagi, and Dublar Char, Katka, and Kochikhali (Islam et al. 2016).

Because of geographical location and topographic characteristics, the sea-level rise had significant effects on the coastal area of Bangladesh (Mahmuduzzaman et al. 2014). The saltwater intrusion was one of the adverse impacts amongst them. Over time, saline water gradually moved towards the inland soil and water as climate change continued. In 1973, there was 83.3 million ha of saline-affected land. But in 2000, 102 million ha and in 2009, 105.6 million ha of land were directly affected by saline water. The rate of increase was 26% over the last 25 years (SRDI 2010).

According to experts, the salinity rate had increased during pre-monsoon and there was a downward trend in the monsoon season. The upward trend usually started at the end of the rainy season. (Hossain et al. 2012).

Researchers conducted a study on 12 coastal districts of Bangladesh to find out the vulnerability level of households in the coastal region. It was delineated that the vulnerability level was higher in the rural area than in the urban area. Households situated in the interior coastal region, faced more threats than those in the exterior region. These households were primarily affected by cyclones rather than any other disasters such as floods or riverbank erosion. There were also several factors that increased the vulnerability level of rural households such as lack of access to a sanitary latrine, a poor social network, a hostile demographic profile, as well as adverse climatic conditions (Toufique and Younus 2013).

10.2.3 Search for Research Trends on Socio-economic Conditions of the Coastal Area

Experts observed that among the houses that were built in the coastal area of Bangladesh, most of them had a non-engineered structure such as one-storied hipped or gable-shaped houses made out of mud, bamboo, straws, tiles, wood, jute sticks, CGI sheets, etc. The structure of these houses was either unstable or weak, and so unable to resist high-speed winds. The main reason for this was lack of knowledge (Alam et al. 2017).

Climate change acts as a potential tool to switch the system of society. It introduced a new community within existing society referred to as "climate migrants." The number of this type of migrants was 25 million in 1995 whereas now it is almost 200 million worldwide. (OECD 2011).

According to Islam (2001), 35 million people live in the coastal area of Bangladesh. If the sea level rises only 1 m, then 11% of the total population would be directly affected, and 60% could be affected by flash floods because of the overflow of the river. Along with 13% of the total agricultural land, 8000 km of roads would be lost. Because of this adverse condition, coastal people would be compelled to migrate to other places within the country. This scenario may increase the risk factors related to health issues as they may be forced to live in unhygienic conditions. The hostile environment has forced marginal people of the coastal area to find a safe and sound place over the last two decades. This trend increased after 2006.

Although there had been a huge loss due to natural disasters and disruption to daily life, sometimes people did not pay enough attention to them. The main reason for this was the cultural beliefs of society. People of the coastal area considered natural disasters as "God's wrath," which was beyond the control of human hands. Because of such faith, the community feels weak, which could have a negative impact on the actions taken in disaster management. On the other hand, culture has taught people to adapt to disasters. (Kulatunga 2010).

10.2.4 Search for Research Trends on Migration and Adaptation Strategy

The activities of the coast communities in protecting themselves from natural disasters can be divided into two categories: pre-event activities and post-event activities. The most highlighted activities before the period of disasters were going to a cyclone center. But strategically, this center was close to the powerful people of the community. Poor people needed to travel a long distance to take shelter. Sometimes they could, sometimes not. These types of people were also socio-economically vulnerable as their literacy rate was low, among them population density was high, and they generally lived in conditions of extreme poverty. Their daily livelihood generally depends on labor activities, fishing, or agricultural practices on other lands. After the disaster, it was found that people's homes, as well as their livelihoods, were destroyed or partially damaged. They were forced to shift from their regular jobs. To cope with this situation, they needed to migrate to other places temporarily or permanently. For migration, most of them chose big cities such as Dhaka. (Mallick et al. 2017)

The economic condition of the central and southeastern coast were greatly affected by the storm surge. Because of the polders and dikes, the central coastal region was riskier than the other regions. Sundarban on the west coast helped to reduce the impact of cyclones and storm surges. Based on socio-economic data, experts reported that capacity, like infrastructure facilities and early warning system, needed to increase in order to reduce the adverse effects of the disaster. (Kabir et al. 2017).

To increase resilience and reduce the level of vulnerability, the government of Bangladesh invested more than US$10 billion over the last three decades (WB 2000).

The Coastal Greenbelt Project, a very ambitious project conducted by the Forest Department of Bangladesh, planting trees along the 9000-km shoreline and creating an artificial mangrove forest environment that helps to reduce the erosion rate and retreat new land from the sea (MoEF 2009a, b). Owing to the dissipated characteristics, it also helped to reduce the height of the wave. The analysis showed that, where the width of the mangrove forest ranges from 100 to 200 m, the height of the wave decreased by about 20–25% from their natural level (MoWR 2000). The government also run the Coastal Embankment Rehabilitation Project (CERP) under which 1300 km of embankment was constructed. In addition, 7500 ha of strip plantation and 665 ha of foreshore plantation were completed.

To reduce the number of deaths in coastal districts, 2583 cyclone shelters, and 924 school buildings were constructed. The height of the pillars of buildings was usually 5 m. These are used for dual purposes. During a period of disaster, it would be used as a safe shelter and in normal times, it would be used as an educational institution. In addition, for the livestock, a 'killa' was built in the new shelters (MoFDM 2009).

An early warning system for floods, cyclones, and storm surges was developed by the government of Bangladesh (GoB). The GoB also launched the National Adaptation Program (NAP) in 2005 to increase the resilience capacity of the inhabitants. The main aim of this program was to identify the root causes of climate change, to analyze the possible impact, and to find out various techniques for coping with this situation (MoEF 2009a).

Bangladesh was regarded as one of the pioneering countries who adopted the program named Community-Based Adaptation (CBA) to increase the capacity of the local community to cope with the unfavorable environment. This program mainly runs at the regional level. It introduced new techniques such as floating gardens for crop culture and vegetation plantation, cage culture for fish production, and cultivation of saline-resistant crops. Because of such activities, food security, as well as nutritional security, would be ensured (Rahman et al. 2009).

10.2.5 Search for Research Gap

The present endeavor traced a large number of research projects that have been conducted in the recent past. The review suggests such research pursuits primarily engaged with the people living within the embankment. So far, no research has been conducted, except for some with inadequate citations. That being said, the present section of the study demonstrates research addressing the coastal settlement of Bangladesh by showing the living dynamic of the people who survive and make desperate attempts to live in a situation where there is a constant threat.

10.3 Methodologies

The whole study is divided into two sections. The first is to observe the migration pattern in the study area and the second is to identify the adaptation strategies that the local residents followed. This study is mainly based on statistical techniques. Data were collected from the field survey and various organizations. The methodologies of this study are divided into four parts: problem identification, study area selection, data collection, and data analysis.

10.3.1 Problem Identification

The impact of climate change on the Bengal delta is now one of the hottest topics and especially for the people who live in the coastal area, who may become refugees because of this change. The intensity and frequency of various hazards will increase in the future, which will makes the lives of those people miserable and compel people to migrate to an urban area or adapt to the hazardous effects. If the population density, population projections, and the amount of land are considered, then it is not feasible to force a large number of people to retreat from the coastal area to the nearby urban area. There are two most recognized words, when it comes to the terms to respond to natural hazards: mitigation and adaptation. Adaptation is considered to be the best technique for the least developed countries such as Bangladesh. The government takes initiatives to formulate various adaptation strategies under the National Adaptation Program of Action (NAPA) and the Bangladesh Climate Change and Action Strategy (BCCAS), which may help people to remain living in their homeland. Adaptation culture for agriculture has already been started (MoEF 2005, 2009a, b). Even then, a significant number of people migrate to improve their current situation. They may not all be environmental migrants, but purposely migrate to big cities for the development of their financial condition. Amongst them, some notably wish to go back to their country roads if they get a chance to do a job that helps to their family to remain solvent.

For that reason, it is required to identify the migration pattern of the coastal people as well as conducting an assessment of present adaptation techniques. It will be helpful to make a bridge between migration and adaptation that will offer future modification of the government's various action strategies.

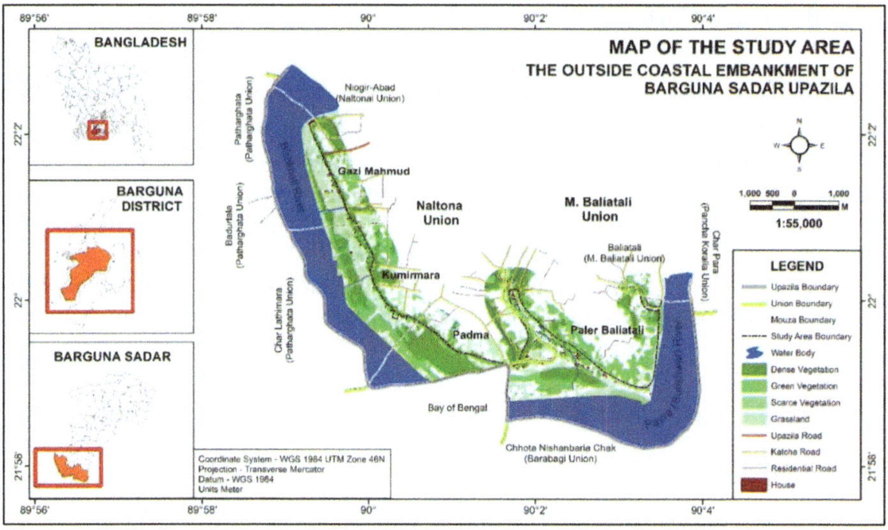

Fig. 10.1 Map of the study area. (Base map source – Ahmed et al. 2011; prepared by the Author)

10.3.2 Selection of the Study Area

10.3.2.1 Study Area

As, this study was conducted outside of the coastal embankment, we selected a place where the embankment is located. Consequently, amongst the 19 coastal districts of Bangladesh, Barguna district is selected for this preliminary study. As time is limited and the budget is constrained, the area coverage is needed to be shortened. In addition, human settlement is needed to fulfill the purposes of this study. From the in-depth preliminary survey, it was observed that, outside of the coastal embankment of Barguna Sadar Upazila, four mouzas that were identified, namely, Gazi Mahmud, Kumirmara, Padma, and Paler Baliatali, where human settlement is found. Amongst them, the first three mouzas fall under the Naltona Union and the remaining are found in the Baliatali Union. After considering all of those things, these four mouzas of Barguna Sadar Upazila were selected for this study (Fig. 10.1).

10.3.2.2 Justification

There are 12 coastal districts of Bangladesh, which are designated as the most vulnerable place to live. One of them is Barguna, which is situated at the mouth of Bishkhali and Payra (Burishwar) river, and the Bay of Bengal is located on the south portion. This area is directly exposed to the sea and regularly faces the wrath of nature. Thus, to conduct this study, this area was purposefully selected.

10.3.3 Data Collection

This study was carried out by using both primary and secondary data. To collect the necessary data and information from the targeted respondents and stakeholders, quantitative and qualitative methods were used. The quantitative indicators regarding livelihood patterns, migration, and adaptation are collected from the selected household. On the other hand, the qualitative survey is conducted covering government and nongovernment officials, decision-makers, community leaders, etc. Secondary data were collected from published and unpublished reports and datasheets from various government and nongovernment offices, journals, etc.

10.3.3.1 Parameters of the Primary Data

The questionnaire, was three categories: livelihood pattern, migration, and adaptation techniques.

10.3.3.2 Survey Instruments

Different kinds of survey instruments have been developed for the different kinds of surveys proposed for this study.

Household-Level Survey

A structured interview schedule (questionnaire) was used to collect the data through the household-level survey. The questionnaire was developed by accumulating various data and information including background characteristics of households, demographic and socio-economic information of household members, migration pattern, adaptation strategy, factors of adaptation strategy, the effectiveness of adaptation techniques, etc. (Fig. 10.2).

Focus Group Discussion

Separate checklists were developed for conducting the focus group discussions (FGDs) based on the participants, focusing on the dynamics and key factors of making the decision to move outside the area of the coastal embankment, the adaptation strategy practiced by respondents, and their factors and effectiveness.

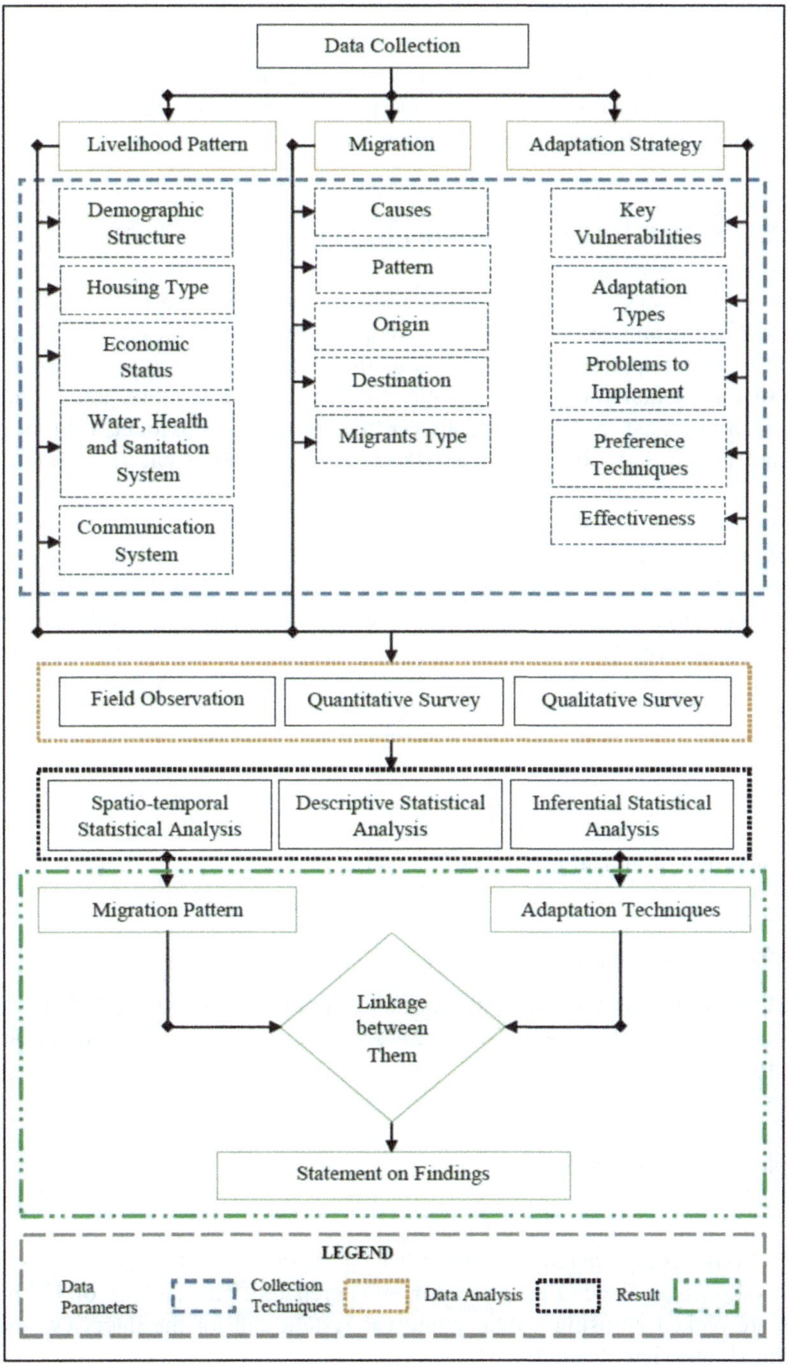

Fig. 10.2 Framework of data collection and analysis

Key Informant's Interview

A separate checklist was used for key informant's interviews (KIIs), mainly to collect information about the initiatives undertaken to reduce the disaster risks, ensure food security, health, and sanitary facilities, etc.

10.3.3.3 Sample Design for Quantitative Survey

Sampling technique and sample size estimation are the key activities undertaken while designing any research study. An adequate estimate (not less and not more) of sample size is important for ethical, scientific, and logistic reasons. The units of analysis for the household-level quantitative survey is the families/households that have been migrated to the outside area of the coastal embankment from the inner area and are currently living there at risk of being affected by natural hazards.

Determination of Sample Size

Assuming a simple random sampling procedure, the sample size is determined by the following formula:

$$n = \frac{pq \times Z^2}{Ne^2}$$

Here,

p = Proportion of indicator = 0.33 (assumed)
q = 1- p
z = 1.96
e = 0.05 (5% level of precision)

According to this formula, the minimum sample size stands at 332. However, this study covers 320 households, as the study area is very tough for the field survey (Fig. 10.3).

Sample Selection Process

According to a simple random sampling procedure, households are selected randomly from a pre-prepared household list. Therefore, at first, a list of households that reside outside the coastal embankment is prepared. Then, a total of 320 households are picked up using a lottery/interval system and for the interview, using a structured interview schedule.

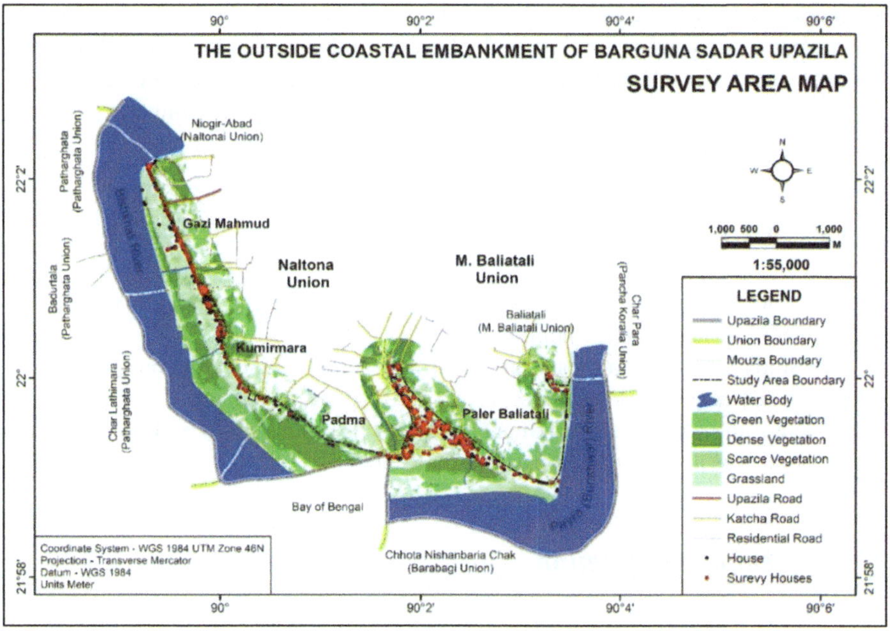

Fig. 10.3 Survey area: map of the study area (Base map source – OSM 2018; Ahmed et al. 2011;prepared by the Author)

10.3.4 Data Analysis

For the data analysis, two methods are followed, namely, quantitative and qualitative methods.

10.3.4.1 Quantitative Method

Several quantitative indicators (social; economic; demographic; health and sanitary; communication and transportation; origin and causes of migration; vulnerability; adaptation techniques; and their types and effectiveness, etc.) are generated through the household-level survey. The targeted respondents of the household-level survey are the vulnerable people who are living outside of the coastal embankment.

10.3.4.2 Qualitative Method

Qualitative data are collected through FGDs and KIIs. The objectives of the qualitative survey are (a) understanding the dynamics and key factors of in making the decision to move outside the area of the coastal embankment, and (b) the adaptation strategy practiced by respondents, and their factors and effectiveness. The respondents who participate in the FGDs are the heads of the households. On the other hand, the KII participants are government and nongovernmental officials responsible for disaster risk reduction in the concerned area, community leaders, local leaders, etc.

10.3.4.3 Data Entry and Analysis Tool

- The collected data were entered into the computer using a "data entry template" by SPSS for data management.
- This study used several descriptive and inferential statistical techniques to analyze the data. For these analyses, Excel 2016 software was used.
- To determine outside embankments, a map was produced. In addition, the sample point was also identified using ArcGIS 10.5.

10.3.5 Data Presentation

All the data analysis was performed using the above-mentioned formula and methods. The study is mainly presented in a descriptive way. Maps, charts, and tables are included where necessary. We also tried to find out the possible reasons for the result.

10.4 Results and Analysis

To figure out the migration pattern and adaptation techniques outside the coastal embankment of Barguna Sadar Upazila, this section is broadly divided into three parts. The first part discusses the lifestyle as well as various facilities and difficulties in the study area. The second part deals with the migration scenario of the study area. This part consists of migration patterns and reasons. The final part shows the various adaptation strategies that local residents of the study area follow.

10.4.1 Socio-economic Conditions

To observe the socio-economic conditions of the study area, the total part is divided into nine categories: likely demographic profile, literature rate, habitation type, livelihood status, the period of working, food supply, water supply, monthly expenditure, and communication system.

10.4.1.1 Number of Population

From the secondary source, it was found that there are 1025 households. Amongst these households 320 were randomly selected for the field survey. It was observed from the primary data that the number of people of the households surveyed is 1340, of which 667 are males and 673 are females, and the sex ratio is 100:101. The average household size is 4.19 (Table 10.1).

Based on primary data it was also observed that the percentage of the active population (whose age is limited between 20 and 50) is 40.72. The percentage of the dependent population is 29.7. Through this analysis, it can be concluded that potential people cut a high portion of the overall population (Table 10.2).

From the analysis, it was found that the number of females (20.67%) is higher than the number of males (19.85%) in the case of the active population. It was identified that for the child, the sex ratio is 100:167, but for the old, the ratio is 170:100. This observation suggested that males live longer than females in the study area (Fig. 10.4).

It was figured out from the census data that the population density of the study area is 504.28 per km^2. The highest density is observed in the Gazi Mahmud area (780.44 per km^2) and the lowest is observed in the Kumirmara area (292.33 per km^2; Table 10.3).

Table 10.1 Total population (amongst 320 households) in the study areas

Sex	Gazi Mahmud	Kumirmara	Padma	Paler Baliatali	Total
Male	130	104	133	300	667
Female	127	99	130	317	673
Total	257	203	263	617	1340
Ratio	0.976923	0.951923	0.977444	1.056667	1.008996

Data Source – Field Survey, 2018

Table 10.2 Percentage distribution of population by age group

Age	Gazi Mahmud	Kumirmara	Padma	Paler Baliatali	Total
0–4	3.89	0.99	1.14	1.78	**1.94**
5–9	14.4	13.79	12.93	12.48	**13.1**
10–14	9.34	10.84	11.03	10.86	**10.6**
15–19	14.79	14.29	15.21	17.67	**16.1**
20–24	11.67	11.82	11.03	9.72	**10.7**
25–29	16.34	14.78	17.87	17.02	**16.7**
30–49	12.84	13.3	14.07	12.8	**13.1**
50–59	5.84	6.9	5.7	7.29	**6.64**
60–64	7.39	7.88	7.22	6.48	**7.01**
65	3.5	5.42	3.8	3.89	**4.03**

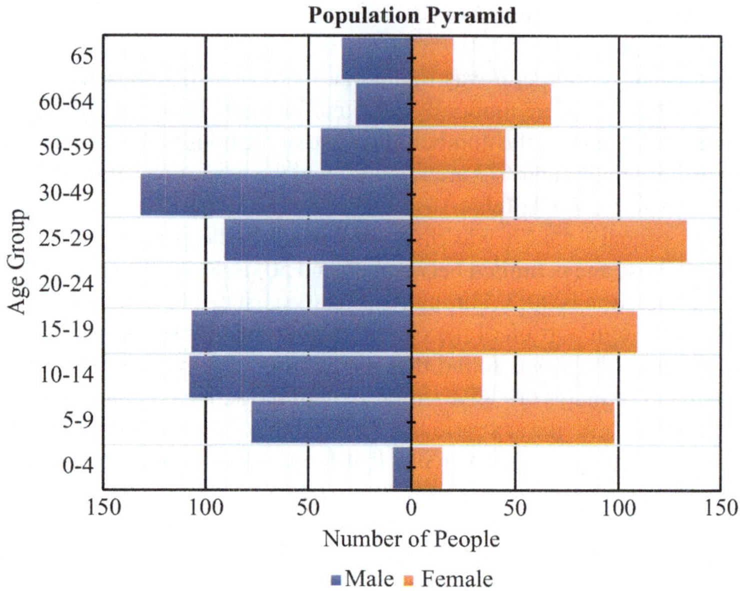

Fig. 10.4 Population pyramid chart of the study area Data Source – Field Survey, 2018

Table 10.3 Population density of the study area

Population	Gazi Mahmud	Kumirmara	Padma	Paler Baliatali	Total
Total	810	619	763	2062	4254
Area (km^2)	1.04	3.51	2.12	1.77	8.44
Density	780.44	292.33	431.61	587.04	504.28

Data Source – Naltona and M. Baliatali Union Parishad Office, Barguna, 2017

10.4.1.2 Literacy Rate

It is evaluated from the primary data that 48.13% of respondents in the study area are literate. The highest rate was found in the Gazi Mahmud area (51.36%), and the Kumirmara area experienced the lowest literacy rate (39.90%). The analysis shows that the female literacy rate (49.03%) is higher than that for males (47.22%) in the study area. The male and female ratio is 100:64 until the Secondary School Certificate, but after this education level, the situation is completely changed. It is observed that the ratio is 100:194 at the Higher Secondary Certificate and graduation levels (Table 10.4).

It was identified that after completing the Secondary School Certificate examination, most of the male members of the family moved away from education as they engaged in various types of work and started to earn money. As their income contributed to developing the economic condition of the family, the family head does not want them to concentrate on their education, and they slowly lose interest. They

Table 10.4 Education level

Gender	Education level	Gazi Mahmud	Kumirmara	Padma	Paler Baliatali	Total
Male	Primary	19	14	18	44	95
	Secondary	28	17	26	50	121
	HSC	16	7	19	50	92
	Graduate	2	0	2	3	7
	Total	**65**	**38**	**65**	**147**	**315**
Female	Primary	23	16	19	40	98
	Secondary	4	10	6	20	40
	HSC	19	14	18	57	108
	Graduate	21	3	21	39	84
	Total	**67**	**43**	**64**	**156**	**330**

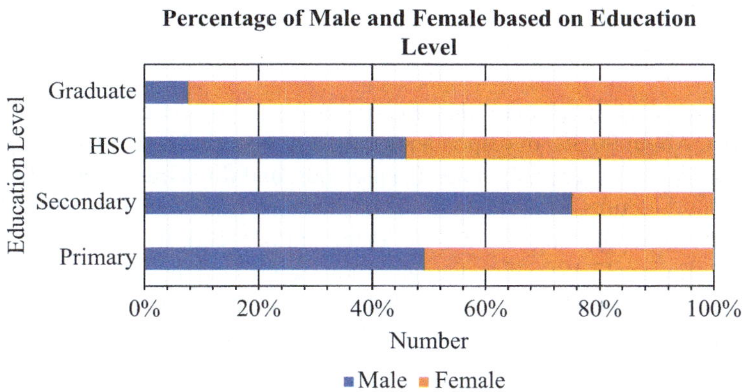

Fig. 10.5 Percentage of males and females based on education level in the study area (Data Source – Field Survey, 2018)

choose to work rather than upgrade their education level. On the other hand, female family members do all household chores along with helping the male members to perform better in their workplace. It was observed that the women in the study area also run a shop near their home. After doing all these activities, they additionally continued their studies (Fig. 10.5).

Another reason for this result has been identified. Natural disasters (namely, cyclones, storm surges, coastal floods, etc.) are one of the common phenomena in the coastal area. Because of these disasters, most of the families had fallen into economic crisis.

To overcome this recession, male members of the family were compelled to take jobs. At that time, they concentrated more on the workplace than on the educational institution. In contrast, female members stayed in the house and continued their studies.

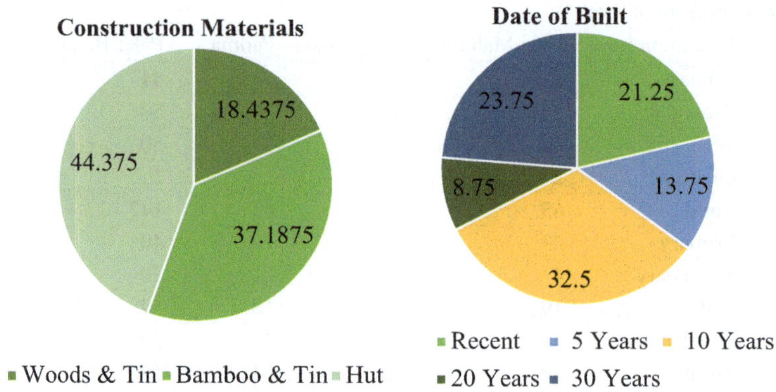

Fig. 10.6 Construction materials and date of building the house (Data Source – Field Survey, 2018)

10.4.1.3 Housing Structure

It was found from the primary data that the one-storied house is prominent in the study area. About 90.3% of respondents have a one-storied house while 9.7% have two-storied houses (generally made from woods, and tin is used as a roofing material). It was identified that 44.38% of the housing consisted of hut-type houses, which were generally built from Nypa Palm and tall reed. It was observed that to construct a house, 37.19% of participants of the study area mainly used bamboo and tin. Very few houses (18.44%) were found that were constructed from wood and tin. It was figured out that mangrove palms and tall reeds are prominent in the study area; hence, most of the respondents use these materials to build a house (Fig. 10.6).

It was found from the primary data that more than half of the houses (53.75%) were built 20 years ago. It is estimated that almost 46% of houses were built in the last 5 years, which means recently migration into the region has increased.

It was observed that only 39.37% of participants think before setting up the house in this area. This statistic suggested that more than 60% of peoples do not have any idea how to build a hazard-resistant house in the outer coastal embankment area. For that reason, the maintenance conditions are very bad in the study area. It was identified that only 8.13% of the houses were found to be in good condition (Fig. 10.7).

It was identified that 88% of participants have a safe sanitation system in their house, whereas 12% used open places as they do not know the disadvantages of unhygienic conditions or being forced to use them.

10.4.1.4 Life and Livelihood Patterns

It was observed that this area is a regular location for tidal water; thus, agricultural practices are impossible. There is also an absence of the activities of commercial (government or private) organizations. Therefore, most of the people in this area are dependent on nature for their income (Fig. 10.8).

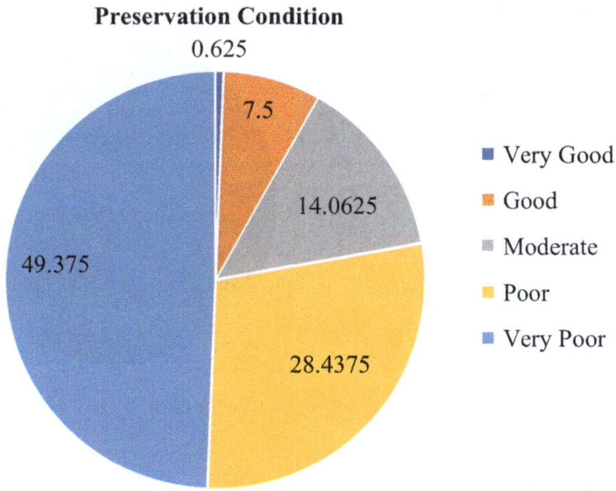

Fig. 10.7 Preservation condition of the houses (Data Source – Field Survey, 2018)

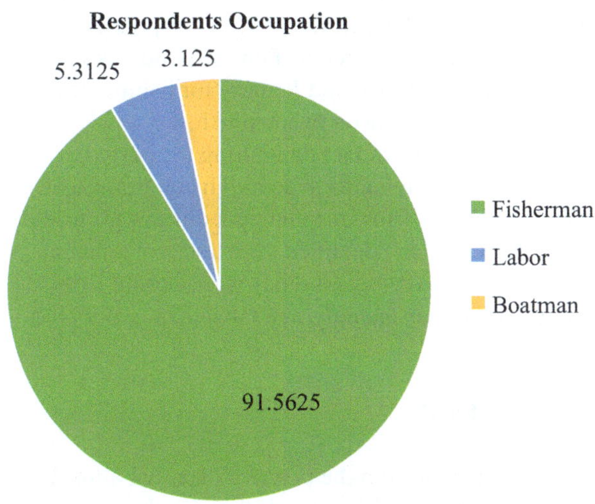

Fig. 10.8 Respondents' occupations. (Data Source – Field Survey, 2018)

From the analysis, it was observed that fishing is the main occupation in the study area. About 91.56% of respondents are engaged with this profession. A few (5.31%) work as boatmen. It was estimated from this finding that 96.87% of participants rely on natural resources for livelihood.

The people of the study area can be identified as very skilled fishermen. In addition, this area is situated near the seashore, which is the preferred place to carry out fishing activities. From the primary data, it was spotted that nobody has their own trawler for catching fish in the sea. They make a contract with the trawler owners. For that reason, they do not get much benefit from their profession.

It was observed from Fig. 10.8 that the remaining 3.13% work as agricultural laborers within the coastal embankment.

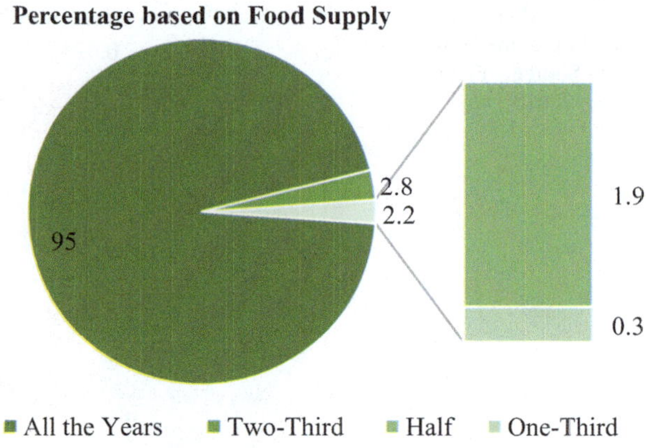

Fig. 10.9 Percentage of respondents based on food supply (per month) **(Data Source – Field Survey, 2018)**

10.4.1.5 Work Schedule

It was figured out from the primary data that the peak time for work in the study area is July to September. About 8.13% of respondents stated that their preferred time for work is August to October. Few respondents (16.88%) mentioned the months of November, December, and January for fishing. It was observed that nobody chooses the pre-monsoon season to catch fish.

It is a well-known fact that during the period of the pre-monsoon season, most of the time, the sea remains in ae rough condition, as well as having frequent warnings about cyclones. Thus, respondents did not go to the open sea to catch fish at that time. It was also identified that the government prohibited fishing practice in October. As there was no other work at that time, they utilize their time repairing fishing equipment, namely, net, trawler, etc., as well as visiting their relatives' houses.

10.4.1.6 Food

It was estimated from the primary data that almost 95% of participants of the study area can fulfill their basic needs through their occupation. All of them have adequate food supplies throughout the year. It was evaluated that 5% of respondents cannot afford food throughout the year for their family. Figure 6.7 shows that amongst them, 2.8% and 1.8% can afford enough food for two-thirds of the year (only working period) and half of the year (peak working period) respectively. It was identified that daily laborers in the study area can afford enough food supply only for one-third of the year (Fig. 10.9).

This area experiences regular flooding by tidal water; thus, people cannot carry out any agricultural practices. For that reason, people depend on the local market to buy food. In addition, the government run various programs in the coastal regions, and in exchange, they supply food. These types of programs are generally known as

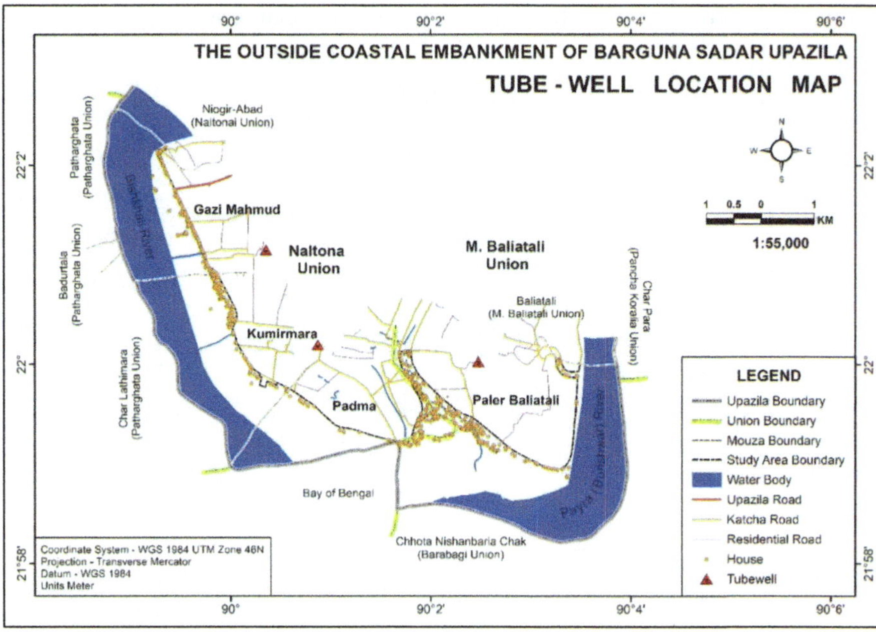

Fig. 10.10 Tube-well location of the study area (Data Source – Field Survey, 2018)

"work for food." There is also the government-run food project, whose main aim is to provide food in the community at a cheap rate.

It was spotted that only 1.56% of respondents in the study area are engaged with agricultural jobs. They do it within the coastal embankment. Most of them borrow the land from the local owners and cultivate various food crops to fulfill the demand.

10.4.1.7 Water for Daily Life

It was observed that the safest source for drinking water is a deep tube-well. More than 97% of respondents used this source for drinking purposes. The astonishing fact is that there were no tube-wells spotted in the study area. Local residents mainly collect water from inside the coastal embankment. It was identified that the number of tube-wells is also insufficient in the study area. For 150–200 households, only one tube-well has been observed. It was evaluated from the respondents' opinion that they have the capability to bore a deep tube-well, but do not know which aquifer zone is free from salinity intrusion. For this reason, they have been unable to do it until now (Fig. 10.10).

It was estimated from the analysis that for every household, the average distance to the tube-well is 2.08 km. The nearest distance was found in the Kumirmara area

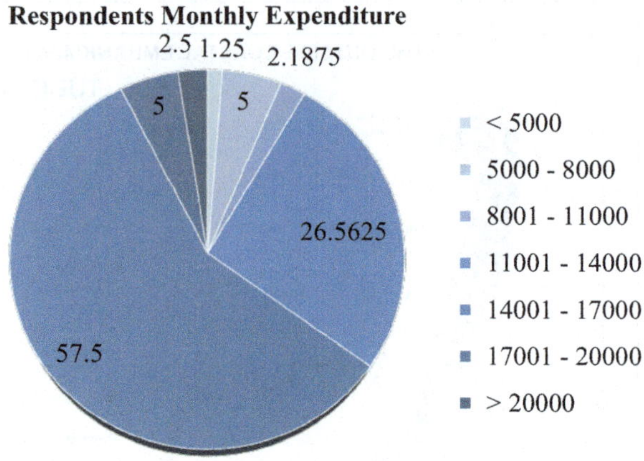

Fig. 10.11 Monthly expenditure range of the respondents (Data Source – Field Survey, 2018)

Table 10.5 Area-wise average expenditure per day

Place	Expenditure (US$/day)
Gazi Mahmud	5.783871
Kumirmara	5.9125
Padma	5.934375
Paler Baliatali	5.916438

Data Source – Field Survey, 2018

(1.43 km), and the longest distance was observed in the Paler Baliatali area (2.35 km) as this area is close to the sea; thus, the tube-well is bored further away from the sea.

It was figured out that almost 2% of respondents use surface water to meet their water demands. As most of the surface water was observed to be in an unhygienic condition as well as having saline intrusion, these participants in the study area regularly faced various types of diseases, namely, diarrhea, prolonged vomiting, and irritation of their eyes, skin, and stomach.

10.4.1.8 Monthly Expenditure

It was estimated that most of the respondents' (57.5%) monthly expenditure ranges from BDT 14,001 to 17,000. Very few participants' (2.5%) monthly expenditure was more than BDT 20,000. Figure 6.8 shows that 8.43% of people's monthly spend range is below BDT 11,000. From the analysis, it was figured out that the expenditure range is high in the study area, because they do not have any money in hand for savings (Fig. 10.11).

According to the Asian Development Bank (ADB) (2016), the people who have purchasing power less than US$1.90 (BDT 159.45) per day, are declared poor people. It is evaluated that the average purchasing power of the study area is US$5.89 (BDT 494.60). The highest average purchasing power is observed in the Padma area (US$5.9; Table 10.5).

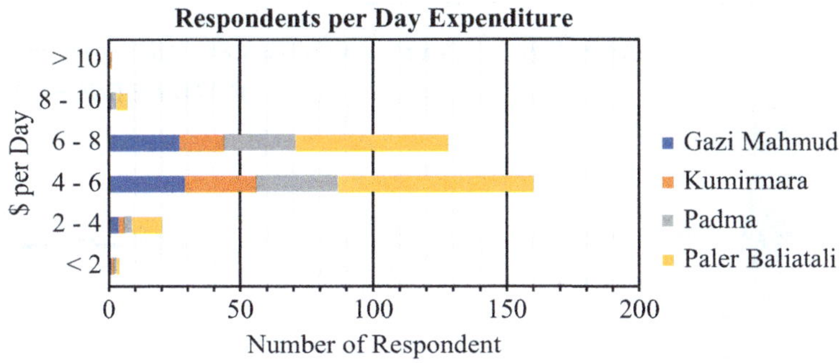

Fig. 10.12 Respondents' expenditure range per day (Data Source – Field Survey, 2018)

It was noticed that only two participants lived below the poverty line. Amongst them, one's purchasing limit is US$1.8 (BDT 151.06) and the other is US$1.6 (BDT 134.28). Their living places are Gazi Mahmud and Paler Baliatali area respectively. All others are identified as being above the poverty line.

The people's purchasing powers of the study area are given in Fig. 10.12.

10.4.1.9 Road Network

It was observed from the primary data that all the roads of the study area were found to be unpaved (locally this type of road is recognized as "katcha" road). In this area, the embankment is generally used as a road. It was identified through observation that the embankment was found to be either broken or in a rough condition. Respondents are forced to use it for their daily commute. This situation is created because of the severe cyclone that occurred in 2007. After that, the responsible authorities have not taken any action to repair it (Fig. 10.13).

It was seen from the field visit that to communicate, there is no vehicle available except a motorbike, but the fare is high. Participants in the study area generally walk to go from one place to another. They use a motorbike only in an emergency.

10.4.2 Migration Scenario

10.4.2.1 Number of Migrants

From the analysis of primary data, it was found that only in-migration occurs rather than out-migration. It is estimated that the percentage of migrants in the last 5 years is 35% (Fig. 10.14).

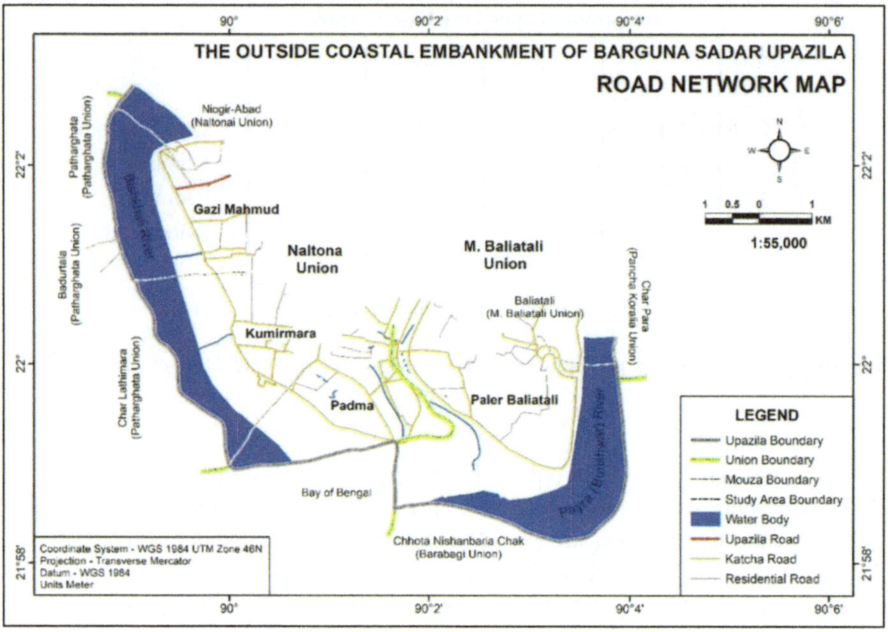

Fig. 10.13 Road network of the study area (Data Source – Field Survey, 2018)

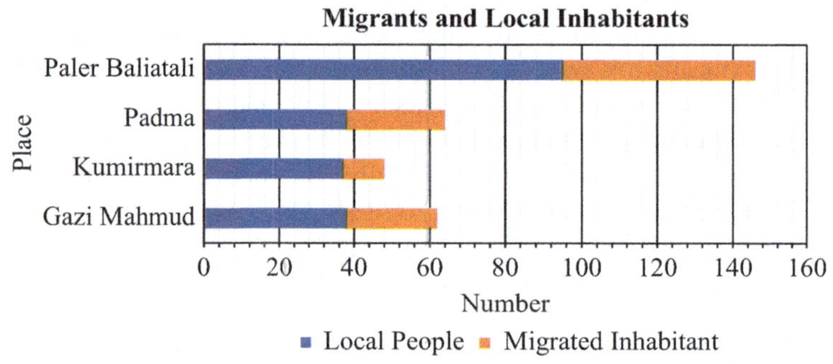

Fig. 10.14 Migrant and local inhabitants in the study area (Data Source – Field Survey, 2018)

10.4.2.2 Origin and Destination Place of Migrants

Respondents who lived outside the embankment, came from the upper part of Naltona and M. Baliatali Union. They are forced to migrate as the erosion rate is higher there. All the people who are affected by riverbank erosion forcefully migrated into the study area permanently, as they have no other place to live.

10.4.2.3 Reasons for Migration

The reasons behind this fact can be broadly divided into two categories: one is push factors (riverbank erosion, environmental degradation, political dispute, and quarrel with the family) and the other is pull factors (working opportunities, availability of new land, and the chance to live with the same class of people).

Push Factors

River Bank Erosion

From the analysis of riverbank shifting of Barguna Sadar Upazila, it was found that the Bishkhali river is highly active in the upper part of the Naltona Union as there is a meander. Through the analysis, it was observed that during the period between 2012 and 2017, Barguna Sadar Upazila lost almost 124 ha of land. In the midst of it, Naltona Union only lost 48.24 ha of land.

From Fig. 10.15, it is figured out that the movement of the shoreline during the period 1980–2017 was between −497.31 m and 714.93 m. The shoreline change is negative at the tip of Naltona Union. Therefore, the people of this region were forced to migrate to another place.

Natural Disasters

It was found from the secondary sources that cyclone Mahasen, which occurred in 2013, fell on the Barguna coast and affected an area of almost 189.68 km^2. The people of that area lost their property and were forced to migrate to a new place (Barguna Sadar Upazila Office, 2018).

Political Dispute

It was observed from the primary data that few respondents were forced to migrate in the study area because of political problems. It was remarked from their opinion that as this is a fringe area, they feel safe there.

Misunderstanding with the Family

It was noticed that several nuclear families quarreled with their families and moved on to this place. The reasons for quarreling are marriage for love, a lack of understanding of the family members, fleeing away because of family pressure, etc.

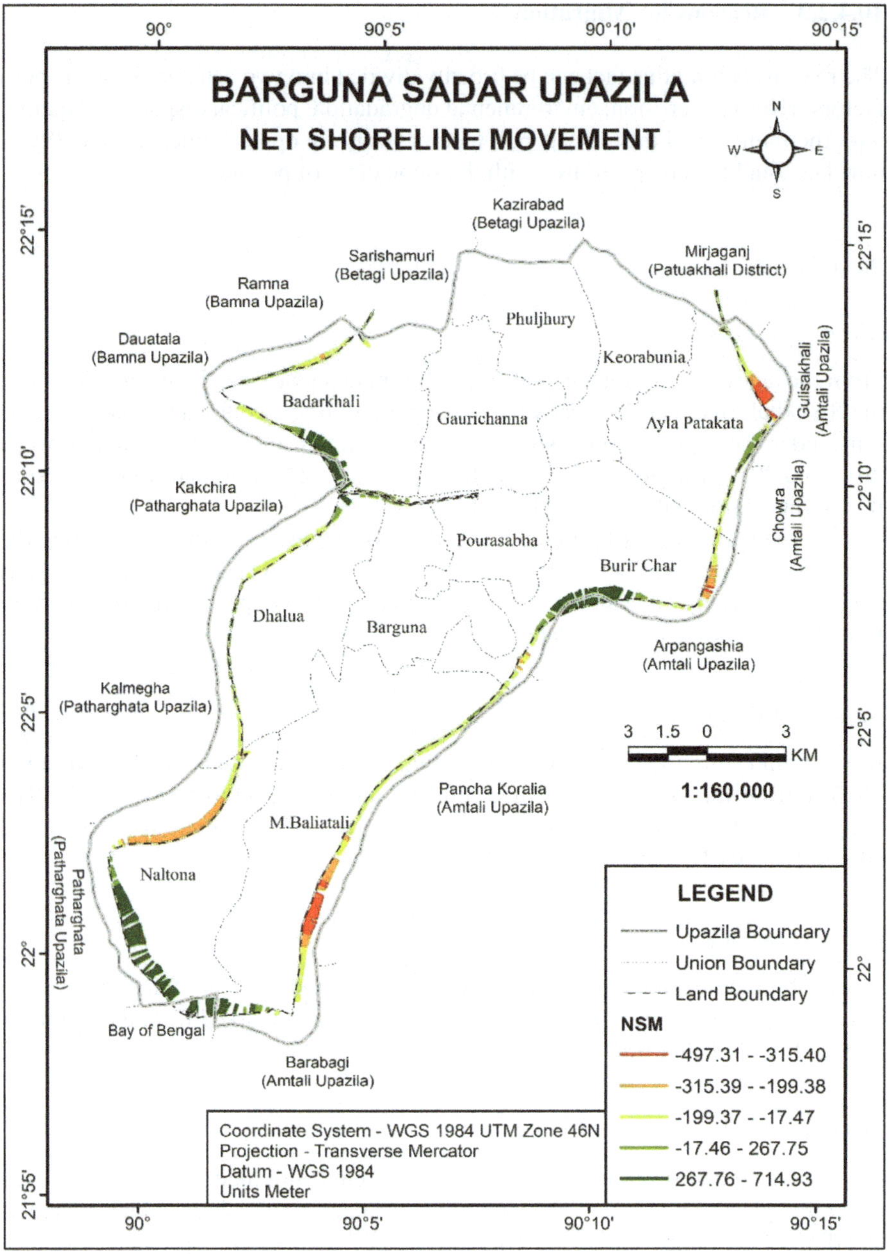

Fig. 10.15 Net shoreline movement of Barguna Sadar Upazila (from 1980 to 2017)

Pull Factors

Working Opportunities

It was seen from the observation that the study area is situated near the seashore, which is designated as a preferable place for fishing. In addition, jobs as boatmen are available here. The people who live here, generally choose either fishing or boatman as an occupation. Thus, the migrant people who lived outside the study area voluntarily come to this place as there are plenty of opportunities for a livelihood.

Availability of New Land

Every year, new land rises in front of the seashore. Anyone can rent this land from the authentic authorities at a cheap rate. In addition, the rate of land is very low on the outside of the coastal embankment, which acts as a pull factor to the migrants.

Opportunity to Live with the Same Class of People

From the analysis, it was figured out that most of the people in the study area are fishermen. Thus, migrants are taking the opportunity to live with the same class of people.

10.4.3 Adaptation Techniques

The people who live outside the coastal embankment of Barguna Sadar Upazila develop various adaptation strategies to cope with the disastrous conditions. It was identified from the primary data that most of the strategies followed by participants are indigenous techniques. It was figured out that based on the respondent's opinion, adaptation methods can be divided into three categories: likely livelihood, habitation, and health.

10.4.3.1 Adaptation for Livelihood

Migration

It is a well-known fact that migration is one of the common adaptation techniques. It was evaluated that most of the respondents of the study area did not leave their house during or after the period of severe disasters as they feel comfortable in their residence. Thus, no out-migration generally occurs. It was observed that people who lived on the northwestern tip of Barguna Sadar Upazila generally migrated into the study area owing to riverbank erosion, as, near the seashore, new land arises every year and there are many opportunities to do various jobs.

Choosing a Compatible Occupation

From the analysis, it was found that the elevation of the study area is very low (only 1.09 m) and has a very gentle slope (0.25^0); thus, this area is regularly flooded by tidal water (Fig. 10.16). For that reason agricultural practices are impossible there. To adapt to this situation, respondents hold jobs in fishing or as boatmen.

Being Involved in Several Occupations

It was estimated that almost 91% of participants of the study area are generally involved with the practice of fishing. During the disaster periods, they are forced to stop this activity. To fulfill the basic needs of the family, at that time, they involve themselves in diversified jobs, namely, day labor, rickshaw puller, mason, etc.

Taking Credit to Combat Rough Conditions

It was found that the respondents lose their property as well as facing financial crisis because of severe natural disasters. To cope with this situation they take out a loan with high interest from the moneylenders and various NGOs. They also forced to sell their barren land as well as homestead land for medical treatment. In addition, they sell domestic animals to meet with family's requirements.

Women and Children Being Engaged with Work

It was observed that few male members of the family were injured by natural disasters and lost their working efficiency. It was also identified from the primary data that several adult members of the family died because of catastrophic disasters. To cover this loss, women and children of those families are forced to engage in risky jobs.

Receiving Support from Relatives

It was experienced that sometimes relatives broaden their shoulders to help the ruined families of the study area. They give them economic support or help them get a job.

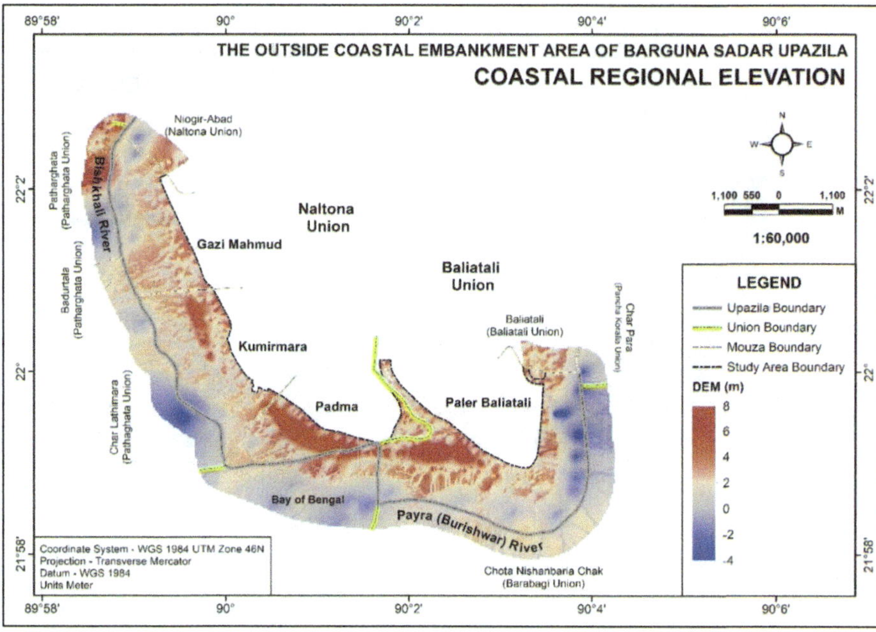

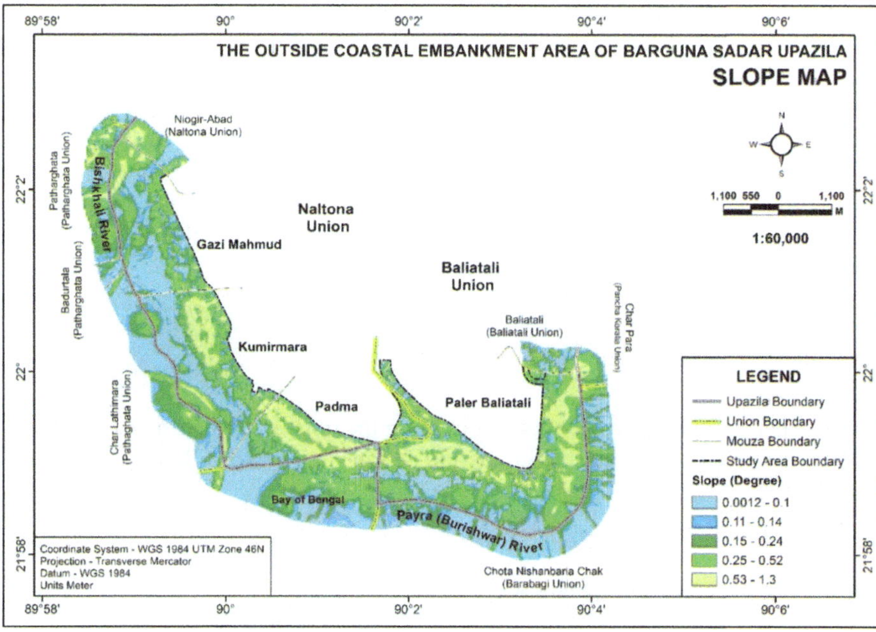

Fig. 10.16 Elevation and slope map of the study area (Data Source – Nasa JPL 2015; Prepared By the Author)

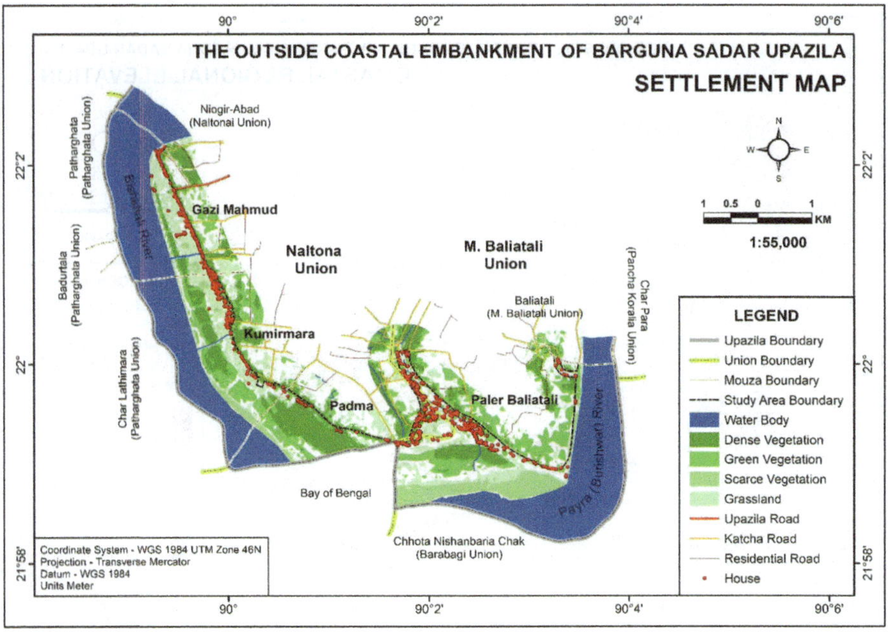

Fig. 10.17 Settlement pattern of the study area (Data Source – Field Survey, 2018)

10.4.3.2 Adaptation to Habitation

Building a Strong House

It was observed that participants tried to build a strong house to protect it from high winds that come from the seashore. For this purpose, they build their house in such a way that the wind can easily pass through it.

Plantation

It was experienced that residents plant trees on the outside of their house to protect themselves from the initial force of the storm surge as well as gusty winds (Fig. 10.17).

Raising the Basement of the House

It was found that almost 95% of respondents raise the basement every year, generally during the dry season, to protect the house from the force of the tidal water wave, taking a lesson from nature.

Modifying the House

It was evaluated from the primary data that when a respondent's house is broken down by natural disasters, they initially modify it to make it habitable. They give their house a tent shape using cloth or polythene.

House Relocation

It was noticed that a few respondents change their house location to avoid unhygienic conditions created by catastrophic disasters.

Embankment Use as a Shelter during the Period of the Disaster

It was figured out that a few respondents take shelter within the coastal embankment during the high tidal period, as the floor of their houses are completely or partially inundated by tide water. In addition, they use this embankment as a shelter during the period of the coastal flood, as well as after a storm surge.

10.4.3.3 Adaptation for Health

Buying Water from the Local Market

It was evaluated from the respondents' opinions that during the period of natural disasters, namely, storm surge, coastal flood, cyclone, etc., all the deep tube-wells go under water and they face the scarcity of fresh drinking water. To cope with this scenario, most of them (97.98%) buy mineral water from the local market.

Temporarily Shifting from the Current Location

Participants face unhygienic conditions because of water-logging, as most of the sanitary latrine are inundated and waste material mixed up with water. Thus, in a short0term period, there is a chance of spread of various kinds of waterborne diseases, and in the long term, leaves, which fall into the surface water, slowly decompose in the water and make the dissolved oxygen level of water low so that it cannot be used for domestic purposes. To adapt to this situation, they temporarily take shelter in the high land of the study area. It was figured out that the government and NGOs supply food and medicine as well as water until the situation comes under control. In addition, local communities try to clean the surface water as quickly as possible.

10.5 Conclusion

From the available literature review, it was observed that the study area is frequently affected by natural disasters. Thus, the people of this region have regularly needed to migrate and adapt to the rages of nature. It was identified from the primary data that the number of people (1340 in 320 households), as well as the population density (504.28 per km^2) of the study area, is very low. In addition, it was noticed that a large number of people were found to be between the age of 20 and–50 (40.72%). Thus, the dependency ratio of the study area is very low (29.7%). It was estimated that the ratio of males to females is almost equal (100:101). Almost half of the total population is literate (48.13%). The one-storied house is most prevalent (90.3%) in the study area. Nypa Palm and tall reeds are the most prominent materials for building a house. Almost all of the respondents (96.87%) depend on natural resources to lead their daily life and their peak time for work is July to September. Through their profession, most of them (95%) feed their families throughout the year. It was calculated from the primary data that the range of monthly expenditure in the study area is high (in 65% of respondents monthly costs are more than BDT 14,000). For that reason, they spend all the money that they earn. The communication system was found to be in very poor condition. The percentage of migrant households is 36.56%. The main reasons for migration are riverbank erosion, miscommunication with family members, a favorable environment for fishing practice, etc. Only in-migration occurred. It was observed from the primary data that local people adapt indigenous strategies to cope with the hostile situation. They mainly adopted the migration strategy, holding available professions, involving women and children in risky jobs, as well as engaging in diversified work in order to fulfill the basic demands, modifying their houses along with planting various trees, buying mineral water at the time of the disaster, and taking shelter on the coastal embankment after the period of natural disaster to protect themselves from unhygienic conditions.

10.6 Recommendation

From this study, recommendations for further study are given below:

- This study strongly recommended that a more in-depth study is needed to find out the effective indigenous techniques of coping with an unfavorable environment.
- As migration is the primitive adaptation strategy, more studies need to be conducted on the origin, causes, and patterns outside of the coastal embankment.
- This study also recommended that to adapt to adverse conditions, several engineering measures need to be undertaken to keep down the exposure level to a lower category, such as building safe houses, sea walls, sluice gates, dikes, etc. In addition, the sensitivity level can be minimized by operating various programs such as improving early warning systems and quickly disseminating them,

formed advanced and well-equipped rescue teams that quickly mobilize in the affected area, and conducting food security programs for the specific community at a local level. as Also, a community's capacity to adapt can be increased by improving the socio-economic status.

References

ADB (Asian Development Bank). 2016. Poverty in Bangladesh. Accessed 30 August 2018 at https://www.adb.org/countries/bangladesh/poverty

Ahmed M, Hasan S, Hossain Z (2011) Mouza Maps. ARC Bangladesh Ltd, 6/5, Rd No 7, Dhaka 1207. Accessed 25 August 2018 at http://gisapps.net/GISMaps/Basemap/Mauza/BD_Mouza_WGS84_BBS.zip

Akter, T. 2009. Climate change and flow of environmental displacement in Bangladesh. Research paper, Unnayan Onneshan, Dhaka, Bangladesh

"Cyclone Mora Batters Bangladesh, Displacement 5,00,000". (30 May 2017) Al Jazeera and News Agencies. Accessed 30 May 2017, at https://www.aljazeera.com/news/2017/05/bangladesh-350000-evacuated-cyclone-hits-170530040833094.html

Alam MR, Kaish ABMA, Zain MFM, Dev SK, Mahzabin MS (2017) Vulnerability assessment and construction recommendations of local houses in the cyclone prone coastal areas of Bangladesh. Int J Disast Risk Reduct 21:118–130

Auerbach LW, Goodbred SL, Mondal DR, Wilson CA, Ahmed KR, Roy K, Steckler MS, Small C, Gilligan JM, Ackerly BA (2015) Flood risk of natural and embanked landscapes on the Ganges – Brahmaputra tidal delta plain. Nat Clim Chang 5:153–157

Dasgupta S, Pandey K, Huq M, Khan ZH, Ahmed MMZ, Mukherzee N, Khan MF (2010) Vulnerability of Bangladesh to cyclones in a changing climate: potential damages and adaption cost. Policy Research Working Papers

DDM (Disaster Management Bureau) (2010) National plan for disaster management. Ministry of Disaster Management and Relief Division, Dhaka

EM-DAT (2018) The International Disaster Database. Center for Research on the Epidemiology of Disasters (CRED). University Catholique de Louvain, Clos Chapelle-Aux-Champs, Bte B1.30.15, 1200 Brussels, Belgium. Accessed at https://www.emdat.be/classification

Harris M (1999) Country report: Bangladesh. IDNDR-ESCAP Regional Meeting for Asia was Held during 23–26 February 1999, Bangkok, Thailand

Hossain MLH, Hossain MK, Salam MA, Rubaiyat A (2012) Seasonal variation of soil salinity in coastal areas of Bangladesh. Int J Environ Sci Manag Eng Res 1(4):172–178

International Organization for Migration (IOM). Retrieved from https://www.iom.int/definitional-issues

IPCC (2007) Climate change 2007: impacts, adaptation and vulnerability. In: Contribution of working group II to the fourth assessment report of the Intergovernmental Panel on Climate Change. Cambridge University Press, Cambridge

Islam N (2001) Urbanization, migration, and development in Bangladesh: recent trends and emerging issues. Center for Policy Dialogue and United Nations Population Fund, Dhaka

Islam MR (2004) Living in the coast: problems, opportunities, and challenges. Workshop paper WP011:13–15

Islam MA, Mitra D, Dewan A, Akhtar SH (2016) Coastal multi-hazard vulnerability assessment along the Ganges deltaic coast of Bangladesh – a geospatial approach. Ocean Coastal Manag 127:1–15

Kabir R, Sakib M, Jahan M, Haque A, Rahman M (2017) Socio-economic vulnerability assessment due to storm surge hazard in Bangladesh coast. 6th international conference on water and flood Management

Khatun M (2013) Climate change and migration in Bangladesh: Golden Bengal to land of disaster. Bangladesh E-J Sociol 10(2):64–79
Kulatunga U (2010) Impact of culture towards disaster risk reduction. Int J Strateg Prop Manag 14(4):304–313
Larour E, Ivins ER, Adhikari S (2017) Should coastal planners have concern over where land ice is melting?. Science Advances 3(11): e1700537.
Mahmuduzzaman M, Ahmed ZU, Nuruzzaman AKM, Sadeque FR (2014) Causes of salinity intrusion in Coastal Belt of Bangladesh. Int J Plant Res 4(4A):8–13
Mallick B, Ahmed B, Vogt J (2017) Living with the risks of cyclone disasters in the South-Western coastal region of Bangladesh. Environments 4(1):13
Mirza M, Monirul Q (2011) Climate change, flooding in South Asia and its implication. Reg Environ Chang 11(1):90–107
MoDMR (Ministry of Disaster Management and Relief) (2014) Comprehensive Disaster Management Program (CDMP II): trend and impact analysis of internal displacement due to the impacts of disaster and climate change
MoEF (Ministry of Environment and Forest) (2005) Bangladesh National Adaptation Program of Action (NAPA). Dhaka, Bangladesh
MoEF (Ministry of Environment and Forest) (2009a) Bangladesh climate change strategy and action plan 2009
MoEF (Ministry of Environment and Forest) (2009b) Bangladesh National Adaptation Program of Action (NAPA). Dhaka, Bangladesh
MoFDM (Ministry of Food and Disaster Management) (2009) Cyclone shelter information for management of Tsunami and cyclone preparedness
MoFDM (Ministry of Food and Disaster Management) (2010) Bangladesh Food Security Investment Forum
MoWR (Ministry of Water Resources) (2000) Final report on improved project planning and implementation methods – 2nd coastal embankment rehabilitation project
NASA JPL (2015) NASA shuttle Radar topography Mission global 3 arc second. Version 3. NASA EOSDIS land processes DAAC, USGS Earth Resources Observation and Science (EROS) Center, Sioux Falls, South Dakota. Accessed 25 August 2018, at https://doi.org/10.5067/MEaSUREs/SRTM/SRTMGL3.003
OECD (Organization for Economic Cooperation and Development) (2011) International migration outlook, Paris
Rahman SH, Alam AKMR, Saadat AHM, Uddin MJ (2009) Climate change and adaptation practices in thirty agro-ecological zones of Bangladesh. OXFAM International, Bangladesh
Sarwar MGM (2011) Vulnerability assessment of the coast of Bangladesh using geographic information system. Doctor of Philosophy Thesis, School of Earth and Environmental Science, University of Wollongong
SRDI (Soil Resource Development Institute) (2010) Saline soils of Bangladesh. SRMAF Project
Toufique K, Younus M (2013) Vulnerability of livelihoods in the Coastal Districts of Bangladesh. Bangladesh Dev Stud 36:95–120
Uzzaman MA (2014) Impact of sea level rise in the coastal areas of Bangladesh: A macroeconomic analysis. J Econ Sus Dev 5(18):105–110
WB (World Bank) (2000) Bangladesh: climate change and sustainable development

Chapter 11
Climate Change Impacts and Mitigation Strategies to Develop the Low Carbon Themes in Bangladesh

Md. Nazrul Islam, Sahanaj Tamanna, Md. Noman, Rathindra Nath Biswas, Sashwot Kharel, Ishtiak Ahmed, Mohammad Hosen Howlader, Kulsum Akhter, Raufun Zamee Evan, Nafiz Imtiaz Chowdhury, Yeasin Arafat, Semonti Kundu, Tasmia Sultana, Washif Ahmed, Afrin Jahan, Fahim Hasan Rashik, Md. Alamin Hasan Shawon, and Jannatul Fardush

Abstract Bangladesh has the limited experiences to reduce the greenhouse gases being spewed out by more developed countries, nor can they hope to reverse what has been done. The country has instead become a hotbed for adaptation strategies and experimentation, where some innovative models are being developed and explored. Aid agencies, donors, and nongovernmental organizations have begun pitching in to help Bangladesh to prepare for a changing climate. Meanwhile, back in more developed countries, where we should be reducing our emissions through what in climate change speak is called mitigation, instead we seem stuck in endless rounds of debates about how to proceed. Throughout the world climate change has become an issue that is afforded less importance given the severity of the problem. This issue of climate change is given rather less value than other crises that exist all around the world. As climate change is a gradual process and as its effects are not immediate, many people all around the world underestimate this issue, whereas others deny the existence of the phenomenon existence. As a result of these efforts,

M. N. Islam (✉) · M. Noman
Department of Geography and Environment, Jahangirnagar University, Savar, Dhaka, Bangladesh
e-mail: nazrul_geo@juniv.edu

S. Tamanna
Bangladesh Environmental Modeling Alliance (BEMA), Non-Profit Research Organization, Dhaka, Bangladesh

R. N. Biswas
Remote Sensing Division, Centre for Environmental and Geographic Information Services (CEGIS), Dhaka, Bangladesh

S. Kharel · I. Ahmed · M. H. Howlader · K. Akhter · R. Z. Evan · N. I. Chowdhury · Y. Arafat · S. Kundu · T. Sultana · W. Ahmed · A. Jahan · F. H. Rashik · M. A. H. Shawon · J. Fardush
North South University, Dhaka, Bangladesh

M. N. Islam, A. van Amstel, *Bangladesh II: Climate Change Impacts, Mitigation and Adaptation in Developing Countries*, Springer Climate,
https://doi.org/10.1007/978-3-030-71950-0_11

Bangladesh has become the first low-income country to adopt low-carbon zone guidelines and a roadmap that, when fully implemented, will reduce greenhouse gas emissions by 17% in the country's industrial zones. In order to prepare for the worsening effects of climate change the Bangladesh government has put forward some plans and policies. But sadly, these plans and policies remain only in this study as they have not been brought in to action effectively. For the human race to mitigate this problem there is no way other than to reduce carbon emissions. Therefore, it is necessary to adopt the low carbon development framework on an international, regional, and national level. It is necessary for the world, as well as for Bangladesh.

Keywords Climate change · Greenhouse gas emissions · International negotiations · Disasters vulnerabilities · Low carbon · Sustainable development · Bangladesh

11.1 Introduction

Since the beginning of the industrial revolution in the nineteenth century, carbon dioxide (CO_2) has been dramatically increased in our earth atmosphere. (Neumann 1985; MoEF 2009; Ghosh et al. 2014; Afroj et al. 2016). This increase in CO_2 from roughly 280 parts per million (ppm) prior to the Industrial Revolution to about 391 ppm today is having a dramatic impact on our climate, warming our climate (Mahmood 2011; Fisher 2013; Fisher and Mohun 2015; Teske et al. 2017; Rahman 2017). The Kyoto Protocol specified six main GHGs that significantly impact the environment: CO_2, methane (CH_4), nitrous oxide (N_2O), hydrofluorocarbons (HFCs), perfluorocarbons (PFCs), and sulfur hexafluoride (SF_6). Amongst them, CO_2 has been considered the most prominent contributor to global climate change (Ghosh et al. 2014; Zhang 2015; Sarkar et al. 2018). For instance, carbon dioxide emissions have increased from 22.15 Gt in 1990 to 36.14 Gt in 2014 (Ahiduzzaman and Islam 2011; World Development Indicators. 2018). The continuous and increasing production of carbon emissions is therefore, a matter of global concern (Yue et al. 2015; Dunnan et al. 2019). Studies exploring carbon emissions of the construction industry have been conducted to gain an insight into the carbon emissions of different industries (Hong et al. 2018; Lu et al. 2013).

Currently, policy building and implementation have depended on the impacts of climate change and global warming as a focal issue (Padilla-Rivera et al. 2018). Luckily, many countries have set ambitious long-term carbon emission reduction targets, e.g., the USA is committed to lower carbon emissions by 17% and 83% below 2005 levels by 2020 and 2050 respectively (Huisingh et al. 2015; Xu et al. 2015); the UK aims to reduce its carbon emissions by at least 80% of 1990 levels by 2050; China is now committed to abating its emissions per unit of economic output by 40–45% of 2005 levels by 2020; India is committed to decrease its emission intensity by 20–25% by 2020; and Brazil is committed to reduce its carbon emissions by 38–42% of BAU levels by 2020 MoEF 2009; (Huisingh et al. 2015; Xu et al. 2015). However, owing to rapid economic growth, fossil fuel emissions have increased from 213 Tg carbon/year in 1990 to about 573 Tg carbon/year in 2009 in South Asia (Boden et al. 2013).

Carbon dioxide is the most dominant GHGs and accounted for 77% of the total global GHG emissions, whereas CH_4, N_2O, and other gases contributed 14%, 8%, and 1% respectively (IPCC et al. 2007; Xu et al. 2015; Dunnan, 2019). Nevertheless, remarkable commercial deeds concern climate change (Backman and Verbeke 2015; Jeswani et al. 2008; Sprengel and Busch 2011; Weinhofer and Hoffmann 2008); rather inconsistently, worldwide CO_2 emissions have risen significantly in the manufacturing zones, even in numerous industrialized states (Oliver et al., 2012; Sarkar et al. 2018). The core cause of this incidence is the dearth of essential inventions to enable the transformation to a low-carbon civilization (Blanford 2009; Tavoni et al., 2012). Despite increasing institutional and regulatory pressures, stakeholders controlling traditional carbon-based technologies are seeking to protect their rents (Neuhoff 2005; Fisher 2013; Fisher and Mohun 2015; Teske et al. 2017) by preserving a compromised regime (Jones and Levy 2007). Instead of appealing eco-friendly inventions to control environmental contamination (Berrone et al. 2013), they make marginal efficiency improvements to existing processes (Pinkse and Kolk 2010) to conform to the dominant practices in their field (Delmas and Montes-Sancho 2010).

The emission of GHGs from Bangladesh was 126.6 $MtCO_2eq$ (metric tons carbon dioxide equivalent per capita) in 2010, which amounted to 0 .3% of global total GHG emissions. In 2005, per capita GHG emission of Bangladesh was 0.6 $MtCO_2eq$ and it is expected to increase up to 1.7 $MtCO_2eq$ in 2025 (Jilani 2012; Fisher 2013; Fisher and Mohun 2015; Teske et al. 2017). Most frequently called upon to react are carbon-intensive firms from the energy and manufacturing sectors because they are most responsible for this form of ecological degradation (Shrivastava 1995; Canadell et al. 2007; Xu et al. 2015). Owing to the vulnerability to climate change Bangladesh now needs to adopt a more effective strategy of low carbon development in order to mitigate this crisis (MoEF 2009; Haq and Rabbani 2011). The surge in the carbon emission research domain encompasses several risks of neglecting important areas of research, which occurs because of the inability to cover the status quo of the domain. As a result of this increasing attention toward global carbon emissions, researchers have begun quantifying the carbon emissions (Lu et al. 2013; Liu and Liang, 2017).

11.1.1 Statement of the Problem

The core cause of global warming is the rise of carbon dioxide in the earth atmosphere. The 2007 Fourth Assessment Report (AR4) by the Intergovernmental Panel on Climate Change (IPCC et al. 2007; Xu et al. 2015) of the United Nations indicated that most of the observed warming over the last 50 years was likely to have been due to the increasing concentrations of GHGs produced by human activities such as deforestation and burning fossil fuels (Fisher 2013; Fisher and Mohun 2015; Teske et al. 2017). The concentration of CO_2 in the atmosphere has increased from a pre-industrial value of about 280 ppm to 391 ppm in 2011. In 2014, the concentration of global CO_2 reached ~ 400 ppm (Sarkar et al. 2018). The continuous and increasing production of carbon emissions is, therefore, a matter of global concern (Ahiduzzaman and Islam 2011; Yue et al. 2015; Dunnan, 2019). The release of CO_2 from burning fossil fuel is becoming an anxiety for ecological species owing to their intensity of concentration in the earth's atmosphere, simultaneous changes in climate, and the direct impact on environments and energy demand (Fleming 2008; Andres et al. 2012; Xu et al. 2015). These environmental alterations along with climatic fluctuations may badly impact human civilization (Fisher 2013; Fisher and Mohun 2015; Teske et al. 2017). This coalescence of statistics on fossil-fuel CO_2 ($FFCO_2$) releases to the air is anticipated to shrink our present comprehension about $FFCO_2$ productions to the atmosphere in support of the Regional Carbon Cycle Assessment and Processes Project (RECCAP). The specific focus on CH4 is motivated by the expectation that it will continue to be the most influential GHG after CO_2 (Gernaat et al., 2015; Saunois et al. 2016).

During this century, earth's average surface temperature rises are likely to surpass the safe threshold of 2 °C above preindustrial average temperature. Moreover, the temperature will be increased significantly at the higher latitudes, by means of situations of moderate danger expecting 2–3 °C temperature escalations by 2090, and 4–5 °C increases in northern Canada, Greenland, and Siberia (Costello et al., 2011). The world mean temperature at present has risen by 0.6 °C since the 1800s. Furthermore, global mean temperature in the last century has increased by 1.5 to 4.5 °C, which intensifies polar ice melting and sea level rise (Fleming 2008: Mahmood 2011). The impacts and vulnerabilities of such climate-induced hazards are greater in Bangladesh, maybe because of its geophysical location, hydrological influence of the monsoons, and regional water flow patterns, low-level resilience of the affected people in terms of technical and financial capacity, and lack of proper arrangement and implementation of policy and institutions to address the challenges (Rahman et al. 2007; Ahiduzzaman and Islam 2011; Rahman 2017). For instance, Bangladesh has been significantly affected by more than 259 natural hazards during the time span 1991 to 2009. More than 80% of the deaths occurred in 1991 in Bangladesh. In Bangladesh, approximately 140,000 people have died, and a significant number were injured in 1991, and a future recommendation for better preparation for climatic events, as well as mitigating the large-scale effects of natural catastrophes (Star 2010). Increasing concentration of GHG emissions is considered a primary cause of these issues (Hansen et al. 1981; Liu et al. 2019). Nearly 80% of global carbon emissions are caused by urban human activities (Heede 2014).

11.1.2 Literature Review

The world GHG emissions have been rising tremendously by anthropogenic intervention, which clearly designates a considerable release of CO_2, CH_4, N_2O, HFCs, PFCs and SF_6) in the earth's atmosphere (Xu et al. 2015; EPA 2014). Global warming at present is +0.60 °C over the last three decades, + 0.80 °C in the last century, and the continued warming in the first half of the twenty-first century is consistent with the recent rate of +0.20 °C per decade (Fleming 2008: Mohajan 2011). The present concentration of GHG in the air has increased since 1750 from a CO_2 of 280 ppm to 450 ppm (Stern 2007, Mohajan 2017). It was well documented that topographic ozone has been increased from air pollution (NOx, CO and others) and is regarded as a vital greenhouse-driving term (Ramanathan and Feng 2009).

In addition, the recognition of chlorofluorocarbons (CFCs) on the stratospheric ozone and the chlorofluorocarbons, or CFCs, which are still utilized in some parts of the world as spray-can propellants and throughout the industrialized world in plastic-foam manufacture, cleaning of industrial components, air conditioning, and refrigeration (Fisher 2013; Fisher and Mohun 2015; Teske et al. 2017; Dunnan, 2019). On the other hand, the absorption of solar radiation by black carbon and some organic matter increases atmospheric heat and tends to expand the atmosphere's greenhouse warming. The five largest emitters of GHG, together accounting for 62%, globally, are China (26%), the United States (13%), the European Union (more than 8%), India (7%), the Russian Federation (5%), and Japan (almost 3%) (Xu et al. 2015; EPA 2014; Dunnan 2019). But increasing this scenario, the emissions come from three regions: Asia, Europe, and the United States, which together accounted for 66 percent of total global emissions in 2012 significantly (EPA 2014; Dunnan 2019). South Asia has also undergone significant changes in the rates of land-use change over the last 20 years, contributing to the net carbon exchange (Ahiduzzaman and Islam 2011; Patra et al. 2013).

Bangladesh is a growing South Asian economy that is maintaining macroeconomic stability despite the political turmoil, structural constraints, and global volatility (World Bank, 2015). For Bangladesh, the concept of low carbon society (LCS) for the energy sector, not only to reduce GHG emission, but also to focus on low-carbon energy mix and better energy efficiency that can meet the challenge of CO_2 emission reduction and improve economic productivity (Fleming 2007; Rahman 2017). In addition, financial institutions develop a number of carbon-based insurance products, derivatives, and structured products in order to manage and transfer risks and costs of carbon emissions and obtain more sustainable profits (Zeng and Zhang 2011; Xu et al. 2015). The new urban agenda and sustainable development goals (SDGs) have a dominant focus on taking urgent and immediate action to combat the impacts of global climate change (SDG Goal 13) to create a more resilient and healthy environment for now and for the future (UN 2015).

11.1.3 Significance of the Study

Climate change is one of the most prominent global issues that has attracted the attention of global academic researchers, policy makers, and other related professionals (Abeydeera et al. 2019). The impacts of climate change have been widely documented in the field of environmental, socioeconomic, and ethical issues that face human civilization from the meaning of population pressure, and environmental pollution. Its consequences and overheads will not be calculated in monetary value, and its impacts will be seriously and asymmetrically spreading worldwide for decades (Fisher 2013; Fisher and Mohun 2015; Teske et al. 2017). The foremost controlling factors of climate change and global warming includes emissions of CO_2 by anthropogenic activities, clearance of forest cover, and burning of fossil fuels (Lu et al. 2013; Sarkar et al. 2018).

Over the past decade, researchers, industry practitioners, and various government authorities across the globe have given greater attention to carbon emission reporting and identification of climate mitigation strategies, resulting in a noteworthy increase in interrelated study journals and works. The study indicated that analyzing past trends of carbon emissions has become a useful method for understanding the current emissions and thereby predicting future emissions (Hammond and Norman 2012; Xu et al. 2015). Accordingly, several studies have been conducted to analyze the past trends of various industrial sectors, including manufacturing (Rahman et al. 2007; Ren et al. 2014) and energy (Ouyang et al. 2015). Moreover, studies exploring carbon emissions of the construction industry have been conducted to gain an insight into the carbon emissions of different industries (Hong et al. 2015). Moreover, global researchers have predominantly focused on exploring operational carbon emissions, which have resulted in a significant increase in embodied carbon (EC) emissions.

Being a minute GHG emitter and as the country most affected by climate change, the Government of Bangladesh (GoB) concentrates on implementing several adaptation and mitigation programs. Considering the advantages of implementing mitigation technologies, the objective of this study is to give an initial vision of a possible LCS scenario in Bangladesh. The LCS scenarios may be possible in future by controlling population, traffic, industry, energy demand, crop production, livestock number, land use patterns, and release of GHS from built-up areas, industrial zones, traffic, and power plants, land use, forestry, and agriculture by 2025 (Sarkar et al. 2018).

The effects of climate change are intensifying throughout the world, and with it, Bangladesh is likely to face severe implications in the future. Bangladesh, with a population of over 150 million is facing a number of challenges, including environmental hazards, socio-political conflicts, development crises, and also the effects of climate change (Fleming 2007; Rahman 2017). Climate change and its associated hazards, including variations in temperature and rainfall, have increased the intensity of flood, drought, cyclone and storm surge, and salinity intrusion that are already affecting the communities, ecosystems, and infrastructure of the country (Amos 2016). The impacts and vulnerabilities of such climate-induced hazards are greater in Bangladesh, and may be due to its geophysical location, hydrological influence of the monsoons and regional water flow patterns, the low-level resilience of the affected people in terms of technical and financial capacity, and the lack of proper arrangement and implementation of policy and institutions to address the challenges (Rahman et al. 2007). GHG emissions are also causing enormous climatic changes, but they are controllable if necessary steps are taken (Hansen et al. 1981; Dunnan, 2019).

11.2 Result and Discussions

11.2.1 Trend of CO_2 Emission in Bangladesh

The concentration of GHGs has increased significantly in the earth's atmosphere owing to anthropogenic intervention since 1750. Although the concentration of all types of GHGs has increased in the atmosphere, the focus is always on the CO_2 emission, as it constitutes a large share of GHG emission (Hansen et al. 1981; Dunnan, 2019). In the Climate Change Synthesis Report 2007, the Intergovernmental Panel on Climate Change (IPCC) has mentioned that the energy sector contributed 25.9% toward global anthropogenic CO_2 emissions (IPCC et al. 2007; Sarkar et al. 2018). The emission of CO_2 by Bangladesh in the context of the global emission is on a very small scale. However, the country's emission scenario has marked a rapid increase in CO_2 emission over time. The CO_2 emission and per capita CO_2 emission in Bangladesh have escalated (Xu et al. 2015). In 2011, the total emission of CO_2 was computed by 57.07 t, which was significantly raised by 140.67% compared with the emission of 15.94 t in 1991 (Fisher 2013; Fisher and Mohun 2015; Teske et al. 2017). This indicates an average yearly increase of CO_2 emission of 6.70% over the period 1991–2011 (Rahman et al. 2007). The per capita emission has also increased by more than two times, which was 0.15 t in 1991 and 0.37 t in 2011 (Hansen et al. 1981; Dunnan, 2019). The CO_2 emission has evolved from various sources such as electricity, transport, and manufacturing. In Bangladesh, highest CO_2 emission was produced by the electricity generation sector followed by the manufacturing sector (Sarkar et al. 2018). In 2012, the electricity sector contributed 47.72% of the total CO_2 emission in Bangladesh whereas manufacturing and construction; residential and commercial and public services; transport; and other sectors were responsible for CO_2 emission of 20.87%, 11.13%, 14.49%, and 5.79% respectively.

11.2.1.1 Carbon Emission from Different Sources

Emission from Vehicles

One of the major sources of air pollution in urban areas of Bangladesh is due to the unburned fuel from two-stroke engine vehicles. Dhaka City is ranked as one of the worst contaminated cities of the world. From the report of the Bangladesh Atomic Energy Commission, every day the emission from automobiles in Dhaka City reveals 100 kg of lead, 3.5 t of SPM, 1.5 t SO, 14 t H_2C, and 60 t (Fisher 2013; Fisher and Mohun 2015; Teske et al. 2017).

Emission from Brick Kilns

It has been found that total annual amount of CO_2 emission for four types of brick kilns from Dhaka, Chittagong, Rajshahi, Khulana, Sylhet, and Barisal are 8.9 t/year, 10.1 t/year, 12.8 t/year, 15.3 t/year, in 2002, 2005, 2007, and 2010 respectively. It shows the increasing rate of CO_2 emission in all of those years simultaneously around Bangladesh.

Emission from Industries

Industry is the sector most responsible for carbon emission all over the world, as well as Bangladesh. Almost 80% of industries directly or indirectly emit carbon in many ways.

11.2.2 Low Carbon Developments

The low carbon development (LCD) approach has become an emerging framework to address the challenge of climate change in the contemporary world (UNEP, 2011; Ahiduzzaman and Islam 2011). The UNFCCC adopted in Rio in 1992 has played a significant role in shaping the concept of LCD over the years. Presently, no globally agreed definition of LCD exists. But many agree on LCD to focus on mitigating of GHG emission in the process of development (Bowen et al. 2011; Asaduzzaman et al. 2013; Fisher 2017; Hunter et al. 2019).

Low carbon development focuses on integrating three aspects of climate change adaptation, mitigation, and development in a coherent way to overcome the global crisis of climate change while driving the country toward economic prosperity and sustainability. It is believed that the integration of the three elements will bring about co-benefits (Fleming 2008; Ahiduzzaman and Islam 2011; Amos 2016). On the one hand, adaptation measures reduce the detrimental effects of climate change and thereby contribute to sustainable development (Fisher 2013; Fisher and Mohun

2015; Teske et al. 2017). On the other hand, development activities lessen vulnerabilities to the impact of climate change by building adaptive capacity. And lastly, mitigation measures reduce the future risk of climate change by reducing carbon emissions (Bowen et al. 2011; Asaduzzaman et al. 2013; Fisher 2017; Hunter et al. 2019). Therefore, it is necessary to understand that these elements are complementary to each other and therefore go hand in hand. The emission of CO_2 in the earth atmosphere has been regarded as the main source of global warming and climate change. In Fig. 11.1, we observed the concentration of global CO_2 in the earth's atmosphere over the last 800,000 years (Fig. 11.2).

Over this period, fluctuations in CO_2 concentrations were observed; however, over the last decade CO_2 emission has been skyrocketing. This spike in CO_2 emission started soon after the industrial revolution (Amos 2016). Before the industrial revolution the atmospheric concentrations of CO_2 did not exceed 300 parts per million (ppm), but after the revolution it shot up and now it is well above 400 ppm (Bowen et al. 2011; Asaduzzaman et al. 2013; Fisher 2017; Hunter et al. 2019). It started breaking CO_2 records in 1950 and it has not stopped since given the size and tremendous heat capacity of the global oceans, it takes a massive amount of heat energy to raise the earth's average yearly surface temperature even a small amount (Fleming 2008). The 2-degree increase in global average surface temperature that

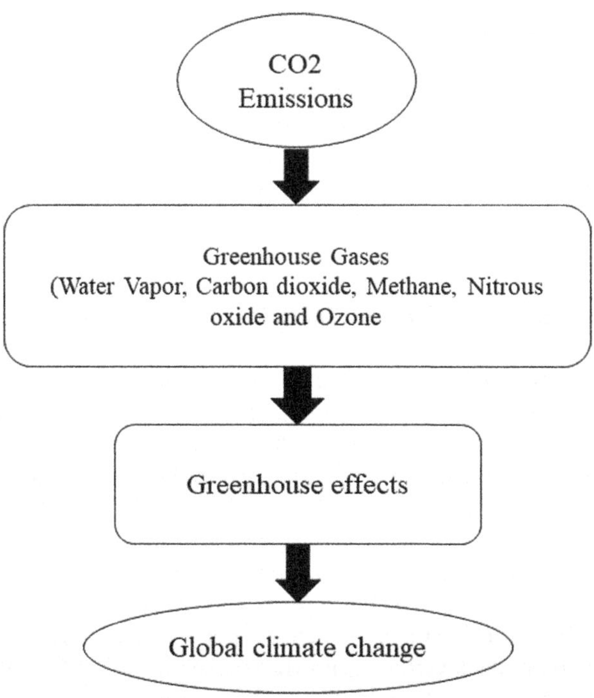

Fig. 11.1 Carbon dioxide is the most dominant greenhouse gas, global warming and climate change

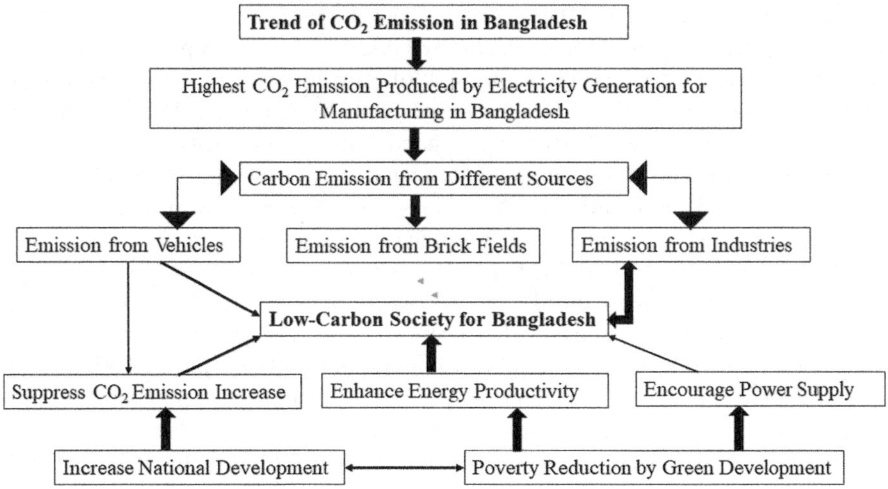

Fig. 11.2 Trend of very small-scale CO_2 emission in Bangladesh in the global context

has occurred since the pre-industrial era (1880–1900) may seem small, but it means a significant increase in accumulated heat (Schurer et al. 2017). It is remarkable that over the last few decades, global temperatures have risen sharply, by 0.6–0.9 °C between 1906 and 2005, and the rate of increase has nearly doubled in the last 50 years Bowen et al. 2011; Asaduzzaman et al. 2013; Fisher 2017; Hunter et al. 2019). Therefore, it is said that the carbon emission has played a major role in increasing the global temperature. Table 11.2 shows that after the industrial revolution carbon emission increased tremendously, and with the increase in carbon emission there has been a rapid increase in the global average temperature (Rahman et al. 2007). (Table 11.1)

11.2.2.1 Low Carbon Society and Its Relevance to Bangladesh

Low-carbon society in a developing country is a combined concept that can develop a sustainable model that can meet economic development, life-style improvements, and mitigate the effects of climate change through technology transfer, using renewable energy, utilizing funding, and finally capacity building, which contributes to the sustainable development of the whole of society (Fisher 2013; Fisher and Mohun 2015; Teske et al. 2017). The function of society will conserve the environment and ensure social as well as economic safety. In that society, equal attention will be paid to environmental protection as well as other socio-economic security.

Bangladesh being one of the most densely populated countries in the world, as well as its economic backwardness, is what makes it one of the most vulnerable countries to climate change. Bangladesh has seen a rapid growth in its GDP for the last few decades, which has allowed the country to reach new heights of economic

Table 11.1 Characteristics of greenhouse gases and the industrial revolution carbon emission increase

Type of greenhouse gases	Source	Removal source	Gas reaction
Carbon dioxide	Burning of fossil fuels	Photosynthesis process	Absorption of infrared radiation
	Deforestation	Ocean	Indirectly affect the ozone concentration in the stratosphere
Nitrous oxide	Burning of biomass	Removal by soil	
	Combustion of fossil fuels	Photolysis in the stratosphere	
	Fertilizers		
Fluorinated gases	Emitted through various industrial processes	Photolysis and reaction with oxygen	
Methane	Burning of biomass	Microorganism uptake	Absorption of infrared radiation
	Rice paddies	Reaction associated with hydroxyl groups	Indirectly affect ozone concentration and water vapor in the stratosphere
	Fermentation by enteric bacteria		Production of CO_2

Source: Modified from Ahiduzzaman and Islam 2011; Yaacob et al. 2020 and adapted from Islam

prosperity and development (Fisher 2013; Fisher and Mohun 2015; Teske et al. 2017; Rahman 2017). This has allowed the country to pull a significant number of people out of poverty, while increasing the living standards of others (Bowen et al. 2011; Asaduzzaman et al. 2013; Fisher 2017; Hunter et al. 2019). This was made possible by Bangladesh's rapid industrialization. With rapid industrialization and urbanization the economy is booming, while the effects of climate change are intensifying, as Bangladesh is now emitting GHGs like never before.

11.2.2.2 Low-Carbon Society for Bangladesh

Bangladeshis are quite powerless to reduce the greenhouse gases being spewed out by more developed countries, nor can they hope to reverse what's been done. But unfortunately the suppression of the CO_2 emission increase and other issues of concern in Bangladesh such as: energy improvement (enhanced energy productivity; low-carbon energy mixture; development of financial progression (encourage power supply with effective energy administration); increased national development opportunities to improve the green economy; poverty reduction by green development; and diminishing of the costs of climate change in the future.

11.2.3 Scenarios of the Problem in Global and Regional Contexts

In the early nineteenth century, when ice ages and other natural changes in paleoclimate had occurred, the history of climate change was first surmised and the greenhouse effect was also first identified. In the late nineteenth century, scientists first argued that human emissions of GHGs could change the climate (Fleming 2008; Rahman 2017). Many theories about climate change were discussed where forces from volcanism to solar variation were involved (Fisher 2013; Fisher and Mohun 2015; Teske et al. 2017). Thomas Edison, precursor of electrical technologies, voiced concern about climate change and support for renewable energy in the 1930s (Electrifying America). It was not until the 1960s that the warming effects of CO_2 gas became increasingly convincing (Hansen et al. 1981; Amos 2016; Dunnan, 2019). During the 1970s, scientists' opinions about human activities that generated pollution undoubtedly favored the warming viewpoint of the Milankovitch Theory of the ice ages, a consonance position that confirmed that GHGs were deeply involved in most climate changes and that human emissions were bringing ascertainable global warming, as a result of improving the reliability of computer models and observational work by the 1990s. Although there are many GHGs such as nitrous oxide, methane, carbon mono-oxide, and SO_2, CO_2 is the main cause of rapid climate change (Fleming 2007; Amos 2016).

Concentration of CO_2 in the atmosphere is at its highest levels in over 800,000 years. Over the period, we see consistent fluctuations in CO_2 concentrations. These periods of rising and falling CO_2 coincide with the onset of ice ages (low CO_2) and inter glacial (high CO_2) (ourworldindata.org). Over this long period the atmospheric concentration of CO_2 did not exceed 300 ppm (Fisher 2013; Fisher and Mohun 2015; Teske et al. 2017). This change occurred only because of the Industrial Revolution and the rise of human emissions of CO_2 from burning fossil fuels. While in over 800,000 years, the concentration of CO_2 did not only not rise above 300 ppm, it is now well over 400 ppm (Rahman et al. 2007; Avi-Yonah and Uhlmann 2008; Boden and Marland 2011; Bo Xu 2012). In Bangladesh, in 2018, CO_2 emissions per capita were 0.56 t. During the time period 1999 to 2018, CO_2 releases per capita rose noticeably from 0.19 to 0.56 t with a maximum annual rate of increase of 15.42% in 2001, declining to 6.64% in 2018 (Table 11.2).

11.2.4 Response of Bangladesh to Climate Change

Bangladesh is one of the most vulnerable nations in the South Asian Region as well as in the world. Although many of the policy makers, professionals, scientists, and bureaucrats are aware of the severity of the issue, the governments over the past decades have not taken this matter a very seriously. The Bangladesh government has recognized climate change as both an environmental and a developmental issue (Delmas and Montes-Sancho 2008; Chowdhury 2013; Dastagir 2015; Donald 2015;

Table 11.2 Average fossil fuel CO_2 emissions and annual growth rates (%) for the decades of the 1990s, 2000s, and the full Regional Carbon Cycle Assessment and Processes Project period (1990–2009)

	Average emissions (Tg carbon/year)	Growth rate (%/ year)	Average emissions (Tg carbon year)	Growth rate (%/ year)	Average emissions (Tg carbon/ year)	Growth rate (%/ year)
Country/ region	1990–2009	1990–2009	1990–1999	1990–1999	2000–2009	2000–2009
Bangladesh	8.207	6.2	5.638	6.0	10.775	6.1
Bhutan	0.113	7.8	0.072	11.4	0.154	8.5
India	319.81	4.7	247.44	5.6	392.18	5.3
Nepal	0.702	5.6	0.514	13.8	0.889	2.3
Pakistan	29.986	4.1	23.019	4.7	36.932	5.7
Sri Lanka	2.368	5.8	1.629	8.9	3.107	2.2
South Asia	361	5.7	278	8.4	444	5.0

Source: Ahiduzzaman and Islam 2011; Patra et al. 2013

Boden and Marland 2017; Dunnan 2019). Throughout the years, Bangladesh has had ten measures and initiatives to better acknowledge the climate change issue in order to overcome future environmental, economic, and social consequences.

The major initiatives taken by Bangladesh are as follows:

- Bangladesh is signatory to the United Nations framework Convention on Climate Change (UNFCCC). The Bangladesh government signed up to the UNFCCCC in 1992 and was approved in 1994.
- It is the Ministry of Environment and Forest (MOEF) that is given the responsibility of coordinating the UNFCCC process in Bangladesh. The government has also formed a National Climate Change Committee 1994 to formulate relevant plans and policies and also to oversee the implementation of obligations under the UNFCCC process (Ahiduzzaman and Islam 2011).
- Under the directives of the government, the Bangladesh Agriculture Research Council (BARC) and leading civil society organizations, e.g., the Bangladesh Centre for Advanced Studies (BCAS), prepared the development of the National Adaptation Programmes of Actions (NAPA) in accordance with the UNFCC. The NAPA basically outlined 15 immediate action plans to address the issue of floods, salinity intrusion, and droughts, which were therefore highlighted as the major consequences of climate change (Amos 2016)
- In 2008, the government adopted the Bangladesh Climate Change Strategy and Action Plan (BCCSAP), focusing on six priorities. It recognized all the climate change-induced hazards including floods, drought, SLR, salinity intrusion, cyclone and storm surge, variations in temperature and rainfall, etc., and their associated impacts on different sectors. The six major themes of BCCSAP as follows:
  - Food security, social protection, and health
  - Comprehensive disaster management

- Infrastructure
- Research and knowledge management
- Mitigation and LCD
- Capacity building and institutional strengthening

- Over the last decades, the Bangladesh government has invested more than US$10 billion to make the country less vulnerable to natural disasters. For this purpose, the initiatives taken by the GoB were establishment of flood control, river bank embankments, construction of resilience cyclone shelters, developing early warning systems to mitigate property loss and damages, to save lives and livelihoods from the vulnerability of natural catastrophes.
- Bangladesh has also created two national climate change trust funds. The Bangladesh Climate Change Trust Fund (BCCTF) was the first climate change trust fund of this country and was established in the financial year 2009–2010. The other one is the Bangladesh Climate Change Resilience Fund (BCCRF), which is a multi-donor trust fund, established in 2010, as a modality for the development partners to sustain Bangladesh in effecting the Bangladesh climate change policy and accomplishment strategy.

11.2.5 Comparison with Other Countries and Global and Regional Contexts

Climate change remains one of the most critical threats that society has ever encountered. Its impacts are already visible and will worsen over time if left unchecked (UNFCC Fact sheet, 2010). Today, the effects of climate change are being felt by countries all around the world. It is distorting various economies and disrupting the livelihood of today's people as well as those of tomorrow (Fleming 2007; Ahiduzzaman and Islam, 2011; Amos 2016). The weather conditions are shifting, sea levels are rising, weather disasters are getting more severe, and GHG emissions are now at their highest recorded levels. Climate change, however, is a global challenge that does not respect national borders. (Bowen et al. 2011; Asaduzzaman et al. 2013; Fisher 2017; Hunter et al. 2019) It is an issue that needs to be addressed collectively and therefore solutions need to be initiated and coordinated at an international level. Therefore, tackling climate change needs concerted intervention from nations around the world (Fleming 2008; Ahiduzzaman and Islam 2011). To strengthen the global response to the threat of climate change, the international community has come together to further address and mitigate this issue. In order to relieve the world of its adverse effects, the international community has taken some major steps throughout the course of history to better acknowledge and mitigate the problem of climate change.

11.2.5.1 United Nations Framework Convention on Climate Change

The United Nations Framework Convention on Climate Change (UNFCCC) is an intergovernmental treaty that was negotiated between 1991 and 1992 and was finally adopted and came into force in 1994, and has been ratified by over 197 countries around the world (Fisher 2013; Fisher and Mohun 2015; Teske et al. 2017). This convention was initiated by the global community in order to reduce the effects of climate change, which were found to have anthropogenic causes (Bowen et al. 2011; Asaduzzaman et al. 2013; Fisher 2017; Hunter et al. 2019). The international community agreed that climate change is man-made; thus, collective efforts made by the UNFCCC in order to mitigate the effects of climate change (Fleming 2008; Iisd.org 2013). The UNFCCC basically acts as a framework for international cooperation to combat climate change by limiting increases in global temperature and the resulting climate change, and coping with impacts that were, by then inevitable (Unfccc.int 2012; Ahiduzzaman and Islam 2011).

The main objectives of the UNFCCC are:

- To stabilize the concentration of GHG concentration in the atmosphere at a level that would prevent the interference with climate change and to ensure the achievement of this goal in a time-bound manner
- To ensure that food production is not affected by climate change
- To ensure that sustainable development is achieved

What the UNFCCC also does is that it mandates the more developed countries to assist the less developed countries, particularly ones that are extremely vulnerable to the effects of climate change. Thus, it mandatory for the Annex 1 members to provide financial assistance as well as to extend support by facilitating technology transfer to developing members (Unfccc.int 2012).

In a nutshell the UNFCCC focuses on four key areas:

- Mitigating (reducing) GHG emissions
- Adapting to climate change
- Reporting national emissions
- Financing climate action in developing countries (Committee on Climate Change, 2017)

11.2.5.2 The Kyoto Protocol

Under the United Nations Framework Convention on Climate Change (UNFCCC), conferences are periodically held to examine the implementation of the convention and also to discuss the future plans and policies to effectively adapt and mitigate the effects of climate change. At the third annual summit, or COP-3, the Kyoto protocol was adopted in 1997. Owing to the complex ratification process, it came into effect in 2005 (Bowen et al. 2011; Asaduzzaman et al. 2013; Fisher 2017; Hunter et al. 2019). The UNFCCC borrowed a very important line from one of the most successful multilateral environmental treaties in history (the Montreal Protocol, in 1987): it bound member states to act in the interests of human safety even in the face of scientific uncertainty. On 12 December 2015, countries under the UN Framework Convention on Climate Change (UNFCCC) adopted the Paris Agreement. After the correlation between human activities and climate change was established, parties to the UNFCCC continue to adopt decisions, review progress, and consider further action through regular meetings of the Conference of the Parties (COP) and the Kyoto protocol came in to effect to reduce human activity and to therefore slow down the process of climate change (Committee on Climate Change, 2017).

The Kyoto protocol is based on the principle "common but differentiated responsibility and respective capabilities." This means that it only binds developed countries, and places a heavier burden on them as they are largely responsible for the current high levels of GHG emissions in the atmosphere, and that individual countries have different capabilities in combating climate change (Fleming 2008; Unfccc.int, 2019). The Kyoto protocol gives more responsibility to the first-world nations, as these countries are seen to be contributing more to the problem of climate change, through their industrial activity (IPCC 2007a, b). The Kyoto protocol basically laid down mandatory targets for the reduction of GHG emissions, which were accepted by leading nations of the world.

The Kyoto protocol also determined what it called commitment periods for various developed countries. These commitment periods set targets for these countries to reduce their carbon emission by certain percentages by a given deadline (Bowen et al. 2011; Asaduzzaman et al. 2013; Fisher 2017; Hunter et al. 2019). The USA and China are among the countries that signed the agreement but never ratified it, because of which the USA and China were not obliged to implement the set standards of carbon reduction. One of the most important features of the Kyoto protocol is the clean development mechanism (CDM) and the carbon trading scheme.

11.2.5.3 Clean Development Mechanism

The CDM encourages developed countries to invest in technology and infrastructure in less-developed countries, where there are often significant opportunities to reduce emissions (Repetto 2001; Ahiduzzaman and Islam 2011). Therefore, it supports sustainable development. Under the CDM, the investing country may demand an effective reduction in emissions as a credit to fulfill its obligations under the protocol. Here, developed countries are given the opportunity to earn extra credits by investing in sustainable development projects in a less developed country.

11.2.5.4 Carbon Trading Scheme

The carbon trading scheme allows participating countries to buy and sell emissions rights and thereby place an economic value on GHG emissions (Aldy and Stavins 2012). The Kyoto protocol basically gives standard targets to countries with which they have to abide. The countries that are able to reduce their carbon emissions by a significant margin are given the opportunity to sell these excess quotas to those countries who have missed their emission targets (Fisher 2013; Fisher and Mohun 2015; Teske et al. 2017). This allows an economic incentive, in which successful countries are rewarded whereas unsuccessful ones are penalized.

11.2.6 Regional Perspective of Climate Change

11.2.6.1 Impacts of Climate Change and Vulnerability in South Asia

The area of South Asia ranges from the high Himalayan peaks of Bhutan and Nepal to the low-lying Delta of Bangladesh and the Indian Peninsula, and to the gem-like islands of Sri Lanka and the Maldives in the Indian Ocean (Fisher 2013; Fisher and Mohun 2015; Teske et al. 2017). Covering climatic areas as diverse as its physical environment, the region is experiencing a number of impacts of climate change, including glacier melting, forest fires, rising sea levels, mountain and coastal soil erosion, and intrusion of saline water (Fleming 2008; Bowen et al. 2011; Asaduzzaman et al. 2013; Fisher 2017; Hunter et al. 2019). The countries of South Asia are susceptible to climate change because of poverty, high population density, and scarcity of resources for adaptation strategies. The region experiences a wide range of extremes: in topography, from the highest mountains to the largest deltas to atolls barely above sea level; in precipitation, from arid land to vast plains prone to intermittent flooding; and in the climate, from glacial to tropical (Assessing the Costs of Climate Change and Adaptation in South Asia).

With shifts in the global climate system expected to cover the next century, South Asia will continue to be vulnerable because of its geography, high population density, and enormous poverty and deprivation in the region (Bowen et al. 2011; Asaduzzaman et al. 2013; Fisher 2017; Hunter et al. 2019). South Asia is predicted to be one of the worst affected regions from global warming and climate change because of geo-physical factors and socio-economic/demographic backwardness. It is vital for these vulnerable countries to take practical steps to adapt to climate change and integrate them into their wider development agenda (Afroj et al. 2016).

Agriculture

Agriculture is one of the most vulnerable sectors in the entire region. Most of the countries in South Asia are highly dependent on this primary sector of the economy. Almost all the countries have an agriculture-driven economy, in which agriculture contributes significantly to the GDP of the countries (Bowen et al. 2011; Asaduzzaman et al. 2013; Fisher 2017; Hunter et al. 2019. It is likely that the surge in temperature will reduce yields of desirable crops while the changing precipitation, which is in turn likely to result in short-term crop failures and long- production declines, therefore compromising food security in the future.

Coastal and Marine Areas

The changes in climatic events that have been observed to have significant impacts on coastal and marine resources in the South Asia region include the rise in land and sea surface temperatures, sea level rises, and increases in rainfall, as well as the intensified life-threatening weather events (e.g., heat waves, intense precipitations, droughts, and cyclones) (IPCC 2007a, b). Being land-locked countries, Nepal, Bhutan, and Afghanistan are spared, whereas others are affected. Among them, the Maldives, Sri Lanka, and Bangladesh are the most vulnerable.

Energy

Climate change changes the nature of energy security in the region, in terms of both electricity supply and demand, thus increasing the vulnerability of the poor by depriving them of the benefits of energy and economic growth (Owusu and Asumadu-Sarkodie 2016). Global warming will increase energy requirements for space cooling while at the same time increasing the energy demand for irrigation. On the other hand, the frequency of extreme events such as storms and sea level rise will in the future increase electrical system failures (Bowen et al. 2011; Asaduzzaman et al. 2013; Fisher 2017; Hunter et al. 2019). Countries dependent on hydro-power may also be severely affected as rivers dry up. Climate change may therefore cause energy shortages, as demand for energy will increase whereas supply may be heated by extreme weathers and natural disasters.

Health

Public health is one of the great concerns of all South Asian countries, and it is highly likely that climate change will bring about new waves of water and vector-borne and other types of contagious viruses. There may be a surge in occurrence of diseases such as cholera, dengue, diarrhea, and malaria. Moreover, there is also a high possibility that viruses such as COVID-19 will appear again and again in the future. Owing to the weak health care system, patients may exceed the capacity of hospitals and can therefore lead to a huge loss of life.

Water

Water is a fundamental resource that is essential for survival as well as for economic livelihoods. It is predicted that in the future, usable water will be scarcer throughout the region. Most of the South Asian countries are dependent on ground water to satisfy their water demands. In the future, it is inevitable that ground water will perish, and water scarcity will persist. Every year in countries such as India, Maldives, Sri Lanka, and Bangladesh salinity intrusion in water is increasing. In countries such as Nepal, Bhutan, and Pakistan the rivers are drying up, therefore decreasing the availability of water for agriculture and domestic utility (Ahiduzzaman and Islam 2011; Chowdhury 2013; Dastagir 2015; Donald 2015; Boden and Marland 2017; Dunnan 2019). Climate change is promoting water shortages and in the future regional tension or conflicts may persist as countries fight to secure their claim on water resources.

11.2.6.2 Regional Response to Climate Change

The most severe impacts of climate change cut across boundaries and do not discriminate among nations. With the impacts of climate change increasing every year, the South Asian Association for Regional Cooperation (SAARC) countries have reaffirmed their commitment to work together to address the common devastating challenges (Fleming 2008; Amos 2016). Throughout the years, the South Asian countries have made efforts to better address the issue so that better adaptation and mitigation strategies can be developed through regional cooperation (Bowen et al. 2011; Asaduzzaman et al. 2013; Fisher 2017; Hunter et al. 2019). Here are some initiatives:

- In 1997, the SAARC environment action plan was initiated
- In 2007, the SAARC Council of Ministers adopted the SAARC Declaration on Climate Change. The Council called all the regional leaders to collectively assess and respond to climate change risks and impacts (Makido et al. 2012).
- In 2008, the SAARC Environment Ministers Dhaka Declaration on Climate Change included a 3-year action plan that urged the international community to promote partnership and provide additional finance to address climate change.
- In their 25th Jubilee Year, the SAARC Summit in 2010 concluded with the Thimphu Declaration on Climate Change, which sets an ambitious goal for South Asia to lead the world in furthering renewable energy, cutting carbon emissions, and reducing poverty while strengthening resilience to climate change (Makido et al. 2012).
- The regional Partners through extensive multilateral dialog have formulated four regional centers to address the issue of the environment, climate change, and natural disasters. The centers are as follows:
  - SAARC Disaster Management Centre (SDMC)
  - SAARC Meteorological Research Centre (SMRC)
  - SAARC Forestry Centre (SFC)
  - SAARC Coastal Zone Management Centre (SCZMC)

Furthermore, the South Asian countries have also made unilateral decisions to better address the issue of climate change at a national level. In order to deal with the issue of climate change effectively, the South Asian partners have mobilized both multilateral initiatives through regional cooperation and unilateral initiatives as part of their national policy. At the same time, with an aim to fulfil their commitments under the UNFCC, most of the South Asian countries have attempted to mainstream climate change into their respective development agendas.

11.2.6.3 Carbon Emission Impact in Bangladesh

Climate change is now a global crisis, affecting almost every country all around the world. Climate change will affect every nation, but the severity among the countries will vary depending on various economic, socio-economic, environmental, geographic, and demographic factors (Ashrafuzzaman and Furini 2019). When considering all the factors, Bangladesh tends to stand in a weaker position compared with other countries. Owing to pre-existing vulnerabilities, Bangladesh is prone to severe consequences in the future (Bowen et al. 2011; Asaduzzaman et al. 2013; Fisher 2017; Hunter et al. 2019). If a similar climatic trend continues then Bangladesh as a country will face economic, socio-economic, and humanitarian catastrophe (Ahiduzzaman and Islam 2011). Many parts of the world experience large-scale climate change extremes and Bangladesh will face the severity of climate change-induced natural calamities including rise in average air temperature, more hot and cold spells, inadequate precipitation during the period of when agriculture plants need it, more rainfall in the rainy season when it already causes floods; glacier melting in the river source areas of Bangladesh by changing the hydrological cycle; more extreme tornadoes and cyclones; sea level rise dislocating societies; revolving of freshwater into saline water and intense extreme storm surge flooding (Ahiduzzaman and Islam 2011; Dastagir 2015). The influence of climate change on Bangladesh will be reinforced by population density, poverty, and the scarcity of natural resources.

Future effects on Bangladesh will be as follows:

- Temperature rise: according to the IPCC's Fourth Assessment Report, all of Asia is likely to warm this century and warming in South Asia is likely to be above the global average at around 3.3 °C (Christensen et al. 2007).
- Increase in flood: flash floods, river floods, rain floods and coastal storm surge floods will increase exponentially in Bangladesh in the coming years.
- Increase in drought: every year, 2.7 million ha of land in Bangladesh are vulnerable to drought. In the coming years it is estimated that this figure will likely double (Tanner 2007).
- Rise in sea level: Bangladesh has been ranked as the third most vulnerable in the world to sea level rise in terms of percentage of the population living in the low elevation coastal zone (McGranahan et al. 2007). In 2050, if the sea level rises by 27 cm, almost seven million will be at a high risk of flooding 58% of these people will be from Khulna, Jhalokati, Barisal, and Bagerhat districts (Mohal and Hossain 2007).
- Increase in winds and cyclones: cyclones are expected to become 10–20% more powerful if sea-surface temperatures rise by 2–4 °C in South Asia; therefore, the number of devastating cyclones will increase (Cruz et al. 2007). Cyclones are expected to have 3% to 12% faster wind speeds by the 2020s, rising to 4% to 20% faster by the 2050s (Tanner 2007).

- River bank erosion: higher volumes of water flowing down rivers owing to climate-related changes such as increased rainfall and summer glacier melting will significantly increase the erosion of land beside Bangladesh's rivers.

11.2.6.4 Carbon Emission Mitigation Strategy in Bangladesh

Owing to this rapid increase in carbon emission, natural calamities such as floods, droughts, cyclones, hurricanes, and typhoons have seen an increase. The sea level is rising, biodiversity is decreasing, and people in coastal areas are being severely affected (Haque et al. 2012). To address the problem of climate change the GoB has also taken some major steps. In an effort to move toward a low carbon society the Bangladesh government has put forward some major plans and policies (Table 11.3).

11.2.6.5 Policy Implications

It is known that the GoB, with the support of its global partners, has already initiated some plans, programs, and activities to address the mitigation of and adaptation to climate change. However, there is still a need for some concrete strategies and actions to reduce CO_2 emissions by power generation, transport, manufacturing, residential, and other sectors (Bowen et al. 2011; Asaduzzaman et al. 2013; Fisher

Table 11.3 Initiatives for reducing CO_2 emission in Bangladesh

Initiatives/programs	Sectors involved	Implementing agency/partner
Bangladesh Climate Change Strategy and Action Plan	Climate	Ministry of Environment and Forest, Government of the People's Republic of Bangladesh
National Renewable Energy Policy	Energy	Ministry of Power, Energy and Mineral Resources, Government of the People's Republic of Bangladesh
National Adaptation Plan of Action	–	Ministry of Environment and Forest, Government of the People's Republic of Bangladesh
National Adaptation Plan Global Support Program	Climate	United Nations Development Program, United Nations Environment Program
Nationally Appropriate Mitigation Activities	Steel	Department of Environment, Bangladesh
Bangladesh-USAID Low Emissions Asian Development Program	Climate, energy, land	ICF International, United States Agency for International Development, United States Environmental Protection Agency, United States Department of State
Enhancing Capacity for Low Emission Development Strategies	Climate, energy, land	United States Agency for International Development, United States Environmental Protection Agency, United States Department of Energy, United States Department of State

Source: Modified and adapted from Ahiduzzaman and Islam 2011; Haque et al. 2012

2017; Hunter et al. 2019). Implementation of the above-mentioned strategies and action plans are essential to accomplish the goal. It can be mentioned that CO_2 emissions will not spontaneously decrease if the country continues to develop its economy without adopting instruments for mitigating climate change (Xu et al. 2012). Thus, mitigation options are important to reduce the GHG emissions, which can promote sustainable energy use, reduction of global warming, as well as environmental sustainability. Some cost-effective mitigation options such as afforestation, sustainable forest management, reduction of deforestation, and cropland management need to be promoted in Bangladesh (Sarkar et al. 2015). There is also a necessity to develop sustainable energy technology, producing electricity by reducing the amount of gaseous carbon compounds, operating hydrogen and electric automobiles, and changing human behavior (Gordon 2007; Harish Jeswani 2008; Erin 2009; Georg 2010; Gernaat 2015). There is also a need for energy-efficient technology, decarbonizing electricity generation, hydrogen and electric vehicles and changing human behavior.

Renewable energy resources constitute one of the most efficient and effective solutions for clean and sustainable energy development in Bangladesh (Ahiduzzaman and Islam 2011). Biomass transformation into energy in the form of liquid, gaseous, and solid pellets or briquettes can be an effective alternative to lessen the pressure on conventional fossil fuels.

11.2.6.6 Mitigation Potential of Climate Change through LCD

With climate change intensifying every year, it is necessary for every country to take action more seriously, as failure to take timely action will have a dire effect on the entire planet. In the present-day scenario, timely action is a must; if proper steps are not taken in due course, then there may be irreversible implications. In order to prevent catastrophic economic and socio-economic devastation comprehensive action must be taken by all countries around the world (Manabe and Wetherald 1975; Haque et al. 2012). Most importantly, to achieve the milestone of limiting temperature rise to less than 2 °C, there is no better way than for all countries to adopt the LCD framework (Fisher 2013; Fisher and Mohun 2015; Teske et al. 2017). Bangladesh being one of the most vulnerable countries with regard to the effect of climate change, it is extremely important for it to be more proactive in establishing and implementing the LCD framework. LCD is a new political and socio-economic pattern that includes other GHGs and achieving new global SDGs and targets.

Low carbon development is a relatively new concept of political and socio-economic development aimed at reducing GHG emissions while achieving global sustainable targets and goals (Haque et al. 2012; Kedia 2016). It is based on the principles of achieving sustainable development goals, reducing GHG emissions, and limiting the rise in the global temperature to below 2 °C (Bowen et al. 2011; Asaduzzaman et al. 2013; Fisher 2017; Hunter et al. 2019). With LCD a country will be able to achieve low power consumption, lower pollution, low GHG emissions, and efficient use in all sectors including agriculture, building, industry, power, and transport (Hunter et al. 2019). The justification for trialing low-carbon progress differs from nation to nation, but for Bangladesh is to:

- Reduce the increase in GHG emission
- Enhance economic growth through efficient energy management
- Expand national development opportunities to enhance a green economy
- Achieve poverty reduction through green development and job creation
- Minimize the effects of climate change in the long run

There are both challenges and opportunities in building a low-carbon society in Bangladesh, but it is not something that cannot be achieved. Being a developing country and a growing economy, reducing carbon usage may be a challenge, but overcoming this challenge will allow Bangladesh to achieve great heights in terms of tackling climate change and achieving sustainability (Karim et al. 1999; Jens 2007; Jos 2012). In order to implement the low-carbon society framework, Bangladesh can implement the following strategies:

- Energy efficiency: bring about improvements in energy efficiency and energy conservation in all sectors (Fisher 2013; Fisher and Mohun 2015; Teske et al. 2017). Therefore, energy efficiency should be encouraged and in time enforced in all sectors of the Bangladesh economy from agriculture, industry, buildings, industry, power, and water to transport (SREDA 2016).
- Low carbon energy: increase investment in low-carbon energy, as the country should use strategies to shift toward nonrenewable sources to meet its country's energy demands. Bangladesh should capitalize more on solar, wind, hydro, biogas, and geo-thermal energy resources.
- Carbon switching: facilitate a shift toward low-carbon fuel in transportation, industry, and buildings. Encourage a shift from highly intensive fossil fuels to bio-energy (Gielen et al. 2019).
- Reduction of non-energy emission: focus on reducing non-energy emissions through afforestation, carbon offsets, and carbon capture and storage.
- Education and awareness: introduce awareness program on energy efficiency and conservation in the curricula of schools, colleges, and universities. It is necessary to plan and implement an effective education and training program model for LCD.
- Carbon trading: establish an efficient carbon trading scheme with subsidies allowance and tax breaks to encourage voluntary restraints on carbon emission by private firms.
- Encourage lifestyle and behavioral change: without community support, a low-carbon society is not achievable. Thus, it is necessary to take drastic action, and community support can be enhanced. The government should take a more proactive role to educate the community with better information on LCD. The government can encourage individuals to:
  - Use public transport more often
  - Walk or bike if the destination is close
  - Buy locally grown food and vegetables
  - Adopt water-harvesting practices
  - Replace compact fluorescent lamp with LEDs

- Investment: increase investment in low carbon technology, including energy efficiency, renewable energy sources, and in green transportation and locomotives. The government should also take steps to reward firms for green innovation; subsidies and tax breaks should be mobilized.

11.2.6.7 Building a Low Carbon Society Through Regional Cooperation

Climate change is a global crisis; mitigation of this problem will not come about unless collective action is taken by regional partners. In terms of dealing with the problem, regional cooperation is a must. In order to make not only Bangladesh but rather the entire South Asian region carbon free, a collective mitigation strategy must be formulated by all the SAARC member countries (Rahman et al. 2007; Avi-Yonah and Uhlmann 2008; Boden and Marland 2011; Bo Xu 2012). The steps that can be taken by the SAARC countries are as follows:

1. Adoption of a regional strategy: it is very important for the SAARC nations to come together to formulate and therefore adopt a long-term emission-limitation strategy for the entire South Asian region. Therefore, the South Asian members should focus on a uniform strategy that will initiate collective action in the region (Mishra et al. 2014).
2. Integration of the strategy: for the effective mobilization of the strategy, it is necessary for the SAARC countries to integrate the regional strategic aim into their overall national policy (Fisher 2013; Fisher and Mohun 2015; Teske et al. 2017). Every country's national policy should reflect the sustainable development outcomes put forward by the agreed regional strategy that was previously agreed upon.
3. Formulation of a green fund: to put the strategy into action it is necessary to exhaust resources, and without it, implementation will be impossible. For effective implementation of the strategy, huge amounts of money must be spent, and sometimes the fiscal budget may not be enough. Thus, the SAARC countries can pool their resources in a particular fund so that every year a certain percentage of GDP can be contributed to this fund by SAARC members (Dutta et al. 2020). This fund can allow countries to fund green projects in the region.
4. Set up technology development centers: technology is one of key elements that play a vital role in the mitigation of climate change. Without technology, mitigation of climate change seems far-fetched. Thus, it is necessary to set up a center for technology development in South Asia to jointly create and deploy clean technologies to meet the adaptation and mitigations needs of the region.
5. Controlling function: lastly, it is important for the SAARC nations to execute the controlling function. At the end of every year, it is necessary to strictly evaluate the effectiveness of each country in terms of attending to their annual objectives and the implementation of the regional strategy to reduce carbon emission.

11.2.6.8 Carbon Tax Implementation in Bangladesh

Bangladesh is a strong actor in the effort to reduce global carbon emission. This is appropriate as it faces a major adverse burden from global climate change. Although per capita carbon emission is low, total carbon emission in Bangladesh is growing. Consequently, as a good global team player, Bangladesh committed to reducing its carbon footprint in its Intended Nationally Determined Contributions (INDC) submissions in 2015 (Ahmed and Khondker 2019). Similar to other taxes, the burden of carbon taxes (CTs) on petroleum energy (i.e., petrol, octane, diesel, furnace oil, and kerosene) in Bangladesh may have an impact on the price and consumption (or sale) of these products directly and indirectly on the sectors and households who use them as input in their production activities or as final consumption. Imposition of CTs will reduce CO_2 emissions compared with a pre-taxed situation owing to lower consumption because of higher CT-inclusive prices (Manabe and Wetherald 1975; Haque et al. 2012). Furthermore, because of the interdependence of the economic system, any effect on particular commodities (or sectors) of the economy is likely to have implications on the prices of other commodities, as well as outputs of other activities.

The implementation of CT can be a suitable way of lessening carbon emission (Avi-Yonah and Uhlmann 2008). Through the price mechanism of the market, the collection of carbon tax makes power producers turn to clean energy and low-carbon products to achieve the purpose of emission reduction. The CT not only lowers CO_2, it also yields revenue. This revenue can be used to invest in clean fuel, clean technology, and infrastructure projects that will help to offset the loss of output from CTs (Fisher 2013; Fisher and Mohun 2015; Teske et al. 2017). The position of Bangladesh in the field of clean energy and energy technology is far away from that of the rest of the world (Islam and Khan 2017). Proper pricing of fossil fuels along with the CT will provide the incentives to reduce the consumption of these fuels and also motivate private investment in clean energy and clean technology (Bowen et al. 2011; Asaduzzaman et al. 2013; Fisher 2017; Hunter et al. 2019). The incidence of CT is progressive in the sense that the cost of living increase is highest for the top ten percentile and lowest for the bottom ten percentile. Additionally, the increase in the cost of living for the poor can be offset through income transfers from additional social protection spending from carbon taxes. Therefore, an appropriate combination of fossil fuel evaluation, CT, and reserves can make CT a win-win policy platform.

11.2.7 Renewable Energy with Low Carbon Emission: An Opportunity for Bangladesh

A downturn in the availability of fossil fuels, their predicted gradual extinction over the next few decades, the need to drastically cut global emissions in order to mitigate climate change (80% reduction by 2050) (Walsham 2010), and finally the issue demanding energy supply to be switched to renewable energy sources.

The use of renewable energy with the help of modern technology has not yet geared up to commercial dimensions. Fortunately, logical policy diffusion on renewable energy has been accelerated to an upper stage (Bowen et al. 2011; Asaduzzaman et al. 2013; Fisher 2017; Hunter et al. 2019). The size and economic potential of the renewable energy resources (e.g., solar photovoltaic (PV), solar thermal power, wind power, biogas) in Bangladesh are quite large and there have been several significant instigations in this sector so far. Owing to the recent energy crisis and limitation of resources, the government has been forced to focus on setting up a guideline along with the observation of current progress in creating renewable energy sources together with the private stockholders (Bowen et al. 2011; Asaduzzaman et al. 2013; Fisher 2017; Hunter et al. 2019). As part of the planning, it is planned to establish an independent institution, the Sustainable Energy Development Agency (SEDA), to be established under the Companies Act, 1994, as a focal point for sustainable energy development and promotion. The SEDA Board will comprise the representatives of stakeholders, including the business community, academics, and/or representatives from the Bangladesh Solar Energy Society, NGOs, as well as financial institutions and implementing agencies. In addition to the domestic approach, many more manufacturing entrepreneurs in the renewable energy sector may inspire indigenous laborers. Owing to some technical confines of local technology, foreign investments can ensure maximum profit using the latest technological knowledge. Bangladesh is one of the worst sufferers of climate change; the countries who are responsible for high carbon emissions should step ahead and encourage the usage of renewable energy sources (Bowen et al. 2011; Asaduzzaman et al. 2013; Fisher 2017; Hunter et al. 2019).

According to the Renewable Energy Policy of Bangladesh (MPEMR 2008), a Renewable Energy Financing Facility will be established that would be capable of accessing public, private, donor, carbon emission trading (CDM), and carbon capitals contingent upon the investments and financial inducements. The Bangladesh Energy Regulatory Commission (BERC) along with the GoB and SEDA, have been planning to create a regulatory framework encouraging generation of electricity from renewable energy. As per the provision of the BERC Act 2003 (BERC, 2003), the organization is currently working on approval of the energy tariff in consultation with a renewable energy scheme with the capability of 5 MW or more. It has been proposed to pay a "green energy" tariff, which gives customers the opportunity to co-pay through their electricity bills. Some private companies such as the Infrastructure Development Company Limited (IDCOL), Grameen Shakti, Bangladesh Rural Advancement Committee (BRAC), and the Rural Services Foundation have already initiated the program and set up their target for the next few years (Khan et al. 2013).

11.3 Present Opportunities and Future Development in Different Energy Sectors

Through the mutual efforts of the GoB and private investors, some significant initiatives have been taken in every field of renewable energy in Bangladesh. The present actions and the prevailing platform are projected to confirm significant development in the green energy system over the next few years. Although the initial set-up demands larger investment, the low operating costs along with minimum maintenance and the increasing demand ensures profitability and proper economic growth for the investors.

11.3.1 Solar Energy

The Rural Electrification Board (REB), the Local Government Engineering Department (LGED), and the Bangladesh Power Development Board (BPDB) are jointly working on scaling up the process of solar photovoltaic technology (PV) systems along with essential procedure progression and crucial alterations (Chowdhury 2018). The uses of PV in the remote or isolated island communities meet their electricity demands, and PV has already been introduced to supply electricity in some rural areas. Nevertheless, the actions in fields excluding SHS are not substantial. The potential for wind power is largely confined to the coastal regions, where wind speeds are high enough. The BCAS launched the Wind Energy Research (WEST) project in October 1995 with the support of the Overseas Development Administration (ODA) Overseas Administration of the United Kingdom (UK) (ODA). The goal of this project was to evaluate the achievability of using wind energy as mechanical and electrical energy. They accumulate and examine the wind speed statistics in seven parts of Bangladesh. The places are extensively disseminated alongside the vast coastline in the districts of Cox's Bazar, Chittagong, Noakhali, Bhola, and Patuakhali. Generally, the best source of renewable energy is solar energy because it does not pollute the environment. Considering the direction of light and heat from the sun, the most favorable spot for the consumption of solar energy lies between the area of 15° and 35° latitude north and south. Fortunately, Bangladesh is situated between 20°43′ north and 26°38′ north latitude and as such Bangladesh is in a very favorable position in respect of the utilization of solar energy (Khan et al. 2013).

The total annual solar radiation in Bangladesh is equivalent to 1010 × 1018 J, considering the average is 1900 kWh/m^2. The present total yearly consumption of energy is about 700 × 1018 J. Therefore, it is evident that utilization of about 0.07% radiation will be able to meet the energy consumption of Bangladesh (Khadem et al. 2007). Currently, the energy consumption in Bangladesh is nearly 0.16 W/m^2 of the terrestrial area, whereas the accessibility is more than 208 W/m^2. Thus, fuel consumption can be diverted from fossil fuel sources since 1990s. According to the report from World Bank, energy access has been increasing at the rate of 3.5% due to installations of 30000 SHSs on average every month. By 2020, the government has aimed to produce 10% of total energy through renewable sources (Ahammed and Azeem 2013). This type of policies promises to result in the exponential growth of SHS ensuring better prospect in various areas in Bangladesh in terms of the size of the SHS market and its economic potential.

11.3.2 Biomass and Biogas

Biogas, mainly derived from animal and municipal waste, could be one of the most auspicious renewable energy assets for Bangladesh. The Bangladesh Council of Scientific and Industrial Research has established bio-gas plants appropriate for Bangladesh. More than 20,000 trees have been planted under an implementation program (Hossain and Khalequzzaman 2004). In a feasibility study prepared for Danish investors on the market potential of Bangladesh, it was indicated that up to 800 MW of electricity could be produced in Bangladesh using organic city waste and poultry litter (Costello 2011; Arabinda Mishra 2014). The total of 12 gasification-based biogas plants corresponding to a capacity of 5 MW are now being deliberated by the donor-financed IDCOL. In 2012, only a portion of the total of 15,000 tons of waste was being reprocessed yearly. About 80% of produced waste is organic, which has a high potential for biogas production. The quantity is estimated to increase to 47,000 tons by 2025 (Bhowmik 2008).

11.3.3 Hydropower Plant

The only hydropower plant in the country is the Karnafuli Hydropower Plant, which has a producing capability of 230 MW. With the exception of Chittagong and Chittagong Hill Tracts, micro-hydro and mini-hydro Bangladesh has limited potential. Hydropower assessments have recognized several potential sites ranging from 10 kW to 5 MW, but no commendable capacity has been installed yet. The Sangu project will be a new project that will generate an installed capacity of 140 MW with an annual capability of approximately 300 GWh per year (Fisher 2013; Fisher and Mohun 2015; Teske et al. 2017). However, in the current context of this project, detailed environmental, social, and economic studies are required. The Matamuhuri development project will be another new project with a capacity of 75 MW per year and an estimated average annual power of 200 GWh per year (Bahauddin and Salahuddin 2012). Foreign investment in this sector will encourage local entrepreneurs to take the next steps in expansion. The government will also contribute when this promising sector shows definite signs of development and a wider range of applicability. The mutual work from the locals, foreign stockholders, and the government would be able to make this a long long-term corporation (Bowen et al. 2011; Asaduzzaman et al. 2013; Fisher 2017; Hunter et al. 2019). The government can contribute to international research by providing expats, and international stockholders can offer the modern machinery to enhance productivity.

11.3.3.1 4. Challenges of Renewable Energy Sectors

Creating opportunities and solving the problems in the renewable energy sectors will be suitable options for establishing a carbon-free society for Bangladesh (Sharif et al. 2018). Moreover, the absence of technological information obstacles in decision making, proper implementation strategy, and lack of an action plan for the investors it is very complex to regenerate the renewable energy in Bangladesh. Bangladesh currently does not have any central database for the primary survey report on the pre-feasibility study, which will be reviewed before any new installation is made on a commercial scale (Fisher 2013; Fisher and Mohun 2015; Teske et al. 2017). Nevertheless, the current environmental goals have not clearly considered the environmental laws and acts of this country, as they were brought into force quite a long time ago. Recently, the GoB has been concentrating only on quick rental power plant (the government allowed the privately owned oil-fired quick rental (QR) power plants to enter the energy market in Bangladesh through tendering), which can only be reflected as a short-term solution to the country's expected energy crisis. Investment in every renewable energy source will be a certain encouragement for private investors, but the initial investment for those projects under the current economic condition will be a challenging job for the government.

11.3.3.2 5. Solutions to the Further Challenge that Await the Country

Bangladesh has succeeded in making an impact to build up awareness among the public, but the further challenge that awaits the country is to make a practical move in every renewable energy sector for the next few years. Bangladesh will benefit in different ways with different approaches to moving toward renewable energy systems (Bowen et al. 2011; Asaduzzaman et al. 2013; Fisher 2017; Hunter et al. 2019). First, the country will be able to transfer loads and reliability from fossil fuel sources and provide assurance and protection for long-term availability. Second, as one of the worst sufferers of climate change, Bangladesh needs to set an example to others, of which the carbon producers should be aware, about the impact on the environment they are creating (Fisher 2013; Fisher and Mohun 2015; Teske et al. 2017). Setting up, for instance, should be an incentive for other countries that are producing substantial amounts of carbon and contaminating the atmosphere. Third, even if the initial investment is high, low operation and maintenance costs will be an incentive for a less developed country such as Bangladesh. Fourth, to keep up with the latest technological inventions in processing industries and energy conversion, the quickest reaction in this regard will empower the country to go for further advancement in the upcoming years.

Finally, to become one of the world's leading leaders in the fight against carbon emissions, Bangladesh must first look at its own strategy and energy sector. Relying solely on fossil fuels certainly will not help to the case for green energy sources (Fisher 2013; Fisher and Mohun 2015; Teske et al. 2017). Bangladesh may be the one of the lowest contributors regarding the amount of carbon emissions, but considering the total percentage of carbon emissions within the country, the result is quite alarming. The tragic distresses of natural disasters in all phases have to be a major obstacle to sustainable development. Bangladesh needs to act in its own interest and take swift action.

11.4 6. Conclusions and Recommendations

Climate change remains one of the most critical threats society has ever encountered. Its impacts are already visible and will worsen over time if left unchecked. Climate is a crisis that has the gravity to push the entirety of humanity toward extinction. The effects of climate change are intensifying each year and it is likely that it will be even worse in the coming years (Karim et al. 2019). Despite the severity of the problem, the issue is under-addressed by the global community. The global community is prioritizing economic prosperity over the livelihoods of our future generations. It is necessary for the international community to give greater importance to this issue and therefore strictly limit carbon emissions while implementing sustainable development and LCD themes (Manabe and Wetherald 1975). Such measures should be taken, in order to preserve the lives of our future generations. Climate change is therefore a global crisis, as it is affecting countries all around the world. Hence, no country is left out, but the developing countries are far worse off. And among these vulnerable countries lies Bangladesh. All around the world, Bangladesh is regarded as one of the most vulnerable countries and yet, although governments have put forward plans and policies, implementation is where the country is lagging behind.

The situation has now worsened to a such a point that many experts are saying that there is less than 5% hope to save the earth if human activities are not slowed down. Now, it has been evident that climate change is the result of anthropogenic activities that are occurring throughout the world. Thus, it is the release of GHGs that is altering the climate of the entire planet dramatically (Fisher 2013; Fisher and Mohun 2015; Teske et al. 2017). The huge release of GHG gases started with the beginning of the industrial revolution, as manufacturing hubs were established throughout the world. With the rise of capitalist markets and gradual shifts toward globalization the release of GHGs has quadrupled. With economic incentives being prioritized, the environment is being compromised for short-term material gains. As economic activities are increasing more and more, environmental problems are overshadowing our planet as climate change is intensified with every minute. As a result, the ice caps are melting, the sea level is rising, long droughts are persisting, weather is being harsher, and therefore new forms of diseases are coming to existence, COVID-19 being one of them.

Similarly, Bangladesh, being one of the largest deltas in the world, is highly vulnerable and is being severely affected. Climate change basically started affecting Bangladesh over a decade ago. Ever since, Bangladesh has been affected one way or the other. Because of climate change, Bangladesh has to face water-related problems every year, which include water shortages and salt water intrusion. Droughts have become common in the northwestern part, whereas flooding often occurs near river banks and beyond. The fertility of the soil is also decreasing, while temperatures are becoming unbearable throughout the country.

References

Abeydeera U, Jayantha W, Samarasinghe T (2019) Global research on carbon emissions: a Scientometric review. Sustainability. 11:3972. https://doi.org/10.3390/su11143972

Afroj, Mahfuza & Rahman, Mahfuzar & Kazal, Mohammad Mizanul Haque & Quamruzzaman, Md. (2016). Carbon trading and its impact in Bangladesh. Middle-East J Sci Res. 24. 2624–2631. doi: https://doi.org/10.5829/idosi.mejsr.2016.24.08.23826

Ahmed S, Khondker BH (2019) Towards a carbon tax for Bangladesh, the financial express, 23 April 2019. Dhaka, Bangladesh

Ahammed F, Azeem A (2013, January) Selection of the most appropriate package of solar home system using analytic hierarchy process model in rural areas of Bangladesh. Renew Energy 55:6–11

Ahiduzzaman M, Islam AKMS (2011) Greenhouse gas emission and renewable energy sources for sustainable development in Bangladesh. Renew Sust Energ Rev 15:4659–4666. https://doi.org/10.1016/j.rser.2011.07.086

Amos J (2016, March 21) BBC.COM. Retrieved from www.bbc.com, https://www.bbc.com/news/science-environment-35867438

Andres, R. J., Boden, T. A., Bréon, F.-M., Ciais, P., Davis, S., Erickson, D., Gregg, J. S., Jacobson, A. Marland, G., Miller, J., Oda, T., Olivier, J. G. J., Raupach, M. R., Rayner, P., Treanton, K. (2012). A synthesis of carbon dioxide emissions from fossil-fuel combustion. Biogeosciences 9:1845–1871

Anthony Costello MA (2011) Managing the health effects of climate change. Lancet 373:1693–1733

Arabinda Mishra KU (2014) Climate risks in the SAARC region: ways to address the social, economic & environmental challenges. New Delhi: The Energy and Resources Institute, New Delhi

Asaduzzaman M, Yunus M, Haque AE, Azad AAM, Neelormi S, Hossain MA (2013) Power from the Sun: an evaluation of institutional effectiveness and impact of solar home Systems in Bangladesh, Final report submitted to the World Bank. Bangladesh Institute of Development Studies, Dhaka. http://dev.sun-connect-news.org/fileadmin/DATEIEN/Dateien/New/Bangladesh_Idcol_Assessment.pdf

Ashrafuzzaman M, Furini GL (2019) Climate change and human health linkages in the context of globalization: an overview from global to southwestern coastal region of Bangladesh. Environ Int 127:402–411

Atiq Rahman A, Alam M, Alam SS, Uzzaman MR, Rashid MM, Rabbani G (2007) Risks, vulnerability and adaptation in Bangladesh, Human development report 2007/2008. Bangladesh Centre for Advanced Studies (BCAS), Dhaka

Avi-Yonah RS, Uhlmann, D. M. (2008) Combating global climate change: why a carbon tax is a better response to global warming than cap and trade. SSRN Electron J 28(3)

Backman CA, Verbeke A (2015) The drivers of corporate climate change strategies and public policy: a new resource-based view perspective. Bus Soc 56(4)

Bahauddin KM, Salahuddin TM (2012) Prospect and trend of renewable energy and its technology towards Climate Change mitigation and sustainable development in Bangladesh. Int J Adv Renew Energy Res 1:156–166

Bank W (2015) World Bank annual report 2015. The World Bank, Washington, D.C.

Berrone P, Fosfuri A, Gelabert L, Gomez-Mejia LR (2013) Necessity as the mother of 'green' inventions: institutional pressures and environmental innovations. Strateg Manag J 34:891–909

Bhowmik (2008) Bangladesh: renewable energy report, Asian and Pacific Centre for Transfer of Technology of the United Nations–Economic and Social Commission for Asia and the Pacific. Retrieved from recap.apctt.org/download.php?p=Admin/publications/5.pdf

Blanford GJ (2009) R&D investment strategy for climate change. Energy Econ 31:27–36

Boden TA, Marland G (2011) Global, regional, and National Fossil-Fuel CO2 emissions. U.S. Department of Energy

Boden TA, Marland, G. (2017) Global, regional, and national fossil-fuel CO2 emissions. Carbon Dioxide Information Analysis Center, Oak Ridge National Laboratory, U.S. Department of Energy

Boden, T. A., Andres, R. J., Marland, G. (2013). Global, regional, and national fossil-fuel CO2 emissions (1751–2010) 2013). USA: Oak Ridge National Laboratory (ORNL), Oak Ridge

Bowen A, Fankhauser, S., Best, S (2011) Low-carbon development for least developed countries. Oxfam Policy and Practice: Climate Change and Resilience 7(2):33–56

Canadell JG, Le Quéré C, Raupach MR, Field CB, Buitenhuis ET, Ciais P, Conway TJ, Gillett NP, Houghton RA, Marland G (2007) Contributions to accelerating atmospheric CO2 growth from economic activity, carbon intensity, and efficiency of natural sinks. Proc Natl Acad Sci 104:18866–18870

Chowdhury SK (2013) Assessment of wind and solar energy resources in Bangladesh. Arab J Sci Eng 38(11):3113–3123

Chowdhury SA (2018) Indicative tariff for utility-scale solar IPP in Bangladesh. United Nations Development Programme (UNDP), Dhaka

Christensen J, Hewitson B, Busuioc A, Chen A, Gao X, Held I, Jones R, Kolli RK, Kwon W-T, Laprise R, Rueda V, Mearns L, Menéndez C, Räisänen J, Rinke A, Sarr A, Whetton P (2007) Regional climate projections. Climate change 2007: the physical science basis. Contribution of working group I to the fourth assessment report of the intergovernmental panel on climate change, pp 847–940

Commission BE (2003) The Bangladesh energy regulatory commission (BERC) act 2003. BERC, Dhaka, Bangladesh

Committee TC (2017) 2017 report to parliament – meeting carbon budgets: closing the policy gap. Committee on Climate Change, United Kingdom

Costello A, Maslin M, Montgomery H, Johnson A, Ekins P (2011) Global health and climate change: moving from denial and catastrophic fatalism to positive action. In: Philosophical transactions, Series a, mathematical, physical, and engineering sciences, vol 369, pp 1866–1882. https://doi.org/10.1098/rsta.2011.0007

Cruz R, Harasawa H, Lal M, Wu S, Anokhin Y, Punsalmaa B, Honda Y, Jafari M, Li C, Ninh N (2007) Asia. Climate change 2007: impacts, adaptation and vulnerability. In: Contribution of working group II to the fourth assessment report of the intergovernmental panel on climate change, pp 469–506

Delmas MA, Montes-Sancho MJ (2008) Organizational responses to environmental demands: opening the black box. Strateg Manag J 29:1027–1055

Delmas MA, Montes-Sancho MJ (2010) Voluntary agreements to improve environmental quality: symbolic and substantive cooperation. Strateg Manag J 31:575–601

Donald Huisingh ZZ (2015) Recent advances in carbon emissions reduction: policies, technologies, monitoring, assessment and modeling. J Clean Prod:1–12

Dunnan Liu XG (2019) What causes growth of global greenhouse gas emissions? Evidence from 40 countries. Sci Total Environ 661:750–766

Dutta UP, Gupta H, Sarkar AK, Sengupta PP (2020) Some determinants of infant mortality rate in SAARC countries: an empirical assessment through panel data analysis. Child Indic Res 13:2093–2116
EPA (2014) Climate change indicators: global greenhouse gas emissions. United States Environmental Protection Agency
Erin Reid MW (2009) Responding to public and private politics: corporate disclosure of climate change strategies. Strateg Manag J 30:1157–1178
Fisher S (2013) Low-carbon resilient development in the least developed countries, Issue paper. International Institute for Environment and Development, London. http://pubs.iied.org/pdfs/10049IIED.pdf
Fisher S (2017) Supporting effective low carbon resilient development: lessons from the political economy of the least developed countries. In: Fisher S, Rai N (eds) The political economy of low carbon resilient development: planning and implementation. Routledge, London, pp 152–169
Fisher S, Mohun R (2015) Low carbon resilient development and gender equality in the least developed countries, Issue paper. International Institute for Environment and Development, London. http://pubs.iied.org/pdfs/10117IIED.pdf
Fleming JR (2007) The Callendar effect. The life and work of guy Stewart Callendar (1898–1964), the scientist who established the carbon dioxide theory of climate change. American Meteorological Society, Boston. ISBN 978-1878220769
Fleming JR (2008) Climate change and anthropogenic greenhouse warming: a selection of key articles, 1824–1995, with interpretive essays. National Science Digital Library Project Archive PALE: Classic Articles. Retrieved 7 October 2019
Georg Weinhofer VH (2010) Mitigating climate change – how do corporate strategies differ? Bus Strateg Environ 19:77–89
Gernaat DEHJ, K. C. (2015) Understanding the contribution of non-carbon dioxide gases in deep mitigation scenarios. Glob Environ Chang 33:142–153
Gernaat D, Calvin K, Lucas P, Luderer G, Otto S, Rao S, Strefler J, Vuuren D (2015) Understanding the contribution of non-carbon dioxide gases in deep mitigation scenarios. Glob Environ Chang 33. https://doi.org/10.1016/j.gloenvcha.2015.04.010
Ghosh BC, Alam KJ, Osmani MAG (2014) Economic growth, CO2 emissions and energy consumption: the case of Bangladesh. Int J Bus Econ Res 3(6):220–227
Gielen D, Boshell F, Saygin D, Bazilian MD, Wagner N, Gorini R (2019) The role of renewable energy in the global energy transformation. Energ Strat Rev:38–50
Gordon Mcgranahan DB (2007) The rising tide: assessing the risks of climate change and human settlements in low elevation coastal zones. Environ Urban 19(1):17–37
Hammond GP, Norman JB (2012) Decomposition analysis of energy-related carbon emissions from UK manufacturing. Energy:220–227
Hansen J et al (1981) Climate impact of increasing atmospheric carbon dioxide. Science 231(4511):957–966
Haque U, Hashizume M, Kolivras KN (2012) Reduced death rates from cyclones in Bangladesh: what more needs to be done? Bull World Health Organ:150–156
Harish Jeswani WW (2008) How warm is the corporate response to climate change? Evidence from Pakistan and the UK. Bus Strateg Environ 17:46–60
Heede R (2014) Tracing anthropogenic carbon dioxide and methane emissions to fossil fuel and cement producers. Clim Chang 122(1–2):229–241
Hong H, Li F, Xu J (2018) Climate risks and market efficiency. J Econ 208. https://doi.org/10.1016/j.jeconom.2018.09.015
Hong J, Shen GQ, Feng Y, Lau W S-t, Mao C (2015) Greenhouse gas emissions during the construction phase of a building: a case study in China. J Clean Prod 103:249–259
Huisingh D, Zhang Z, Moore J, Qiao Q, Li Q (2015) Recent advances in carbon emissions reduction: policies, technologies, monitoring, assessment and Modeling. J Clean Prod 103. https://doi.org/10.1016/j.jclepro.2015.04.098

Hunter GW, Sagoe G, Vettorato D, Jiayu D (2019) Sustainability of low carbon city initiatives in China: a comprehensive literature review. Sustainability 2019(11):4342

IISD (2013) IISD annual report 2012-2013. Winnipeg, Canada: the International Institute for Sustainable Development (IISD)

IPCC (2007a) AR4 Climate Change 2007: synthesis report. The Intergovernmental Panel on Climate Change, Geneva

IPCC (2007b) Climate Change 2007: impacts, adaptation and vulnerability. Contribution of working group II to the fourth assessment report of the Intergovernmental Panel on Climate Change. Intergovernmental Panel on Climate Change, Geneva, Switzerland

IPCC, Parry M, Canziani O, Palutikof J, van der Linden P, Hanson C (2007) IPCC fourth assessment report: working group II report. In: Impacts, adaptation and vulnerability

Islam S, Khan MZ (2017). A review of energy sector of Bangladesh. Energy Procedia. 110. 611–618. doi: https://doi.org/10.1016/j.egypro.2017.03.193

Jens Hesselbjerg Christensen BH (2007) Regional climate projections. In: Climate change 2007: the physical science basis. Cambridge University Press, Cambridge

Jilani T (2012) Low carbon society development towards 2025 in Bangladesh. Kyoto

Jeswani H, Wehrmeyer W, Mulugetta Y (2008) How warm is the corporate response to climate change? Evidence from Pakistan and the UK. Bus Strateg Environ 17:46–60. https://doi.org/10.1002/bse.569

Jones CA, Levy D (2007) North American business strategies towards climate change. Eur Manag J 25:428–440

Jos Oliver GJ-M (2012) Trends in global CO_2 emissions: 2012 report, background studies. Netherlands Environmental Assessment Agency; European Commission, The Hague

Karim et al (1999) In: Huq S, Karim Z, Asaduzzaman M, Mahtab F (eds) Climate change vulnerability of crop agriculture: a case study in vulnerability and adaptation to climate change for Bangladesh. Kluwer Academic Publishers, Dordrecht, pp 39–54

Karim, Md & Karim, Ridoan & Islam, Md & Muhammad-Sukki, Firdaus & Bani, Nurul & Muhtazaruddin, Mohd. (2019). Renewable energy for sustainable growth and development: an evaluation of law and policy of Bangladesh. Sustainability. 11. 1–30. doi: https://doi.org/10.3390/su11205774

Kedia S (2016) Approaches to low carbon development in China and India. Adv Clim Chang Res:213–221

Khadem, Md Shafiuzzaman Khan & Hussain, Muhtasham & Ullah, Saeed Mahmud & Bhowmik, N & Mazumder, Rezaul & Ghosh, Himangshu & Mariam, Lubna. (2007). Solar and wind energy resource assessment (SWERA) - Bangladesh

Khalequzzaman AH (2004) Utilization of renewable energy in Bangladesh. Utilization of renewable energy in Bangladesh Prepared for symposium of Association of Academies of Sciences in Asia Dalian. Symposium of Association of Academies of Sciences, Dalian

Khan I, Alam F, Alam Q (2013) The global climate change and its effect on power generation in Bangladesh. Energy Policy 61:1460–1470. https://doi.org/10.1016/j.enpol.2013.05.005

Liu L-J, Liang Q-M (2017) Changes to pollutants and carbon emission multipliers in China 2007–2012: an input-output structural decomposition analysis. J Environ Manag 203:76–86. https://doi.org/10.1016/j.jenvman.2017.07.061

Liu D, Guo X, Xiao B (2019) What causes growth of global greenhouse gas emissions? Evidence from 40 countries. Sci Total Environ 661. https://doi.org/10.1016/j.scitotenv.2019.01.197

Lu Y, Stegman A, Cai Y (2013) Emissions intensity targeting: from China's 12th five year plan to its Copenhagen commitment. Energy Policy 61:1164–1177. https://doi.org/10.1016/j.enpol.2013.06.075

Mahmood SA (2011) Impact of climate change in Bangladesh: the role of public administration and government's integrity. J Ecol Nat Environ 4(8):223–240

Makido Y, Dhakal S, Yamagata Y (2012) Relationship between urban form and CO2 emissions: evidence from fifty Japanese cities. Urban Clim 2:55–67. https://doi.org/10.1016/j.uclim.2012.10.006

Manabe S, Wetherald RT (1975) The effects of doubling the CO2 concentration on the climate of a general circulation model. J Atmos Sci 32(3):3–15
Mcgranahan G, Balk D, Anderson B (2007) The rising tide: assessing the risks of climate change and human settlements in low elevation coastal zones. Environ Urban 19:17–37. https://doi.org/10.1177/0956247807076960
Mishra A, Liu S (2014) Changes in precipitation pattern and risk of drought over India in the context of global warming. J Geophys Res Atmos 119. https://doi.org/10.1002/2014JD021471
MoEF (2009) Bangladesh climate change strategy and action plan, Ministry of Environment and forest. Government of the People's Republic of Bangladesh, Dhaka, 76pp. Available at https://cmsdata.iucn.org/downloads/bangladesh_climate_change_strategy_and_action_plan_2009.pdf
Mohajan H (2017) Greenhouse gas emissions, global warming and climate change. Jamal Nazrul Islam research Centre for Mathematical and Physical Sciences (JNIRCMPS). University of Chittagong, Chittagong
Mohajan HK (2011) Greenhouse gas emissions increase global warming. Int J Econ Polit Integr 1(2):21–34
Mohal N, Hossain M (2007) Investigating the impact of relative sea level rise on coastal communities and their livelihoods in Bangladesh. Institute of Water Modelling (IWM) and Center for Environmental and Geographic Information Services (CEGIS), Dhaka
MPEMR (2008). Government of Bangladesh. (2008). Renewable energy policy of Bangladesh. Retrieved from Ministry of Power, Energy and Mineral Resources: www.powerdivision.gov.bd/pdf/REP_
Neuhoff K (2005) Large-scale deployment of renewables for electricity generation. Oxf Rev Econ Policy 21:88–110
Neumann J (1985) Climatic change as a topic in the classical Greek and Roman literature. Clim Chang 7(4):441–454
Oliver J, Janssens-Maenhout G, Peters J (2012) Trends in global CO_2 emissions; 2012 report
Ouyang X, Lin B, Lin B (2015) An analysis of the driving forces of energy-related carbon dioxide emissions in China's industrial sector. Renew Sustain Energy Rev:838–849
Owusu PA, Asumadu-Sarkodie S (2016) A review of renewable energy sources, sustainability issues and climate change mitigation. Civil & Environmental Engineering
Padilla-Rivera A, Amor B, Blanchet P (2018) Evaluating the link between low carbon reductions strategies and its performance in the context of climate change: a carbon footprint of a wood-frame residential building in Quebec, Canada. Sustainability 10. https://doi.org/10.3390/su10082715
Patra PK, Canadell JG, Houghton RA, Piao SL, Oh N-H, Ciais P, Manjunath KR, Chhabra A, Wang T, Bhattacharya T, Bousquet P, Hartman J, Ito A, Mayorga E, Niwa Y, Raymond P, Sarma VVSS, Lasco R (2013) The carbon budget of South Asia. Biogeosciences 10:513–527
Pinkse J, Kolk A (2010) Challenges and trade-offs in corporate innovation for climate change. Bus Strateg Environ 19:261–272
Rabbani SH (2011) Climate change and Bangladesh: policy and institutional development to reduce vulnerability. J Bangladesh Stud 13:1–10
Rahman KM (2017) Climate Change governance: Bangladesh perspective. Int J Hort Agric 2(1):1–8. https://doi.org/10.15226/2572-3154/2/1/00110
Rahman M, Kashem M, Bank, Bangladesh, Dhaka, Bangladesh (2007) Carbon emissions, energy consumption and industrial growth in Bangladesh: empirical evidence from ARDL cointegration and granger causality analysis. Energy Policy 110:600–608. https://doi.org/10.1016/j.enpol.2017.09.006
Ramanathan V, Feng Y (2009) Air pollution, greenhouse gases and climate change: global and regional perspectives. Atmos Environ:37–50
Rehan Dastagir M (2015) Modeling recent climate change induced extreme events in Bangladesh: a review. In: Weather and climate extremes, pp 49–60

Ren S, Yin H, Chen XH (2014) Using LMDI to analyze the decoupling of carbon dioxide emissions by China's manufacturing industry. Environ Dev:61–75
Repetto R (2001) The clean development mechanism: institutional breakthrough or institutional nightmare? Policy Sci:303–327
Sarkar MSK, Sadeka S, Sikdar MMH, Badiuzzaman (2015) Energy consumption and CO2 emission in Bangladesh: trends and policy implications. Asia Pacific J Energy Environ 2:175–182
Sarkar MSK, Sadeka S, Sikdar MMH, Badiuzzaman (2018) Energy consumption and CO2 emission in Bangladesh: trends and policy implications. Asia Pacific J Energy Environ 5(1):41–48. https://doi.org/10.18034/apjee.v5i1.249
Saunois M, Jackson R, Poulter B, Canadell J (2016) The growing role of methane in anthropogenic climate change. Environ Res Lett 11:120207. https://doi.org/10.1088/1748-9326/11/12/120207
Schurer AP, Mann ME, Hawkins E, Tett SFB, Hegerl GC (2017) Importance of the pre-industrial baseline in determining the likelihood of exceeding the Paris limits. Natural Climate Change:563–567
Sharif SI, Anik MAR, Al-Amin M, Siddique MAB (2018) The prospect of renewable energy resources in Bangladesh: a study to achieve the National Power Demand. Energy Power:1–6
Shrivastava P (1995) The role of corporations in achieving ecological sustainability. Acad Manag Rev 20:936–960
Sprengel DC, Busch T (2011) Stakeholder engagement and environmental strategy – the case of Climate Change. Bus Strateg Environ 20:351–364
SREDA (2016) Energy efficiency and conservation master plan up to 2030. Sustainable and Renewable Energy Development Authority, Dhaka
Star D (2010) Bangladesh among top 10 climate-risk nations. Available. Retrieved from Daily Star: http://www.thedailystar.net/newDesign/news-details.php?nid=165007
Stavins JE (2012) The promise and problems of pricing carbon. J Environ Dev:152–180
Stern N (2007) Stern review: the economics of Climate Change. HM Treasury
Tavoni M, De Cian E, Luderer G, Jan, Steckel J, Waisman H (2012) The value of technology and of its evolution towards a low carbon economy. Clim Chang 114. https://doi.org/10.1007/s10584-011-0294-3
Teske S, Morris T, Nagrath K (2017) 100% renewable energy for Tanzania: access to renewable and affordable energy for all within one generation. Report prepared by Institute for Sustainable Futures for Bread for the World, October
Thomas Tanner AH (2007) Piloting climate risk screening in DFID Bangladesh. University of Sussex, Institute of Development Studies
UN (2015) The United Nations. Retrieved from Sustainable Development Goals. https://www.un.org/sustainabledevelopment/s ustainable-development-goals/
UNEP (2011) UNEP 2011 annual report. United Nations Environment Programme, Kenya
UNFCCC. (2012). United Nations framework convention on climate change. Retrieved from United Nations Framework Convention on Climate Change: https://unfccc.int/resource/docs/publications/unitingonclimate_eng.pdf
Walsham M (2010) Environment, climate change and migration in Bangladesh. International Organization for Migration
Weinhofer G, Hoffmann V (2008) Mitigating climate change - how do corporate strategies differ? Bus Strateg Environ 19:77–89. https://doi.org/10.1002/bse.618
World Development Indicators (2018) (2019, June). Retrieved from The World Bank Group.: http://datatopics.worldbank.org/world-development-indicators/#:~:text=World%20Development%20Indicators%20(WDI)%20is,country%20comparable%20data%20on%20development.&text=The%20World%20Development%20Indicators%20is,and%20the%20fight%20against%20poverty
Xu B, Wennersten R, Brandt N, Sun Q (2012) A projected turning point in China's CO2 emissions: an environmental Kuznets curve analysis. Int J Glob Warm 4(3/4):317–329
Xu B, Wennersten R, Brandt N, Sun Q (2015) A projected turning point in China's CO2 emissions: an environmental Kuznets Curve analysis. Int J Glob Warm 4(3/4):317–329

Xu XF, Tian HQ, Chen GS, Liu ML, Ren W, Lu CQ, Zhang C (2012) Multifactor controls on terrestrial N2O flux over North America from 1979 through 2010. Biogeosciences 9:1351–1366. https://doi.org/10.5194/bg-9-1351-2012

Yaacob N, Yazid M, Muhamad, Maulud K, Basri N (2020) A review of the measurement method, analysis and implementation policy of carbon dioxide emission from transportation. Sustainability. 12:5873. https://doi.org/10.3390/su12145873

Yu Y, Jin Z-x, Li J-z, Jia L (2020) Research on the impact of carbon tax on CO2 emissions of China's power industry. J Chem

Yue TX, Zhao MW, Zhang XY (2015) A high-accuracy method for filling voids on remotely sensed XCO2 surfaces and its verification. J Clean Prod 103:819–827

Zeng S, Zhang S (2011) Literature review of carbon finance and low carbon economy for constructing low carbon society in China. Low Carbon Econ:15–19

Zhang Xu QT (2015) Research focus and trend of energy development and climate change. Renewable Energy Resources 8:1214–1218

Chapter 12
Governance for Sustainable Development to Combat the Impact of Climate Change in a Medium-Sized City in Bangladesh

A. K. M. Mahmudul Haque, Abdullah Al-Maruf, and Tasnim Nazira Rida

Abstract Bangladesh is struggling with an increasing trend of environmental degradation due to the adverse shocks of climate change. Rajshahi, a northwestern city of Bangladesh, is mostly vulnerable to drought, heat waves, air and water pollution, etc. Although various earlier studies regarding environmental governance are mostly focused on mega cities such as Dhaka, studies on medium-sized cities are less frequently highlighted. Thus, achieving sustainable government through environmental governance in the medium-sized city is an urgent issue for academia and policy makers alike. Local environmental governance plays a significant role in attaining local sustainable development. In Bangladesh, the local government authorities perform a wide range of development activities in the cities. To understand local sustainable development, the environmental impact of these development activities must be explored. The main purpose of this chapter is to investigate the overall status of environmental governance for promoting sustainable development considering environmental issues in development programs. This investigation is empirical research mainly based on primary data as collected from the government officials and field staff of the Rajshahi City Cooperation, and various stakeholders of the city, e.g., residents and civil society, through questionnaire surveys and key informant interviews, including a case study. The central finding indicates that the state of environmental governance for sustainable development at the Rajshahi City Cooperation is functionally poor. Despite having a policy framework for prioritiz-

A. K. M. M. Haque (✉)
Department of Political Science, University of Rajshahi, Rajshahi, Bangladesh
e-mail: akmmahmudul@ru.ac.bd

A. Al-Maruf
Department of Geography & Environmental Studies, University of Rajshahi, Rajshahi, Bangladesh

T. N. Rida
Department of Philosophy, University of Rajshahi, Rajshahi, Bangladesh

M. N. Islam, A. van Amstel, *Bangladesh II: Climate Change Impacts, Mitigation and Adaptation in Developing Countries*, Springer Climate,
https://doi.org/10.1007/978-3-030-71950-0_12

ing environmental issues in the development activities of the city, a huge gap still remains between policies and practices. The findings of the study may assist the policy makers and concerned authorities in formulating policies and taking initiatives to ensure environmental governance to strengthen sustainable development in the urban areas of Bangladesh.

Keywords Environmental governance · Sustainable development · Medium-sized city · Climate change · Environmental degradation · Rajshahi City Cooperation (RCC)

12.1 Introduction

Local governance is the key to inclusive and sustainable development at the local level, as it remains the main entry point to accessing public authorities and state institutions. In Bangladesh, urban local governance institutions are entrusted with the responsibilities of ensuring environmental sustainability at the urban level through implementation of rules, regulations, and strategies of sustainable development. However, the state of environmental sustainability at the urban level is in question (Barkat et al. 2015; Rahim 2019). The scale and pace of urban environmental degradation is rapidly increasing; hence, saving the environment is now a major concern for academia and policy makers, particularly for the major cities of the country. This situation is caused by the parlous state of local governance, in which current local governance institutions are unable to address environmental issues properly. Urban local government bodies perform a wide range of regular and development activities such as waste management; drainage maintenance; traffic management; construction of roads, bridges, culverts, housing, potable water supply, and irrigation, flood control, recreation centers, and markets, etc. Such activities are needed to enhance the status of the environmental and socio-economic conditions of the urban dwellers, explicitly the poorest residents in a poor land (Uddin 2013). To build sustainable development, and to ensure environmental sustainability, urban legal frameworks and the sustainable development goals (SDGs) guidelines reveal a set of environmental protection parameters for analyzing the adverse environmental consequences of the regular and development programs by the authority concerned and to adopt appropriate measures to reduce such adverse consequences of climate change to acceptable levels (Sharafuddin and Rahman 1994; UNDP 1997; Usui 2000). However, the interference of unplanned development activities in local urban ecosystems is increasing and taking its toll in urban areas of Bangladesh. Good policies are there in this context, yet these are not working properly. Development is chipping away at the environment. One of the reasons for the gap between policy and implementation in development is that the rules and regulations concerned are not properly enforced by the authorities to consider sustainable development issues in routine activities, and planning and implementation of development projects. Therefore, proper study is required to assess the extent to which urban bodies obey rules and regulations in considering environmental issues in their development projects to ensure sustainable development. This study is an in-depth investigation to explore the state of urban local government in integrating environmental issues into the planning and implementation of development activities.

12.2 Key Concepts

12.2.1 Governance

The term "governance" has a multidimensional meaning and approaches. Simply, "governance" refers to the process of decision making and decision implementation. In the economic, social, environmental, and political disciplines, the term "governance" has multiple interpretations (Harman 2008). In general, governance means strategy, process, and institutions for managing a country's affairs at all levels. It is the means of articulating interest and exercising the legal rights of the citizens and groups where widely accepted components of good governance generate policy with the consensus orientation, accountability at all levels, transparency beyond expectations, effectiveness and efficient progressing status, free from corrupt practices, and establishing the mechanisms of the rule of law (Karim et al. 2018).

12.2.2 Environmental Governance

Environmental governance is an important instrument of governance systems that influence environmental outcomes. The term 'environmental governance' is used to describe how decisions about the environment are made and who makes such decisions. It includes the formal and informal institutional arrangements for resource and environmental decision-making and management. It includes and extends beyond the state to involve the private sectors and civil society organizations (Barkat et al. 2015; Rahim 2019). Thus, it involves a wide range of institutions, social groups, processes, interactions and traditions, all of which influence how power is exercised, how public decisions are made, how citizens become engaged or disaffected, and who gains legitimacy and influence and achieves accountability. According to the Asian Development Bank-ADB (2012) "environmental governance is the manner in which people exercise authority over nature. It concerns the actors – within each level of government, between elected and appointed officials, and among 'traditional' private and non-governmental bodies – and the power that they exercise to make decisions about the disposition of natural resources and benefits that flow from the environment."

12.2.3 *Local Environmental Governance*

Local environmental governance shifts the decision-making power to the grassroots level from the central governments. It is a democratic system based on decentralization composed of diverse actors engaged in environmental protection and management, and in sustainable economic activities at local, provincial, or district levels (Lima 2010). Local environmental governance materializes through the existence of a permanent and active task force composed of the state, civil society, and private sector (direct or indirect participation) to manage the regional and local environmental issues and natural resources based on principles of sustainable development, rationality, and environmental management. This consortium must look for projects, programs, policies, and rules that balance and combine coherently at least three interest areas: economic, social, and environmental, maximizing the benefits to the population (and its minorities) with minimum stress to nature (Lima 2010). Local environmental governance is important for bringing back power to local communities in the global fight against environmental degradation.

12.3 Sustainable Development and Combating the Impact of Climate Change

Sustainable development is a concept that appeared for the first time in 1987 with the publication of the Brundtland Report. Brundtland pointed to sustainable development as development that meets the needs of the present without compromising the ability of future generations to meet their own needs (Brundtland 1997 cited in Bello and Dola 2014). Sustainable development warns of the negative environmental consequences of economic growth and globalization, which tried to find possible solutions to the problems caused by industrialization and population growth. There are three dimensions of sustainable development, i.e., environmental, social, and economic.

From the environmental perspective, sustainability prevents nature from being used as an inexhaustible source of resources and ensures its protection and rational use. Through environmental sustainability, a healthy environmental system is maintained for current and future generations (Bello and Dola 2014). Environmental conservation, investment in renewable energies, saving water, supporting sustainable mobility, and innovation in sustainable construction and architecture, etc., contribute to achieving environmental sustainability on several fronts (Acciona 2018). Development of people, communities, and cultures to help to achieve a reasonable and fairly distributed quality of life, healthcare, and education across the globe, and the fight for gender equality, are different aspects that form the basis of social sustainability (Acciona 2018). In the context of the economy, sustainability refers to a futuristic approach to maintaining economic welfare (Pearce et al. 2010). It focuses on equal economic growth that generates wealth for all, without harming the environment. Investment and equal distribution of the economic resources strengthen the other pillars of sustainability for a complete development (Acciona 2018). From the local governance perspective, all three dimensions are required to be carried along equally balanced for comprehensive development.

Now it is very obvious to expect an impact on the climate of a region where the environmental components are affected by the decisions of the authorities. Simply, environmental governance of any region has all the power to standardize climatic scenery of the location and its surroundings (Sharafuddin and Rahman 1994; UNDP 1997; Usui 2000). Long-term mismanagement of development planning, town planning, waste management system, river exploitation, sanitation, industrial management, even the public's behavior, can easily bring about adverse changes in the climate. In this case, multiple factors can be aroused, i.e., carbon emission or emission of any other greenhouse gases (Barkat et al. 2015; Rahim 2019). The effect of climate change in Rajshahi is clearly visible. This drought-prone region of the northern part of Bangladesh often faces a water crisis in the dry season. The ground water reserve is under enormous threat (Rahman 2015). On the other hand, storms, high temperatures, and cold waves are very familiar inside the city. These disasters have achieved acceleration in the frequency of happening and causing untold sufferings for the localities (Haque and Salehin 2019). This chapter tries to justify the environmental governance, or in this case, mal-governance, and forecasts the effect of this mishap on the sustainability of the city's development and climate.

12.4 Research Methodology

A mixed methods approach (both qualitative and quantitative data) has been used in this study. The empirical findings are based on primary data, which were collected through questionnaire surveys, key informant interviews (KIIs), and observation techniques. Survey interviews were conducted with 10 government officials and 60 field staff of the Rajshahi City Corporation (RCC) who have involved in implementing environmental policies, as well as 120 stakeholders of the city such as residents and members of civil society in 2015–2016.

A number of policies have direct and indirect relationships with consideration of environmental issues in development activities. Among them some important policies have been quoted below.

- Activities that have adverse effects on public health, including development activities, will be eradicated from the country to develop a healthy environment in rural and urban areas (Government of Bangladesh, The National Environment Policy 1992b).

- No industrial unit or project shall be established or undertaken without obtaining an environmental clearance certificate in the manner prescribed by the rules, from the DG (Directorate General) Government of Bangladesh, The Bangladesh Environment Conservation Act 1995, Amended in 2002).
- Necessary steps will be taken to protect the environment and public health from the adverse impact of all sorts of waste materials (Government of Bangladesh, National Environment Management Action Plan 1992a).
- The Location Clearance Certificate and Environment Clearance Certificate are compulsory for all industrial and other projects (Government of Bangladesh, The Bangladesh Environment Conservation Rules 1997).
- The Corporation may, and if so required by the Government shall, prepare and implement development plans for such periods and in such a manner as may be specified (Government of Bangladesh, The Local government (City Corporation) Act 2009). The Corporation shall make adequate arrangements for the removal of refuse from all public streets, public latrines, urinals, drains, and all buildings and land vested in the Corporation, and for the collection and proper disposal of such refuse (Government of Bangladesh, The Local government, City Corporation Act 2009).

12.5 Results and Discussions

In order to explore the practical implications of the sustainable development policies concerned, the aspect of environmental issue consideration in the development activities of the RCC was investigated. In this regard, the level of environmental issue consideration in the planning and implementation process, the monthly meetings of the RCC, the activities of standing committees, people's perception about environment-related committees, etc., were investigated. Simultaneously, a case study was conducted to achieve a more comprehensive understanding about the matter of environmental issue consideration in different activities and policies. These case studies are selected development projects that have already been completed by the RCC. The study findings have been presented below.

12.5.1 Consideration of Environmental Issues in Planning

The planning process of development projects requires primary consultation with a broad base of groups: beneficiaries, persons affected by the project and community leaders, elected representatives, disadvantaged groups, and various development agencies in the area (Rabbani 2015). In other words, local people are to be involved in determining the relationship between socio-economic development and the environment. This process is helpful in developing effective methods of "people involvement" in project design, management, operation and maintenance, monitoring and

Table 12.1 Consideration of environmental issues in planning

Variables	Responses	Percentage (%)
Does the RCC consult with you during the planning of development projects?	Yes	0.0
	No	86.8
	No comment	13.2
Do you think that RCC considers environmental issues during the planning of development projects?	Zero consideration	39.4
	Moderate consideration	59.4
	Full consideration	1.1
Do you give any information to the RCC about improving the environment of your area?	Yes	13.9
	No	74.4
	No comment	11.7

(Source: Field Survey 2016)

evaluation, keeping the environmental issues at the center of the planning (Chowdhury et al. 2016).

Table 12.1 indicates that most of the respondents (86.8%) report that the RCC does not consult with them at the time of planning of development projects, whereas 13.2% of the respondents abstained from answering. No one said that the RCC consults with them in this regard. On the other hand, 39.4% of respondents believe that the RCC does not consider environmental issues during the planning of development projects, whereas 59.4% say that the RCC gives environmental issues moderate consideration. Only 1.1% of the respondents say that the RCC fully takes environmental issues into account when planning development projects.

In response to another question, the majority (74.4%) of the respondents replied that they do not give any information to the RCC authorities about improving the environment of their area. Only 13.9% of the respondents stated that they provide information to the RCC in this regard. Only 11.7% of respondents refrained from answering, which is shown in Table 12.1.

The study findings also illustrate that, in spite of having guidelines, the RCC does not consider environmental issues fully during the planning of development projects. In this regard, the study shows that a lack of mandatory provisions within the legal framework for environmental assessment in local government development projects makes such a course of conduct possible and easy for local councils. Although BECA 1995 and ECR 1997 contain provision for undertaking environmental assessment for proposed development projects in all sectors, they do not provide specific guidelines or requirements for the local government bodies on conducting and reviewing the environmental assessment of non-industrial projects.

The study also finds a lack of awareness of key players about environmental management. During the KIIs, the RCC's executive engineer noted, "We have resource constraints. We do not have enough manpower. While making the plans, we have to follow the Development Project Proforma (DPP) of the Planning Commission that does not contain mandatory provision for environmental assessment. We just answer a question of the DPP saying that this project will not create any adverse impact on the environment and we send the proposal to the higher authority for approval. They do not seek further information in this regard. Besides, we need to plan a project within a very short time. So, it is very difficult to integrate people's opinions during planning of projects. Moreover, ordinary citizens are not much interested to participate in the planning process without financial incentives."

Another key informant (an RDA Town Planner) indicated "there is no special budget allocation for Environmental Assessment (EA) activities of development projects. So, continuous enforcement and improvement of EA becomes difficult. Besides, to conduct EA of development projects, we need highly experienced consultants with high salaries. But we do not have sufficient funds to appoint expert EA consultants."

12.5.2 Consideration of Environmental Issues During Project Implementation

A sustainable development strategy emphasizes the involvement of stakeholders in the development management process. In practice, stakeholders can be involved in the project implementation process in two ways: indirectly, as beneficiaries or information sources; or directly, as members of the committees related to project implementation.

At the City Corporation level, each project has a committee from the Works Division, named the Urban Infrastructure Development and Conservation, which is responsible for implementation of planning. This committee monitors implementation in order to ensure that the project has no adverse impact on environment or biodiversity. This committee of the Works division is assigned to conduct the environmental and social assessment of projects. It will collect data and provide a progress report to concerned higher authorities, giving special consideration to the environment. Hence, data have been collected to obtain information about the role of the Works Division regarding environment issues in practice.

The field survey reveals that 52.8% of the respondents have heard about such committees of the works division, whereas 47.2% have not. On the other hand, when asked, a large majority of the respondents (80.6%) reply that they do not know the duties and responsibilities of such committees or any other related authority. Only 19.4% of respondents have some idea about the role and responsibilities of such committees or any other authority related to the implementation of projects. In response to another question, 94.4% say that committee members never discuss environment issues with them, whereas only 5.6% of the respondents reply positively in response to this question (Table 12.2).

During the KIIs, the ex-Chief Engineer of the RCC said that committees related to project implementation do not actively play their assigned role. Most of the members of these committees are not very aware of environmental rules and regulations.

Table 12.2 Consideration of environmental issues in project implementation

Variables	Responses	Percentage (%)
Do you know anything about committees related to implementation of development projects?	Yes	52.8
	No	47.2
Do you know the duties and responsibility of such committees or other authorities with regard to project implementation?	Yes	19.4
	No	80.6
Did any member of such committees discuss with you about consideration of environment issues during implementation of development projects?	Yes	5.6
	No	94.4

(Source: Field Survey 2016)

Besides, members of these committees are mainly selected by the Mayor and local-level influential leaders of the ruling party. Thus, they emphasize mainly their personal and party interests, rather than considering environmental issues, in implementation of the projects. Moreover, ordinary people are not very aware of environment issues, nor are they interested in participating in environment-related discussions.

12.5.3 Consideration of Environmental Issues in Monthly Meetings

Under the Local Government (City Corporation) Act 2009, the Corporation will prepare and implement development plans in such a manner as may be specified. The Act also states that the Corporation will organize monthly meetings, at least once a month, in which the Mayor, Councilors (female and male) and Chief Executive Officer (CEO) will participate, to discuss overall development issues of the Corporation. This forum acts as a policymaking body.

To investigate the consideration of environmental issues in this meeting, Councilors and the CEO were asked whether or not they discuss environmental issues in the monthly meetings. They replied that they discuss the general activities of the Corporation in the monthly meetings, based on priority. They do not have any discussions about environmental issues at every meeting. Sometimes, if they need to consider the environmental issues in project management, however, they discuss this matter in a special meeting.

For sake of the study, a high official of the RCC was asked whether or not they are keeping proper track of climatic change within their region. In response, they failed to give any relevant answer. In fact, they held a very pragmatic concept of being concerned as little as possible with the climate issue.

12.5.4 Consideration of Environmental Issues in Ward Meetings

According to the policy framework, each Ward Councilor will hold a Ward meeting with residents of the Ward at least twice a year, to discuss environmental issues, identify problems, prioritizing problems, identifying schemes, prioritizing schemes, and short listing the projects in an open and exclusive manner. After the Ward meeting, further participatory planning processes will be carried out at the City Corporation.

In this context, data were collected from a field survey. Table 12.3 shows that just over half of the respondents (51.2%) state that Ward Councilors occasionally arrange Ward meetings to discuss development issues of the Ward, whereas 48.8% respondents do not agree with this statement. When they are asked about their participation in Ward meetings, only 13.9% of respondents replied that they participate in such meetings, but 86.1% have never participated in Ward meetings. In response to another question, a large number of respondents (76.67%) report that they have never taken part in the Ward meeting to discuss environmental issues. Only 3.33% replied positively and 20.1% did not answer this question.

During a KII, an Engineer from the Project Planning Unit of the RCC said that, at the time of project selection, planning, and implementation, priority is given to making the citizens happy by undertaking physical and social development rather than environmental development. He added that, during selection of the projects, the Mayor and other politicians mainly emphasize the people's demands and the political commitments given to the voters during the last election. Thus, environmental considerations are not given priority. People always evaluate the politicians on the basis of what infrastructural development they have completed rather than on the development of the environment.

Table 12.3 Consideration of environmental issues in Ward meetings

Variables	Responses	Percentage (%)
Does the Ward Councilor occasionally arrange Ward meetings to discuss development issues of your Ward?	Yes	51.2
	No	48.8
Have you ever participated in such a Ward meeting?	Yes	13.9
	No	86.1
Does the Ward meeting discuss environmental issues during project selection and other regular and development activities?	Yes	3.3
	No	76.6
	No comment	20.1

(Source: Field Survey 2016)

12.5.5 People's Perceptions of Environment-Related Committees

In accordance with sections 50 (1) and 50 (2) of the Local Government (City Corporation) Act 2009, the RCC is required to form 14 Standing Committees to carry out its duties. With the prior approval of the Minister, the Corporation may also constitute additional Standing Committees. A Standing Committee is required to consist of not more than six members, elected by the Councilors from amongst themselves. No Councilor is allowed to be the member of more than two Standing Committees at the same time. However, the Mayor is a member of all the Standing Committees. A Standing Committee elects one of its members as Chairman and another as Vice-Chairman. The Department Head concerned acts as the Member-Secretary of each Standing Committee.

Presently, there are 18 Standing Committees in the RCC; however, the following Committees are related to environmental governance in RCC:

(a) Standing Committee on Finance and Establishment
(b) Standing Committee on Waste Management
(c) Standing Committee on Education, Health, and Family Planning
(d) Standing Committee on Urban Planning and Development
(e) Standing Committee on Urban Infrastructure Development and Conservation
(f) Standing Committee on Environmental Development

Standing Committees have been entrusted with the responsibilities of considering environment-related issues while discharging their own responsibilities. To explore the Standing Committees' role in environmental governance, data have been collected in the field survey (Table 12.4).

Most of the respondents (83.3%) have never heard about the Committees on Waste Management, Urban Planning and Development, and Environmental Development listed above, and a large number (85.6%) of people are not informed

Table 12.4 People's perception of environment-related committees

Variables	Responses	Percentage (%)
Have you heard anything about the committees on waste management, urban planning and development, environment development, etc.?	Yes	16.7
	No	83.3
Do you know about the duties and responsibilities of these committees?	Yes	14.4
	No	85.6
Do you think that these committees are successfully performing their duties and responsibilities in the environmental development of the city?	Yes	30.6
	No	42.2
	No comment	27.2

(Source: Field Survey 2016)

about what they do. Only 16.7% and 14.4% of respondents respectively knew about these Committees and what they do. In response to another question, 30.6% of respondents thought that these committees were successfully performing their duties, but 42.2% of respondents do not think so. Most of those who had never heard anything about the Committees said that they were, or were not, performing well: strictly speaking, this is illogical and calls the reliability of these responses into question. Yet a good number of respondents (27.2%) refrained from answering; some of these may have been those who had never heard of the environment-related committees.

A previous survey also revealed that the Standing Committees are not effective in the true sense. The members of the Standing Committees were not well aware of their responsibilities. They lacked awareness of environmental issues in the management of other issues and development activities (Rabbani 2015). A key informant (Head of a Civil Society Organization) said that, although there were Standing Committees in the RCC, these only functioned on paper. Most of the Committees were not active.

To ensure sustainability of development management with better service delivery, Standing Committees must become effective. Currently, they barely exist and almost no one knows what, if anything, they do. Both Golam Rabbani's survey of the Standing Committee members in 2010 and the field survey for this report in 2015–2016 point to this same conclusion. Yet their role can be very important as an environmental oversight of all the actions of the Council.

12.5.6 Impacts of Development Activities

Urban development undertaken by City Corporations, such as construction of public or private buildings and roads, raises environmental issues. However, the field survey shows that, when construction of roads or buildings take place, the RCC or their contractors leave building materials such as bricks, sand, metal rods, etc., in the roads for long periods, obstructing the free movement of vehicles and pedestrians. The waste of development work is, in some cases, kept at, or dumped into, roadsides, open spaces, and drains, creating environmental hazards. In interviews, 66.4% of respondents stated this, whereas 33.6% of respondents stated otherwise. A huge amount of solid waste is generated during road excavation for water supply and sanitation, drain construction or repair, and other development work, by the RCC.

The only way that responses to questions 1 and 2 in Table 12.5 can be read consistently is to say that RCC's road excavation waste is kept on the roadside for a short time and eventually dumped into the drains: although 30.5% of the respondents did not answer the second question. Physical observation also reveals that,

Table 12.5 Impact of development activities

Variables	Responses	Percentage (%)
Do you think that construction materials of any buildings or infrastructure development are kept on the road and dumped into the drains?	Yes	66.4
	No	33.6
Do you think that waste from road excavation is stored or kept along the roadside for long periods of time?	Yes	8.2
	No	61.3
	No comment	30.5
Do you think that the RCC's open drains create adverse environmental impacts?	Yes	97.2
	No	2.8

(Source: Field Survey 2016)

when the RCC undertakes any road excavation work, the solid waste from such works is not removed quickly in many cases.

By leaving waste at the side of the road or dumping it into drains, the RCC is following the practice of private sector construction in Rajshahi: private home and office builders do the same thing. Yet the RCC also takes no action against the owners of buildings or others who do not keep or dispose of such materials properly. There is a Magistracy Department in the RCC, headed by a Magistrate. This Department could hear complaints and impose penalties under the Building Materials Act 1952. However, they rarely do so.

A key informant (an RCC Executive) thinks that the City Corporation is mainly run by the Mayor and his Councilors, who are politicians. All of them are elected by the voters of the city. Therefore, they are always afraid of losing support among the voters if the City Corporation is perceived as harassing private development. Construction is one of Rajshahi's few successful private industries and it makes a large amount of money. A large amount of construction money can be given now, to make the City Corporation inspectors go away quietly, or withheld later at election time from politicians or parties who increase the costs of construction by insisting on proper disposal of its waste.

The RCC constructed 118 km of completed and 162 km of partially completed drains in the city, through the implementation of drainage development projects, in 2009–2015. Yet still more than 85% of drains are open. Open drains are risky for pedestrians and significant sources of environmental pollution, as they attract insects, create bad smells, and run into the river. More than 97.2% of respondents say that these open drains create adverse environmental impacts in the city, whereas only 2.8% do not agree with this statement.

In interviews, the Chief of the Conservancy Division of the RCC tried to justify this situation, saying that open drains are easier to clean than closed drains. However, "cleaning" these open drains usually means mucking them out and dumping the sludge, often containing human and animal waste, onto the side of the road, which is more polluting than the drains themselves!

12.5.7 Case Studies: Environmental Compliance in the Development Projects of the RCC

Local government bodies in Bangladesh, such as *Paurashava* and City Corporations, design and implement a large number of development projects every year. The RCC is no exception in this regard. It has planned and implemented a total of ten development projects between 2010 and 2015 using central government's allocation of Annual Development Program (ADP) funds. Among them, five projects have been studied to assess the impact of the RCC's environmental governance. All projects are infrastructure development projects that were implemented by the RCC using funds of the Ministry of LGRD and Co-operatives/Local Government Division under the ADP.

What is most striking from the case studies is that the consideration of environmental issues in the planning remains standard; however, implementation of the projects was far from the necessary activities and action are needed. There is the Development Project Proorma (DPP). Yet the DPP seems in practice only a formality in which the local government makes self-justifying assertions about a positive environmental impact. Sustainability has little to do with the environment in these DPP statements: it means only that the building will stand up and the road will remain in place for a long time. In each case of the case study, there was a great deal of serious and easily predictable environmental harm that could result. Committees are set up on paper to monitor environmental impact for the RCC development projects, but know little, do a limited amount, and are virtually unheard of by stakeholders: they are merely party sinecures for the ruling actors. There is no popular participation at all in environmental assessment: nor is there any demand for it from the people.

12.6 Conclusions and Recommendations

Integration of environmental issues in development activities by the local governments is crucially important for ensuring sustainable development in urban areas to combat the adverse impact of climate change. In Bangladesh, there is a robust policy framework to govern the environmental issues at an urban level, but there is a gap between the policies and their practices in reality. The study finds that adequate assessment of adverse environmental impacts of development projects is hardly done by the RCC at the stage of planning and implementation of the projects. However, the limited actions of the RCC exactly follow the environmental rules and regulations in this respect. The study finds a gap between the policies and practices, as a result of which the process of environmental governance is being hampered considerably. Behind this gap, a lack of responsible performance by the public officials, a lack of proper knowledge of the elected representatives on environmental issues and laws, a lack of commitment by policymakers, the political influence in decision making and implementation, a lack of effective planning, and the knowledge gulf between the people and the authorities are mainly responsible.

The study indicates insufficient attempts by the RCC to ensure public participation in the environmental decision-making and implementation process and to raise the knowledge and awareness level of community people. The local community is also not very aware of environmental issues. Apart from that, the role of civil society in environmental governance is also not strong enough to put pressure on the city corporation. A previous study also revealed that lack of enthusiasm and awareness of the policymakers are major causes of poor policy implementation in environmental governance (Nishat 2015). It is clear that the environment cannot be protected by only the government offices. The people are an indispensable part of environmental governance in fighting the negative impact of climate change, but they are left out.

In addition, where there is so little concern about the environmental impact of the regular development works conducted by the local authority, it is presumable that there will be a shortage of data regarding climate change. Therefore, the sustainable development of the city will be under threat because the climate change will pose economic, social, and political predicaments that will challenge the successful implementation of the Sustainable Development Goals (SDGs). Multiple disasters and hazards, which are being posed by the changing climate, will inevitably impact the overall sustainability of the city.

The study suggests the following recommendations to strengthen sustainable government through the local government.

- Amend the Bangladesh Environment Conservation Act (BECA) to prepare clear guidelines by the DoE for environmental assessment of non-industrial projects.
- Produce local environmental assessment guidelines to better reflect local conditions, including the laws, institutions, standards, and procedures. It would promote information flow, awareness, and interdepartmental cooperation.
- Public consultation must be a part of the statutory process of environmental assessment. It would make the environmental assessment effective and meaningful.
- Conduct environmental assessment only to ensure sustainability of the project, not to make the donor agencies and government happy.
- Appoint expert staff to review the quality of the environmental impact assessment, the design of mitigation measures, and the quality of monitoring data.
- The RCC's Magistracy Department should be strict and more active to ensure appropriate punishment for the environmental polluters in accordance with environmental legislations.
- Ensure proper monitoring of environmental activities including waste management and environmental assessment of development activities of the city.
- Strong commitment of both politicians and bureaucrats for proper enforcement of environmental compliance legislations. Steps must be taken to remove pervasive corruption in this regard.
- Selection of the projects should not be to fulfill political commitments. Before selection, the adverse environmental impacts of such projects should be studied properly.

- Selection of the members of the PIC/Standing Committee should be done on the basis of the knowledge required. Proper training should be arranged for them.
- Adverse impacts of the development activities on the environment should be properly mitigated in a timely manner.

References

Acciona (2018) What is sustainable development and how to achieve it? Available at https://www.acciona.com/sustainable-development. Accessed 20 July 2018

Barkat A et al (2015) Local governance and decentralization in Bangladesh: politics and economics. Pathak Shamabesh, Dhaka

Bello A, Dola K (2014) Sustainable development and the role of local governance: experience from Malaysian model regions. Int J Humanities Soc Sci 4(1):268–280

Brundtland G (1997) Our common future. Report of the 1987 World Commission on Environment and Development'

Chowdhury JU et al (2016) Integrating environmental considerations into economic policy making processes: modalities for environmental assessments: country studies on flood loss reduction in Bangladesh, urban development and environmental protection in Shanghai, China and fishery resources development and management in Samao. Integrating Environmental Considerations Into Economic Policy Making Processes: Background Readings, Vol. IV, United Nations. Economic and Social Commission for Asia

de Lima IB (2010) Deforestation in Amazonia and Local Environmental Governance. International University of Japan

Government of Bangladesh (1992a) National Environment Management Action Plan (NEMAP), 1992. Dhaka

Government of Bangladesh (1992b) The National Environment Policy, 1992. Dhaka

Government of Bangladesh (1995) The Bangladesh environment conservation act 1995 (Amended 2002). Dhaka

Government of Bangladesh (1997) The Bangladesh Environment Conservation Rules 1997. Dhaka

Government of Bangladesh (2009) The Local government (City Corporation) Act 2009. Dhaka

Haque M, Salehin A (2019) Climate change in Bangladesh: effect versus awareness of the local agencies of Rajshahi city. Soc Sci J, Rajshahi University 22

Harman SJ (2008) The relationship between good governance and environmental compliance and enforcement. E-book library [online]. Available at https://www.peacepalacelibrary.nl/ebooks/files/C08-0084-Harman-Relationship.pdf. Accessed 2 June 2020

Karim BA, Abdul Z, Karim ZA, Nasharuddin MN (2018) Corruption and foreign direct investment (FDI) in ASEAN-5: a panel evidence. In: Economics and finance in Indonesia, Faculty of Economics and Business, University of Indonesia, vol 64, pp 145–156

Nishat A (2015) Policy and legal premise for environment and natural resources management in Bangladesh: issues concerns trends and institutional framework. UNDP Bangladesh Policy Dialogue Series No-11

Pearce DW et al (eds) (2010) Sustainable development: economics and environment in the third world. Edward Elgar Publishing Limited, London

Rabbani G (2015) Environmental governance: policies and practices at local government level in Bangladesh. Inst Bangladesh Stud J (Rajshahi University) 35

Rahim A (2019) Governance and good governance: a conceptual perspective. J Public Admin Gover 9. Available at 133. https://doi.org/10.5296/jpag.v9i3.15417. Accessed 2 June 2020

Rahman MS (2015) Climate change possesses a threat to livelihood in Rajshahi. The Daily Independent, 15 June. Available at http//:m.independentbd.com/arcpprint/detail/3072/2015. Accessed 15 June 2018

Sharafuddin AM, Rahman A (1994) Environmental education and awareness, environment and development in Bangladesh (Part 1). University Press Limited, Dhaka

Uddin M (2013) Towards good urban local governance in Bangladesh: lessons learnt from Japanese local government system to overcome the challenges. Lex Localis, Available at 11.933–953. https://doi.org/10.4335/11.4.933-953. Accessed 2 June 2020

UNDP (1997) Governance for sustainable human development program. United Nations Development Program, New York

Usui Y (2000) Governance legal order, social integration: reviewing new governance approaches in EU studies. Available at http://www.nuis.ac.jp/~usui/Usui_2000_Journal_of_NUIS.htm. Accessed 29 May 2010

Chapter 13
Climate Change Impact on Urban Cities in Bangladesh: Waste Management and Pollution Control Challenges

A. H. M. Monzurul Mamun and Mallik Akram Hossain

Abstract Municipal waste management is a challenging environmental issue in many cities around the world. Municipal solid waste (MSW) continues to increase in thriving cities in developing countries due to rapid increases in urban populations along with sustained economic growth. MSW is generated from various sources such as households, shopping centers, health care facilities, industry, development activities, and others. Cities in Bangladesh face multiple challenges in mitigating the problems associated with MSW. Among others, the handling of MSW contributes to climate change by producing greenhouse gases. This chapter reviews the contemporary literature on MSW and climate change impacts both in Bangladesh and beyond. Focusing on a typical municipal environment, this chapter attempts to shed light on the dynamics of municipal waste, including the relationship between the socioeconomic status of the residents and waste generation. Finally, climatic impact from MSW is investigated using scientific evidence. Solutions to mitigate the climate change impact originating from MSW are recommended in the context of Bangladesh cities.

Keywords Municipal environment · Greenhouse gases · Climate change · Waste management · Pollution control

A. H. M. M. Mamun (✉)
Department of Geography and Environment, Pabna University of Science and Technology, Pabna, Bangladesh
e-mail: hasnat@pust.ac.bd

M. A. Hossain
Department of Geography and Environment, Jagannath University, Dhaka, Bangladesh
e-mail: mallik.a@geography.jnu.ac.bd

M. N. Islam, A. van Amstel, *Bangladesh II: Climate Change Impacts, Mitigation and Adaptation in Developing Countries*, Springer Climate,
https://doi.org/10.1007/978-3-030-71950-0_13

13.1 Introduction

Urban areas are one of the major contributors to waste generation. Rapid urbanization in developing countries has accelerated the production of municipal solid waste (MSW). In Bangladesh, large cities generate an overwhelming proportion of MSW in the country (Alamgir and Ahsan 2007 cited from Kabir 2016), with Dhaka city contributing the most. MSW in Bangladesh cities typically includes food waste, paper, textiles, rubber, plastic, glass, metals, soil, and wood (Islam 2016a).

MSW management in Bangladesh cities is deplorable, even as the amount of MSW continues to increase day by day. MSW management in South Asian and South East Asian cities are constrained by many problems associated with high population growth, rapid industrialization, and unabated urbanization. Poor management of MSW may have adverse effects on the environment and urban health. Among other environmental issues, climate change has emerged as a major concern of policymakers, and greenhouse gas (GHG) emissions from MSW are a contributor to climate change. Chowdhury et al. (2014) reported that Dhaka city (consisting of Dhaka North City Corporation and Dhaka South City Corporation) generate approximately 1.6 million tons of municipal waste per year, which emit approximately 1 million tons of GHG annually. Therefore, the linkage between MSW management and climate change is obvious. The following sections focus on issues related to municipal waste, a distinct municipal environment, and climate change issues arising from MSW.

13.2 Municipal Waste and Associated Issues

Waste is regarded as the byproducts and end products of production and human consumption (Mamun and Monzurul 2017; Singh et al. 2014). Household solid waste (HSW) accounts for the majority (55–80%) of MSW (Ye et al. 2020). Population growth, economic development, and rapid urbanization have caused increased rates of MSW generation, from 0.68 billion tons per year in 2000 to 1.3 billion tons in 2010 and a projected 2.2 billion tons by 2025 and 4.2 billion tons by 2050 (Islam 2016a; Oribe-Garcia et al. 2015). In Bangladesh, municipal authorities are mainly responsible for providing an effective waste management (WM) system to inhabitants of cities. However, they often face problems beyond their abilities (Sujauddin et al. 2008) due to organizational, financial, and multidimensional system barriers (Guerrero et al. 2012; Burnley 2007). Poor WM practices and associated public health and climatic impacts remain severely problematic in many developing countries for a variety of reasons (Marshall and Farahbakhsh 2013; Konteh 2009). Municipal solid waste management (MSWM) is one of the most costly urban services, typically requiring 1% of the gross national product and 20–40% of municipal revenue in developing countries (Aliu et al. 2014). It is a highly visible municipal service that involves large expenditures in terms of investment, operational, and environmental costs (Faccio et al. 2011). The municipalities of Bangladesh are facing problems due to physical planning issues, governmental systems and policies, administrative and managerial procedures, and people's attitudes and behavior toward MSW (Hasan 1998).

Climate change is a serious global environmental concern because of its impact on the lives and livelihoods of human beings. Research findings confirm that the main climatic parameters, such as air temperature and precipitation, have changed in recent decades. Climatic changes are intertwined with the increase in greenhouse gases. GHG emission is considered to be one of the most significant environmental impacts of waste management (Lackner and Jospe 2017). The global urban population is expected to be 5 billion by 2030, which will be accompanied by an alarming increase in MSW (Puppim de Oliveira 2018). Significant environmental problems, such as pollution and anthropogenic GHG emission, are associated with rapidly increasing MSW in Bangladesh. Insanitary landfills are used to manage almost 80% of global MSW because they are the cheapest MSWM system (Kumar and Sharma 2013). In Bangladesh, MW has high organic content, which is considered to be biodegradable and linked to anthropogenic methane (CH_4) emissions unless managed properly (Mamun and Hossain 2012). Mismanagement, inefficient and inadequate service, and lack of strategy make the MSWM situation hazardous and threatening. Low collection efficiency and unhygienic disposal practices such as crude dumping, inappropriate transportation, and insanitary land filling have contributed to GHG emissions (Fig. 13.1). CO_2, CH_4, N_2O, and perfluorocarbons are potent GHGs emitted from WM activities. In municipal areas, energy used by fossil fuel combustion results in most CO_2 emissions. A great amount of energy is consumed when a product is manufactured and then discarded.

MSW management generates GHG emissions in two ways: direct emissions (from landfills, waste-to-energy plants, recycling, and collection of the waste) and indirect emissions. Indirect GHG emissions that are associated with the extraction and processing of primary resources or fossil fuels versus those associated with recycling or incineration operations and the use of raw materials for plastic, paper, metals, etc. (Moora et al. 2017).

Fig. 13.1 GHG emission sources from an MSWM system in Bangladesh. (Modified from Mamun and Hossain 2012)

A more potent GHG is CH_4, which is 34 times stronger than CO_2 (Islam 2016a). CH_4 is produced when biodegradable or organic waste decomposes in an anaerobic environment, such as a landfill. A study in 2008 found that European countries generate 71% of their CH_4 from landfills used for dumping MW. One ton of biodegradable waste produces 200 to 400 m^3 of landfill gas, of which 50% are GHGs, including CH_4 (Mamun and Hossain 2012). N_2O results from the decomposition of organic matter; the small amount of perfluorocarbon emissions has a significant global warming potential. Anthropogenic activities, such as MSWM, increase the concentration of GHGs in our atmosphere and consequently influence global warming. Global annual GHG emissions of 52 $GtCO_2$ eq. were recorded in 2010 due to increased contributions of CH_4 from MSW (Islam 2016b; Stocker et al. 2014). Fig. 13.2 illustrates the linkage between MSWM and GHG emissions.

The recycling of biodegradable and organic waste, such as by composting, decreases the total amount of waste to be landfilled and therefore significantly reduces landfill area requirements (Mamun and Monzurul 2014). This WM process also shows some energy recovery and environmental conservation potential by producing biogas and organic fertilizer or compost as end products. In developed countries, such as Germany and the United States, the emissions from waste management were reduced drastically after the introduction of mechanical biological treatments, source separation, recycling, and composting (Dehoust et al. 2010; Weitz et al. 2002). Tremendous progress has been observed in the practices and technologies of WM (e.g. collect, treat, recycle, and recover) over the past several decades, with an aim to improve public health conditions in urban communities and minimize the adverse environmental impacts of MSWM.

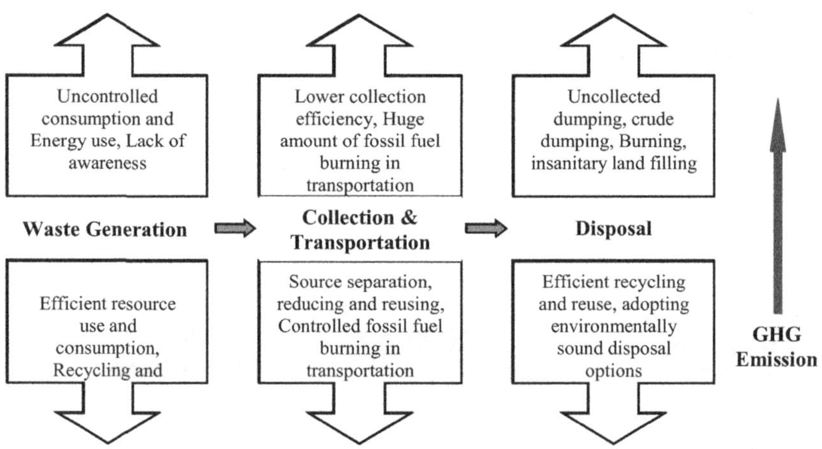

Fig. 13.2 Linkage between MSWM and GHG emissions in Bangladesh. (Modified from www.epa.gov)

13.3 Household Waste Management in a Typical Municipal Environment

This section identifies the factors affecting waste generation and investigates the awareness of household-level waste management practices in the Pabna municipality of Bangladesh (Mamun and Monzurul 2017). The study was conducted in the Monsurabad residential area of the Pabna municipality (24°00′ N and 89°15′ E) in northwestern Bangladesh, 219 km away from the capital city (Dhaka). It aimed to identify the micro-level socioeconomic and demographic factors affecting the source of waste generation outside of the large cities that are typically studied (Dhaka, Chittagong, Khulna, and Rajshahi) (Hasan 1998; Sufian and Bala 2006; Afroz et al. 2007; Sujauddin et al. 2008; Bhuiyan 2009; Bari et al. 2012a, b; Hossain et al. 2014), even though they contain only 32.03% of total urban population of this country (BBS 2012). The municipal authority of Pabna was formed in 1876. It serves as the home of 144,000 people (BBS 2012) and is one of the major economic hubs of the region. Monsurabad residential area is the only planned and fastest-growing residential area of this municipality. The number of total households is 346, as counted during a reconnaissance survey. The main tool for data collection was a structured questionnaire designed in the local language. For the Monsurabad residential area, the appropriate sample size was determined to be 125 using the methodology developed by Cochran (1977), which has been widely used in other studies (Gomez et al. 2008; Gallardo et al. 2012; Fei-Baffoe et al. 2014). The sample proportion was considered to be 0.15 based on a pilot study prior to the questionnaire survey. The value of the standard variate was taken to be 1.96 per the table of area under the normal curve at a given confidence level of 95%; the acceptable error was 0.05 as the estimation was considered to be within 5% of the true value.

To collect the residential per capita solid waste generation, plastic bags were supplied to each household during the questionnaire survey. Collected waste from each household was segregated from the plastic bags (Enayetullah et al. 2005; Sujauddin et al. 2008; Otoma et al. 2013), weighed, and recorded every day from each surveyed household. Different income groups were identified on the basis of each household's monthly income by surveying and following the methods of relevant studies (Sujauddin et al. 2008; Otoma et al. 2013; Pirani et al. 2015). In addition to household income, the number of residents per dwelling, the household owner's age and education level, length of time at current location, and average hours spent by parents at home per day were selected as determinants of per capita solid waste generation based on a reconnaissance survey and review of relevant literature (Afroz et al. 2007; Sujauddin et al. 2008; Otoma et al. 2013; Aliu et al. 2014; Pirani et al. 2015).

The determinants represent the socioeconomic and demographic characteristics of households and may influence the quantity of waste generation (Sujauddin et al. 2008). To analyze the degree of correlation of the determinants with per capita household solid waste generation (HSWG), Pearson's product moment correlation analysis was performed using Statistical Package for Social Science (SPSS).

Significantly correlated determinants with per capita HSWG were then evaluated with least square regression analysis to identify their potential effects on per capita HSWG. Information regarding managerial aspects, such as the respondent's knowledge on waste segregation and willingness to cooperate with household waste management, were also examined and interpreted. The distribution of the selected determinants among the sample population indicated that the average size of the surveyed households was 4.76, which is in line with the Population and Housing Census of 2011 published by Bangladesh Bureau of Statistics (BBS 2012). Approximately 44% of the surveyed households had a size of 4 persons per dwelling. The average age of the head of the household was around 50 years, and about 30% were between 40 and 50 years of age. The mean years of schooling of the household heads was 11.68, and approximately 30% of them had an education level of 12th class. About 44% of the families lived in their households for a period ranging from 6 months to 5 years; they were mostly renters. Approximately 30% of the families who were homeowners had lived in their households for more than 20 years. The income groups were identified as 0–20,000 bdt, 20,001–40,000 bdt, 40,001–60,000 bdt, and more than 60,000 bdt (1 US$ = 80 bdt). The range of 20,001–40,000 bdt was the most common (48%) among the income groups. The average hours spent by parents at home per day was 15–18 hours (including all activity) for 37.5% of the interviewed households. No information regarding hours spent by parents at home was available for 4% of the sample population. Figure 13.3 shows the distribution of the determinants (independent variables) among the sample population.

Residential (household) waste comprises approximately 30% of the total municipal waste stream in Asia (World Bank 1999). The results of the survey indicate that solid waste generation in the Monsurabad residential area of Pabna Municipality is approximately 1.8 kg per household per day. During survey work, the collected wastes from each household were carefully segregated using the methodology of some empirical studies (Enayetullah et al. 2005; Sujauddin et al. 2008). Table 13.1 presents the household solid waste categories and composition.

The composition of household solid waste is quite diverse in category; organic waste accounts for approximately 76% of the municipal waste composition in Bangladesh (Bari et al. 2012b). The household waste generation rate in Monsurabad residential area of 0.38 kg/day/person is in line with the estimated per capita waste generation rate of six major urban areas of Bangladesh (Enayetullah et al. 2005; Sujauddin et al. 2008). The generation of solid waste per household per day is found to be positively correlated with the size of the household ($r_{xy} = 0.526, p < 0.01$), year of schooling of household owner ($r_{xy} = 0.532$, $p < 0.01$), and monthly income of household ($r_{xy} = 0.829$, $p < 0.01$); the correlations with household owner's age, period of residence at current location, and hours spent by parents at home were statistically insignificant. Table 13.2 shows the correlations between the selected determinants and household solid waste generation.

The correlation between HSWG and year of schooling can be explained by an increase of living standard and income due to the higher education level, which results in more consumption and more waste (Sujauddin et al. 2008). The

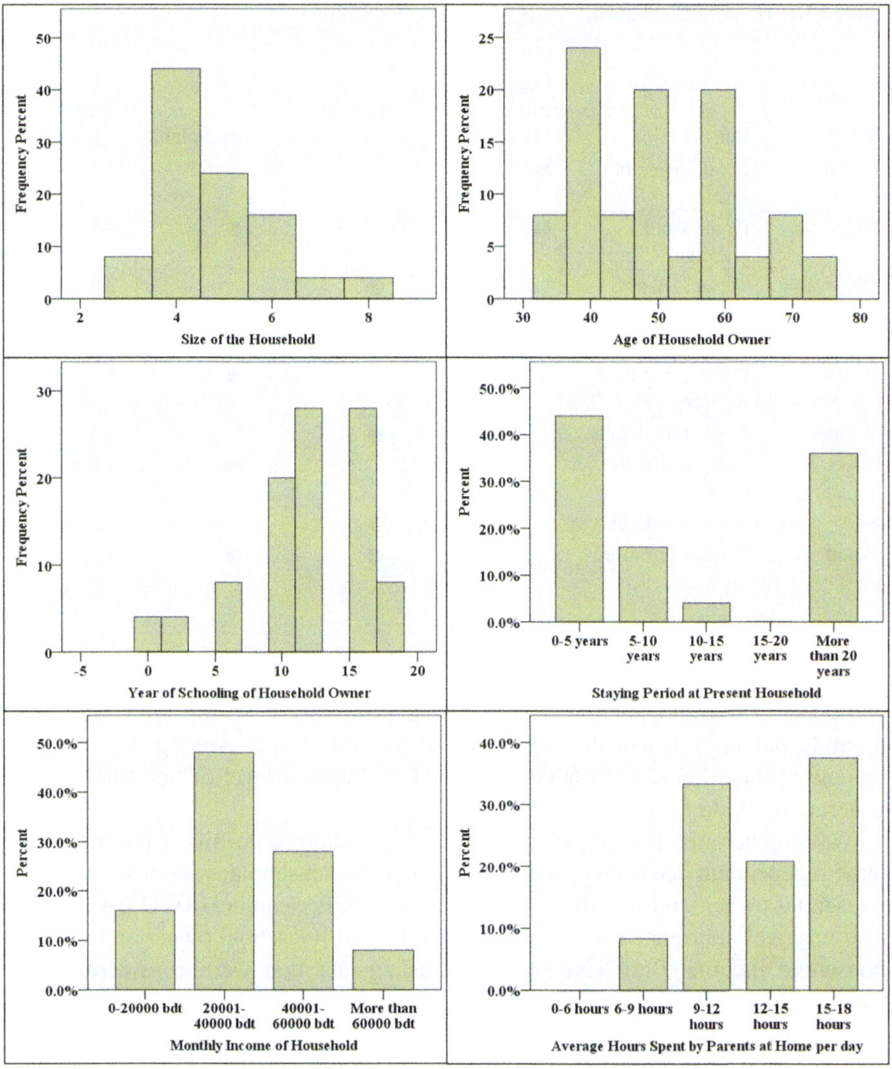

Fig. 13.3 Distribution of the determinants (independent variables) among the sample population

correlation analysis also showed a positive correlation ($r_{xy} = 0.434$, $p < 0.05$) between year of schooling of household owner and monthly income of household. As expected, family size did affect the quantity of household waste at a household because more members result in more consumption. However, in our study sample, the household size did not affect the per capita HSWG (Table 13.2). Per capita HSWG was positively correlated with the year of schooling of the household owner ($r_{xy} = 0.580$, $p < 0.01$) and household monthly income ($r_{xy} = 0.715$, $p < 0.01$). Other

Table 13.1 Household solid waste categories and composition

Type	Item	Category: Organic / Bio-degradable / Green	Inorganic / Disposable	Recyclable	Composition (%)
Green	Food, Vegetable, Fruits	●			72.4
Paper	Paper, Book, Packages	●		●	4.1
Disposable paper	Tissue, Diaper, Container/ Pouch		●		2.7
Metal	Can, Jar, Tin, Metal			●	.3
Plastic	Plastic, Polythene, Rubber			●	5.3
Textile	Rugs, Jute, Cloth		●		1.6
Glass	Glass stuff, Ceramic			●	.5
Soil	Rock, Brick, Dirt		●		11.3
Wood	Broken Furniture		●	●	1.8

indicators were insignificant. This analysis also revealed that per capita HSWG and waste per household per day had a strong positive correlation with household monthly income (statistically significant at the 0.01 level). Among the identified income groups, 'more than 60000 bdt' had the highest level of household waste generation (Table 13.3).

Waste generation per household per day was found to positively correlate with three independent determinants, namely size of the household, year of schooling of household owner, and monthly household income. Per capita HSWG per day was also positively correlated with year of schooling of household owner and monthly household income. Both categories of waste generation were considered to be dependent variables in the linear regression analysis with the correlated independent determinants (Table 13.4).

The significantly fitted linear regression models revealed that the three independent determinants explain 77% of the variance in household waste generation per day, whereas 60% of the variance of per capita HSWG per day can be explained by monthly household income and year of schooling of household owner. In both models, monthly household income was revealed to be the most significant determinant, with p values of 0.000 and 0.001, respectively. The results of bivariate simple regression with household waste generation per day and per capita HSWG where monthly household income was the independent variable indicate that monthly income has an explanation of variance of 68% ($p = 0.000$) in the case of waste generation per household and 51% ($p = 0.000$) of in the case of per capita HSWG.

Like most urban areas of Bangladesh, the municipal authority of Pabna is fully responsible for municipal waste management; currently, it is in a challenging

Table 13.2 Correlations between selected determinants and household solid waste generation

		Size of the Household	Age of Household Owner	Year of Schooling of Household Owner	Period of Residence at Current Household	Household Monthly Income (bdt)	Average Hours Spent by Parents at Home per day	Waste per Household (kgs)	Household Waste Generation per capita per day (kgs)
Size of the Household	*r*	1	0.187	0.159	0.003	0.391	−0.089	**0.526**[a]	0.009
	p		0.372	0.446	0.988	0.053	0.671	**0.007**	0.965
Age of Household Owner	*r*	0.187	1	0.116	0.210	0.077	0.264	0.202	0.148
	p	0.372		0.580	0.313	0.713	0.203	0.332	0.479
Year of Schooling of Household Owner	*r*	0.159	0.116	1	0.000	**0.434**[b]	−0.262	**0.532**[a]	**0.580**[a]
	p	0.446	0.580		0.999	**0.030**	0.206	**0.006**	**0.002**
Period of Residence at Current Household	*r*	0.003	0.210	0.000	1	−0.097	0.120	0.097	0.120
	p	0.988	0.313	0.999		0.646	0.567	0.646	0.569
Household Income (bdt)	*r*	0.391	0.077	**0.434**[b]	−0.097	1	−0.023	**0.829**[a]	**0.715**[a]
	p	0.053	0.713	**0.030**	0.646		0.914	**0.000**	**0.000**
Average Hours Spent by Parents at Home per day	*r*	−0.089	0.264	−0.262	0.120	−0.023	1	0.027	0.089
	p	0.671	0.203	0.206	0.567	0.914		0.897	0.672
Waste Generation per Household (kgs)	*r*	**0.526**[a]	0.202	**0.532**[a]	0.097	**0.829**[a]	0.027	1	**0.839**[a]
	p	**0.007**	0.332	**0.006**	0.646	**0.000**	0.897		**0.000**
Household Waste Generation per capita per day (kgs)	*r*	0.009	0.148	**0.580**[a]	0.120	**0.715**[a]	0.089	**0.839**[a]	1
	p	0.965	0.479	**0.002**	0.569	**0.000**	0.672	**0.000**	

[a]. Correlation is significant at the 0.01 level (2-tailed)
[b]. Correlation is significant at the 0.05 level (2-tailed)
p = Sig. (two tailed).

Table 13.3 Solid waste generation based on monthly income of the surveyed households

Household monthly income groups	Average per capita HSWG (kg)	Average amount of household waste per day (kg)
0–20,000 bdt	0.21	0.87
20,001–40,000 bdt	0.35	1.56
40,001–60,000 bdt	0.50	2.37
More than 60,000 bdt	0.52	3.33

Table 13.4 Results of multiple regression analysis between independent determinants and household solid waste generation

Model	R	R^2	Adjusted R^2	F	P
1	.879	.773[a]	.741	23.88	.000
2	.775	.601[b]	.564	16.554	.000

[a]Dependent: household waste generation per kg per day
Independent: (Constant), size of the household, year of schooling of household owner, monthly household income.
[b]Dependent: per capita HSWG per day
Independent: (Constant), year of schooling of household owner, monthly household income.

situation due to limited resources, technology, and manpower (Mamun and Hossain 2012). The effectiveness of any household waste management program requires effective support from the households; a well-informed and concerned public can significantly facilitate program implementation and ensure success (Sujauddin et al. 2008). This study investigated the awareness of waste management practices at the household level to generate regional-level comprehensive information. For the surveyed area, we found that 83.2% of households received municipal waste management services, with a service-level satisfaction of 57.7%. Table 13.5 shows the distribution of solid waste management aspects among surveyed households.

The investigation revealed that 72.0% of the surveyed households were in favor of receiving waste management services and 75.6% of them were willing to make a monthly payment. The preferred payment was approximately bdt 50 for 55.9% of the families. Before 9.00 AM was the preferred waste collection period from the households. The surveyed families preferred their waste dumping place to be the dustbin (52.0%) or lawn (40%), which differs from some empirical studies (Sujauddin et al. 2008) but is the system that was in use for a long period of time. This common practice also influenced the awareness regarding self-role in household solid waste management, which was only collection for 76.0% of respondents. In all, 64.0% of respondents were knowledgeable about source separation or segregation, and about half of them (47.5%) identified it as a tool for reducing pollution. The waste management system of Pabna is characterized by an outdated collecting and dumping practice with no involvement of local or community-based organizations.

Table 13.5 Distribution of solid waste management aspects among surveyed households

Variable	Frequency	Percentage
1. Receive services from the municipality		
Yes	104	83.2
No	8	6.4
No Information	13	10.4
Total	125	100.0
2. Satisfaction regarding present service		
Satisfied	60	57.7
Not satisfied	44	42.3
Total	104	100.0
3. Willingness to receive service from the municipality/local waste management initiative		
Yes	90	72.0
No	5	4.0
No information	30	24.0
Total	125	100.0
4. Rationality of payment for service		
Rational	68	75.6
Not rational	17	18.9
No information	5	5.5
Total	90	100.0
5. Willingness to pay for service (bdt)		
0–20	21	30.9
21–40	2	2.9
41–60	38	55.9
61–80	0	0.0
81–100	6	8.9
>100	1	1.4
Total	68	100.0
6. Preferable time for waste collection		
Before 9.00 AM	60	48.0
9.00 AM–10.00 AM	23	18.4
10.00 AM–11.00 AM	11	8.8
11.00 AM-12.00 PM	0	0.0
12.00 PM–1.00 PM	0	0.0
1.00 PM–2.00 PM	0	0.0
2.00 PM–3.00 PM	0	0.0
After 3.00 PM	28	22.4
No information	3	2.4
Total	125	100.0
7. Preferable place for dumping		
Open space	5	4.0
To the collector	5	4.0
Lawn	50	40.0

(continued)

Table 13.5 (continued)

Variable	Frequency	Percentage
Dustbin	65	52.0
Total	125	100.0
8. Self-role in household solid waste management		
Collection	95	76.0
Collection and segregation	15	12.0
Recycle	10	8.0
Dispose	5	4.0
Total	125	100.0
9. Preference for waste segregation		
Yes	80	64.0
No	45	36.0
Total	125	100.0
10. Why use segregation?		
Waste reduction	12	15.0
Cost reduction	21	26.3
Pollution minimization	38	47.5
Composting	9	11.2
Total	80	100.0

13.4 Climatic Impact of Municipal Waste Generation

In populous cities of developing countries, HSW has become a serious concern due to rapid population growth, uncontrolled and unplanned urbanization, industrialization, and lifestyle transformation (Ye et al. 2020). These factors have resulted in tremendous generation of MSW in urban areas of Bangladesh. In a typical municipal environment of Bangladesh, HSW generation is proportional to the population, education, and income of the people. Additional factors such as climatic conditions and social/public attitudes also may affect the amount and composition of municipal waste (Nilanthi et al. 2010 cited in Rahman et al. 2010), GHG emissions from the urban waste sector and associated climate change consequences are now receiving global attention, as the increased generation of CH_4 from MSW has encouraged individuals worldwide to implement initiatives for a sustainable MSWM system (Islam 2016a). The waste sector contributes 5% of anthropogenic GHG emissions worldwide (Hammed et al. 2018; IPCC 2006). The municipal area described in this chapter is characterized by 41 tons of MSW generation per day (Mamun and Monzurul 2014), and the municipal authority has an MSWM collection efficiency of less than 55% (Enayetullah et al. 2005). Open drains for municipal discharge, roadside spaces, and potholes are locations for uncollected disposals. Crude dumping may be practiced as the only disposal method. The whole system is unhygienic, and a variety of environmental problems (including GHG emissions) are prevalent.

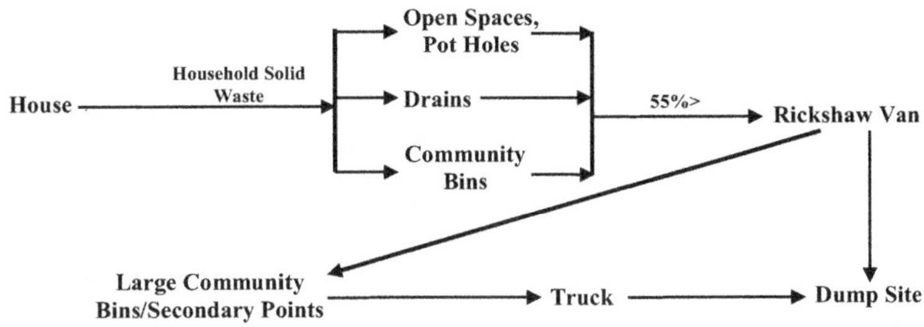

Fig. 13.4 General MSWM system practiced in most municipalities in Bangladesh

Figure 13.4 shows the traditional MSWM system practiced in most municipalities in Bangladesh.

The deplorable situation of the municipal authority is characterized by crude dumping along with increasing rates of waste generation, inadequate waste collection, no utilization of organic waste, no or very limited informal inorganic recycling, lack of awareness among citizens, and no community participation. These factors result in environmental pollution and GHG emissions. With these unhygienic practices, the organic or biodegradable portion of MSW is totally dumped either by the municipal authority through collection or by the individual during disposal. Both are subjected to anaerobic decomposition, which results in CH_4 emission. CH_4 emissions from crude dumping sites were estimated to be approximately 26,800 tons for Dhaka city in the year 2010 (Rahman et al. 2010), 2350 tons for Rajshahi city in 2010 (Mamun and Hossain 2012), and 255 tons for Pabna municipality in 2013 (Mamun and Monzurul 2014), based on IPCC tier 1 calculations (IPCC 2006). When applying IPCC tier 1 for Bangladeshi municipalities, degradable organic carbon was considered to be 15% for a year of MW deposition, the fraction of actual degradable organic carbon was 77%, methane correction factor was 0.4 for unmanaged landfill sites, the fraction of CH_4 in the generated landfill gas was 0.5, and the recovered methane and oxidation factor was 0 due to a lack of burning on dumping sites (Rahman et al. 2010). By assuming the generated per capita HSW was totally dumped and using the (Intergovernmental Panel on Climate Change (IPCC) tier 1 considerations of cited studies, it was estimated that an individual in the Pabna municipality generates 4.26 kg of CH_4 annually by waste disposals.

To improve the waste management system laws and regulations, awareness building, financial incentives and support, improvements in collection and disposal, optimum utilization of resources and manpower, waste minimization, separate collection of organic and inorganic wastes, and regular maintenance of municipal waste infrastructures can be performed by the municipal authority through public–private partnerships. Annual per capita CH_4 emissions can be reduced to a negligible level by adopting certain municipal waste managerial practices, such as source separation of biodegradable waste, curbside and/or door-to-door waste collection with 100% collection efficiency, and cost-effective manual composting of biodegradable/green waste (Mamun and Monzurul 2014). Figure 13.5 presents the generalized waste management plan for municipal areas of Bangladesh.

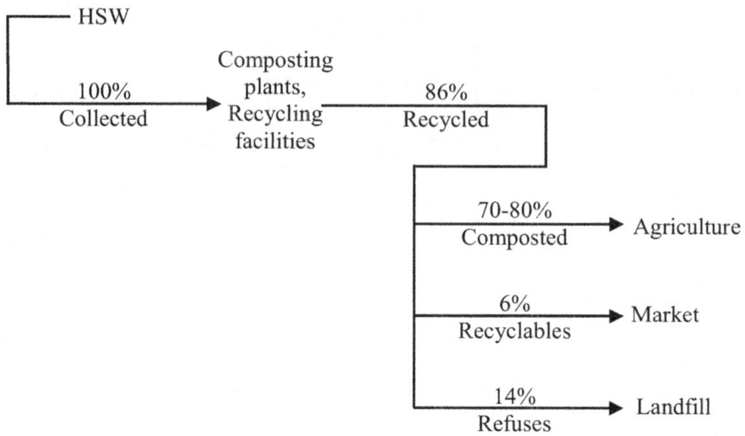

Fig. 13.5 Generalized waste management plan for municipal areas of Bangladesh. (Modified from Mamun and Hossain 2012)

13.5 Conclusion

Low-carbon and environmentally sustainable urban development greatly depends on the community (Islam 2016b). The potential GHG emissions from MSW in Bangladesh can be reduced by adopting feasible techniques, properly implementing plans, and establishing an effective and sustainable waste management system. GHG emission reductions, such as by avoiding crude dumping of biodegradable waste and composting to decrease chemical fertilizer use, are cost-effective and environmentally sustainable waste management approaches. Effective implementation of composting for biodegradable waste may reduce CH_4 emissions by 400 kg per ton (Mamun and Hossain 2012). In the context of global warming and climate change, local governments should rethink MSWM approaches in urban areas of Bangladesh. Furthermore, public education campaigns, adequate training and facilities for the conservancy workers, proper sanitary landfill site selection, and community-based organization involvement in MSWM should be ensured to achieve an improved MSWM system and a healthier municipal environment with reduced GHG emissions.

References

Afroz R, Hanaki K, Hasegawa-Kurisu, k. (2007) Willingness to pay for waste management improvement in Dhaka city, Bangladesh. J Environ Manag 90(1):492–503. https://doi.org/10.1016/j.jenvman.2007.12.012

Alamgir M, Ahsan A (2007) Municipal solid waste and recovery potential: Bangladesh perspective. Iran J Environ Health Sci Eng 4(2):67–76

Aliu IR, Adeyemi OE, Adebayo A (2014) Municipal household solid waste collection strategies in an African megacity: analysis of public private partnership performance in Lagos. Waste Manag Res 32(9):67–78. DOI: 1177/0734242X14544354

Bari QH, Hasan KM, Haque ME (2012a) Solid waste recycling in Rajshahi city of Bangladesh. Waste Manag 32(11):2029–2036. https://doi.org/10.1016/j.wasman.2012.05.036

Bari QH, Hasan KM, Haque ME (2012b) Scenario of solid waste reuse in Khulna city of Bangladesh. Waste Manag 32(12):2526–2534. https://doi.org/10.1016/j.wasman.2012.07.001

BBS (Bangladesh Bureau of Statistics) (2012) Population and Housing Census 2011. Statistics and Informatics Division, Ministry of Planning, Government of the people's republic of Bangladesh. Available at: http://www.bbs.gov.bd/PageWebMenuContent.aspx?Menukey=243

Bhuiyan SH (2009) A crisis in governance: urban solid waste management in Bangladesh. Habitat Int 34(1):125–133. https://doi.org/10.1016/j.habitatint.2009.08.002

Burnley SJ (2007) A review of municipal solid waste composition in the United Kingdom. Waste Manag 29:2779–2786. https://doi.org/10.1016/j.wasman.2006.06.018

Chowdhury DAH, Mohammad N, Haque MRU, Hossain DT (2014) Developing 3Rs (reduce, reuse and recycle) strategy for waste Management in the Urban Areas of Bangladesh: socio-economic and climate adoption mitigation option. IOSR J Environ Sci Toxicol Food Technol (IOSR-JESTFT)., e-ISSN: 2319-2402,p- ISSN: 2319-2399 8(5):09–18

Dehoust G, Schueler D, Vogt R, et al. (2010) Climate protection potential in the waste management sector, federal environment agency (Umweltbundesamt). Available at: https://www.umwelt-bundesamt.de/sites/default/files/medien/461/publikationen/4049.pdf. Accessed 5 June 2020

Enayetullah I, Sinha AHMM, Khan SSA (2005) Urban solid waste management scenario of Bangladesh: problems and prospects. Waste Concern Technical Documentation, Dhaka. Available at: http://www.wasteconcern.org/publication/waste%20Survey_05.pdf

Faccio M, Persona A, Zanin G (2011) Waste collection multi objective model with real time trace-ability data. Waste Manag 31:2391–2405. https://doi.org/10.1016/j.wasman.2007.06.013

Fei-Baffoe B, Nyankson EA, Gorkeh-Miah J (2014) Municipal solid waste management in Sekondi-Takoradi metropolis, Ghana. J Waste Manag:1–9. https://doi.org/10.1155/2014/823752

Gallardo A, Bovea MD, Colomer FJ, Prades M (2012) Analysis of collection systems for sorted household waste in Spain. Waste Manag 32:1623–1633. https://doi.org/10.1016/j.wasman.2012.04.006

Gomez G, Meneses M, Ballinas L, Castells F (2008) Characteristics of urban solid waste in Chihuahua, Mexico. Waste Manag 28(12):2465–2471. https://doi.org/10.1016/j.wasman.2007.10.023

Guerrero LA, Maao G, Hogland W (2012) Solid waste management challenges for cities in developing countries. Waste Manag 33:220–232. https://doi.org/10.1016/j.wasman.2012.09.008

Hammed TB, Wandiga SO, Mulugetta Y, Sridhar MKC (2018) Improving knowledge and practices of mitigating greenhouse gas emission through waste recycling in a community, Ibadan, Nigeria. Waste Manag 81:22–32. https://doi.org/10.1016/j.wasman.2018.09.044

Hasan S (1998) Problems of municipal waste management in Bangladesh: an inquiry into its nature. Habitat Int 22(2):191–202. https://doi.org/10.1016/s0197-3975(97)00039-8

Hossain HMZ, Hossain QH, Monir MMU, Ahmed MT (2014) Municipal solid waste (MSW) as a source of renewable energy in Bangladesh: revisited. Renew Sust Energ Rev 39:35–41. https://doi.org/10.1016/j.rser.2014.07.007

IPCC (2006) IPCC guidelines for national greenhouse gas inventories. In: Eggleston HS, Buendia L, Miwa K, Ngara T, Tanabe K (eds) Prepared by the National Greenhouse gas Inventories Programme. IGES, Japan

Islam KMN (2016a) Municipal solid waste to energy generation in Bangladesh: possible scenarios to generate renewable Electricity in Dhaka and Chittagong City. J Renewable Energy. https://doi.org/10.1155/2016/1712370

Islam KMN (2016b) Greenhouse gas footprint and the carbon flow associated with different solid waste management strategy for urban metabolism in Bangladesh. Sci Total Environ. https://doi.org/10.1016/j.scitotenv.2016.12.022

Kabir MR (2016) Municipal solid waste management system: a study on Dhaka north and South City corporations. J Bangladesh Inst Planners 8(2015 (Printed in December 2016)):35–48

Konteh FH (2009) Urban sanitation and health in the developing world: reminiscing the nineteenth century industrial nations. Health Place 15(1):69–78. https://doi.org/10.1016/j.healthplace.2008.02.003

Kumar A, Sharma MP (2013) Estimation of GHG emission and energy recovery potential from MSW landfill sites. Sustain Energy Technol Assessments 5:50–61. https://doi.org/10.1016/j.seta.2013.11.004

Lackner KS, Jospe C (2017) Climate change is a waste management problem. Issues: Sci Technol XXXIII(3) Spring

Mamun AHMM, Hossain MA (2012) Green waste management and climate change: case of Rajshahi City Corporation. Oriental Geogr 53:55–72

Mamun AH, Monzurul M (2014) An approach towards sustainable city environment: critical analysis of municipal solid waste management of Pabna municipality of Bangladesh from the perspective of resource recovery and environmental conservation. Sci Res J 2:33–37

Mamun AH, Monzurul M (2017) Characteristics of household solid waste and its Management in Pabna Municipality. Rajshahi Univ J Environ Sci 6:10–17. Rajshahi, Bangladesh

Marshall RE, Farahbakhsh K (2013) System approaches to integrated solid waste management in developing countries. Waste Manag 33:988–1003. https://doi.org/10.1016/j.wasman.2012.12.023

Moora H, Urbel-Piirsalu E, Voronova V (2017) The climate change impact of possible municipal Solid waste management scenarios in Estonia. Linnaeus Eco-Tech 78. https://doi.org/10.15626/Eco-Tech.2010.009

Nilanthi JG, Pattnaik S, Reddy MV (2010) Assessment of municipal solid waste Management in Puducherry (Pondicherry), India. Resour Conserv Recycling 54(8):512–520

Oribe-Garcia I, Kamara-Esteban O, Martin C, Macarulla-Arenaza AM, Alonso-Vicario A (2015) Identification of influential municipal characteristics regarding household waste generation and their forecasting ability in Biscay. Waste Manag 39:26–34. https://doi.org/10.1016/j.wasman.2015.02.017

Otoma S, Hoang H, Hong H, Miyazaki I, Diaz R (2013) A survey on municipal solid waste and residents' awareness in Da Nang city, Vietnam. J Mat Cycles Waste Manag 15:187–194. https://doi.org/10.1007/s10163-012-0109-2

Pirani SI, Al-Khatib IA, Halaweh R, Arafat MA, Arafat HA (2015) Household level determinants of residential solid waste generation rates: a study from Nablus-Palestine. J Mat Cycles Waste Manag 17:725–735. https://doi.org/10.1007/s10163-014-0304-4

Puppim de Oliveira JA (2018) Intergovernmental relations for environmental governance: cases of solid waste management and climate change in two Malaysian states. J Environ Manag 233:481–488. https://doi.org/10.1016/j.jenvman.2018.11.097

Rahman S M, Shams S, Mahmud K (2010) Study of solid waste management and its impact on climate change: A case study of Dhaka City in Bangladesh. Proceedings of ICEAB10

Singh J, Laurenti L, Sinha R, Frostell B (2014) Progress and challenges to the global waste management system. Waste Manag Res 32(9):800–812. https://doi.org/10.1177/0734242X14537868

Stocker TF et al (2014) Climate change 2013: the physical science basis. Cambridge University Press, Cambridge, UK/New York

Sufian MK, Bala BK (2006) Modelling of urban solid waste management system: the case of Dhaka city. Waste Manag 27(7):858–868. https://doi.org/10.1016/j.wasman.2006.04.011

Sujauddin M, Huda MS, Rafiqul Haque ATM (2008) Household solid waste characteristics and management in Chittagong, Bangladesh. Waste Manag 28:1688–1695. https://doi.org/10.1016/j.wasman.2007.06.013

Weitz K, Thorneloe SA, Nishtala SR et al (2002) The impact of municipal solid waste management on greenhouse gas emissions in the United States. J Air Waste Manag Assoc 52:1000–1011

World Bank (1999) What a waste: solid waste management in Asia. Washington DC, pp 43

Ye Q, Anwar MA, Zhou R, Asmi F, Ahmad I (2020) China's green future and household solid waste: challenges and prospects. Waste Manag 105:328–338. https://doi.org/10.1016/j.wasman.2020.02.02

Chapter 14
Challenges for Char Dwellers from Riverine Floods Due to Climate Change: A Case Study from Northern Bangladesh

Md. Shamim Hossain and Md. Habibur Rahman

Abstract Bangladesh is a riverine country that is geographically located in a disaster-prone area where flooding is a very common. Riverine islands are extremely vulnerable to floods and are where char dwellers suffer the most. Flooding has short-term and long-term impacts on their livelihoods, earnings, and lifestyles. Therefore, this chapter analyzes the impacts of climate change on river floods and scrutinizes the adaptation techniques of char dwellers to river floods. The study follows mixed methods of research, using both qualitative and quantitative methodology, with a convenient sampling method. This study also correlates social capital theory with the objectives of the research. In the last 50 years, Bangladesh monsoon rainfall has increased by 2.65 millimeters per year, and the monthly average river discharge of Brahmaputra has increased by about 8%. Furthermore, the temperature of the country is likely to increase by 1.4–1.6 °C by 2050. These changing patterns of rainfall, river discharge, and temperature show the possible impacts of climate change and accelerate the intensity and frequency of river floods. Approximately 58% of char dwellers are considered to be seasonal migrants. To cope up with floods, char dwellers depend on the availability of their resources, practicing individual adaptation strategies, making soil mounds to protect fruits and vegetables, designing flood-tolerant paddy, and cultivating fast-growing fish (carp,

M. S. Hossain (✉)
Department of Geography and Environmental Science, Begum Rokeya University, Rangpur, Bangladesh
e-mail: shamimgeodu@gmail.com

M. H. Rahman
60 Decibels, Inc., USA

The Daily Law News (Dainik Ain Barta), Dhaka, Bangladesh

Department of Public Administration, Dr. Wazed Research and Training Institute, Begum Rokeya University, Rangpur, Bangladesh
e-mail: habibpacs@gmail.com

M. N. Islam, A. van Amstel, *Bangladesh II: Climate Change Impacts, Mitigation and Adaptation in Developing Countries*, Springer Climate,
https://doi.org/10.1007/978-3-030-71950-0_14

tilapia). They are receiving economic and logistical support from the concerned authorities. Moreover, they collaborate socially when encountering floods. The existing bonding, bridging, and linking mean that char dwellers cooperate more with their neighbors than with administration, donors, and other stakeholders. The study suggests that social capital is a new dimension since the 1970s to cope with any kind of natural disaster, including floods.

Keywords Char dwellers · Riverine floods · Climate change · Adaptation techniques · Bangladesh

14.1 Introduction

Climate change affects long-term trends in temperature, moisture, atmospheric pressure, airstream, rainfall, atmospheric particle count, and different meteorological measurements in an area (Rahman 2013). The impacts of climate change are not confined to only one country or region; rather, it has effects across the globe. Researchers argue that the marine ecosystem has been changed by anthropogenic climate change as well as atmospheric and oceanographic conditions (Sumaila et al. 2011). The impacts of climate change increase the migration of people in developing countries. People of Bangladesh are changing their mitigation and adaptation processes for climate-induced challenges (Bose 2016).

The flow and water quality of rivers have been changed by the impacts of climate change. In addition, socio-economic conditions are heavily affected by patterns of climate change. Monsoon flows are expected to increase from climate change after evaluating the results from the 2050 and 2090 models (Whitehead et al. 2015, pp. 1–6). Although Bangladesh emits relatively low amounts of global greenhouse gases, it is one of the most affected area from climate change (Rawlani and Sovacool 2011, pp. 845–863). The impacts of climate change and flooding have negative impacts on livelihood, agriculture, livestock, poultry, water, health, sanitation, and food security in northern and other parts of Bangladesh, according to German NETZ[1] and the Bangladesh Centre for Advance Studies (The Daily Star 2020).

Geographically, Bangladesh is situated in a natural disaster area (Roy et al. 2014, pp. 59–63). It is flood-prone because of its location on the Ganges Delta, which is fewer than 5 meters above the mean sea level; 80% of Bangladesh is classified as floodplain (Brammer 1990). Flood is one of the climate-induced disasters by which people are heavily affected (Rawlani and Sovacool 2011, pp. 845–863).

Bangladesh has experienced four unique floods and two dangerous floods from 1954 to 2010 (Ali 1996, cited in Rawlani and Sovacool 2011, pp. 845–863; Chowdhury et al. 1993, cited in Rawlani and Sovacool 2011, pp. 845–863; Haque

[1] NETZ, the Partnership for Development and Justice, is a German international NGO that is implementing voluntary and humanitarian activities in Bangladesh and providing assistance to marginalized people (The Daily Star 2020).

1997, cited in Rawlani and Sovacool 2011, pp. 845–863). Bangladesh is considered to be the sixth most flood-affected state due to frequent floods across the globe (UNDP 2004, cited in Azad et al. 2013, pp. 190–199). One third of Bangladesh is highly affected by floods every 10 years (DMB 2008, cited in Azad et al. 2013, pp. 190–199). Generally, floods often occur in July and August each year (Brammer and Khan 1991 cited in Azad et al. 2013, pp. 190–199).

Public health is heavily affected by floods with regard to the unavailability of basic foods and services. As a result, people experience a variety of diseases (Rawlani and Sovacool 2011, pp. 845–863). Floods destroy agricultural crops and damage personal property. Communicable and non-communicable diseases and loss of human lives are associated with floods (Azad et al. 2013, pp. 190–199).

Some districts of northern Bangladesh, such as Gaibandha, Kurigram, Rangpur, and Dinajpur, have been categorized as disaster-prone areas where flood and river bank erosion are regular problems. Char dwellers are extremely vulnerable to floods. The inhabitants of char lands[2] face much economic and personal loss from the floods (Mallick and Hossain 2019). Thus, char dwellers desperately need to adapt to floods. Therefore, northern Bangladesh is an important area to study the impacts of and adaptation techniques for floods. Char dwellers elsewhere in the country can also benefit from this research on how to cope with floods. Environmentalists, geographers, academicians, students, policy makers, gender specialists, feminist scholars, and domestic and international researchers may be interested in the present study. The objectives of the study are to analyze the impacts of climate change on river floods and scrutinize the adaptation techniques of female inhabitants of char lands to river floods.

14.2 Research Methods

This study used both qualitative and quantitative research methodology in which researchers collected primary data and information from the field. Researchers also used direct observations during the data collection process to cross-check data analysis. This study analyzed data in a persuasive[3] approach for maintaining correlations with the theoretical framework and objectives of the research. The present study includes 460 char dwellers of 37 different char lands in the five districts of Rangpur, Kurigram, Gaibandha, Dinajpur, and Lalmonirhat within 15 days from May 16, 2020 to May 30, 2020 (Map 1).

This study used a convenient sampling method of five districts out of the eight districts in the Rangpur Division. Most respondents were between 19 and 70 years of age. Primary data were collected from char dwellers of particular char lands using open-ended and close-ended questions. This paper uses a purposive sampling technique to attain the desired respondents within a short time in order to gather

[2] Also called riverine islands.

[3] Persuasive is a type of academic writing that develops a logical argument (EDU 2019).

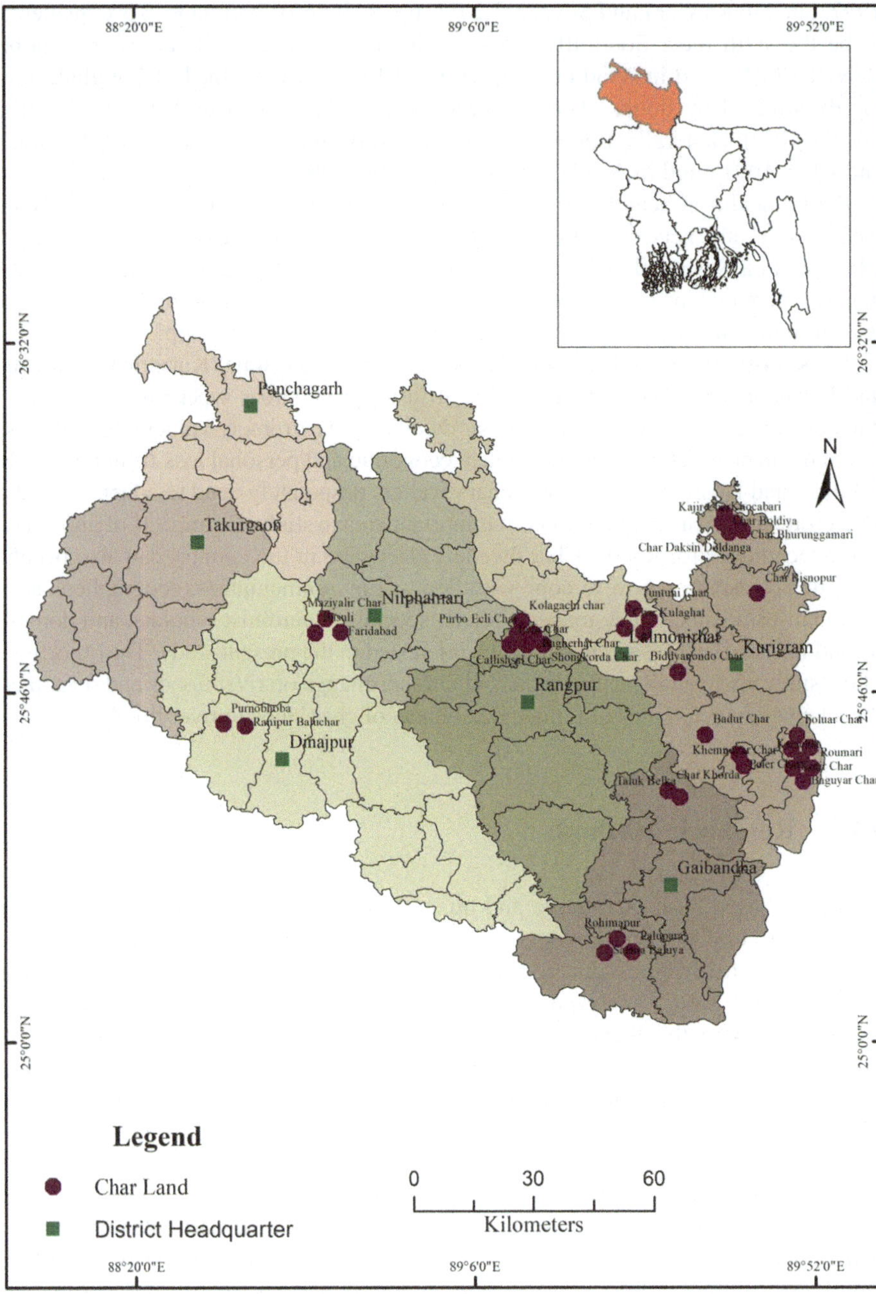

Map 1 Char Lands of Rangpur Division, Bangladesh

authentic data. Secondary data were gathered from peer-reviewed journals and publications, books, newspapers, articles, periodicals, websites, and other sources.

14.2.1 Study Area

The northern part of Bangladesh includes Rajshahi Division and Rangpur Division. This study was conducted in the five districts of Rangpur Division: Rangpur, Kurigram, Gaibandha, Dinajpur, and Lalmonirhat (Map 1). Rangpur Division is located between 25°50′N and 89°00′E; it is 1634.37 square kilometers, with a population density of 848 per square kilometer (Bangladesh National Portal 2020). The Division has a humid subtropical monsoon climate with high temperatures, considerable humidity, and heavy rainfall.

Almost every year, a flood hits the northern part of Bangladesh because of temperature increases, changes in rainfall patterns (long periods of steady rainfall persisting over several days, and high-intensity rainfall of short duration) during the monsoon. Moreover, the Brahmaputra river basin is located there and receives four major tributaries: the Dudkumar, the Dharla, the Teesta, and the Hurasagar (FFWC 2017). Therefore, it is the perfect ground for studying the adaptation techniques of char dwellers for river floods.

14.3 Literature Review

Some international scholars have researched the impacts and adaptations associated with climate change and floods. Researchers have reviewed literature relevant to this study, including the adaptation strategies of char dwellers to floods. Fatma Denton, Nigel W. Arnell and Simon N. Gosling, B.K. Roy, M.B. Ullah, M. H. Rahman, Shakibul Islam, Abul Kalam Azad, Khondker Mokaddem Hossain, and Mahbuba Nasreen have investigated the topic.

14.3.1 International Adaptive Strategy for Climate Change and Floods

Denton (2002, pp. 10–20) assessed that gender inequalities are insidious in developing states for some causal factors, although it varies from country to country and continent to continent for several reasons. Most African women remained poor and in vulnerable condition where they contributed 80% to the agricultural sector. Thus, women should also be considered in future developments to ensure sustainable development as floods affect them adversely along with men.

Arnell and Gosling (2016) discussed the implications of climate change for global river flood risk. This study highlighted four indicators of flood vulnerability. There was a change in the return period and scale of flood climaxes. The current 100-year flood was predicted to occur at least twice as frequently across 40% of the world in 2050. Thus, 450 million flood-prone people and 430,000 square kilometers of flood-prone cropland would be exposed to twice the flood frequency and global flood risk, which would increase by approximately 187%. There is also strong regional variability, such as in Asia, and considerable variability among climatic models.

14.3.2 Bangladesh Adaptation Strategies of Char Dwellers Due to Climate Change and Floods

Roy et al. (2014, pp. 59–63) studied adaptations to climate change in the char lands of the Sirajganj district in Bangladesh. The objectives of the study were to explain the climatic parameters, such as temperature and rainfall patterns, and adaptations to climate change in a farming community using focused group discussions, direct observations, and primary data collected from the field. Researchers found that children and elderly people are the most vulnerable groups in char lands; they face many problems, such as unsafe drinking water and river bank erosion. Farming communities have some adaptation techniques, such as repairing embankments and roads, to cope with climate change.

Islam (2016, pp. 169–179) assessed the coping techniques of char land people for flood calamities in the Kurigram district of Bangladesh. The objectives of that research were to explore the preparedness programs, coping techniques, flood emergency response of the local people, and the causes and history of disasters in Kurigram. That study revealed the poor economic conditions of char lands people, but governmental and nongovernmental organizations (NGOs) provide assistance to the vulnerable groups.

Currently, the women of char lands are especially being deprived of some fundamental human rights and healthcare services. They often walk long distances to obtain healthcare services. A study by Unnayan Shamannay[4] revealed that 70% of pregnant women are unable to obtain healthcare services, whereas 85% have no access to postnatal medical care in char lands, especially in Nilphamari and Lalmonirhat districts (The Daily Star 2009). Women are in the most vulnerable position because they are closely attached to the household chores, including cooking, childcare, and water collection (Nasreen 1995 and 2009, cited in Azad et al. 2013, pp. 190–199). The status and capability of women are evaluated by the patriarchal structure of society (Azad et al. 2013, pp. 190–199).

[4] Unnayan Shamannay is a non-profit research organization (Usshamunnaybd, n.d.).

Women are also vulnerable to domestic violence and harassment at shelters during floods. One study examined the flood-prone vulnerabilities and problems encountered by women in northern Bangladesh using mixed methods of research. The objectives of the study were to examine the vulnerabilities and livelihoods of marginalized women during and after a flood in Sirajganj of Rajshahi Division. The study indicated that 25% of the respondents suffered from physical injuries and 85% were affected by various diseases (Azad et al. 2013, pp. 190–199). People are keeping various types of crops and seeds to cope up with natural calamities e.g. floods. Thus, they try to reduce their poverty and to earn money for their daily lives (Kader 2007).

Our study focuses on the adaptation strategies of char dwellers for river floods in Rangpur, Kurigram, Gaibandha, Dinajpur, and Lalmonirhat districts. Previous studies did not follow any theory or model to examine the data from the viewpoint of environmental sciences and geography; our study uses "social capital theory" to evaluate the data in a scientific manner. We found a research gap on the adaptation of char dwellers to river flood in our literature reviews, with regard to objectives of the research, methodology of the studies, and the findings of the research. Therefore, the objectives of the present study are to analyze the impacts of climate change on river floods and scrutinize the adaptation techniques of char dwellers to river floods.

14.4 Theoretical Underpinnings

Researchers have already defined some phenomena included in the study, including char lands, climate change, river flood, and adaptation. Char lands are riverine islands; generally, they are short-term sand islands that are shallow and different from the mainland. Char dwellers are the most vulnerable groups to floods and erosion (Roy et al. 2014, pp. 59–63). Char land is one of rarest assets for Bangladeshi people because resource limitations are common everywhere (Kader 2007). Nearly 6.5 million people live in the northern part of Bangladesh along the Brahmaputra (RDCD 2015); this study focuses on the char lands of Rangpur, Kurigram, Gaibandha, Dinajpur, and Lalmonirhat.

Climate change refers to any type of alteration to the climate (either an increase or a decrease) over the time due to natural variability (e.g. solar variability, astronomical effects, tectonic processes, volcanic eruptions), anthropogenic activities (e.g. greenhouse gas emissions, deforestation, urbanization), or both (Pielke 2004).

River floods are common phenomenon in Bangladesh. During the monsoon season, river floods occur annually because of intensive river inflow and excessive rainfall, which causes the water levels in the rivers to rise and fall (FFWC 2020). River floods often occur from the spilling of major rivers and their tributaries. The distributaries rise and fall slowly over 10–20 days or more and can cause widespread harm to property and loss of lives (Mirza 2002). Common causes of river floods include excessive rainfall, long duration of rainfall, melting of snow, steep slopes, impermeable rocks,[5] saturated soils, compacted or dry soils, urbanization,[6] and deforestation.[7]

Adaptations may be carried out in response to or in anticipation of changes in conditions to survive. They enhance resiliency and enforce changes to protect against unpleasant impacts (Ahmed et al. 1999). Disaster often occurs naturally; it cannot be stopped by human beings. Therefore, adaptation strategies are necessary for survival. Developing countries such as Bangladesh need to focus more on adaptation than mitigation because adaptation techniques are not very costly. An adaptation technique of char dwellers (social capital) helps them to cope with floods in the char lands of Rangpur, Kurigram, Gaibandha, Dinajpur, and Lalmonirhat.

The term *social capital* was used for the first time by two theorists, Bourdieu and Coleman, who are the pioneers of the social capital theory (Hauberer 2011). Social capital refers to trust, norms, and networks that enhance social collaboration and harmonization for mutual wellbeing (Pelling 1998). Social capital is a network or linkage among people who are living and working together for mutual benefit (Putzel 1997). Social linkages are associated with a particular social group that is facing a specific developmental problem, such as local vulnerability to floods. Social capital reflects both the quantity of social cooperation[8] and the quality of this organization[9] (Pelling 1998). Additionally, adaptation techniques may be adopted by individuals and societies that are closely linked to the form and availability of their resources. Thus, vulnerability and adaptation strategies are manipulated by livelihood, community structure, social group, household structure, age, ethnicity, historical time, and physical or psychological health (Chambers 1989; Burton et al. 1993).

Adams (1995) stated that access to data and information shapes peoples' relationships with disasters, such as floods. Information is also vital to local decision-making; the individuals who hold information stay in central positions. Moreover, access resources or negotiated political structures are likely to be best placed to obtain resources and enhance security for themselves or others. Conversely, vulnerable families have been shown to continue their dependency on individual adaptations during floods. Therefore, underdeveloped civil society, weak local government, and failure of the participatory methods can be inclusive or representative of the most marginalized and vulnerable. There is also a failure of participatory methods to fully incorporate the poorest, most marginalized, and socially isolated members of local communities. Char dwellers, especially women, were acknowledged to have passive and active resistance to imposed models of community. Consequently, social capital and vulnerability are associated with the loss of an opportunity to

[5] It does not allow water to pass through.

[6] Towns and cities have more impermeable surfaces.

[7] Deforestation is a process to remove of trees, which reduces the amount of water intercepted and increases runoff.

[8] Responses to cope with environmental hazards take place individually within households and collectively between households.

[9] It refers to the inclusiveness, transparency, and accountability of decision-making institutions.

develop local communities (Pelling 1998). Additionally, adaptation strategies to climate change are continuously being restructured through social relationships.

There is an opportunity to incorporate a social capital framework in climate change. Local communities have begun to investigate the role of institutions, culture, and political change in individual and collective adaptations to flooding. Bridging, bonding, and linking are some key indicators of social capital theory. Bonding ties are common interests among the same groups, such as ethnic or religious groups. Social relationship exchanges can be used for bridging opportunities in a community. Linking is a sub-category of bridging; cross-group boundaries occur in a vertical direction, such as between social classes or community groups and donors (Pelling 1998, 2002; Adger 2003; Putnam 2000; Woolcock and Narayan 2000). Therefore, the triangular association among bridging, bonding, and linking capital in a social classification can help us understand the direction and speed in which adaptation unfolds. Strong bonding ties are more closely associated with survival than development. Anyone can manage to recover from natural disaster and conflict using strong bonding (Pelling 2003). Social capital is a process that correlates social networks and norms in the production of adaptive capacity among the collectives. It has brought about material adaptations with the help of mobilization and teamwork to raise the level of river embankments, invest in children's education, and increase familial resiliency to future socio-economic risks. With the aim of generating access to resources for the future material interventions, an individual can build her or his social capital link (Pelling and High 2005).

14.5 Impacts of Climate Change on River Floods in Bangladesh

Due to its geographical location, climatic patterns, and confluence of three large rivers (the Ganges, the Brahmaputra, and the Meghna), Bangladesh is one of the most vulnerable states in the world in terms of climate change. The country is characterized by a sub-tropical monsoon climate and is located at the foot of the Himalayas, which is also the highest precipitation zone in the world. This rainfall is caused by the influence of the southwest monsoon and has resulted in higher frequencies of disastrous floods in the country (Huq et al. 1996). Cherrapunji in India is the highest rainfall area in the world, and it is located a few kilometers northeast of the Bangladesh border (FFWC 2018). Therefore, river floods have changed dramatically with the overall impacts of climate change.

Bangladesh has been experiencing the adverse impacts of global warming and climate change, such as summer becoming warmer and wetter, irregular monsoons, early rainfall, heavy rainfall over short time periods, increased river flow, and inundation during heavy rain. Therefore, river floods increased due to the frequency, intensity, and recurrence factors. Hossain et al. (2012) reported that snows of the

Table 14.1 History of Catastrophic Floods in Bangladesh

Year	Flood-Affected Area		Socio-economic Impact of Flood	
	Area (km^2)	Percent (%)	Deaths	Damage (Billions of US$)
1974	52,600	36	2000	NA
1987	57,300	39	2055	1.0
1988	89,970	61	6500	1.2
1998	100,250	68	1100	2.8
2004	55,000	38	700	6.6
2007	62,300	42	1100	1.1
2017	61,979	42	145	NA

Source: Hossain et al. 2012; Rawlani and Sovacool 2011, pp. 845–863; DDM 2017

high Himalayans are melting; deep monsoon rainfalls are occurring over the Himalayans and the Assam Hills due to climate change, which are accelerating the river floods in Bangladesh. An analysis of the last 50 years of rainfall data showed that monsoon rainfall in Bangladesh increased by 2.65 millimeter per year and monthly average discharge of Brahmaputra increased about 8% in the last 50 years. These changing patterns of rainfall and flooding scenarios in Bangladesh are probable impacts of climate change (Ali et al. 2013).

Every year, floods occur in Bangladesh. Consequently, people are killed and homes, vegetation, and roads are destroyed. The floods have caused great devastation in Bangladesh throughout history, especially in the years 1974, 1987, 1988, 2004, 2007, and 2017 (Table 14.1). Moreover, Bangladesh faces the cumulative effects of floods due to the change of climate.

When water overflows the banks of rivers, it results in inundation. Severe floods result in water entering adjacent residential and commercial areas. During the historical floods of Bangladesh, the flow of river water always exceeded the dangerous level and lasted for days (Table 14.2).

Model-based predictions of future climate change for Bangladesh reveal that there is an increasing pattern in both mean annual temperature and rainfall. Results of General Circulation Model (GCM) simulations indicate that both mean annual and seasonal temperatures of Bangladesh are likely to increase on the order of 1.0–1.1 °C and 1.4–1.6 °C by 2030 and 2050, respectively. GCM also reveals that mean annual rainfall in Bangladesh will increase 6% by 2050. The winter dry season is predicted to become significantly drier due to negative increases of rainfall, whereas the wet season will see a 6–8% increase of rainfall between 2030 and 2050, respectively (Table 14.3). The intensification of the hydrological cycle will increase, causing heavy rainfall in the coastal regions and resulting in widespread flooding of Bangladesh at the end of 21st century due to global warming and climate change (Christensen et al. 2007).

Table 14.2 Water level (in meters) during 2017, 1998, and 1988 floods at stations along the Brahmaputra Basin

River	Station	Danger Level	Peak Level			Days above Danger Level		
			2017	1998	1988	2017	1998	1988
Dharla	Kurigram	26.50	27.84	27.22	27.25	12	30	16
Teesta	Dalia	52.40	53.05	52.20	52.89	6	NA[a]	8
Jamuneswari	Badarganj	32.16	33.61	33.00	32.80	8	6	5
Ghagot	Gaibandha	21.70	22.55	22.30	22.20	15	51	17
Brahmaputra	Chilmari	24.00	24.87	24.77	25.04	14	22	15
Jamuna	Bahadurabad	19.50	20.84	20.37	20.62	25	66	27
Jamuna	Sirajganj	13.35	14.87	14.76	15.12	33	48	44
Jamuna	Aricha	9.40	10.16	10.76	10.58	12	68	31
Old Brahmaputra	Jamalpur	17.00	17.01	17.47	17.83	1	31	8
Lakhya	Narayanganj	5.50	5.74	6.93	6.71	10	71	36

Source: FFWC 2017
[a]Not available

Table 14.3 Temperature and Rainfall Change Scenarios for Bangladesh Based on GCM

Year	Period	Increases of Temperature (°C)	Increases of Rainfall (%)
2030	Mean Annual	1.0	5
	Summer (June–August)	0.8	6
	Winter (December–February)	1.1	−2
2050	Mean Annual	1.4	6
	Summer (June–August)	1.1	8
	Winter (December–February)	1.6	−5

Source: Rashid and Paul 2014

14.6 Adaptation Techniques of Char Dwellers to River Floods

Char dwellers have followed different types of adaptive capacities to cope with floods. This section focuses on the following issues of char dwellers adaptive capacities, including material adaptations regarding educational qualification and knowledge; the availability of individual resources, such as savings money and storage of foods; maintenance of movable and personal properties; the role of institutions considering knowledge of children about disaster management; access to information along with flood forecasting information; weak local government together with inactiveness of hotline numbers; using different types of tools and vehicles for continuing communications and stable networks; practicing individual adaptation in the way of seasonal migration from origin; structural changes to their residences; safety training and capability of swimming; getting social cooperation and collective adaptation with economic and logistics support; taking long-term strategy for saving crops and fishes; cultivating paddy, farming fishes, farming cattle (recapturing strategies from cattle diseases); eating balanced diets and ensuring health status of char dwellers along recovering from diseases; addressing women's health risks during floods; obtaining healthcare for women with regard to menstruation and pregnancy; ensuring women's psychological health after marital breakup; preference of apparel during floods; women's participation in decision making; taking short-term and long-term adaptive strategies; social collaboration when encountering floods; and reviewing the connection between social capital and flood adaptation.

14.6.1 Material Adaptation

Material adaptation simply focuses on educational qualification and acquisitioning of flood management knowledge. Generally, education plays a vital role in coping with natural disasters. The literacy rates of selected districts differ and are below 53%, whereas the country's literacy rate is 73.9% for those 15 years and older (Alamgir 2019). The literacy rate for both men and women (7 years and older) is 52.42% in Dinajpur (BBS 2015a), 42.52% in Kurigram (BBS 2015b), 46.09% in Lalmonirhat (BBS 2015c), 48.6% in Rangpur (BBS 2015d), and 42.81% in Gaibandha (BBS 2015e).

Because the literacy rate of this region is below average, it is assumed that people also lag behind in social awareness and capacity building for resilience. Therefore, their adaptation knowledge is affected. Adaptation techniques for disasters sometimes depend on educational background, and people of different educational background live in the char lands. Some of them are illiterate, whereas others are literate. Lack of job opportunities, economic underdevelopment, and natural disasters hinder their education. Illiterate or less educated people often face some challenges in adapting themselves to floods due to lack of adequate knowledge; for example, they are not used to receiving opportunities from Information and Communication Technology (ICT) and other media. Educated people utilize their knowledge to adapt themselves to floods by easily connecting with ICT.

People also are spending money to build up embankments and protect themselves from floods. Again, some respondents initiate some plans to repair local barrage to save their personal properties and themselves. Stakeholders repair embankments, roads, and bridges of the flood-affected area. This is another good indicator of material adaptation for flood.

14.6.2 The Role of Institutions

Institutions can enhance the knowledge of children about disaster management and preparedness for forthcoming natural disasters. Disseminating disaster management knowledge to children is essential because children can play pro-active roles as a volunteer. However, the authority is not addressing the issue accordingly. Large portions (87.7%) of children of char lands do not receive any lessons from formal and informal educational institutions regarding coping with and protecting themselves from disasters. Only 12.3% of children received some knowledge (e.g., flood adaptation techniques, disaster response) from conventional (i.e., school) and non-conventional (e.g., grandparents, neighborhood) educational institutions about disaster management and prevention. Some students of char lands drop out of educational institutions during sudden floods. When those learners continue their studies, they face unorganized sessions. After the floods, they cover the syllabus with extra classes and some additional facilities provided by educational institutions. The authorities should provide more support and take a pro-active role in avoiding any kind of session disorganization and dropout from educational institutions due to floods.

14.6.3 Availability of Individual Resources

People need to store foods and other relevant grocery items in preparation for floods, in addition to saving money for flood prevention strategies. Char dwellers of Rangpur, Kurigram, Gaibandha, Dinajpur, and Lalmonirhat save money and store foods before floods in order to survive. On average, each family saves 5680 BDT[10] and stores different types of food items before floods, such as rice, molasses, pulse, Khai,[11] biscuits, medicine, potatoes, salt, sugar, rice powder,[12] flour, wheat, and maize. Basically, they store food items that can be preserved for a long time. Nearly 61% of char dwellers are able to store the necessary food items for their family in order to fulfill their basic needs during and after floods. However, 38.90% of the respondents were not able to store basic food items for their economic constraints and other factors for survival.

Maintenance of movable and personal properties depends on the availability of individual resources. People need to preserve and protect their movable and personal properties before disasters (e.g., flood) to avoid economic loss and future uncertainty. Therefore, char dwellers of Rangpur, Kurigram, Gaibandha, Dinajpur, and

[10] Bangladesh Taka

[11] A Bengali fried rice food item.

[12] A Bangladeshi food item named Chalergura

Lalmonirhat use some local adaptive strategies. They keep their properties on a high platform, on the roof of their house or building, in a temporary storage room at a temporary flood/cyclone shelter or neighbors' house, on the road or on boats, etc. Sometimes, they move their properties to their parents' home, relatives' building, or flood-free areas. In some cases, they take their personal properties with them to the flood shelter.

Resources and economic solvency sometimes differentiate maintenance capacity. Some char dwellers are unable to protect their movable and personal properties due to personal constraints or sudden floods. A few char dwellers can protect some portion of their personal properties, whereas the rest of the property is destroyed by the flood.

14.6.4 Access to Information

Access to information is imperative to ensure flood forecasting. Forecasting news reduces the vulnerability and economic loss for a certain community in the time of flood. In all, 54.6% of char dwellers obtain flood forecasting information from different sources and authorities, including radio, television, local meteorological departments, social media platforms, and other local sources. Radio (38.2%) was more accessible to them, followed by television (31.3%). They get nearly 70% of information from radio and television (Table 14.4).

Miscommunication and inactiveness of the concerned authority about forecasting news could be devasting for a community that is unable to migrate before the start of a flood. For example, 45.4% of char dwellers do not get sufficient flood forecasting information before the flood occurs. However, they apply indigenous knowledge to migrate from their homes, cope with floods, and following other strategies.

Table 14.4 Flood forecasting information providers

Source	Percent
Radio	38.2
Television	31.3
Community	15.3
Other[a]	10.7
Social Media	3.7
Local Meteorological Department	0.8
Total	**100**

Source: Field Survey

[a]Includes announcements from local mosques and local people, etc.

14.6.5 Weak Local Government

A weak local government may fail to implement flood-related projects, connect a hotline, or provide utility of its services. The Department of Disaster Management (DDM) uses two numbers: 1090 to disseminate disaster forecasting news and 999 to provide service to the citizens in emergency situations (DDM 2020). However, 90% of char dwellers do not know about the 1090 hotline number and 85% of char dwellers are not familiar with the 999 hotline number.

These hotline numbers do not serve the interest of marginalized people for disaster preparedness and emergency service. Most char dwellers do not use these hotline numbers to get support, due to a lack of awareness programs about them, absence of communication, deficiency of coordination from local authorities, lack of knowledge about ICT,[13] not living in a well-facilitated area, illiteracy, and being unaware of the hotline numbers.

14.6.6 Networks and Communication

People are using different types of tools to maintain technological networks and communicate within shared identity groups. Communication is a must to cope with any disaster. Hence, 60% of char dwellers try to continue their communication by using cell phones during floods, whereas 40% of char dwellers use computers and other methods[14] to resume their communication (Fig.14.1).

Char dwellers use various kinds of vehicles during floods to continue their reciprocal communication and resettlement. Continuation of communication using vehicles is another opportunity and medium to resume families' daily activities and

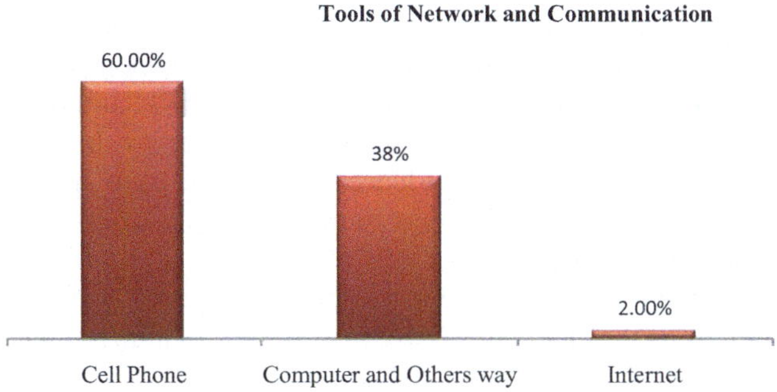

Fig. 14.1 Network and communication tools. (Source: Field Survey)

[13] Information and Communication Technology

[14] Verbal communication and messengers

Fig. 14.2 Uses of Vehicles during Floods. (Source: Field Survey)

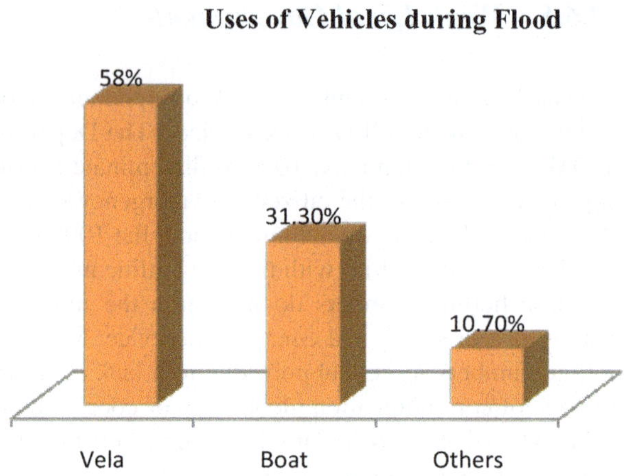

to save human resources. In all, 58% of the char dwellers use *vela*[15] for moving one place to another place and continuing their families' daily activities, whereas 31% of the char dwellers utilize boats for carrying out their important chores and moving items during floods. The rest of the families used other types of vehicles[16] to resume their activities during floods (Fig.14.2).

14.6.7 Individual Adaptations

People use different types of individual adaptations to cope with floods, including migration and structural changes to their residences. For example, 58% of char dwellers migrate from their local area to survive floods, taking shelter at parents' or relatives' homes, roads, embankments, educational institutions, shelter centers, boats, etc. Char dwellers may seasonally migrate from their residence more than twice in their lifetime due to flood-induced riverbank erosion. However, 42% of families do not migrate from their residences, remaining to secure valuable property and for other reasons.[17] Most families have to stay due to their socioeconomic conditions, as they have no option to leave their home, have opportunities to do agricultural activities at their home, have nowhere to go, or do not have enough money to buy land in a new place.

Structural change to the residence is another individual adaptation. To adapt to floods, households of the char lands have to alter the structures of their residences. For example, 59% of the households in char lands change their residence structure

[15] A way of floating using banana trees

[16] Rickshaw, van car, wheelbarrow, bicycle etc.

[17] They believed that if they migrated from their homes, then their personal movable properties might be stolen and destroyed.

before and after floods in order to cope with the flood. To live with floods, they build their houses on a high platform or high land based on previous experiences. However, 41% of the families do not alter their residence structurally. Almost every year, flooding hits the chars and affects char dwellers. Therefore, they have to repair the damaged parts of their houses after the end of the flood and natural disasters. However, most families do not alter the structure of their residence due to economic and resource constraints.

Safety training and capability to swim are important individual adaptations. Training is significant for safety to reduce casualties and economic loss from any kind of disaster, such as floods. However, only 3.20% of char dwellers received training (e.g., first aid training) for personal safety from NGOs, volunteer, and other organization. Unfortunately, 96.8% of char dwellers have not received any personal safety training from desired stakeholders. Those without any personal safety training try to cope with the flood by applying their indigenous knowledge. Women, children, and elderly people are vulnerable to floods, so they have to take precautionary measures to cope.

14.6.8 Social Cooperation and Collective Adaptation

Social cooperation highlights economic and logistical support to strengthen economic and logistic capacity for char dwellers. Economic and logistic support contribute to the positive impacts for char dwellers in the northern part of Bangladesh. Almost 80% of char dwellers receive economic support from different stakeholders to address the impacts of floods. They receive relief, loans, and housing from different stakeholders and concerned authorities. Furthermore, char dwellers receive some assistance from different projects, such as Kabikha and Kabita. In all, 46.7% of char dwellers obtained some adaptive assistance and economic support from the government of Bangladesh, whereas 26.7% received adaptive aid and economic support from NGOs. Sometimes, char dwellers work together to address their problems collectively (Fig.14.3).

Fig. 14.3 Service Providers. (Source: Field Survey)

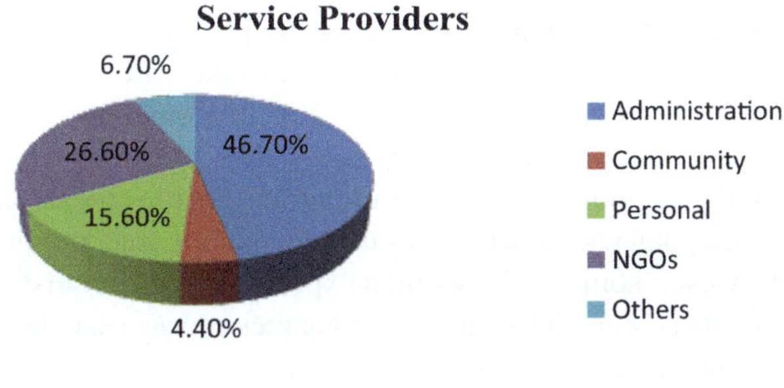

14.6.9 Balanced Diet and Health Status of Char Dwellers

Marginalized people who are living below the poverty line are severely affected by malnutrition (BBC 2019). Many char dwellers are not able to eat a balanced diet because of their economic status, striving to buy necessary foods rather than a balanced diet. For example, 55% of char dwellers lacked knowledge about a balanced diet, whereas 45% were aware of a balanced diet.

People may suffer from various diseases from contaminated water due to flood-induced impacts. For example, 42% of char dwellers are affected by different types of diseases during floods, such as cholera, diarrhea, skin diseases, dysentery, and other water-borne diseases. They recover from these diseases by taking medicine and following medical guidelines provided by physicians. However, 58% of char dwellers were not affected by any kind of diseases because of their healthy immune systems and consumption of a balanced diet.

14.6.10 Women's Health Risks during Floods

In all, 87.60% of the respondents believed that women are in the most vulnerable position during floods because they face different kinds of health risks. The health risk varies from district to district in the northern part of Bangladesh, but may include water-borne diseases, malnutrition, and other problems. However, 12.40% of the respondents believed that women were not most vulnerable position as were children and elderly people.

14.6.11 Women's Healthcare

Women try to ensure their well-being by maintaining healthcare during menstruation and pregnancy. In all, 24.90% of the char land women experienced menstruation during a flood, but few had sanitary napkins available. Most used dry cotton cloths, torn pieces of clothes, and local methods during menstruation. Others contracted infectious diseases, skin diseases, and urinary tract infections from unsafe water. Moreover, they tried to maintain extra physical awareness with reference to hygiene.

Family members were willing to take care of pregnant women and newborn babies and their mothers. In all, 21% of women were pregnant or gave birth during a flood. During their pregnancies, they got help from their husbands and other family members, including providing healthy meals, assisting with daily chores, consulting with doctors about their health, and providing medicine if necessary. However, some of the women expressed their dissatisfaction regarding their family members' care during their pregnancies and accused them of having non-cooperative attitudes.

14.6.12 Women's Psychological Health

Some women feel psychological pressure and trauma due to family problems and marital discord. Violence against women is rare in the char lands of northern part of Bangladesh, which is a positive situation in terms of societal, political, and feminist points of views in the patriarchal society of Bangladesh. Very few women (5%) have experienced marital breakup or other relationship problems. Quarrels are often about contributing to household chores, buying groceries for the family, rearing children, buying cosmetic products, using slang language, creating mental pressure for a dowry, and other trivial matters. Occasionally, they are engaged in verbal or physical attacks. However, 95% of women in the chars did not face any marital problems during floods. They resolved any problems using dialogue, cooperation, showing respect to each other, and mutual understanding.

14.6.13 Women's Individual Adaptations

Women wear different kinds of apparel during floods—mainly movable and comfortable apparel. Typically, women of char lands wear different types of dresses according to their economic capacity. A saree is a very common clothing choice for the women of Bangladesh, with 67% of the women in the chars preferring to wear a saree[18] in the time of floods instead of other types of dresses. Only 17% of the women preferred three-piece dresses during floods for their convenience and for other reasons (Fig.14.4).

The ability to swim is a positive method of coping during floods. More than two thirds (89%) of women of the char lands reported knowing how to swim and using this ability for their personal safety during floods. However, 11% of the women did not know how to swim and felt insecure during floods.

14.6.14 Women's Participation in Decision Making

An interactive and participatory approach is very important when facing any natural disasters, especially floods. However, women's participation in decision making to cope with floods is not sufficient in the char lands. In fact, 69% of the women do not have any opportunity to join policy-making dialogue about flood prevention. However, 31% of women in the chars can access some policy-making discussions about flood prevention strategies, such as repair of embankments and roads, storage of foods, and agricultural activities.

[18] A kind of apparel for women which makes them feel comfortable

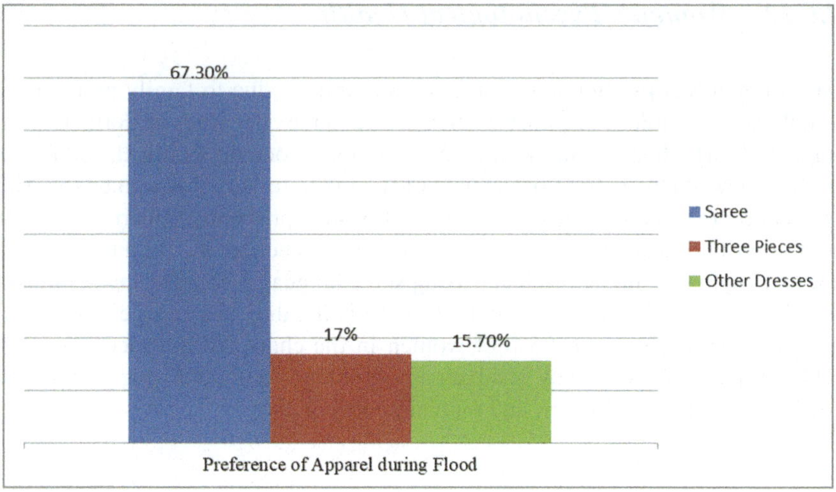

Fig. 14.4 Preference of Apparel during Flood. (Source: Field Survey)

Fig. 14.5 Flood Support Initiatives. (Source: Field Survey)

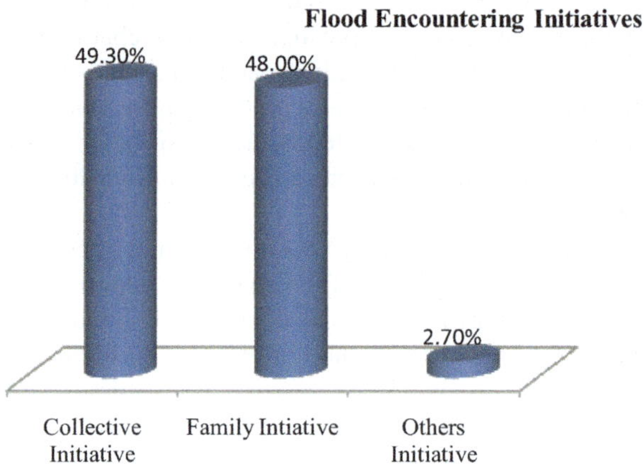

14.6.15 Social Collaboration

Encountering a flood is a good exercise for social collaboration, although it is difficult to attain. Human beings cannot control floods, but they can reduce their vulnerability and economic losses through pre-planning and disaster management techniques. About half (49%) of char dwellers received collective support to adapt to floods, whereas 48% survived the flood with only family support (Fig.14.5).

14.6.16 Short-Term Adaptation Strategies

Char dwellers take shelter in different places as a short-term strategy, such as relocating to flood shelter centers when they receive a flood forecast. Approximately 70% of char dwellers took shelter at different places, such as a temporary shelter center (school, Upazilla office, mosque, etc.), on roads, on embankments or high platforms, or at relatives' homes during floods. However, 30% of char dwellers did not relocate during the flood due to the lack of a shelter center. Some char dwellers had the opportunity to protect themselves at the shelter center. However, they did not want to relocate because the shelter center was very far from their char lands, communication was not good, and the previous experience of char dwellers indicated that the shelter center flooded very early. Hence, they remained in their residences.

14.6.17 Long-Term Adaptation Strategies

Char dwellers are using different long-term strategies to save crops, fruits, paddy, fishes, and cattle. People of the char lands usually cultivate different crops and fishes to support their local adaptive capabilities. Most of the char dwellers lead their lives by doing primary activities. Households of char lands cultivate vegetables, fruits, paddy (a stable food of Bangladesh), and fish to fulfill their basic needs. In all, 65.5% of char dwellers are used to cultivating different vegetables and fruits to fulfill their basic and economic demands. They cultivate seasonal vegetables, such as jute leaves, bottle gourd, pumpkin, basil, water amaranth, cucumber, potol (pointed gourd), Indian spinach (Poishakh), nuts, lady fingers, barb (*Vigna sinensis*), balsam (Karala), kachu (edible root), potato, onion, red, spinach, peacord, brinjal, green chili, radish, tomato, chalkumra, and kalmishaak, among others. They also cultivate various fruits, such as papaya, kamranga (*Averrhoa carambola*), guava, banana, walk (chalta), latkon, wood apple, bangi, mango, black beery, pineapple, jack fruit, lichi, and pomegranate. Char dwellers try to save their fruits and vegetables using some pre-flood strategies, such as picking fruits and vegetables before floods, selling, and making soil mounds to protect fruits and vegetables from floods. However, a few households could not save their fruits or vegetables due to a sudden flood. Furthermore, 34.5% of char dwellers do not cultivate vegetables and fruits due to resource constraints, unwillingness to perform agricultural activities, and land crises.

Char dwellers are following long-term strategies to cultivate paddy to fulfill their food demand. People often cultivate paddy to grow rice because of the country's agricultural structure and land. Most char dwellers (80.8%) cultivate paddy (e.g. IRRI,[19] Swarna etc.), wheat, or corn, among others, to lead their livelihood and earnings. Char dwellers try to save their crops by taking measures such as cultivating

[19] International Rice Research Institute

a flood-tolerant paddy (e.g. Swarna Sub-1), picking crops before floods, selling and storing in a safe place. However, a few households could not save their crops due to sudden floods, and 19.2% of char dwellers cannot cultivate any kind of vegetable.

Char dwellers are following long-term strategies to farm fishes. For example, 78.9% of char dwellers are involved with farming fishes to earn income, especially rapidly growing carp fish (silver carp, grass carp, mrigal carp, catla carp, and rui, etc.), tilapia, and poti. They catch the fish before floods and sell them, although sometimes this is not possible due to sudden flood. Some (21.1%) households do not farm any fish.

Char dwellers are following long-term strategies to farm cattle as an income source. For example, 78% of char dwellers have farmed cattle to survive and for better solvency. Cattle must have access to adequate and appropriate food (e.g., crop residues such as hay, roughage, green grasses, maize, tree foliage, etc.) and have to remain far away from flood areas. Hence, during floods, char households need to feed livestock. To protect their cattle, they put the cattle on boats, high platforms, or embankment, or relocate them to a relative's house or temporary shelter. However, 22% of char dwellers do not farm cattle.

In all, 52.5% of cattle were affected by different diseases during floods, such as fever, weakness (due to shortage of food), diarrhea, measles, pox, skin disease, and cough. Cattle recovered from the disease after treatment from a veterinarian, although a few cattle died from disease during the floods. However, 47.5% of cattle were not infected by any disease during the floods due to some precautionary measures and other indigenous methods of treatment.

14.6.18 Association between Social Capital and Flood Adaptation

Social capital theory is a collective approach to reduce vulnerability to floods among the community in certain regions. Bonding, bridging, and linking are three important indicators of social capital theory. The balance among these three ties can explain flood adaptation speed and direction (Fig. 14.6).

Likely, bonding ties are helpful to survive a flood. Bonding is a social relationship shared between groups. Confronting flood with support from different groups is obviously helpful. Therefore, 52% of necessary support came from the community or neighbors during floods, whereas 20% of char dwellers received support from family members. A smaller portion of support came from administration (Table 14.5).

Bridging are social relationships of exchange between groups. The exchange of necessary resources and information is very significant to cope with floods. Consequently, more than 50% of cases of sharing during floods came from neighbors or the community, whereas a quarter came from administration, NGOs, and relatives (Table 14.6).

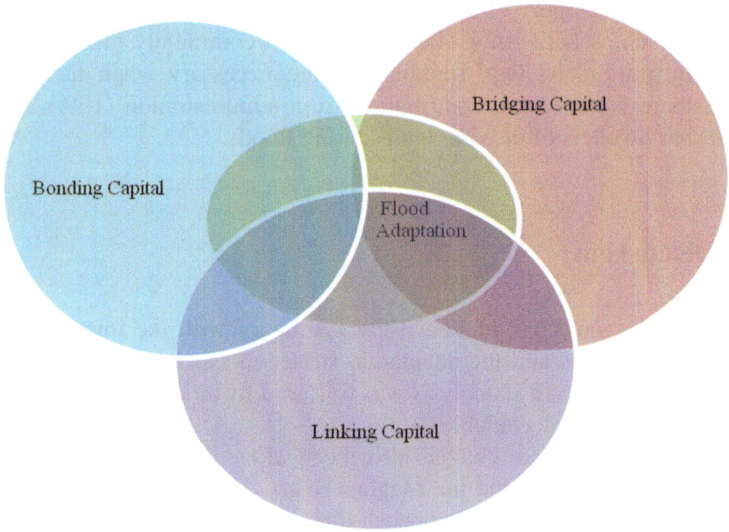

Fig. 14.6 The Relationship between Social Capital and Flood Adaptation. (Source: Field survey)

Table 14.5 Support groups during floods

Support group	Percent (%)
Neighbors / Community	52
Family	20
Administration	13
NGOs	10
Relatives	5
Total	**100**

Source: Field Survey

Table 14.6 Sharing groups during flood

Sharing groups	Percent (%)
Neighbors / Community	53
Family	19
Administration	12
NGOs	10
Relatives	6
Total	**100**

Source: Field Survey

Linking, a sub-category of bridging, is a social association that traverses group boundaries, such as between social stratifications/community groups and donors (e.g., administration, NGOs). Linking ties are necessary when facing a flood. However, there is very little contribution from administration (15%) and NGOs (10%) to char dwellers of the char lands during floods.

14.7 Discussion

This study endeavored to link social capital theory with the impacts of climate change on river floods and the adaptation strategies of char dwellers in northern Bangladesh. The theoretical analysis was followed by an analysis of indicators for the study's variables and available data.

The patterns of climate change have been altered for the confluence of the Ganges, the Brahmaputra, and the Meghna rivers. The impacts of climate change and geographical location of Bangladesh directly affect the flood situation in this region. Some environmentalists argue that the monsoon rainfall of Bangladesh and monthly average discharge of Brahmaputra have increased.

Social capital theory refers to a system in which people are willing to support each other for common interests and benefits. Human beings are not free from natural disaster and calamities; they live in a society where they need to cope with natural disasters for survival. Social capital theory discourages isolationism and focuses instead on collectivism followed by transparency, inclusiveness, and accountability of decision-making institutions.

Material adaptation indicates infrastructure development of local people to cope with floods and augment resilience capacity. Educational qualification is a relevant issue to material adaptation. The quality of education and the literacy rate of this region lag behind the overall literacy rate of Bangladesh. Therefore, they face some problems in adapting to floods.

The availability of individual resources allows people to prepare themselves to face natural disasters. Char dwellers usually save money and store foods before floods to survive. In addition, they protect their movable and personal properties for future economic solvency and savings. Access to information (e.g., flood forecasts) ensures local decision-making processes where people can obtain relevant data about natural disasters and floods from radio, television, social media, and local meteorological departments.

The local government and civil society play vital roles in disseminating information on natural disasters. Sometimes, these announcements do not occur where there is weak local government or a failure of participatory methods. Thus, the Department of Disaster Management has introduced two unique hotlines numbers—1090 (Disaster Forecasting News) and 999 (National Emergency Services)—to support common people. However, these hotline numbers do not serve the interest of char dwellers, who are not receiving necessary support from these hotline numbers.

Sometimes, community structures differentiate vulnerability and adaptation strategies in their collective effort to reduce economic losses and casualties. Different stakeholders extend economic and logistic support to vulnerable groups and flood-affected areas. Basically, vulnerable families use individual adaptation techniques when needed for their survival. People use different types of adaptive strategies in response to floods and natural disasters, such as saving crops and fishes, cultivating flood-tolerant paddy, farming climate-friendly fishes, and so on. Individual adaptation depends on the availability of a person's resources. Although they have limited resources (e.g., vela, boat), people try to utilize their resources properly to adapt to floods. These marginalized people keep matches, necessary medications, and some dry food on hand for their personal safety. Moreover, to reduce the loss of their resources, they incorporate some indigenous knowledge, such as building high platforms and constructing temporary store rooms.

This study also endeavored to connect the present theory with classification of bonding, bridging, and linking. People are using bonding, bridging, and linking capital to cope with floods and other difficulties caused by climate change.

14.8 Conclusion

Floods affect the socio-economic condition of human beings. This study analyzed the impacts of climate change on river floods and the adaptation strategies of char dwellers in northern Bangladesh. Vulnerable communities need to collaborate with their neighbors, government, and NGOs as well in order to cope with floods because Bangladesh is geographically located in a flood-prone area and in the Ganges-Brahmaputra-Meghna river basin. Monsoon rainfall in Bangladesh has increased 2.65 mm each year based on an analysis of the last 50 years of rainfall data. GCM indicates that the mean annual rainfall of Bangladesh will increased by 8% and 6% respectively by 2030 and 2050.

The study revealed that the increased volume of rainfall caused by climate change has intensified the river floods during monsoons in Bangladesh. Some parts of the country have experienced river floods almost every year with considerable damage, including loss of human lives, property, livestock, crops, plants, and infrastructure, as well as deterioration of health conditions due to waterborne diseases. This research highlights that material adaptations are correlated with educational qualification, which often determine the adaptive capacity of char dwellers in char lands. They do not apply proper flood resilience and prevention knowledge during floods due to less education and material adaptation crisis. Char dwellers depend on the availability of their resources to save money and store sufficient food to prepare for floods. They protect their movable and personal properties before transferring to a shelter center. They use different media, such as radio, television, and social networks, to obtain information on floods and natural disasters.

Local governments do not fulfill the interest of marginalized people. The Department of Disaster Management and the government should find a solution to the inactiveness of hotline numbers (1090 and 999) that deliver the forecasts of disasters and provide emergency services. The concerned authorities should implement personal training programs for the char dwellers and provide sufficient medicine to the char land people. The concerned authority needs to take positive steps to reduce mental and psychological health risks of women in order to enhance their wellbeing and healthcare. Moreover, DDM should take feedback and suggestions from women during policy formulation on flood adaptation.

The authorities need to work together to communicate and disseminate adaptation knowledge by providing information on how to access infrastructural development and internet facilities. In such a way, people can quickly obtain accurate information about floods. The study suggests that social capital theory can be an approach to obtain proper adaptive capacity during floods. Community assistance can be critical in coping with floods with regard to bonding, bridging, and linking capital. Therefore, the concerned authorities should formulate an approach to address these challenges in order to help char dwellers adapt to riverine floods.

Acknowledgement Alhamdulillah, all praises to Almighty Allah for giving His blessing and the potency to complete this chapter successfully and we remember the name of Hazrat Mohammad (Peace be upon him). A number of professionals, researchers, colleagues, and friends have helped us in the writing of this chapter. First of all, we owe unreturnable debt to the environmentalists, gender specialists, and geographers who published their precious works on climate change and flood adaptation. Many colleagues have generously provided their insights and comments on our chapter, notably Dr. Asib Ahmed (Assistant Professor, Department of Geography and Environment, University of Dhaka), Biddut Kumar Ghosh (Assistant Professor, Department of Geography and Environmental Sciences, Begum Rokeya University (BRUR), Rangpur), Iffat Ara Badhan (Lecturer, Department of Electrical and Electronic Engineering, BRUR), undergraduate students (the Department of Geography and Environmental Science, BRUR), Khadizatul Kobra Urmy (Lecturer, Department of English, Shahjalal University of Sciences and Technology, Sylhet), and Md. Al Amin (Research Officer, Clinical Governance and System, Nutrition and Clinical Services Division, Dhaka). They have provided tremendous support and important feedback with valued input on this chapter. Moreover, the authors are grateful to their teachers, mentors, family members, classmates, and well-wishers who always support to them. Last but not the least, the authors are thankful to the char lands' women of northern Bangladesh.

References

Adams J (1995) Risk. University College London Press, London

Adger WN (2003) Social capital, collective action and adaptation to climate change. Econ Geogr 79(4):387–404

Ahmed AU, Alam M, Rahman AA (1999) Adaptation to climate change in Bangladesh: future outlook. In: Vulnerability and adaptation to climate change for Bangladesh. Springer, Dordrecht, pp 125–143

Alamgir M (2019) 100 pc literacy still a far cry [Online]. Available from https://www.thedailystar.net/frontpage/literacy-rate-in-bangladesh-2019-100-pc-still-far-cry-1796734. Accessed 07 Sept 2019

Ali MS, Mahjabin T, Hosoda T (2013) Impact of climate change on floods of Bangladesh and introducing flood intensity index to characterize the flooding scenario. Journal of Engineering 4(1):23–34

Arnell NW, Gosling SN (2016) The impacts of climate change on river flood risk at the global scale. Clim Chang 134(3):387–401

Azad AK, Hossain KM, Nasreen M (2013) Flood-induced vulnerabilities and problems encountered by women in northern Bangladesh. International journal of disaster risk science 4(4):190–199

Bangladesh National Portal (2020) Rangpur division: Rangpur division at a Glance [Online]. Available from http://www.rangpurdiv.gov.bd/site/page/7be6dde1-18fd-11e7-9461. Accessed 28 May 2020

BBC (2019) Borpete Khete Peleo Keno Opostite Voghe Bangladeshira? [Online]. Available from https://www.bbc.com/bengali/news-50683916. Accessed 6 Dec 2019

BBS (2015a) Population and housing census 2011, Zila Report: Dinajpur, Bangladesh Bureau of Statistics (BBS), Statistics and Informatics Division, Ministry of Planning. Accessed 24 May 2020

BBS (2015b) Population and housing census 2011, Zila Report: Kurigram, Bangladesh Bureau of Statistics (BBS), Statistics and Informatics Division, Ministry of Planning. Accessed 24 May 2020

BBS (2015c) Population and housing census 2011, Zila Report: Lalmonirhat, Bangladesh Bureau of Statistics (BBS), Statistics and Informatics Division, Ministry of Planning. Accessed 24 May 2020

BBS (2015d) Population and housing census 2011, Zila Report: Rangpur, Bangladesh Bureau of Statistics (BBS), Statistics and Informatics Division, Ministry of Planning. Accessed 24 May 2020

BBS (2015e) Population and housing census 2011, Zila Report: Gaibandha, Bangladesh Bureau of Statistics (BBS), Statistics and Informatics Division, Ministry of Planning. Accessed 24 May 2020

Bose PS (2016) Vulnerabilities and displacements: adaptation and mitigation to climate change as a new development mantra. Area 48(2):168–175

Brammer H (1990) Floods in Bangladesh: geographical background to the 1987 and 1988 floods. Geogr J:12–22

Burton I, Kates RW, White GF (1993) The environment as Hazard, 2nd edn. Guildford Press, London

Chambers R (1989) Editorial introduction: vulnerability, coping and policy

Christensen JH, Hewitson B, Busuioc A, Chen A, Gao X, Held R, Jones R, Kolli RK, Kwon WK, Laprise R, Magaña Rueda V (2007) Chapter 11: regional climate projections. In: Climate change, 2007: the physical science basis. Contribution of working group I to the fourth assessment report of the intergovernmental panel on climate change. Cambridge University Press, Cambridge, pp 847–940

DDM (2017) Disaster situation report: 2017. Department of Disaster Management, Government of the People's Republic of Bangladesh, [Online], Available from http://www.ddm.gov.bd/. Accessed 28 May 2020

DDM (2020) Department of disaster management. [online], Available from http://www.ddm.gov.bd/. Accessed 24 May 2020

Denton F (2002) Climate change vulnerability, impacts, and adaptation: why does gender matter? Gend Dev 10(2):10–20

EDU (2019) Types of academic writing [Online]. Available from https://www.sydney.edu.au/students/writing/types-of-academic-writing.html. Accessed 3 Oct 2019

FFWC (2017) Reports: Annual flood report 2017. Flood Forecasting and Warning Center, Bangladesh Water Development Board [Online]. Available from http://ffwc.gov.bd/index.php/reports/annual-flood-reports. Accessed 28 May 2020

FFWC (2018) Reports: annual flood report 2018. Flood Forecasting and Warning Center, Bangladesh Water Development Board [Online]. Available from http://ffwc.gov.bd/index.php/reports/annual-flood-reports. Accessed 28 May 2020

FFWC (2020) Definitions: Floods. Flood Forecasting and Warning Center, Bangladesh Water Development Board [Online]. Available from http://ffwc.gov.bd/index.php/definitions. Accessed 28 May 2020

Häuberer J (2011) The founding concepts of social Capital-Bourdieu's theory of capital and Coleman's rational-choice approach to social capital. In: Social capital theory. VS Verlag für Sozialwissenschaften, pp 35–51

Hossain MA, Reza MI, Rahman S, Kayes I (2012) Climate change and its impacts on the livelihoods of the vulnerable people in the southwestern coastal zone in Bangladesh. In: Climate change and the sustainable use of water resources. Springer, Berlin, Heidelberg, pp 237–259

Huq SU, Ahmed AU, Koudstaal R (1996) Vulnerability of Bangladesh to climate change and sea level rise. In: Climate change and world food security. Springer, Berlin, Heidelberg, pp 347–379

Islam S (2016) Coping strategies, preparedness and emergency response of char-land peoples in Flood Disaster: A case study in Kurigram, Bangladesh

Kader MA (2007) Charland and right of the landless [Online]. Available from https://www.thedailystar.net/news-detail-9034. Accessed 27 Oct 2007

Mallick D, Hossain Z (2019) Climate change, food security and our vulnerability [Online]. Available from https://www.thedailystar.net/environment/climate-change/news/climate-change-food-security-and-our-vulnerability-1800865. Accessed 17 Sept 2019

Mirza MMQ (2002) Global warming and changes in the probability of occurrence of floods in Bangladesh and implications. Glob Environ Chang 12(2):127–138

Pelling M (1998) Participation, social capital and vulnerability to urban flooding in Guyana. Journal of International Development: The Journal of the Development Studies Association 10(4):469–486

Pelling M (2002) Assessing urban vulnerability and social adaptation to risk: evidence from Santo Domingo. Int Dev Plan Rev 24(1):59–76

Pelling M (2003) The vulnerability of cities: natural disasters and social resilience. Earthscan

Pelling M, High C (2005) Understanding adaptation: what can social capital offer assessments of adaptive capacity? Glob Environ Chang 15(4):308–319

Pielke RA Jr (2004) What is climate change? Energy & Environment 15(3):515–520

Putnam RD (2000) Bowling alone: The collapse and revival of American community. Simon and schuster

Putzel J (1997) Policy arena: accounting for the 'dark side' of social capital: reading Robert Putnam on democracy. Journal of International Development: The Journal of the Development Studies Association 9(7):939–949

Rahman MS (2013) Climate change, disaster and gender vulnerability: a study on two divisions of Bangladesh. American Journal of Human Ecology 2(2):72–82

Rashid H, Paul B (2014) Climate change in Bangladesh: confronting impending disasters. Lexington Books, London

Rawlani AK, Sovacool BK (2011) Building responsiveness to climate change through community based adaptation in Bangladesh. Mitig Adapt Strateg Glob Chang 16(8):845–863

RDCD (2015) Chars Livelihoods Programme (CLP), Rural development and co-operatives division, Government of the People's Republic of Bangladesh [Online]. Available from https://rdcd.portal.gov.bd/sites/default/files/files/rdcd.portal.gov.bd/publications/edcdc30e_a754_4758_aa2b_17cf3c5b82d7/Success%20Story%20of%20CLP.pdf. Accessed 31 May 2020

Roy BK, Ullah MB, Rahman MH (2014) Climate change impact in Charlands in central area of Bangladesh: assessing vulnerability and adaptation by the farming communities. Journal of Environmental Science and Natural Resources 7(2):59–63

Sumaila UR, Cheung WW, Lam VW, Pauly D, Herrick S (2011) Climate change impacts on the biophysics and economics of world fisheries. Nat Clim Chang 1(9):449–456

The Daily Star (2009) Basic healthcare services absent in Char lands [Online]. Available from https://www.thedailystar.net/news-detail-80796. Accessed 22 Mar 2009

The Daily Star (2020) German NGO NETZ provides support to 20,000 families in north Bangladesh [Online]. Available from https://www.thedailystar.net/events/news/german-ngo-netz-provides-support-20000-families-north-bangladesh-1900366. Accessed 7 May 2020

Usshamunnaybd (n.d.) Unnayan Shamannay: About Unnayan Shamannay., Available from http://www.usshamunnaybd.org/. Accessed 12 June 2020

Whitehead PG, Barbour E, Futter MN, Sarkar S, Rodda H, Caesar J, Butterfield D, Jin L, Sinha R, Nicholls R, Salehin M (2015) Impacts of climate change and socio-economic scenarios on flow and water quality of the Ganges, Brahmaputra and Meghna (GBM) river systems: low flow and flood statistics. Environ Sci: Processes Impacts 17(6):1057–1069

Woolcock M, Narayan D (2000) Social capital: implications for development theory, research, and policy. World Bank Res Obs 15(2):225–249

Chapter 15
Climate Change Impact and Adaptation Strategies in Bangladesh to Strengthen Regional Cooperation

Md. Nazrul Islam, Sahanaj Tamanna, Md. Noman, Effat Akhter, Md. Maruf Hasan, Mohammad Ashikur Rahman Rifat, Eram Islam Sakib, B. M. Yousuf, Sany Mohammad Khaled, Rabeya Akhter Akangkha, Ibrahim Sufi Mohammad Riaz, Anik Barua, Nafisa Mubashsara, Sheikh Mohammad Abrar Jahin, and Fahim Hossain

Abstract There is a need for global, regional, and local-level mitigation and adaptive strategies to address the reality of climate change impact. Adaptation to climate change is a critical issue in the Southeast Asian region. Geospatial, socioeconomic, and demographic conditions make South Asia to be one of the most vulnerable regions to climate change. The affected regions of Bangladesh have formed an alliance to study, adapt, and mitigate such calamities. Academic and policy research is increasing sharply as a result of the growing evidence on inevitable impacts in the region. Quick and effective actions for South Asia can make significant changes for the people of this region, who necessarily depend on agricultural and natural resources to make a living. South Asian countries 'as a union' can better negotiate with international assemblages and design joint coping mechanisms. For that reason, more efforts are required to drive future responses, including such aspects as vulnerability and adaptation assessments, methodology and approaches for modelling and data gathering, institutional approaches, financial and technical support, awareness and training, and networking and information management. Adaptation is a knowledge-intensive undertaking, which requires the provision of access to

M. N. Islam (✉) · M. Noman
Department of Geography and Environment, Jahangirnagar University, Savar, Dhaka, Bangladesh
e-mail: nazrul_geo@juniv.edu

S. Tamanna
Bangladesh Environmental Modeling Alliance (BEMA), Non-Profit Research Organization, Dhaka, Bangladesh

E. Akhter · M. M. Hasan · M. A. Rahman Rifat · Eram Islam Sakib · B. M. Yousuf · S. M. Khaled · R. A. Akangkha · I. S. Mohammad Riaz · A. Barua · N. Mubashsara · S. M. Abrar Jahin · F. Hossain
North South University, Dhaka, Bangladesh

M. N. Islam, A. van Amstel, *Bangladesh II: Climate Change Impacts, Mitigation and Adaptation in Developing Countries*, Springer Climate,
https://doi.org/10.1007/978-3-030-71950-0_15

relevant and usable knowledge to ensure effective implementation and outcomes. Thus, there is a need for more regional collaborations and cooperation to reduce the diverse effects of climate change in South Asian countries. This chapter reviews the overall situation of climate change in Bangladesh and also explains various adaptation strategies from different dimensions.

Keywords Climate change · Natural hazards · Adaptation strategies · Coping mechanism · Regional negotiation · Bangladesh

15.1 Introduction

Climate change is one of the most significant challenges to global economic development (Ali 1999; Anbumozhi et al. 2012). Over the past three decades, the number of recorded natural disasters has more than doubled in number, from around 200 a year to more than 400 (Rodriguez et al. 2009; Parvin et al. 2010; Kabir 2019; Monirul and Alam 2017). Nine out of every 10 disasters are now climate-related, with increasingly significant repercussions across economies, societies, and natural ecosystems (Thomas and López 2015). The Intergovernmental Panel on Climate Change (IPCC) projected a global mean temperature increase of 1.1 °C to 6.4 °C by 2100, which is likely to affect storms and floods, as well as increase the sea level due to the thermal expansion of the oceans and the melting of ice sheets and glaciers (IPCC 2007; Islam 2008). Recent research predicts a global sea level rise between 50 cm and 190 cm from 1990 to 2100 (Vermeer and Rahmstorf 2009). The adverse impacts of the natural hazards affecting socio-economic conditions need to be abated for long-term development (Pande 1992; Ali 1999). They will affect all sectors but will have a disproportionate impact on millions of poor rural people of developing countries (Black et al. 2013; Mohammad 2015; Bhuyan et al. 2018). Countries are very vulnerable to climate change for various reasons, including geo-climatic conditions, socio-economic status, demographic, and dependence on agriculture and rural sectors for livelihoods. Food, textiles, biodiversity, water resources, coastal ecosystems, human health, and land degradation are considered to be highly vulnerable for South Asia due to climate change (Islam et al. 2009; Amin and Rahman 2014; Bhuyan et al. 2018). Bangladesh and Nepal are among the most vulnerable developing countries of South Asia (Thornton et al. 2014; Hoque and Haque 2016), with Bangladesh being more vulnerable due to sea and river flooding. In India, agricultural lands are hampered by floods and droughts; in Nepal, glacier outburst floods lead to temporary displacement and disruption of livelihoods (Bartlett et al. 2010; Hossain et al. 2014; Kilroy 2015). In Pakistan, extreme weather events are causing water-related disasters (Rahman et al. 2015). Besides climate vulnerability, these four countries experience similar climate change impacts with their transboundary natural resources, including rivers and mountains (Vij et al. 2017).

The Indo-Gangetic Plain (IGP), Indus delta, and Bangladesh flood plains are geographically uncommon systems of regional importance that have produced food and many resources; they have been the backbone of cultural and economic development in the region for centuries (Wang et al. 2017a, b; Bhuyan et al. 2018). The region is also called the "third Pole" and the water tower of Asia, as it stores a large volume of water in the form of ice and snow and regulates the flow of 10 major river systems in the region (Morton 2011; Wang et al. 2017a, b). Bangladesh has experienced many types of natural disasters due to its geographic and geologic settings every year (Carter 1991). In the last 30 years, the country has experienced nearly 200 climate-related disasters including drought, extreme temperatures, floods, and storms (Hay and Mimura 2010). More than 80% of the land is situated on the deltas and floodplains of the Ganges, the Brahmaputra, and the Meghna (GBM) rivers (Siddiqi 2012; Chaudhuri and Mishra 2016). The overall economic development of the country has been adversely affected by climate change, especially high temperatures, sea-level rise, cyclones and storm surges, salinity intrusion, and heavy monsoon downpours (Amin and Rahman 2014; Hossain et al. 2014; Bhuyan et al. 2018). The vulnerable low-elevation coastal zone of 54,000 km^2, or approximately 40% of the country, is facing the consequences of growing environmental pressure, resulting in increasing demand for water, insufficient maintenance of existing embankments and other environment protection measures, and unplanned urbanization and industrialization (Ali 1999; Karim and Mimura 2008; Rejaur Rahman 2015). Since 1989s the frequent and severe floods, cyclones, storms, tidal surges, sea level increases, salt water intrusions, and river and coastal erosions force many rural people to migrate to the coastal cities, thus creating new environmental problems in the country's urban slums (Thompson and Walsham 2010; Abedin et al. 2012; Dasgupta et al. 2015).

To reduce the impacts of climate change, adaptation policy approaches are defined as the ways in which policies are designed and implemented (Dessai and Hulme 2004). Despite many international environmental conventions, treaties, agreements, protocols, and legislations—such as the UN Conference on Human Environment (1972), Our Common Future (1987), the Kyoto Protocol (1992), the Earth Summit (1992), Johannesburg Summit (2002), Bali Conference (2007), and Poznan Conference (December, 2008), among others— human society is still facing severe mitigation and adaptation related challenges responses. (Fig. 15.1; Ali 1999; Islam et al. 2009). Those responses of adaptation to climate change may be used to mitigate vulnerability and are designed to take advantage of new opportunities that may come as output of climate change (Burton 2009). In the Fourth Assessment Report, the IPCC (2007) recognizes that some adaptation is occurring, but on a very limited basis, and affirms the need for extensive adaptation across nations and economic sectors to address impacts and reduce vulnerability (Smith et al. 2008; Islam et al. 2009). Bangladesh has developed some facility for dealing with the impacts of climate change at the national level. Policy response options have been mobilized to reduce vulnerability to natural variability and most recently to climate change in particular coastal areas in Bangladesh (Ayers 2007).

In 2005, Bangladesh developed the National Adaptation Program of Action (NAPA). In 2008, Bangladesh prepared Bangladesh Climate Change Strategy and

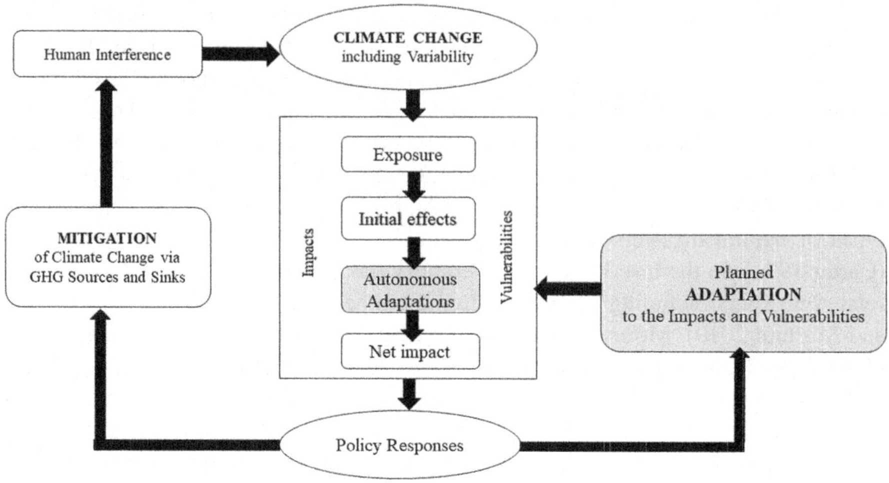

Fig. 15.1 Places of adaptation to climate change. (Adopted and modified from Smit et al. 1999)

Action Plan (BCCSAP) to strengthen its six pillars: Food security, social protection, and health; Comprehensive disaster management; Infrastructure; Research and knowledge management; Mitigation and Low-carbon development; and Capacity building and institutional strengthening (MoEF 2009). Bangladesh has also formulated a Coastal Zone Policy (CZPo) in 2005, although it paid very little attention to Sea Level Rise (SLR) (CZPo, Coastal Zone Policy 2005), as well as the challenges of the coastal cities and their infrastructure (Hoque and Haque 2016). South Asian countries and their development partners have invested significant portions of their financial budgets and much effort into formulating climate policies, strategies, and action plans and implementing numerous adaptation and mitigation projects to increase the capacity of small farm holders to contain climate impacts and reduce greenhouse gas (GHG) emissions. However, there have been few attempts to assess the status of policies, projects, and other efforts or to identify gaps and needs for guiding future policies and projects (Black et al. 2013; Islam et al. 2015; Wang et al. 2017a, b).

15.1.1 Statement of the Problem

There are few studies investigating how climate policy approaches are designed and implemented in South Asia (Smith et al. 2008; Islam et al. 2009; Saito 2013; Butler et al. 2016). There are also barriers to developing long-term and flexible adaptation strategies (Le Dang et al. 2014; Biesbroek et al. 2013). Bangladesh has total 64 districts, including 24 districts located in coastal zones and mainland from where people are displaced. Because of sea level rise, almost 160 million of people in Bangladesh are directly vulnerable to climate change impact and displacement (Rabbani 2009). Twenty-eight percent of the population of Bangladesh lives in the coastal regions of the country (Ali 1999; Dasgupta et al. 2015; Parvin et al. 2016). Climate change may have dramatic consequences for several regions. Fragile countries with limited capacity to adapt are most vulnerable; Bangladesh is one of them. The effects of climate change will aggravate many of the natural hazards faced by Bangladesh every year, including all of the natural hazards currently leading to climate displacement, such as flooding, tropical cyclones, storm surges, salinity intrusion, and riverbank and coastal erosion and SLR (Islam et al. 2009; Abedin et al. 2012; Black et al. 2013; Dasgupta et al. 2015). IPCC has indicated that climate change and global warming are likely to lead to an increase of rainfall, more flash floods and large-area floods, earlier melting of snow packs and melting of glaciers, frequent and intense droughts, intense tropical cyclones, rising sea levels, frequent and intense storm surges, intense inland rainfall, and stronger winds (IPCC 2007; Dasgupta et al. 2015). It is anticipated that 13% of Bangladesh's coastal land will be subsumed by sea level rise by 2080 (Pender 2008; Chaudhuri and Mishra 2016). The GBM system is likely to lead to further riverbank erosion, resulting in mass displacement in the mainland areas of Bangladesh by heavier and erratic rainfall (Ali 1999; Islam et al. 2009; Amin and Rahman 2014; Chaudhuri and Mishra 2016). Erratic rainfall also leads to exceeding and breaching of embankments, resulting in widespread flooding on both sides of river banks. In addition, it is likely to cause more droughts (especially in the drier northern and western regions of Bangladesh, which record significantly less rainfall), causing the destruction of crop yields and severe disruptions to livelihoods (IPCC Third Assessment Report 2001; Ahsan et al. 2011). In the hill region of Bangladesh, frequent erratic rainfall creates landslides, which trigger human exodus. As the glaciers of the Himalayas continue to melt, there will be higher river flows in the warmer months of the year, followed by lower river flows and enhanced saline intrusion after the glaciers have shrunk or disappeared (Ali 1999; Mohammad 2015).

Climate change will obviously affect the country, so Bangladesh has to start thinking about this major problem affecting the world very soon. Measures have to be taken right away, and officials should come up with adaptation strategies. Without adaptation strategies, it will be very hard for Bangladesh, as a vulnerable country, to adapt to climate changing rapidly (Ahsan et al. 2011; Amin and Rahman 2014; Hoque and Haque 2016). One of the key approaches is to strengthen regional cooperation. Bangladesh is surrounded by other countries—Myanmar, India, and Nepal—that also have to adapt to climate change. International cooperation, in which resources and knowledge are shared, can be really beneficial to Bangladesh, which is currently not taking proper measures (Atta-Ur-Rahman 2015; Alam 2018; Ahmed 2018). Economic standards and technology might be interfering with a strategy implementation. Thus, with regional cooperation, climate change adaptation will be a lot easier (Islam et al. 2009; Black et al. 2013). Bangladesh and its people have adapted over generations of floods, droughts, and cyclones by involving communities in planning, construction, and management. Climate change investments should thus be made at the community level (Ahsan et al. 2011; Hossain et al. 2014).

15.1.2 Literature Review

Global climate change and associated climate extremes and disasters have increased uncertainty (IPCC 2012) in the livelihoods of people in developing countries. They pose various threats to human life and security (Burton et al. 2006; IPCC 2007; Alam et al. 2017). The South Asia region is endowed with rich natural resources and contains global biodiversity hotspots that provide various ecosystem services, including water, biodiversity, soils, natural beauty, recreational opportunities, and wilderness—all of which financially support many in the population (Tiwari and Joshi 2014). However, the Headly Center for Climate Prediction and Analysis estimated that ocean levels in Asian nation may rise 40 cm (15 in.) by 2080 (Streatfield and Karar 2008; Islam et al. 2009). The climate models indicate that temperature may increase in Asian nations by 0 °C to 1.5 °C by 2030 (IPCC 2018) (Table 15.1).

According to the Third Assessment Report of the IPCC, South Asia is the most vulnerable region of the world to climate change impacts (McCarthy et al. 2001). The international community also recognizes that Bangladesh ranks high in the list of most vulnerable countries on earth (ADB 2009; CSR 2010; World Bank 2018; Ahmed 2018). Bangladesh is one of the top-ten nations largely at risk of global climate change, according to the German Watch International Climate Risk Index (CRI) 2011 report. Natural hazards coupled with climate change will create complex development challenges for remote rural communities (Twigg and Bhatt 1998; Ayersa et al. 2014; Hoque and Haque 2016). The WB (2005) estimated that 97% of the total area and 99% of the population of Bangladesh are at risk from multiple hazards such as cyclones, floods, droughts, earthquakes, and tornados. Water levels have increased by a minimum of 5 cm a year at Hiron point, 1.4 cm at Cox's Bazar, and a 0.9 cm at Khepupara, according to the Asian Nation Water Development Board (Ann 2010). Because these economies greatly rely on agriculture, natural

Table 15.1 The climate models indicate that temperature may increase in Asian nations by 0 to 1.5 °C by 2030 (IPCC 2018)

Year	Temperature change (°C) mean			Rainfall change (%) mean			Sea Level Rise (cm)
	Annual	DJF	JJA	Annual	DJF	JJA	
2030	1	1.1	0.8	5	-2	6	14
2050	1.4	1.6	1.1	6	-5	8	32
2100	2.4	2.7	1.9	10	-10	12	88

Source: Adopted from the Bangladesh NAPA Document (GOB 2005)

resources, forestry, and fisheries, an increased risk of floods and droughts would decrease production in these sectors and exacerbate poverty (Fischer et al. 2005).

As documented by the IPCC, adaptation to climate change is among the most fundamental challenges facing human society over the coming decades (ISET 2008). Mitigation of GHGs and adaptations to live in the changing environment are two anthropogenic ways to tackle climate change (Füssel and Klein 2006; Byomkesh et al. 2011; IPCC 2018). However, adaptation to climate change did not receive much attention in the early years of international climate change studies, where there was more focus on mitigation and impacts (Black et al. 2013; Kates 2014), although adaptation has recently been covered more extensively and has an important place in the fourth assessment report of the IPCC (2007). Climate change adaptations are adjustments of humans or natural systems in response to real or expected climatic effects, which moderate harms or exploit beneficial opportunities (Parry and Canziani 2007). Developing strategies to adapt to climate change implies that people must protect themselves against unavoidable new and exogenous threats to their health and properties (Collins 2008; Abedin et al. 2012). Many developing countries have already experienced weather events, such as floods, droughts, heat waves, and tropical cyclones, that are more frequent or intense than previous experiences (Dai et al. 2004; Yamin et al. 2005; Ministry of Finance 2005; Yohe et al. 2008; Ministry of Forests 2009); the resulting impacts point to the consequences on the environment, production systems, and livelihoods from future climate variability and change (Hoque and Haque 2016). Developing countries have specific steps for adaptation because of high vulnerabilities; thus, they will bear much of the global burden of climate change (Mertz et al. 2009; Abedin et al. 2012).

15.1.3 Climate Change Is an Evolving Physical Condition

Climate change is an evolving physical situation with very dire and inauspicious consequences, particularly for Bangladesh. It is a major and long-term constraint permeating all sectors of the economy, adversely affecting the wellbeing of men, women, young, old, people on the coast, and even people in the inlands, hills, and plains (Smith et al. 2008; Islam et al. 2009; Abedin et al. 2012). Climate change in Bangladesh is no longer a future problem; rather, it is among the countries that are already being severely affected (Islam et al. 2009; Ahsan et al. 2011). The country frequently suffers from floods, tropical cyclones, droughts, salinity intrusion, scarce drinkable water, ecosystem effects, riverbank erosion, high-velocity stream flows, storm surges, and sea level rise. These phenomena have resulted in the displacement of millions of people—"environmental refugees"—from coastal regions by creating huge adverse impacts on the livelihoods and long-term health of a large proportion of the population (Karim and Mimura 2008; Ahsan et al. 2011). Bangladesh is already recognized as one of the most climate-vulnerable countries in the world. Along with experiencing repeated natural disasters, it has also faced damage to infrastructure and economic assets, which challenge the country's ability to achieve the high rates of economic growth needed to sustain these reductions in poverty (Islam et al. 2009; Amin and Rahman 2014).

15.1.4 Background to the Study Area and Issues

It is clearly known that climate change has harmed communities, societies, and countries all over the world. One of the most significant issues is global warming (Smith et al. 2008; Islam et al. 2009), which is having disastrous impact on nature, society, and communities worldwide (Ahsan et al. 2011; Shahnila Islam 2016). Excessive emissions of GHGs have caused a greenhouse effect and associated temperature increases since the beginning of the industrial age (Byomkesh et al. 2011). Additionally, several human activities contribute to climate change, including deforestation and population growth. For example, deforestation may occur to clear lands for agriculture, in an effort to feed the growing number of people. Furthermore, the large population has been responsible for GHG emissions causing the greenhouse effect, as well as sea level rises (Abedin et al. 2012).

Due to its geographical location, population density, poverty, flat and low-lying landscape, and lack of institutional systems, Bangladesh is highly vulnerable to natural disasters. However, its social, physical, and economic statuses are very similar to other climate change–affected countries in the world (Hoque and Haque 2016). Almost 80% of Bangladesh lands are floodplains, with a major portion of the country (except the northwestern highlands) being flooded during the rainy season. Events due to climate change, such as high temperatures, cyclone and storm surges, and salinity intrusion, have impeded the overall economic progress of the country to a great extent (Karim and Mimura 2008; Denissen 2012; Amin and Rahman 2014). Global temperatures and sea levels are increasing, with abnormal climate changes around the world, making it a more imperative issue than ever (BCCSAP 2009). In Bangladesh, climate change has hampered human life and economic development, as well as displaced human beings (Fig. 15.2). Due to global warming, the country has experienced unusual changes in seasons and faces unexpected rains, dry spells, temperatures, and other symptoms of changes in global weather patterns.

Bangladesh's vulnerability to climate change lies mainly in its population density and geography, with a large part of the country consisting of low-lying coastal areas and expansive floodplains. The population of the country is increasing while forests are decreasing (Alam et al. 2009; Hussain et al. 2016). An increasing world population as well as harmful industries in developed countries are the main causes of climate change. The severity of storms, droughts, rainfall, floods, and other natural disasters has been increasing in developing countries such as Bangladesh due to climate change (Black et al. 2013; Dasgupta et al. 2015). Climate change has hampered every sector of the economy, especially in agriculture. Natural phenomena, such as SIDR (Sidr, with a maximum wind speed of 240 km/h, tidal waves up to five

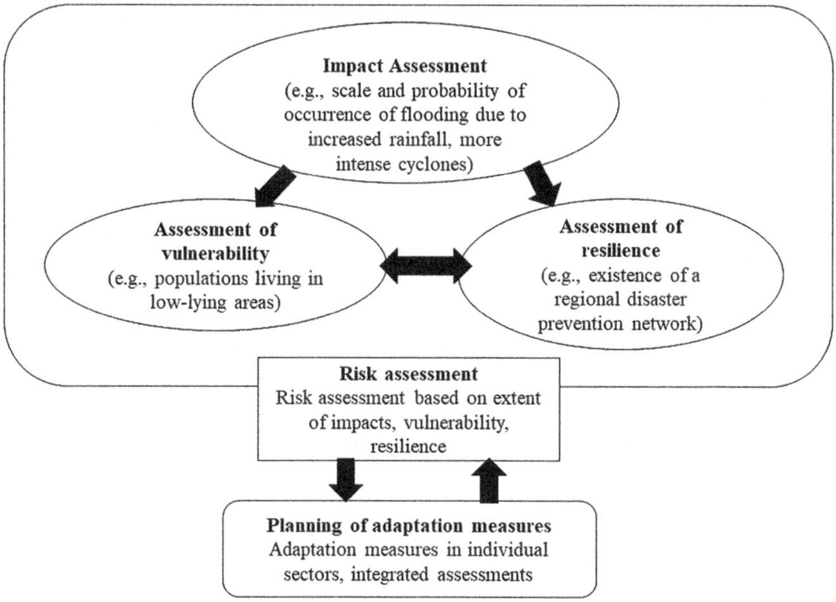

Fig. 15.2 Relationships between adaptation measures and assessments of impacts, vulnerability, and resilience

meters high, and storm surges at a maximum of 10 meters in some areas, struck the southwest coast of Bangladesh on 15 November 2007) and AILA (The cyclonic storm "Aila" hit the south western part of Bangladesh (Khulna Division) and West Bengal in India on 25th May 2009), cause great damage to humans and properties throughout the country (Islam et al. 2009). Due to lack of resources and other causes, Bangladesh does not have the capacity to ensure that appropriate measures are taken to mitigate the damage (Ahsan et al. 2011; Ahamed 2018).

The coastal areas of Bangladesh and the coastal people are most affected by climate change; they are losing ponds, lakes, dams, and forestry due to natural disasters (Hussain et al. 2016). Researchers estimate that 54 types of fish have already become endangered due to climate change and natural disasters. The weather patterns have also changed because of climate change. Human life, animal life, and food production are affected by temperature increases, frequent floods, and cyclones. At least 30 rivers, including the Padma and the Teesta, have dried up and other rivers have already lost their natural attributes because of a lack of water flow (Smith et al. 2008; Islam et al. 2009; Saito 2013; Butler et al. 2016). The northern area of Bangladesh is almost a desert due to insufficient water flow from the rivers. Geological and biological changes have occurred in this area (Shamsuddoha and Chowdhury 2009; Hossain et al. 2014). To mitigate the effects of climate change, Bangladesh needs more economical and technical support to collaborate with other developed countries.

15.1.5 South Asian Association for Regional Cooperation

The South Asian Association for Regional Cooperation (SAARC) is an economical and geopolitical organization of eight countries located in South Asia that was established in 1985. The countries are India, Pakistan, Bhutan, Nepal, Bangladesh, Sri Lanka, and Afghanistan. The South Asian nations decided to form a regional organization to develop economic conditions and address climate change consequences (Black et al. 2013; Amin and Rahman 2014; Chaudhuri and Mishra 2016). SAARC formed to fulfill some objectives, such as to promote the welfare of South Asian people, improve their quality of life, accelerate economic growth, implement social programs, and use cooperation for mutual benefits. Persistent problems, such as population growth, poverty, climate change, healthcare, terrorism, floods, and droughts also were areas of improvement for SAARC. In South Asia, the construction of a common framework to implement SAARC policies is a priority. In response to the impacts of climate change (Smith et al. 2008; Islam et al. 2009; Saito 2013; Butler et al. 2016), SAARC has integrated policies and regulations amongst the countries to build strategic plans and policies on prevention and protection (Smith et al. 2008; Islam et al. 2009; Chaudhuri and Mishra 2016). SAARC aims to find a common and sustainable methodology to counter the climate change adaptation limitations in designated four countries in below (Table 15.2).

15.1.6 Association of South East Asian Nations

The Association of South East Asian Nations (ASEAN) was established in 1967 in Bangkok. Myanmar, Laos, Thailand, Brunei, Cambodia, Indonesia, Malaysia, Philippines, Singapore, and Vietnam are the associated countries (Smith et al. 2008;

Table 15.2 National climate action plans, coordinating institutions, and funding mechanisms

Country	National climate action plans	National focal agency	Funding mechanism
Bangladesh	NAPA (2005) Bangladesh Climate Change Strategy and Action Plan (2009)	Climate Change Unit, Ministry of Environment and Forests	Climate Change Trust Fund Bangladesh Climate Resilience Fund
Bhutan	NAPA (2006, 2012)	Climate Change Unit, NEC	None
Nepal	NAPA (2010) LAPA (2011) Strategic Program for Climate Resilience	Climate Change Management Division, Ministry of Environment	None
India	National Action Plan for Climate Change INDCs (2015)	Climate Change Division, Ministry of Environment and Forests	None

Source: Islam (2009), Atta-Ur-Rahman (2015), Alam et al. (2017), Ahmed (2018)

Islam et al. 2009). This organization was formed to work on establishing economic growth, cultural development, regional stability, training, and safety. Climate change impacts and adaptations are major concerns of these regional cooperation organizations, as all the countries have large populations and economic activity that is concentrated on the coastlines (Abedin et al. 2012; Bhuyan et al. 2018). These countries often announce mitigation targets. The existing commitment of the ASEAN member states to international climate policy provides a good foundation for joint regional climate policy formulation and action (Smith et al. 2008; Islam et al. 2009; Saito 2013; Butler et al. 2016). All ASEAN member states have ratified the Kyoto Protocol and signed the Paris Agreement; nine out of 10 countries have ratified the Paris Agreement, with Myanmar expected to do so in the near future (Black et al. 2013; Bhuyan et al. 2018). At least half of the ASEAN member states also reacted publicly to President Donald Trump's announcement in 2017 that the United States would withdraw from the Paris Agreement, criticizing it directly and/or reiterating their own country's commitment to climate action (Smith et al. 2008; Islam et al. 2009; Saito 2013; Butler et al. 2016).

15.2 Overview of Climate Change Adaptations

The IPCC's Fourth Assessment Report (2007) estimated that climate change impacts are expected in upcoming decades, such as rising temperatures, high-intensity storms, frequent heat waves, accelerated loss of biodiversity, and natural calamities. Poverty rates and economic growth will be hampered (Black et al. 2013; Amin and Rahman 2014). The current policies, strategies, and mechanisms to reduce climate change impact in developing countries in Southeast Asia are inadequate, specifically in densely populated river deltas that are more vulnerable to climate hazards.

15.2.1 Climate Change Adaptation and Mitigation

Climate change in IPCC usage refers to any change in climate over time, whether due to natural variability or as a result of human activity. This usage differs from that in the Framework Convention on Climate Change, where climate change refers to a change of climate that is attributed directly or indirectly to human activity that alters the composition of the global atmosphere and that is in addition to natural climate variability observed over comparable time periods (Mcgranahan et al. 2007; Parvin et al. 2010; Islam 2016). Adaptation is the adjustment in natural or human systems in response to actual or expected climatic stimuli or their effects, which moderates harm or exploits beneficial opportunities. Vulnerability is the degree to which a system is susceptible to, and unable to cope with, adverse effects of climate change, including climate variability and extremes. Vulnerability is a function of the character, magnitude, and rate of climate change and variation to which a system is exposed, the sensitivity and adaptive capacity of that system (IPCC 2007; Smith et al. 2008; Islam et al. 2009; Ahsan et al. 2011; Hoque and Haque 2016). Adaptation to the impacts of climate change has not been attracting the same attention and interest as the issue of mitigation; thus, it is relatively less developed than mitigation as a policy response. This is evident from the Kyoto Protocol, in which the main concern is the reduction of greenhouse gas emissions, particularly from developed countries (Smith et al. 2008; Islam et al. 2009).

15.2.2 Assessment of Climate Change Adaptation

Early research on adaptation mainly focuses on the assessment of climate change impacts in general, with limited analysis of adaptation measures (Feenstra 1998; Ahsan et al. 2011). The limitations of the assessment of biophysical aspects alone is progressively recognized with more analysis on vulnerability being seen as an important component of any attempt to define the magnitude of climate change (Füssel and Klein 2006; Black et al. 2013). Vulnerability assessment provides a starting point for the determination of effective means of promoting remedial action to limit impacts by supporting coping strategies and facilitating adaptation (Kelly and Adger 2000). Identification of particularly vulnerable nations or regions can act as an entry point for both understanding and addressing the processes that cause and exacerbate vulnerability (Smith et al. 2008; Islam et al. 2009; Saito 2013; Butler et al. 2016). As a factor in determining the level of vulnerability, adaptive capacity is often used when assessing the potential to adapt to future climate change, although it is important to remember that whether or not such adaptive capacity is drawn upon to bring about adaptation will be dependent upon a range of uncertain variables (Vincent 2007; Black et al. 2013). Knowing other adaptive capacities is thus crucial in determining the impacts of climate change and necessary adaptation responses.

15.2.3 Vulnerability and Adaptation in Southeast Asia

Dynamic characteristics are found in South Asian regions, including tropical rainforests, monsoon climates, and high population density (Ahmed 2006; Denissen 2014). Despite the diversity in politics, economy, and culture, the region shares a common challenge in the face of climate change: it is one of the most vulnerable regions in the world. Its vulnerability is attributed to a number of reasons: geographical (tropical climate, long coastlines, etc.), demographical (highly dense populations in coastal and delta areas), and economical (necessary economic activities in delta areas for livelihood) (Smith et al. 2008; Islam et al. 2009; Saito 2013; Butler et al. 2016). According to the IPCC Fourth Assessment Report, warming is likely to be above the global mean in Southeast Asia (Islam 2009; Atta-Ur-Rahman 2015; Alam 2018; Ahmed 2018). Precipitation in boreal winter is likely to increase in the southern parts of Southeast Asia, whereas summer precipitation is likely to increase in most parts of Southeast Asia. In Southeast Asia, tropical cyclones are increasing, with extreme rainfall and winds. In addition to the impacts of climatic change and variability, the region also faces increasing vulnerabilities arising from rapid urbanization, degradation of resources, and unsustainable development (Abedin et al. 2012; Chaudhuri and Mishra 2016). Adequate adaptation is already underway in response to the expected outcomes of climate change in this region. Although the region has made significant progress in adapting to climate change, many steps remain to be done. One priority is to strengthen the overall adaptive capacity through greater efforts to raise public awareness; more research to better understand climate change and its impacts, especially at the local level; increased policy and planning coordination; and mainstream adaptations in development planning (Smith et al. 2008; Islam et al. 2009; Saito 2013; Butler et al. 2016).

15.2.4 Observed Climate Trends and Variability

On the basis of surface air temperature, previous and current climate variability can be characterized in South Asia. The observed increases in some parts of Asia during recent decades ranged between less than 0.1 °C to 0.3 °C per decade (Smith et al. 2008; Islam et al. 2009). Interseasonal, interannual, and spatial variability in rainfall trends have been observed during the past few decades across all parts of Asia. Decreasing trends in annual mean rainfall were observed in Indonesia and Philippines, whereas annual mean rainfall exhibits increasing trends along the western coast of the Philippines (Table 15.3).

There is new evidence on recent trends, particularly on the increasing intensity and frequency of extreme weather events in Asia over the last century and into the twenty-first century. South Asia El-Nino effects have been increasing this century, along with other extreme weather events. Significantly longer heatwave durations have been observed in many countries of Asia, as indicated by pronounced warming trends and several cases of severe heat waves (Islam 2009; Atta-Ur-Rahman 2015; Alam 2018; Ahmed 2018). Intense precipitation events have occurred in many parts of Asia, causing floods, landslides, and debris flow. In contrast, some countries have reported a decreased frequency of precipitation (Smith et al. 2008; Islam et al. 2009; Saito 2013; Butler et al. 2016. The increasing frequency and intensity of droughts in many parts of Asia can be attributed to increased temperatures, particularly during the summer and normally drier months, and during ENSO events (Karim and Mimura 2008) (Table 15.4). The intensity and frequency of tropical cyclones have developed in last few decades from the Pacific Ocean. The damage caused by intense cyclones has risen significantly in the affected countries, particularly the Philippines, Vietnam, and Cambodia.

Table 15.3 Key past and present climate trends and variability in Southeast Asia

Country	Change in temperature	Change in precipitation
General	0.1 °C to 0.3 °C increase per decade reported between 1951 and 2000.	Decreasing trend between 1961 and 1998. Number of rainy days have declined throughout the region
Indonesia	Increase of 1.04 °C–1.40 °C per century.	Decline in rainfall in southern and increase in northern region.
Malaysia	Warming trends (0.18 °C per decade) from 1951 to 1996.	More frequent the El Niño-Southern Oscillation (ENSO) warm-phase episodes since 1977 significantly influenced rainfall in Malaysia.
Philippines	Increase in mean annual, maximum, and minimum temperatures by 0.14 °C between 1971 and 2000.	Increase in annual mean rainfall since 1980s and in number of rainy days since 1990s, increase in interannual variability of onset of rainfall.
Singapore	Increase by about 0.3 °C per decade between 1987 and 2007.	Decrease in annual rainfall in the past three decades.
Thailand	Increase of 1.04 °C–1.80 °C per century.	Decreasing annual rainfall for the last five decades.
Vietnam	Increase of 1.0 °C per century.	Decrease in monthly rainfall during July–August and increase in September–November.

Source: Adopted and modified from Pereira (2010), IPCC (2011), Abedin et al. (2012); Dasgupta et al. (2015), and Chaudhuri and Mishra (2016)

Table 15.4 Observed changes in extreme events and severe climate anomalies in Southeast Asia

Extreme climatic events	Key trends
Cyclones/typhoons	On an average, 25 cyclones cross the Philippines Area of Responsibility (PAR) with about eight to nine landfalls each year; this is an increase of 4.2 in the frequency of cyclones entering PAR during the period 1990–2010
Intense rains and floods	Increased occurrence of extreme rains causing flash floods in Vietnam; landslides and floods in 1990 and 2004 in the Philippines; and floods in Cambodia in 2000.
Droughts	Droughts normally associated with ENSO years in Myanmar, Laos, Philippines, Indonesia, and Vietnam; droughts in 1997 to 1998 caused massive crop failures and water shortages and forest fires in various parts of Philippines, Laos, and Indonesia.
Heatwaves	Increase in hot days and warm nights and decrease in cold days and nights between 1961 and 2004.

Source: Ali (1999), Islam (2009), Atta-Ur-Rahman (2015), Alam et al. (2017), Ahmed (2018)

15.3 Climate Change Impacts Observed in Bangladesh

Bangladesh, ass one of the least developed countries, and felt the impact of climate change since the beginning of the industrial age (Smith et al. 2008; Islam et al. 2009; Saito 2013; Butler et al. 2016) because of its geographical position, population density, and underdeveloped infrastructure (Amin and Rahman 2014). Many climatic events have exacerbated the poor health of people in the northern part of

Bangladesh due to the intrusion of salt in water, effects on the surrounding habitat, and damage to infrastructure (Karim and Mimura 2008). This section presents some examples of such climatic events and their impacts on the environment.

15.3.1 Floods

The most common water-related natural hazard in a deltaic floodplain such as Bangladesh is flooding. Most of Bangladesh lies in the delta of three of the largest rivers in the world (GBM): the Brahmaputra, the Ganges, and the Meghna (Bhuyan et al. 2018). These rivers make a collective peak discharge in flood season of 180,000 cubic meters per second (the second highest in the world after the Amazon) and bear about 2 billion tons of sediment every year. With a low and flat topography, the country is less than 5 meters above sea level; thus, it is susceptible to river and rain-water flooding, as well as to tidal flooding during storms in lower-lying coastal areas (Smith et al. 2008; Islam et al. 2009). One quarter of the country was inundated in recent years. Such floods occur as waters from the hilly upstream rush to the plains at a high velocity, destroying standing crops and causing loss of life and substantial damage to infrastructure, housing, agriculture, and livelihoods. During severe floods, the poorest and most vulnerable residents suffer the most because their houses are located in the most susceptible locations.

Particularly in the Haor areas, flash floods cause extensive damage to crops and property. Each year, the eastern part of the foothill regions face extreme damage to boro rice production from early floods (April–May). Damage to roads and railway embankments, bridges, and buildings parallel to river channels occurs during exceptionally high flash floods (Amin and Rahman 2014. Sediment deposited in channels reduces the drainage capacity of minor rivers, culverts, and irrigation and drainage canals. Flood embankments next to certain eastern rivers, particularly the Khowai, are split by floods virtually every year. Arable lands and lands adjacent to foothill streams sometimes are buried under sand (Smith et al. 2008; Islam et al. 2009; Saito 2013; Butler et al. 2016). Climate change is causing increased monsoon precipitation, higher transboundary water flows, and riverbank erosions during high floods. In the last 25 years, Bangladesh has experienced severe, successive, and damaging floods, as shown in Table 15.5.

15.3.2 Tropical Cyclones and Storm Surges

Every 3 years, Bangladesh is hit by severe tropical cyclones because the coastal zones of the country are increasingly vulnerable to climate change–driven cyclonic storm surges. Not only has the frequency of cyclones along the Bay of Bengal increased as a response to rising sea surface temperatures, but cyclonic intensity has also increased, with a corresponding increase in surge height in newly flooded

Table 15.5 Climate change impacts in Bangladesh at a glance

Types	Description of impact	Sources
Salinity	There are 2.8 million hectares of land, including deltaic floodplains and offshore islands, affected by salinity and poor-quality water in Bangladesh.	Haque (2006), Khan et al. (2011), and Dasgupta et al. (2015)
Sea level rise	Statistics indicate that by the year 2050, a 45-cm rise of sea level may submerge 10–15% of the land and displace more than 35 million people from the coastal regions.	MoEF (2009)
Climate refugees and migration	Approximately 16 to 26 million people are likely to migrate long-term from areas affected by inland flooding, storm surges, and riverbank erosion in Bangladesh.	Dasgupta et al. (2015) and Kabir and Baten (2019)
Tropical cyclones and storm surges	Records of the last 200 years show that at least 70 major cyclones hit the coastal belt of Bangladesh and during the last 35 years nearly 900,000 people died due to catastrophic cyclones	PDO-ICZMP (2004:14)
Floods	Approximately 34% of the land area is flooded for about 5–7 months every year	Islam (2004 cited Annya Chanda Shimi 2010)
Drought	Bangladesh has been affected by droughts approximately 20 times in the past 50 years	Chowdhury (2002) and Habiba and Sha (2011)
Biodiversity loss	Up to 30% of animal and plant species of the Sundarbans may be wiped out by a global temperature rise of 2.7 °C–4.5 °C	Nishat and Chowdhury (2019)
High temperatures	Temperatures could increase 2.4 °C higher than the current level by 2100	Jahan and Quddusi (2014)
Desertification	The ratio of cultivable land to rural population (acre/person) has decreased in the northwestern area by 23.2% compared with a decreased ratio of 17.2% in the whole of the country.	Choudhury et al. (2006)
Food production	IPCC estimates that, by 2050, rice production in Bangladesh could decline by 8% and wheat by 32% (against a base year of 1990), which could lead to serious food insecurity.	BCCSAP (2009)

Source: Islam (2009), Atta-Ur-Rahman (2015), Alam et al. (2017), and Ahmed (2018)

shorelines; furthermore, the surge height along the continental shelf has decreased as a consequence of sea level rise (Smith et al. 2008; Islam et al. 2009; Amin and Rahman 2014). These storms are accompanied by high winds of more than 150 kph, causing extensive damage to houses and high loss of human life and livestock in coastal communities. For example, the tropical cyclones in 1970 and 1991 killed approximately 500,000 and 140,000 people, respectively (Table 15.6). The storm surges are higher in Bangladesh than in neighboring countries. Cyclonic activity in the Bay of Bengal has become more frequent, causing rougher seas that make it difficult for fishermen and small craft to sail.

Table 15.6 Impacts of SIDR on crops and livestock in some villages of Pirojpur

Crops	Pre-SIDR	Post-SIDR	Livestock	Pre-SIDR	Post-SIDR	Reduction
Coconut	81	74	Cow	294	242	1%
Betel nut	86	79	Goat	237	202	14%
Vegetables	88	80	Buffalo	112	89	21%
Pulses	5	25	Chicken	1218	426	10%
Chili	30	26	Duck	306	279	8.82%
Rice	97	86	Pigeon	50	38	24%

Table 15.7 Drought-affected areas by cropping season

	Size of drought-affected area by severity class (in million ha)					
Crop season	Very severe	Severe	Moderate	Slight	Unaffected	Non-T. Aman rice
Pre-Kharif	0.407	1.17	4.78	4.11	2.12	–
Kharif (T. Aman rice only)	0.348	0.75	3.19	2.96	0.69	4.73
Rabi	0.447	1.73	2.96	4.24	3.18	

During SIDR, saline water damages the crop lands and reduces production. Damage to river embankments reduces fish and vegetables production. Approximately two million people have lost income due to the impact of SIDR.

15.3.3 Droughts

Droughts in Bangladesh are primarily a seasonal agricultural phenomenon that affect huge amounts of crops, causing hardships for poor agricultural laborers and others who cannot find work (Smith et al. 2008; Islam et al. 2009; Saito 2013; Butler et al. 2016). Bangladesh receives an excessive amount of water during monsoons, causing floods, and insufficient water in the dry season, causing droughts (Table 15.7). Monga (unemployment-related seasonal hunger) is often a complication, mostly in the months leading up to the November–December rice harvest. Severe drought conditions in the northwestern regions, which have lower rainfall compared with the rest of the country, can lead to total crop failure (Amin and Rahman 2014; Bhuyan et al. 2018).

The combined effects of rising temperatures, greater precipitation, severe flooding, occasional seasonal droughts, and loss of cultivable land in coastal areas resulting from climate change have caused declines in rice production of 3.9% every year, or an accumulative total of 80 million tons (Smith et al. 2008; Islam et al. 2009; Saito 2013; Butler et al. 2016). Drought caused the growth of continuous broken cracks on the dried-up topsoil, burnt-out yellowish foliage in the vegetation cover (top yellow syndrome; particularly in betel nut trees and bamboo groves), and loss of soil structure, which results in the topsoil transforming into a dusty layer (Amin and Rahman 2014). Pre-monsoon droughts commonly affect major crops, including

HYV Boro, Aus, wheat, pulses, sugarcane, and potatoes. Post-monsoon droughts affect Aman crops. Drought caused by climate change occurred 19 times between 1960 and 1991. Extreme droughts hit the country in 1951, 1961, 1975, 1979, 1981, 1982, 1984, 1989, and 1995 (Table 15.8).

15.3.4 *Heavy Rainfall*

Countries face inconsistent rainfall behavior. Rainfall during and after a monsoon was found to be increased, while pre-monsoon and winter rainfall has decreased. Heavy rainfall causes serious damage, including soil erosion, water logging in cultivated land, drinking water contamination, the spread of many water-related

Table 15.8 Most vulnerable drought hotspots districts and upazillas' in Bangladesh

		Vulnerability	Ranking		Combined	Severity
District	Upazila	Rabi	Kharif-1	Khari-2	ranking	ranking*
Naogoan	Niamatpur	7	5	3	15	1
Rajshahi	Tanore	6	2	11	19	2
Nawabganj	Nchole	9	7	4	20	3
Naogoan	Porsha	4	3	16	23	4
Naogoan	Sapahar	5	4	17	26	5
Thakurgoan	Baliadangi			30	30	6
Dinajpur	Hakimpur	18		12	30	7
Naogoan	Patnitala	12	18	1	31	8
Nawabganj	Shibganj	11	8	14	33	9
Joypurhat	Panchbibi	36		2	38	10
Rangpur	Badarganj			39	39	11
Rajshahi	Godagari	13	11	15	39	12
Naogoan	Mahedebpur	8	31	6	45	13
Nawabganj	Gomastapur	16	9	22	47	14
Rajshahi	Shah Makhdum	25	13	10	48	15
Rajbari	Goalanda			50	50	16
Rangpur	MithaPukur			51	51	17
Panchagrah	Tentulia			53	53	18
Thakurgaon	Thakurgoan			54	54	19
Rajshahi	Durgapur		29	25	54	20

Source: Islam (2009), Atta-Ur-Rahman (2015), Alam et al. (2017), Ahmed (2018)

The drought that occurred during 1978–79 caused massive harm. In 1995, a drought occurred during the late Kharif (post-monsoon drought) period and caused a loss of 377,000 tons of Aman production

*Combined ranking means the sum of the vulnerability ranking of Rabi, Kharif-1 and Kharif-2 cropping pattern in the case study area. But the severity ranking shows the ranking of the less drought vulnerable hotpots area. It means that low severity rank means drought vulnerable hotpots area is less and high severity ranking means the large number area affected by the drought vulnerability

diseases (Adger et al. 2007), landslides and avalanche damage, loss of life, damage to buildings and property, and loss of livelihoods due to longer periods of flooding (Depledge and Lamb 2005).

15.3.5 Riverbank Erosion

Riverbank erosion is a major threat to Bangladesh. According to the National Water Management Plan, at least 10,000 hectares of land are destroyed each year due to riverbank erosion (Government of Bangladesh 2010). The losses caused by erosion are usually slow and gradual. However, they are more destructive in the long run than other sudden and devastating calamities. This situation is even more worrisome considering that approximately 1 million people are affected by the erosion of riverbanks each year (Working Group on Equity and Justice, 2007 cited in IOM, 2009). Given the magnitude of this phenomenon and the number of people that are at risk, riverbank erosion has the potential to displace large numbers of people in the future. Nearly 130,000 people are displaced each year due to riverbank erosion (Mollah and Ferdaush 2015).

15.3.6 Heat Indices

The heat index considers the effects of humidity as well as temperature on the human body. This unified effect is causing serious threats to human health because of climate change. Climate variability and thus the phenomenon of heat waves have increased throughout Bangladesh, especially during summer (Smith et al. 2008; Islam et al. 2009). Winters have become shorter, whereas the summer season is prolonged and hotter, thus affecting the lives of people who are occupied in outdoor activities in scorching sun. In 2003, approximately 62 people died because of heat waves throughout the country. Severe heat waves affect human health and individual performance, causing heat cramps, heat exhaustion, discomfort, heat stroke, and fatigue (Bhuyan et al. 2018). Heat index increases can lead to death for humans and animals.

15.3.7 Salinity Increases

Sundarban is the largest mangrove in the world, located in the outer GBM deltas. The Sundarban ecosystem is affected by lower flow of upstream water. The alarming decrease in water flow down the rivers is causing high salinity in both the water and soil of Sundarban, causing a massive change in faunal composition of the forest (Dasgupta et al. 2015). The number of large timber-producing trees, such as Sundari, are decreasing at a rate proportionate to the increase in salinity (Dasgupta et al. 2015). The latest report from World Conservation Monitoring warned that long-term ecological changes are taking place in Sundarbans due to the eastward migration of the Ganges, abandonment of some distributaries, past diversion of water, and withdrawals for irrigation.

15.3.8 Agriculture

The vast majority of the population in Bangladesh depends on the agricultural sector for their livelihood, as it employs approximately 60% of the labor force. Only 1.2 million hectares of cultivated land in Bangladesh are available for production (Amin and Rahman 2014; Bhuyan et al. 2018). The sector is extremely vulnerable to higher temperatures, changing rainfall patterns, increased flooding and droughts, and rising salinity in coastal areas that affect crop yields and crop production (Hoque and Haque 2016). The IPCC estimates that, by 2050, rice production in Bangladesh could decline by 8% and wheat by 32% (against a base year of 1990), which could cause serious food shortages in Bangladesh (BCCSAP 2009) (Table 15.9).

15.3.9 Health

Climate change and the resulting natural disasters are likely to increase the incidence of water-borne and vector-borne diseases. Saline intrusion of fresh drinking water will also lead to increased health hazards (Abedin et al. 2012). In fact, the threat that floods pose in terms of increasing sicknesses and deaths from diarrhea, malaria, cholera, dengue fever, and the like, could eventually lead to exorbitant levels of spending on preventable public health epidemics.

15.4 Comparison with Other Countries and Regional Perspective

India and Pakistan have both faced major heat waves, with temperatures hot enough to soften roads. Jacobabad in Pakistan saw a temperature of 51 °C, making it the hottest place on earth (Smith et al. 2008; Islam et al. 2009; Saito 2013; Butler et al. 2016). In India, the highest temperature occurred within the northern state of Rajasthan, reaching 50.8 °C. For many, heat waves turn deadly. One heat wave in India claimed 184 lives within the Jap province alone. Delayed monsoon rains intensify water-scarce conditions across South Asia (Chaudhuri and Mishra 2016). By 2100, three out of four individuals on earth may be subject to a minimum of 20

Table 15.9 Adaptation measures and requirements for crop cultivation under climate change

Adaptation Measures	Requirements	Comment
Bear loss (no adaptation) Loss of production Loss of assets		Hypothetical, highly unlikely to take place.
Share losses Crop insurance Cooperative management Governmental subsidies	Additional investment in terms of premium. Agreement for sharing the output. State allocation for offering subsidies. Adequate legal and institutional framework.	Provisions to be made. Political motivation is required.
Modify the threats Preparedness (early warning) Awareness and training Investment for structural measures	Research and training Extension, media campaign Investments (anticipatory) Crop calendar adjustment Opting for less susceptible crops	Farmers are already practicing it, based on ancestral behavior/ knowledge. Manifold opportunities are plausible; barrier removal and implementation could be less costly. High-priority option.
Prevent adverse effects Structural measures	Large investment Political wiliness Long-term planning	Investment intensive option. Financial constraints might hinder implementation process.
Change land use Alternative cropping Abandon crop agriculture	Innovation through research, investment Means of survival, skills for alternative employment	Unless alternative employment opportunities are created, it is not likely to be accepted socially.
Change location Relocate to less vulnerable places	Free cultivable land	Heavily constrained due to unavailability of fallow cropland.

days of excessive heat each year if temperature changes are not addressed (Dasgupta et al. 2015). The issue is especially acute in megacities, where searing heat is exacerbated by dense concrete structures and mingles with cyanogenic pollution, putting the lives of urban dwellers in danger. Worsening urban heat may result in a worldwide health care crisis that will cost billions worldwide.

Nepal is also a vulnerable country with fragile geology, steep topology, inadequate human resources, and a poor economy. Nepal has very diverse climatic conditions. Global circulation model (GCM) projections indicate that the temperature over Nepal may increase between 0 5 °C and 2.0 °C with a multi-model mean of 1.4 °C by the 2030s and between 3.0 °C and 6.3 °C, with a multi-model mean of 4.7 °C, by the 2090s. GCM outputs suggest that extraordinarily hot days (the hottest 5% of days in the period from 1970 to 1999) are projected to increase by up to 55% by the 2060s and up to 70% by the 2090s. GCM outputs suggest that extremely hot nights (the hottest 5% of nights in the period from 1970 to 1999) are projected to increase by up to 77% by the 2060s and 93% by the 2090s.

Bhutan has an outstanding natural environment, with rugged mountains and deep valleys that comprise a rich and diverse ecosystem. Bhutan is the only country in the world with a constitution mandating that there always be 60% forest cover; very strong legal instruments have been put in a place to protect its rich biodiversity (Smith et al. 2008; Islam et al. 2009). However, this biodiversity faces many challenges, both manmade and natural. Climate change poses a grave danger to sustainable development, as well as the lives and livelihoods of its people (Trenberth et al. 2007; Andreasen et al. 2010; López 2015). Glaciers are retreating rapidly and dramatically increasing catastrophic glacial outburst floods. In Bhutan, 60% of the people depend on subsistence agriculture for their livelihoods, but agriculture is greatly affected by climate change. The monsoon season has become very erratic; some mountain springs have ceased to flow and farmers have to use other means of water for cultivation in the lean season. An increasing number of flash floods, landslides, and wild storms also threaten crops, property, and critical infrastructure. Bhutan is heavily reliant on hydroelectricity (Smith et al. 2008; Islam et al. 2009). Long-term threats to water security, agriculture, food security, and human settlement need to be addressed by mitigating GHG emissions, protecting forest lands, and adopting lower-emission approaches in all sectors. To remain carbon neutral, and thus safeguard the lives and livelihoods of its people, Bhutan needs the international community.

The impact of climate change on Singapore is alarming, as the country is heating twice fast as the rest of the world. By 2100, the maximum daily temperature is expected to reach 35–37 °C. Singapore is more affected because it is surrounded by the sea and always has high humidity, which causes health concerns for its residents. Singapore's industrialization is a major contributor to its rapid temperature increase. Air conditioners in its large buildings are also emitting heat to the outside and increasing temperature. Singapore must reduce the burning of fossil fuels and its carbon dioxide emissions, as well as prevent rapid development in urban areas. The construction of roof or wall gardens may help, although the research does not show that these greening initiatives reduce the urban heat island effect. To fight climate change, Singapore established the Interministerial Committee on Climate Change.

Myanmar is at high risk of climate change events. For example, in 2008, Cyclone Nargis claimed the lives of 140,000 people and displaced another 800,000 people. According to a German think tank who published a risk index 2018, Myanmar is among the top three countries at high risk for climate change impacts, including flooding cyclones and droughts. Many people in Myanmar rely on farming to provide food and income, so their livelihoods are at risk (Smith et al. 2008; Islam et al. 2009). Aid agencies have been working since 2008 to improve the country's ability to survive catastrophic weather incidents. In Myanmar, monsoons decreased in the 1990s, after which the heat and drought indices increased. After 1977, the annual rainfall began to decrease. In the 1980s and 1990s, several abnormal synoptic situations were recorded (Stern 2006; Solutions 2012; Saito 2013; Tiwari and Joshi

2014). The sea surface temperature is also increasing, which causes storms, cyclones, and hurricanes. In 2011, Myanmar had its heaviest rainfall, with its highest temperature recorded in 2010. Thus, it appears that Myanmar is already facing the consequences of climate change.

15.5 Adaptation Plan Against Climate Change Impacts in Bangladesh

In Bangladesh, adaptation policies for climate change are included under NAPA (2005 and 2009) and BCCSAP (2009). Adaptation strategies are moving toward mainstreaming as a main objective (Smith et al. 2008; Islam et al. 2009; Saito 2013; Butler et al. 2016). Many organizations are involved in adaptation policies, including the Ministry of Water Resources, Ministry of Agriculture, Ministry of Environment and Forests, Ministry of Planning, and the Ministry of Finance.

In 2009, the BCCSAP prepared a document describing the implementation challenges of such policies (Smith et al. 2008; Islam et al. 2009). Instead of transboundary coordination, BCCSAP and NAPA have emphasized scalability. The policies have specific guidelines for floods, but comparatively little attention is paid to microlevel planning in villages (Andreasen et al. 2010; Amin and Rahman 2014). BCCSAP and NAPA collaborate with other management policies in a reflexive way. Certain sectors are prioritized in BCCSAP and NAPA, including water sector, agricultural production, infrastructure, disaster response, food security, and energy. To make strong and target-based connections between sectors, annual development plans (the sixth and seventh 5-year plans) have included adaptation strategies. In addition, a gender action plan connects the gender aspects and climate-affected sectors (Smith et al. 2008; Islam et al. 2009; Saito 2013; Butler et al. 2016). Bangladesh's government has budgeted $45 million to address adaptation to climate change via NAPA (2005) and BCCSAP (2009) (Fatima et al. 2018; Bhuyan et al. 2018).

15.5.1 The National Adaptation Programme of Action in 2005

The Ministry of Environment and Forest adopted NAPA in 2005, which aggregates all appropriate stakeholders to understanding the current state of affairs via four subnational workshops and one national workshop (MOEF 2005). The objective of NAPA was to address environmental issues and natural resource management by highlighting sustainable development goals (BCAS 2008). NAPA stated that an unstable economy, insufficient infrastructure, lack of institutional framework, and overdependency on natural resources turned into Bangladesh into one of the most adversely affected countries in the world (Smith et al. 2008; Islam et al. 2009; Saito 2013; Butler et al. 2016). NAPA aimed to mitigate the adverse effects of climate change using coping mechanisms and practices.

In 2005, NAPA identified the adverse impacts of climate change, including higher evapotranspiration and low rainfall occurring from scarcity of freshwater, drainage congestion due to higher levels of water, frequent flood and riverbank erosion, and migration in coastal zones because of salinity intrusions in water and soil. However, the links to climate displacement were not defined in concrete terms (Black et al. 2013; Mohammad 2015; Bhuyan et al. 2018). The document also stated that possible long-term effects of the "adaptation to agriculture systems in areas prone to enhanced flash flooding" would be that "people might get a means to continue with farming, instead of reducing to cities after the flood." To tackle climate change, NAPA has considered medium-term and long-term issues for adaptation. It places importance on food security, energy security, water security, and livelihood security issues (including human rights and health care), as well as respect for local communities' resource management (Haque 1996).

15.5.2 Bangladesh Climate Change Strategy and Action Plan in 2009

The Bangladesh government formulated BCCSAP in 2009 to address mitigation and adaptation capacity and reliance on climate change for the decade of 2009–2018. To address possible challenges and variations in climate change conditions, the 10-year program plan was used (BCCSAP 2009). BCCSAP addresses all natural hazards, such as cyclones, storm surge, floods, high rainfall, and salinity intrusion, and assesses their impacts in all sectors.

BCCSAP features a set of measures with six major themes:

- Food security, social protection, and health
- Comprehensive disaster management
- Infrastructure
- Research and knowledge management
- Mitigation and low carbon development
- Capacity building and institutional strengthening

The BCCSAP noted that, "Many people of different zone of the country have to shift one day to other places. The process of migration of climate change-affected people, both inside and outside the country, needs to be monitored closely and adequate institutional support should be provided for their proper resettlement."

In addition to these two mechanisms, other traditional fund mechanisms exist to address the issue of climate change, as discussed in the next two sections. The Bangladesh Climate Change Trust Fund (BCCTF) was created by a legal mandate of the Bangladesh Climate Change Trust Act in 2010 based on national budget revenue. In addition, the Bangladesh Climate Change Resilient Fund (BCCRF), also known as the Multi-donor Trust Fund (MDTF), was created to provide funds from the country's development partners.

15.5.3 Bangladesh Climate Change Trust Fund

The Bangladesh Climate Change Trust Fund (BCCTF) is the first ever national climate fund established by a least developed country (LDC) and is an example to other countries for institutionalizing national climate finance. The BCCTF funds programs and projects from the national budget to help communities recover and become resilient to climate change impacts. Operational since 2010, the fund is currently managed by the Bangladesh Climate Change Trust and (BCCT) and the government and has allocated Tk2900cr during the last six fiscal years until 2014–2015 (Ministry of Water Resources 2005; Haque et al. 2013a; Sarker et al. 2013; Fatima et al. 2014). From 2009 to 2012, a US$100 million allocation was provided in each year's budget for the BCCTF. Of this, 66% of the total budget was allotted for the implementation of BCCSAP, whereas 34% was spent as a fixed deposit for emergencies. The interest accrued on the 34% fixed deposit was spent for project implementation. The BCCTF funds were used for nongovernment projects in 10% and financial public sectors. However, it was not mandatory to spend all annual funds within the same year (Haque et al. 2013a, b).

15.5.4 Bangladesh Climate Change Resilience Fund

The government created the BCCRF to address climate change using funds from international development partners of Bangladesh. In 2008, a UK–Bangladesh Climate Change Conference was held in Dhaka. The international development partners expressed a desire to establish a financial mechanism to reduce climate change impacts in Bangladesh (Smith et al. 2008; Islam et al. 2009). Thus, in 2010, BCCRF was established by the Bangladesh Development Forum and was managed by the Government of Bangladesh (GoB). For the next 3 years, the World Bank provided fiduciary backup to BCCRF over the GoB. During this period, BCCRF gained $200 million USD from many development partners. From this fund, 90% was spent by ministries and the other 10% was collected by Palli Karma Sahayak Foundation (PKSF) to help NGO activities (Hoque 2007).

15.5.5 Sectoral Adaptation Policies

Since 1990, Bangladesh has created various adaptation policies to mitigate climate change impacts. Although Bangladesh is highly vulnerable to climate change, the coastal zone has only one sectoral policy on climate change (BCAS 2010). By including short-, medium-, and long-term perspectives in 1999, the National Water Policy (NWP) was drawn up. The NWP was organized by the National Water Management Plan (NWMP) in 2001. Despite the various effects of climate change on water resources, the NWMP did not include climate issues at all. After that, in 1995, the National Environmental Management Action Plan (NEMAP) was published, although it did not expound on climate change policy. In the same way, the National Forest Policy (NFoP) and National Land Use Policy (NLUP) did not adequately address climate change issues, nor did the Poverty Reduction Strategy (PRS). However, more recently, PRS noted the adverse impacts of climate change in sustainable development and expressed the need for integration/mainstreaming (Black et al. 2013; Mohammad 2015; Bhuyan et al. 2018). Adaptation measures for policy areas and adaptation projects are found in the NAPA (BCAS 2010).

15.5.6 Institutional Mechanisms

In Bangladesh, many departments are developing institutional mechanisms for climate change. The Ministry of Environment and Forest (MoEF) is coordinating the formulation of policies and adaptations in climate change. The Department of Environment (DOE) and Climate Change Unit (CCU) is maintaining programs at the community level with the help of NAPA and BCCSAP. The National Steering Committee on Climate Change (NSCCC), headed by the Minister of MOEF, works with stakeholders of all climate-affected sectors (Table 15.10). The National Environment Committee is expected to mainstream climate change into national development planning. Climate Change Focal Points (CCFP) in various ministries have been working through five technical groups constituted for adaptation, mitigation, technology transfer, financing, and public awareness (Country Summary Report 2010).

15.6 Regional Adaptation Responses and Initiatives

To mitigate climate change impacts, efforts give rise to opportunities in international affairs. Climate change has affected the whole world, ranging from a mere change in temperature to the extinction of an entire species for that very same reason. The only way to reverse and survive is to cooperate with neighboring countries (Smith et al. 2008; Islam et al. 2009). Bangladesh has joined alliances with its neighboring countries, such as in The South Asian Association for Regional Cooperation Alliance (SAARC). SAARC was created in 1985 as a joint treaty between Afghanistan, Bangladesh, Bhutan, India, Maldives, Nepal, Pakistan, and Sri Lanka. The purpose of its creation is to promote the welfare of the peoples of South Asia and to improve their quality of life; to increase economic growth, social progress, and cultural development in the region; to provide all individuals the opportunity to live in dignity and to realize their full potentials (Black et al. 2013; Mohammad 2015; Bhuyan et al. 2018); to promote and strengthen self-driven living among the countries of South Asia; to contribute to mutual trust, understanding, and

Table 15.10 Community-led adaptation actions of the Bangladeshi coastal communities

Adaptation sector		Description of adaptation	Sources
Livelihood	Changing rice farming to non-rice farming	Different varieties of crops are farmed, such as cultivation of jute, wheat, plums, and different types of pulses	Sarkar et al. (2013)
	Increasing involvement in a variety of income sources	Money is earned by wage labor, small business, construction work, and livestock farming	Pouliotte et al. (2009) and Abedin et al. (2013)
	Selling land and taking loans	Poor households often temporarily adapt to extreme climate events by selling land and taking loans	Alam (2002), Alam (2003), and Pouliotte et al. (2009)
	Gender dimensions	Women are forced to take difficult jobs outside their comfort zones	Abedin et al. (2013)
Human habitation	Raising homesteads on plinths	Low-lying coastal and island inhabitants often raise their homesteads on plinths, much higher than those used by mainland people, to mitigate the severe effects of coastal flooding	Alam (2002) and Bhuyan et al. (2018)
	Planting of trees	Trees are planted around houses to reduce the intensity of storm surges	Alam (2002) and Bhuyan et al. (2018)
Health	Household coping strategies	Season-specific household-level strategies are used to prevent sickness and diseases caused by extreme heat, cold, and precipitation	Haque et al. (2013a, b) and Bhuyan et al. (2018)

Source: Adopted and modified from Islam (2009), Atta-Ur-Rahman et al. (2015), Alam et al. (2017), Ahmed (2018), and Bhuyan et al. (2018)

solving of one another's problems; to promote active mutual helping hand in the economic, social, cultural, technical, and scientific fields; to develop communication media among themselves in the same language with common interests; and to support each other with similar goals.

It is in the interest of all countries to be aware of the risks and prepare for them; the overarching purpose of this study is to support ASEAN and its member states in this area. South Asia's geographical location with various archipelagoes, complex border lines, and long coastal area will affect interstate relationships in this region (Smith et al. 2008; Islam et al. 2009). The ASEAN countries are facing several challenges by climate change. SAARC and ASEAN are assigning tasks according to the member countries' institutional capabilities. Countries in an area share the same water bodies, forests, ecological and natural resources, and culture. At the same time, countries are varied in terms of resource subsidizing, economic development, skills, and institutional capacities. Therefore, they have different capacities to tackle a problem like climate change (Black et al. 2013; Mohammad 2015; Bhuyan et al.

2018). Regional cooperation aids in exploiting the problems that exist between the nations in a region to achieve outcomes in terms of development, adaptation, and prevention (ML Parry et al. 2007; Andreasen et al. 2010). This is true for the South Asian region, given the shared vulnerabilities of the countries to climate change as well as the possibility of exploiting the similarities of the countries in dealing with the problems. Bangladesh has started installing compressed natural gas instead of traditional fuels to greatly reduce the effects of fossil fuels on the atmosphere and climate (Table 15.11), and they have seen good results from it. Because fossil fuels are one of the biggest drivers of climate change, people need to be educated about their impacts and the consequences they have on the climate, as well as taught about different adaptation possibilities. There are significant efforts toward planning and researching adaptations for climate change in Southeast Asia (Resurreccion et al. 2008; Smith et al. 2008; Islam et al. 2009). The policy and research approaches to adaptation in the region can be classified into the following five categories:

- National efforts to meet the obligations of the UNFCCC
- Assessment of climate change impacts and vulnerabilities
- Community-based adaptation and integrated approaches
- Disaster risk reduction and climate change adaptation
- Economic analyses and adaptation research

Policies may mention other specific themes for strengthening adaptation strategies in Southeast Asia, such as human approaches to climate change. These themes include migration, social security mechanisms, livelihood security for small-scale fishers and farmers, strengthened resilience to health-related impacts, and governance of adaptation across scales.

There are many dissimilarities between the developed world and the South Asian region. Some issues need to be addressed to ensure regional adaptive capacity, such as the following:

- Income, inequality, poverty, literacy, and regional disparities
- Capacity and governance of public institutions and public finance
- Availability or adequacy of public services, including education, health, social protection, and social safety nets
- Capacity for economic diversification, especially at local levels

Adaptation spans from scientific assessment (climate change projections, impact assessment, and vulnerability analysis) to policy facilitation (capacity-building, policy formulation, and planning) and response implementation (piloting, demonstrating, and full-scale implementation of adaptation actions) (Smith et al. 2008; Islam et al. 2009; Saito 2013; Butler et al. 2016). It is a knowledge-intensive undertaking, which requires provision of access to relevant and usable knowledge to ensure effective implementation and outcomes. Knowledge is needed to address uncertainties associated with climate change and its impacts, and the societal responses to render adaptations (Smith et al. 2008; Islam et al. 2009). Thus, it is important to manage and share the scientific, technological, and socio-economic insights of the adaptation process. The goal of knowledge management and sharing

Table 15.11 Priority response areas to advance climate change adaptation in the Asia Pacific

Theme	Priority areas
Data collection, management, and dissemination	Establishment and maintenance of observation facilities
	Collection and compilation of climatic, social, and biophysical data;
	Improvement of information and data management, sharing, and networking.
Biophysical impact studies	Basic physiological and ecological studies on the effects of changes in atmospheric conditions.
Socioeconomic aspects	Social vulnerability to multiple stressors due to climate change and environmental change;
	Identification and characterization of vulnerable communities and groups;
	Capability to diversify local economies, livelihoods, and coping strategies beyond tackling the natural systems;
	Migration as adaptation strategy and the support systems and social networks;
	Institutions and mechanisms supporting social security;
	Economic analysis of climate change impacts and adaptation interventions.
Sectoral and cross-sectoral assessments	More proactive, systematic, and integrated approaches to adaptation in key sectors (agro-technology, water resource management, integrated coastal zone management, pathology, and disease monitoring and control; etc.)
	Sectoral interaction, such as between irrigation and water resources, agricultural land use and natural ecosystem, water resources and cropping, water resources and livestock farming, water resources and aquaculture, water resource and hydropower, sea-level rise and land use, sea-water invasion and land degradation.
Extreme event and critical threshold, and disaster risk reduction	Impacts of extreme weather events such as disasters from flood, storm surges, sea-level rise, heatwaves, plant diseases, and insect pests;
	Critical climate thresholds for various regions and sectors;
	Linkage and integration of responses on climate change adaptation and disaster risk reduction.
Policy responses, implementation, and integration	Mainstreaming science of climate change impacts, adaptation, and vulnerability in policy formulation;
	Governance across scales and stakeholder groups;
	Raising public awareness of climate change and its impact;
	Communication of research findings to different stakeholder groups;
	Private sector involvement;
	Integrated and balanced responses of adaptation and mitigation.

Source: Adopted and modified from Islam (2009), Atta-Ur-Rahman et al. (2015), Alam et al. (2017), and Ahmed (2018)

is to meet the knowledge needs for adaptation policy setting, planning, and implementation; contribute to improving the understanding and assessment of impacts, vulnerability, and adaptation; and make informed decisions on practical adaptation actions, with credible scientific, technical, and socioeconomic knowledge (Smith et al. 2008; Islam et al. 2009; Saito 2013; Butler et al. 2016).

Building an informed understanding of the capacity to adapt to the effects of climate change, while seemingly a straightforward task, is actually one that is affected by various challenges. Knowledge, information, and data are needed with respect to social and economic changes that might manifest in the future, such as population growth and development and the depreciation and aging of existing infrastructure (Smith et al. 2008; Islam et al. 2009). Scenario planning is one methodology that can support enhancement of adaptive capacity to an uncertain future. The approach is often useful for development of long-term strategies in an uncertain future, but also a suitable methodology in developing scenarios for the short term.

The IPCC, through its Fourth Assessment Report, pointed out several key specific research-related priorities for advancing the understanding of adverse impacts of climate change in Asia. Similar needs apply in the Southeast Asian region, which will require strengthening the academic and research institutions to conduct innovative research on the response of human and natural systems to multiple stresses at various levels and scales. Several subsequent regional studies further identified knowledge gaps, priority research aspects, and critical policy responses on climate change adaptation (Wang et al. 2018). Major priorities for Southeast Asia when moving forward on adaptation responses are summarized in Table 15.12.

15.7 Conclusions and Recommendations

Over the last 100 years, South Asia has gradually become vulnerable to consistent trends in global warming. Studies have shown that temperatures are increasing at higher rates in high-altitude areas compared with other areas (ICIMOD 2010). These high temperatures result in a reduction of permafrost and glacial area by a gradual melting of the ice sheets. Rainfall makes up the largest proportion of total precipitation in South Asia regions, but great snowmelts and short winters are also occurring. These variations result in natural hazards, as well as river and ecosystem changes. Immediate research is needed to develop climate-resilient strategies. Without adaptation and mitigation steps in national developmental programs, the progress in South Asia's economic development will be slowed by climate change.

At the community level in South Asia, climate change is viewed as a responsibility of developed countries because they are emitting high volumes of GHGs. This perception can be changed by strong political leadership. Climate change is a global problem that needs a local solution. All stakeholders, from individuals to institutions, must give attention and political support to climate change programs. These steps, when combined with pro-climate government support, will provide excellent opportunities for advocacy and awareness about climate change and human

Table 15.12 Needs for research and systematic observations in relation to climate change adaptation in South Asian countries

Issues	Needs
Research	Develop regional data for vulnerability and adaptation;
	Strengthen the research capacity and observation capability in climate, environment, natural resources, and land-use and cover change;
	Strengthen the research capacity in understanding the impacts of climate change, and developing appropriate adaptation strategies and measures;
	Tailor research and systematic observation to better understanding of the impacts of and adaptation to climate change in water resources, coastal zones, resources, agriculture, forests, biodiversity, fisheries, and human health;
	Modernize the technology (equipment and capacity) used in the current networks;
	Coordinate efforts regionally and internationally.
Observation	Currently no comprehensive marine/oceanographic observation programs in place;
	Upgrade meteorological and hydrological monitoring programs;
	Rehabilitate and expand the existing station networks for more representative monitoring of weather, climate, and other environmental variables;
	Upgrade and expand climate observation networks at the national level and improve the contributions to the global observing systems through development, utilization, and accessibility of databases;
	Lack of observation stations result in limited and unreliable climate data;
	Strengthen existing stations for data to enhance understanding of impacts of future climate change on agriculture, marine ecosystem, land use and forestry, biodiversity, waste and water resources;
	Improve maintenance of observation equipment coverage of systems through the provision of more resources (financial and technical).
Institutional strengthening	Establishment of strong and effective institutions to manage observation systems, including development of human resources and information technology;
	Invest resources to support observation programs;
	More programs that facilitate the exchange of experts from developing countries' institutions to those of developed countries;
	Lack of availability/accessibility to good quality data and poor research facilities and opportunities to undertake research.

Source: Adopted and modified from Islam (2009), Atta-Ur-Rahman (2015), Alam et al. (2017), Ahmed (2018)

development goals. Institution building, professional capacity, and technological knowledge can increase by collectively channeling funds and reducing the obstacles to socioeconomic development. Governments must provide environment and legal backing to translate policies, programs, and projects into action at the grass-roots level.

Climate change mitigation policy, especially adaptation, will be an important part of the development policies of the country. Responsible decision-making on global change issues requires scenario-based reasoning. The alternative to scenarios, which explicitly stipulate assumptions about key uncertainties, is to make implicit, unexamined assumptions about future conditions that resemble the present. This not likely to produce better decisions. The main alternative approaches proposed for global-change decisions are seeking decision strategies that are robust for major uncertainties, constructing scenarios of desired future conditions, and examining how to get there; these approaches do not avoid the need to specify relevant future trends. The changes to global-change scenario practices proposed here are likely to make scenarios more useful for decision-making. They are not likely, however, to make scenarios less controversial or less subject to attacks, such as those leveled against the IPCC and the US National Assessment. Global-change scenarios are controversial because they are powerful public framers of the issue, and because they act as proxies for the need to take action. Political actors with strong views on the desired actions have an interest in attacking scenarios that challenge their preferred progression (Smith et al. 2008; Islam et al. 2009; Saito 2013; Butler et al. 2016). Consequently, opponents of emission restrictions attack high-growth scenarios as biased or unrealistic and highlight low-growth ones, while proponents of limits do the opposite. It is easy to attack one scenario in isolation, exaggerating its predictive intent, describing it as "speculative" or "unscientific," and digging into its details to find elements that appear implausible or erroneous.

Climate change is a worldwide problem that needs global action to find a solution. Because South Asia is highly vulnerable, the region should take urgent steps to collaborate with development partners. South Asian countries, including Bangladesh, India, Nepal, and Bhutan, must work together to tackle these unavoidable impacts.

References

Abedin MA, Habiba U, Shaw R (2012) Health: impacts of salinity, arsenic and drought in South-Western Bangladesh. In: Environment disaster linkages (community, environment and disaster risk management), vol 9. Emerald Publishing, Bingley, pp 165–193

Abedin MA et al (2013) Gender and climate change: impacts and coping mechanisms of women and special vulnerable groups. Springer, London/New York, pp 33–44

ADB (2009) Building climate resilience in the agriculture sector of the Asia and the Pacific. Asian Development Bank, Mnila

Adger NA et al (2007) Climate change 2007: impacts, adaptation and vulnerability: working group II contribution to the Intergovernmental Panel on Climate Change: summary for policymakers. IPCC, Geneva

Ahmed AU (2006) Bangladesh climate change impacts and vulnerability. Climate Change Cell, Department of Environment, Dhaka

Ahmed B (2018) Who takes responsibility for the climate refugees? Intl J Clim Change Strateg Manage 10:5–26

Ahsan R, Karuppannan S, Kellett J (2011) Climate migration and urban planning system: a study of Bangladesh. Environ Just 4(3):163–170

Alam E (2002) Post cyclone adjustment process: the case of Sandwip, Bangladesh. University of Dhaka, Dhaka

Alam E (2003). Post-cyclone adjustment process: Basic needs perspective. Oriental Geographer 47:47–60

Alam E (2018) Climate impacts and adaptation strategies of the Bangladeshi coastal communities. Am J Environ Sci 14(4):195–202

Alam MZ et al (2009) Production of bioethanol by direct bioconversion of oil-palm industrial effluent in a stirred-tank bioreactor. J Ind Microbiol Biotechnol 36(6):801–808

Alam E et al (2017) Climate change impacts on the coastal zones of Bangladesh: perspectives on tropical cyclones, sea level rise, and social vulnerability. Springer Climate:145–166

Ali A (1999) Climate change impacts and adaptation assessment in Bangladesh. Clim Res 12:109–116

Amin SMA, Rahman A (2014) Opportunities and challenges of urban and Peri-urban agriculture to face climate change: a critical analysis of policy and urban governance of Dhaka City. In: The security of water, food, and energy & live ability of cities. Springer, New York, pp 365–382

Anbumozhi V et al (2012) Climate change in Asia and the Pacific: how can countries adapt? Saga, New Delhi

Andreasen MH et al (2010) Understanding climate change from below, addressing barriers from above: practical experience and learning from a community-based adaptation project in Bangladesh. Action Aid Bangladesh, Dhaka

Ann L (2010) Introduction: In saving point Reyes national sea-shore 1969–1970. An oral history of citizen action in conservation. Oral History Transcript/1993. Berkeley, CA: Bancroft Library Regional History Office and Nabu Press.

Annya Chanda Shimi GA (2010) Impact and adaptation to flood: a focus on water supply, sanitation and health problems of rural community in Bangladesh. Disaster Prev Manag 19(3):298–313

Atta-Ur-Rahman AN (2015) Hazard, vulnerability and risk: the Pakistan context. In: Disaster risk. Springer, Sendai

Ayers SH (2007) Climate change impacts and responses in Bangladesh. Clim Change

Ayers J, Huq S, Wright H, Faisal AM, Hussain ST (2014) Mainstreaming climate change adaptation into development in Bangladesh. Clim Dev 6(4):293–305. https://doi.org/10.1080/17565529.2014.977761

Bartlett M, Briones M, Neilson R, Schmidt O, Spurgeon D, Creamer R (2010) A critical review of current methods in earthworm ecology: from individuals to populations. Eur J Soil Biol 46:67–73. https://doi.org/10.1016/j.ejsobi.2009.11.006

BCAS (2010) Bangladesh Centre for Advanced Studies Report 2010. Bangladesh Centre for Advanced Studies, Dhaka, Bangladesh

BCCSAP (2009) Bangladesh climate change strategy and action plan (BCCSAP). In: Dhaka, Bangladesh: economic relations division. Ministry of Finance

Bhuyan MDI, Islam MM, Bhuiyan MEK (2018) A trend analysis of temperature and rainfall to predict climate change for North-Western region of Bangladesh. Am J Clim Chang 7:115–134

Biesbroek GR et al (2013) On the nature of barriers to climate change adaptation. Reg Environ Chang 13:1119–1129

Black R, Kniveton D, Schmidt Verkerk K (2013) Migration and climate change: toward an integrated assessment of sensitivity. In: Disentangling migration and climate change. Springer, Dordrecht, pp 29–53

Burton EL (2009) The Earthscan reader on adaptation to climate change. Earthscan, London

Burton M, Evans R, Sanders A (2006) Are special measures for vulnerable and intimidated witnesses working? Evidence from the Criminal Justice Agencies, London, UK

Butler JRA et al (2016) Priming adaptation pathways through adaptive co-management: design and evaluation for developing countries. Clim Risk Manag:1–16

Byomkesh T, Nakagoshi N, Dewan A (2011) Urbanization & green space dynamics in Greater Dhaka, Bangladesh. Land Ecol Eng 8:45–58

Carter WN (1991) Disaster management – a disaster management hand book. Asian Development Bank, Manila

Choudhury A, Quadir D, Ahmed A (2006) Socio-economic and physical perspectives of water related vulnerability to climate change: results of field study in Bangladesh. Sci Cult 71
Chaudhuri G, Mishra NB (2016) Spatio-temporal dynamics of land cover and land surface temperature in Ganges-Brahmaputra delta: a comparative analysis between India and Bangladesh. Appl Geogr 68:68–83
Chowdhury SD (2002) Turnarounds: A stage theory Perspective. Canadian Journal of Administrative Sciences / Revue Canadienne des Sciences de l'Administration 19:249–266. https://doi.org/10.1111/j.1936-4490.2002.tb00271.x
Collins AE (2008) Health security or climate change adaptation? What do we really mean by disaster risk reduction in times of disease? In: Proceedings of the international disaster and risk conference, IDRC DAVOS 2008. IDRC, Davos
CSR (2010) Country summary report, Bangladesh (2010) project for increasing stakeholder utilization of GAR 11, Preparation Study for Asia. UNISDR
Dai A, Trenberth KE, Qian T (2004) A global dataset of palmer drought severity index for 1870–2002: relationship with soil moisture and effects of surface warming. J Hydrometeorol 5:1117–1130
Dasgupta S, Hossain MM, Huq M, Wheeler D (2015) Climate change and soil salinity: the case of coastal Bangladesh. Ambio 44:815–826. https://doi.org/10.1007/s13280-015-0681-5. Royal Swedish Academy of Sciences
Denissen AK (2012) Climate change & its impacts on Bangladesh, the National Commission for International Cooperation and Sustainable Development Foundation (NCDO) 63, 1092. Amsterdam, Mauritskade
Denissen A-K (2014) Climate change & its impacts on Bangladesh, 25 April. Retrieved from http://www.ncdo.nl/artikel/climate-change-its-impacts-bangladesh, http://www.cgdev.org/files/1422836_file_Future_Storm_Surge_Disasters_FINAL.pdf. Accessed 25 Apr 2014
Denton F, Wilbanks TJ, Abeysinghe AC, Burton I, Gao Q, Lemos MC, Masui T, O'Brien K, Warner K, Bhadwal S, Leal W, Ypersele J.-P, Wright SB (2015) Climate-resilient pathways: Adaptation, mitigation, and sustainable development. https://doi.org/10.1017/CBO9781107415379.025
Depledge J, Lamb R (2005) Caring for climate: a guide to the climate change convention and the kyoto protocol. Climate Change Secretariat (UNFCCC), Bonn
Dessai S, Hulme M (2004) Does climate adaptation policy need probabilities? Clim Pol 4(2):107–128
Fatima R, Wadud A, Coelho S (2014) Human rights, climate change, environmental degradation and migration: a new paradigm. Issue No. 8. Migration Policy Institute (MPI), Washington, DC, USA
Fatima Y, Kazmi S, King S, Solomon S, Knight S (2018) Positive placement experience and future rural practice intentions: findings from a repeated cross-sectional study. J Multidiscip Healthc. 2018 Nov 5;11:645-652. doi: 10.2147/JMDH.S178138. PMID: 30519033; PMCID: PMC6233952
Feenstra RC (1998) Integration of trade and disintegration of production in the global economy. J Econ Perspect 12(4):31–50. https://doi.org/10.1257/jep.12.4.31
Fischer G et al (2005) Socio-economic and climate change impacts on agriculture: an integrated assessment, 1990–2080. Philos Trans R Soc B 360:2067–2083
Füssel H-M, Klein RJ (2006) Climate change vulnerability assessments: an evolution of conceptual thinking. Clim Chang 75(3):301–329
Habiba U, Sha R (2011) Drought risk reduction through a socio-economic, institutional and physical approach in the northwestern region of Bangladesh. Environmental Hazards 10:121–138
Haque M (1996) Climate change: issues for the policy makers of Bangladesh, pp 13–14
Haque SA (2006) Salinity problem and crop production in coastal regions of Bangladesh. Department of Soil Science, Bangladesh Agricultural University, Mymensingh, pp 59–78
Haque MA et al (2013a) Health coping strategies of the people vulnerable to climate change in a resource-poor rural setting in Bangladesh. BMC Public Health 13:1–11

Haque M et al (2013b) An assessment of climate finance governance-Bangladesh. Transparency International Bangladesh, Dhaka
Hoque, R. (2007). Bangladesh's emergency challenges the country's judiciary. Development and cooperation 34(11):426
Hoque MZ, Haque ME (2016) Impact of climate change on crop production and adaptation practices in coastal saline areas of Bangladesh. Int J Appl Res 2(1):10–19
Hossain MS, Roy K, Datta DK (2014) Spatial and temporal variability of rainfall over the South-West Coast of Bangladesh. Climate 2:28–46. https://doi.org/10.3390/cli2020028
Hussain M, Alak P, Azmz I (2016) Spatio-temporal analysis of land use and land cover changes in Chittagong city corporation. Bangladesh Int J Adv Remote Sens GIS Geogr 4:56–72
ICIMOD (2010) Climate change in the Hindu Kush-Himalayas, the state of current knowledge. International Centre for Integrated Mountain Development, Kathmandu, Nepal
IPCC (2001) TAR climate change 2001: the scientific basis. The Intergovernmental Panel on Climate Change, Geneva
IPCC (2007) Intergovernmental Panel on Climate Change, Working Group II: impacts, adaptation and vulnerability, fourth assessment report. Intergovernmental Panel on Climate Change, Genava
IPCC (2011) Summary for Policymakers. In: Edenhofer O, Pichs-Madruga R, Sokona Y, Seyboth K, Matschoss P, Kadner S, Zwickel T, Eickemeier P, Hansen G, Schlömer S, von Stechow C (eds) IPCC special report on renewable energy sources and climate change mitigation. Cambridge University Press, Cambridge, United Kingdom and New York, NY, USA
IPCC (2012) Managing the risks of extreme events and disasters to advance climate change adaptation. Cambridge University Press, New York
IPCC (2018) Special report: global warming of 1.5 °C. IPCC, Geneva
Islam MR (ed.) (2004) Where land meets the sea: A profile of the coastal zone of Bangladesh. The University Press Limited, Dhaka
Islam AKMN (2008) SIDR cyclone in Bangladesh, 'End of the course project' submitted to the National Institute of Disaster Management (NIDM), New Delhi, for 4th online course on comprehensive DM
Islam SN (2016) Deltaic floodplains development and wetland ecosystems management in the Ganges – Brahmaputra-Meghna Rivers Delta in Bangladesh. Sustain Water Resour Manag 2:237–256
Islam AKMN, Sultan S, Afroz (2009) Climate change and South Asia: what makes the region most vulnerable? MPRA Paper 21875. University Library of Munich, Munich
Islam GM, Islam AK, Shopan AA, Rahman MM, Lazar AN, Mukhopadhyay A (2015) Implications of agricultural land use change to ecosystem services in the Ganges delta. J Environ Manag 161:443–452
Jahan F, Quddusi K (2014) Climate change and Bangladesh: geographical and socio-economic impacts. Dhaulagiri Journal of Sociology and Anthropology 7. https://doi.org/10.3126/dsaj.v7i0.10439
Kabir M, Abdul Baten M (2019) Community mangrove aqua-silviculture (CMAS Culture): An innovation and climate resilient practice by the Sundarbans mangrove forest dependent rural communities of Bangladesh. Int. J Environ Clim Chang 9(1): 1–16. https://doi.org/10.9734/ijecc/2019/v9i130093
Karim MF, Mimura N (2008) Impacts of climate change and sea-level rise on cyclonic storm surge floods in Bangladesh. Glob Environ Chang 18:490–500
Kates RW (2014) Cautionary Tales: adaptation and the global poor. Clim Chang
Kelly PM, Adger WN (2000) Theory and practice in assessing vulnerability to climate change and facilitating adaptation. Clim Chang 47:325–352
Khan A et al (2011) Drinking water salinity and maternal health in coastal Bangladesh: implications of climate change. Environ Health Perspect 119(9):125–137
Kilroy G (2015) A review of the biophysical impacts of climate change in three hotspot regions in Africa and Asia. Reg Environ Chang 15:771–782

Le Dang H et al (2014) Farmers' perceptions of climate variability and barriers to adaptation: lessons learned from an exploratory study in Vietnam. Mitig Adapt Strateg Glob Chang 19:531–548
López VT (2015) Global increase in climate-related disasters. Manila, Asian Development Bank
McCarthy JJ et al (2001) Climate change 2001: impacts, adaptation, and vulnerability. Cambridge University Press, Cambridge
Mcgranahan G et al (2007) The rising tide: assessing the risks of climate change and human settlements in low elevation coastal zones. Environ Urban:17–37
Md Humayain Kabir MA (2019) Sustainability of climate change adaptation practices in South-Western coastal area of Bangladesh. J Sustain Develop 12(5):1
Md Rejaur Rahman HL (2015) Climate change in Bangladesh: a Spatio-temporal analysis and simulation of recent temperature and rainfall data using GIS and time series analysis model. Theor Appl Climatol 128 (1–2)
Mertz O et al (2009) Adaptation to climate change in developing countries. Environ Manag 43:743–752
Mimura JH (2010) The changing nature of extreme weather and climate events: risks to sustainable development. Geomat Nat Hazards Risk 1:3–18
Ministry of Finance (2005) Coastal zone policy. Ministry of Finance, Dhaka
Ministry of Forests (2009) Bangladesh climate change strategy and action plan 2019. Dhaka, Ministry of Environment and Forests
Ministry of Water Resources (2005) Coastal Zone Policy, 2005. Ministry of Water Resources, Dhaka
ML Parry et al (2007) Climate change 2007: impacts, adaptation and vulnerability. Contribution of Working Group II to the Fourth Assessment Report on Climate Change. Cambridge University Press, New York
MOEF (2009) Bangladesh climate change strategy and action plan 2009. Ministry of Environment and Forestry, Dhaka
Mohammad N (2015) Climate change and displacement in Bangladesh: issues and challenges. Clim Change Adapt:177–194
Monirul GM, Alam KA (2017) Climate change perceptions and local adaptation strategies of hazard-prone rural households in Bangladesh. Clim Risk Manag 17(C)
Mollah TH, Ferdaush J (2015) Riverbank erosion, population migration and rural vulnerability in Bangladesh (a case study on Kazipur upazila at Sirajgonj district). Environ Ecol Res:125–131
Morton K (2011) Climate change and security at the third pole. Survival 53(1)
NAPA (2005) National adaptation programme of action (NAPA) 2005. Economic Relations Division, Ministry of Finance, Dhaka, Bangladesh
Nishat A, Chowdhury JK (2019) Influence of climate change on biodiversity and ecosystems in the Bangladesh Sundarbans: increasing livelihood security. Springer, New York
Pande (1992). Disaster Relief Trisul. Trisul J III(2):33–37
Parvin S, Wang C, Li Y, Nishino N (2010) Effects of inoculation with lactic acid bacteria on the bacterial communities of Italian ryegrass, whole crop maize, Guinea grass and Rhodes grass silages. Anim Feed Sci Technol 160(3–4):160–166. https://doi.org/10.1016/j.anifeedsci.2010.07.010
Parvin GA et al (2016) Land use change in southwestern coastal Bangladesh: consequence to food and water supply. Springer, Berlin
Parry ML, Canziani OF (2007) Climate change 2007: impacts, adaptation and vulnerability. IPCC, Brussels
Pender J (2008) Community-led adaptation in Bangladesh. Forced Migrat Rev. Citing research by the UK Institute of Development Studies, p 54
Pereira ES, Pimentel PG, Duarte LS, Villarroel ABS, Regadas Filho JGL, Rocha Junior JN (2010) Intestinal digestibility of protein of adapted forages and by-products in Brazilian North-East by three-step technique. Rev Bras Saúde Prod Anim 11(2):403–413
Pouliotte J et al (2009) Adaptation and development: livelihoods and climate change in Subarnabad, Bangladesh. Clim Dev 1(1):31–46

Rabbani M (2009) Climate forced migration: a massive threat to coastal people in Bangladesh. Climate Action Network-South Asia Newsletter, Dhaka

Rahman S et al (2015) Climate extremes and challenges to infrastructure development in coastal cities in Bangladesh. Weather Clim Extrem 7:96–108

Resurreccion BP et al (2008) Climate adaptation in Asia: knowledge gaps and research issues in South East Asia. ISET-International, Kathmandu

Rodriguez JM et al (2009) Annual disaster statistical review 2008 – the numbers and trends. Centre for Research on the Epidemiology of Disasters, Université catholique de Louvain, Brussels

Saito N (2013) Mainstreaming climate change adaptation in least developed countries in South and Southeast Asia. Mitig Adapt Strateg Glob Chang 18(6):825–849

Sarker MAR et al (2013) Assessing the determinants of rice farmers' adaptation strategies to climate change in Bangladesh. Intl J Clim Change Strateg Manage 5:382–403

Sarkar M, Kabir S, Begum RA, Pereira JJ, Jaafar AH, Saari MY (2013) Impacts of and adaptations to sea level rise in Malaysia. Asian J Water Environ Pollution 11: 29–36

Shahnila Islam NC-D (2016) Structural approaches to modeling the impact of climate change and adaptation technologies on crop yields and food security. Glob Food Sec:63–70

Shamsuddoha M, Chowdhury T (2009) Opportunities of developing tourism industry in Bangladesh. J Tour Stud Res Tour 8(8):22–26

Siddiqi MS (2012) A situation analysis of climate change adaptation initiatives in Bangladesh. The Asia Foundation, Bangladesh, Dhaka

Smith TF et al (2008) Case study of adaptive capacity: systems approach to regional climate change adaptation strategies. Sydney, Sydney Coastal Councils Group

Smit B, Burton I, Klein R, Street R (1999) The science of adaptation: a framework for assessment. Mitigation Adaptation Strategies Global Change 4:199–213. https://doi.org/10.1023/A:1009652531101

Solutions D (2012) Climate displacement in Bangladesh. Displacement Solutions, Geneva

Stern N (2006) The economics of climate change. Cambridge University Press, London

Streatfield PK, Karar ZA (2008) Population challenges for Bangladesh in the coming decades. J Health Popul Nutr:261–272

Thornton PK et al (2014) Climate variability and vulnerability to climate change: a review. Glob Chang Biol 20(11):3313–3328

Thomas V, López R (2015) Global increase in climate-related disasters. In: Thomas V, López R (eds) Global increase in climate-related disasters (November 2015), Asian Development Bank economics working paper series no. 466, Available at SSRN: https://ssrn.com/abstract=2709331 or. https://doi.org/10.2139/ssrn.2709331

Thompson M, Walsham G (2010) ICT research in Africa: need for a strategic developmental focus. Inf Technol Dev:16. https://doi.org/10.1080/02681101003737390

Trenberth, Jones P, Ambenje, Bojariu R, Easterling D, Tank K, Parker D, Rahimzadeh F, Renwick J, Rusticucci M, Soden, Zhai P, Folland C (2007) Observations: surface and atmospheric climate change. In: Climate change 2007: the physical science Basis. Contribution of working group I to the fourth assessment report of the intergovernmental panel on climate change, chapter: 3. Cambridge University Press, Cambridge, United Kingdom/New York, NY, USA

Tiwari, PC, Joshi, B (2014) Environmental changes and their impact on rural water, food, livelihood, and health security in Kumaon Himalaya. Intl J Urban Regional Stud Contemp India:1–12

Twigg J, Bhatt MR (1998) Understanding vulnerability: south Asian perspective. Intermediate Technology Publications, London

Vermeer M, Rahmstorf S (2009) Global sea level linked to global temperature. Proc Nat Acad Sci 106:21527–21532

Vij S et al (2017) Climate adaptation approaches and key policy characteristics: cases from South Asia. Environ Sci Pol 78:58–65

Vincent K (2007) Uncertainty in adaptive capacity and the importance of scale. Glob Environ Chang 17(1):12–24

Wang H et al (2017a) Towards a universal model for carbon dioxide uptake by plants. Nature Plants 3:734–741

Wang SW, Lee WK, Son Y (2017b) An assessment of climate change impacts and adaptation in South Asian agriculture. Intl J Clim Change Strateg Manage 9(4):517–534

Wang G et al (2018) Climate change impacts and forest adaptation in the Asia–Pacific region: from regional experts' perspectives. J For Res 30:277–293

World Bank (2018) Bangladesh disaster risk and climate resilience program. World Bank, Dhaka

Yamin F et al (2005) Vulnerability, adaptation and climate disasters: a conceptual overview. Inst Develop Stud (IDS) Bull 36:1–14

Yohe GW et al (2008) Reducing poverty and hunger in Asia- climate change in the context of Asia: pro-poor adaptation, risk management and mitigation strategies. International Food Policy Research Institute, Washington, DC

Chapter 16
Climate Change and Living with Floods: An Empirical Case from the Saghata Union of Gaibandha District, Bangladesh

Biddut Kumar Ghosh and Koyel Rani Sarker

Abstract Flood is a recurring natural disaster in Bangladesh due to the intersection of the Ganges, Brahmaputra, and Meghna (GBM) Rivers and their numerous distributaries and tributaries across the country. Bangladesh is considered to be a vulnerable county with regard to climate change. Bangladesh has experienced an increased number of hydro-meteorological disasters caused by climate change, including a loss of lives and property. The country is adopting survival mechanisms to respond to the worst situations caused by natural disasters from climate change. This empirical study illustrates the effects of the 2017 flood in Gaibandha District, Bangladesh and adaptation mechanisms in response to this event. Quantitative and qualitative data were used in this empirical study of the mitigation processes for the 2017 flood's effects and the adaptation strategies of flood victims. Because they continue to be affected by floods, the victimized households became a marginalized group with unstable income; many have been displaced into different industries or other locations with better availability of income opportunities; some victims migrated to Dhaka, Chottogram, and Sylhet City for work. The study assesses the policies for providing sufficient support and assistance to flood victims. Adaptation strategies for climate change should be introduced on a large scale via South Asian regional planning with regard to transboundary water management to reduce the effects of floods.

Keywords 2017 flood · Climate change · Mitigation process · Adaptation strategies · Victim households

B. K. Ghosh (✉) · K. R. Sarker
Department of Geography and Environmental Science, Begum Rokeya University, Rangpur, Bangladesh

M. N. Islam, A. van Amstel, *Bangladesh II: Climate Change Impacts, Mitigation and Adaptation in Developing Countries*, Springer Climate,
https://doi.org/10.1007/978-3-030-71950-0_16

16.1 Introduction

Bangladesh experiences floods every year. This hydro-meteorological hazard has dual impacts on the Bangladeshi people: Although it results in fertile land, it also inundates the surrounding land to damage crops, homesteads, and infrastructure and endanger the lives of people and their domestic animals (Rasheed 2008). Bangladesh is highly vulnerable to floods geographically (Ghosh and Mahbub 2014) because it is located in the downstream section of the Ganges-Brahmaputra-Meghna (GBM) river basin (Rasheed 2008) and has 80% floodplain land (Brammer 1990). Every year, recurrent floods inundate an average of 20.5% of Bangladesh, or 3.03 million hectares (BDRCS 2019). The huge floods of 1988 and 1998 inundated 61% and 68% of Bangladesh, respectively (Miah 2004). Generally, the floods in Bangladesh start with flash floods in the northern and eastern hill streams in April and May, followed by monsoon floods in the middle and downstream section of the GBM rivers in June to September, varying from one part to another part of Bangladesh from one year to another (Rasheed 2008). Bangladesh experienced two floods in 2017 (BDRCS 2019), endangering the north, northeast, and middle parts of Bangladesh (BDRCS 2017).

Low-lying areas of Rangpur region were inundated by the 2017 flood as a result of heavy floods in the Indian State of Assam, which were caused by heavy rain in the upper stream of the Brahmaputra River, according to the Flood Forecasting and Warning Centre. As a consequence, the Brahmaputra, Jamuna, and Padma rivers were also showing a rising trend. More than 3000 people in the Gaibandha District were marooned in remote char areas of Gaibandha Sadar, Fulchhari, and Saghata Upazilas in July 2017 as water levels rose in the Brahmaputra River (The Daily Star 2017a). Farmers in the Rangpur region (Dinajpur, Rangpur, Kurigram, Lalmonirhat, Bogura, and Gaibandha Districts) were especially vulnerable as 1,69,335 hectares of land were inundated and 5.5 lakh tons of Aman rice were destroyed (The Daily Star 2017b).

Historical documents indicate that the northern part of North Bengal is less vulnerable to floods in August. However, the 2017 flood broke the records, and a subsequent 2019 flood followed almost the same pattern as the 2017 flood in Bangladesh. Thus, climate change issues seem to be responsible for the abnormal occurrence of floods in Bangladesh and South Asian countries in general.

16.2 Rationale of the Study

This study examines how the 2017 flood-affected households returned to their normal lives by some adaptive mechanisms. Basically, all floods cause some unavoidable complex issues. Riverbank erosion is one type of issue that becomes more active during the monsoon period in Bangladesh (Ghosh and Mahbub 2014), causing societal and economic problems (Ghosh 2016). The 2017 flood affected 31 districts all over Bangladesh, 1200 unions of 183 upazilas of those districts. Almost 33% of Bangladesh's land (Hossain 2017) and approximately 6.77 million people were affected by the 2017 flood—50,042 fully affected households (HHs) and 1,305,257 partially affected HHs (Nirapad 2017c). Dinajpur, Kurigram, Lalmonirhat, Rangpur, Naogaon, Sirajganj, Jamalpur, Bogura, Tangail, Netrokona, Sylhet, Sunamganj, Gaibandha, Thakurgaon, and Panchagarh were the districts most affected by the 2017 flood (BDRCS 2017). In Gaibandha District, 262 villages in 42 unions were affected. The most affected upazilas of this district were Gaibandha Sadar, Fulchhari, Sundarganj, and Saghata. A total of 252,103 people in 34,956 HHs were extremely affected and 205,553 people took shelter in a flood protection area in Gaibandha District, Bangladesh (BDRCS 2017). Climate change has been linked to the abnormal floods, which have social, economic, and employment impacts (Fig. 16.1). Figure 16.2 shows the impacts of floods on victims at different scales, as well as the emergency responses needed from different institutional bodies to recover from monsoon floods in Bangladesh.

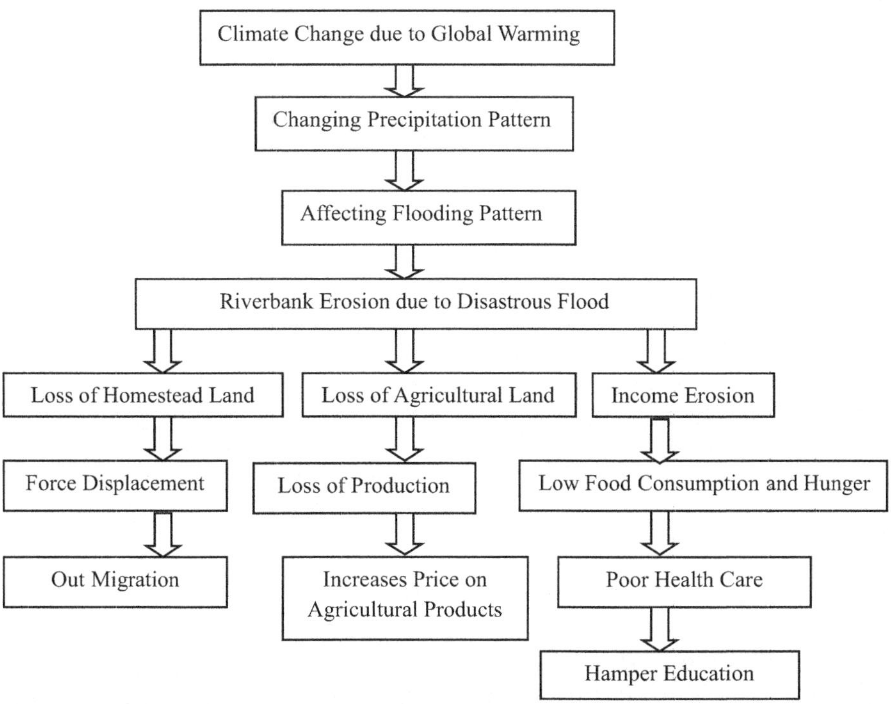

Fig. 16.1 Conceptual framework of climate change linked to livelihood vulnerability. (Source: Uddin and Basak n.d.)

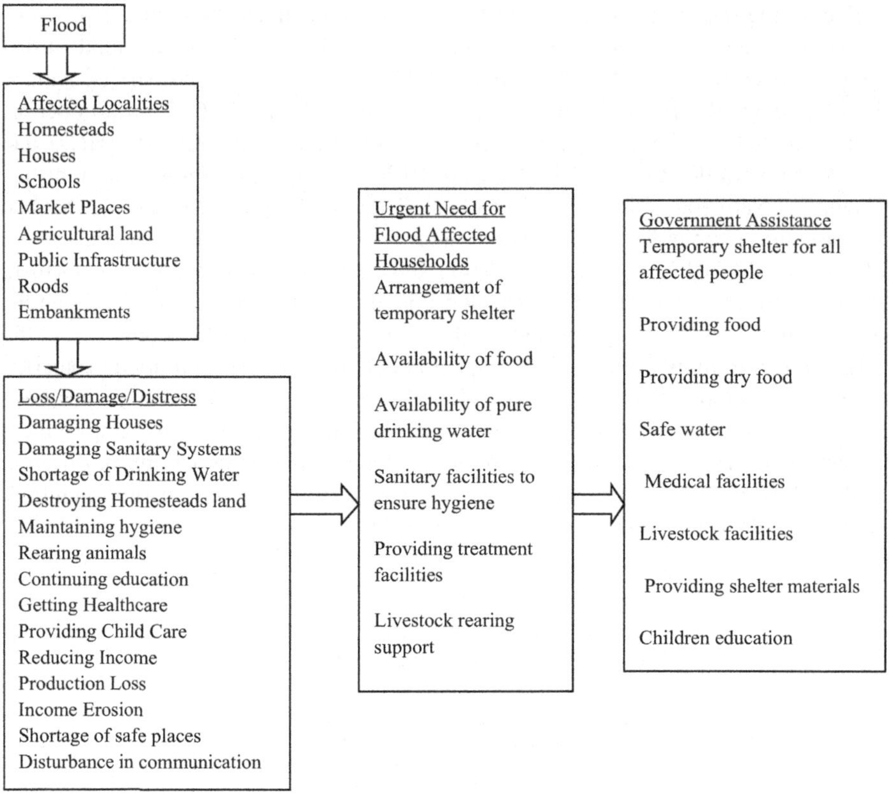

Fig. 16.2 Schematic diagram of impacts on and assistance for flood victim households. (Source: Nirapad 2017c)

16.3 Objectives of the Study

The study explores the challenges and adaptive strategies used by the victims of the 2017 flood in Gaibandha District, Bangladesh. The objectives of the study were to understand the socio-environmental challenges of the 2017 flood and describe the adaptive mechanisms used by the flood victims to recover from the disaster.

16.4 Study Area and Methodology

16.4.1 Selection of the Study Area

Saghata Union (part of Saghata Upazila in Gaibandha District, Bangladesh) was selected as the study site. This area was extremely affected by the monsoon flood in 2017. Gaibandha District is located on the right bank of the Jamuna River; the study area is adjacent to the Jamuna River (Fig. 16.3) and becomes inundated every year

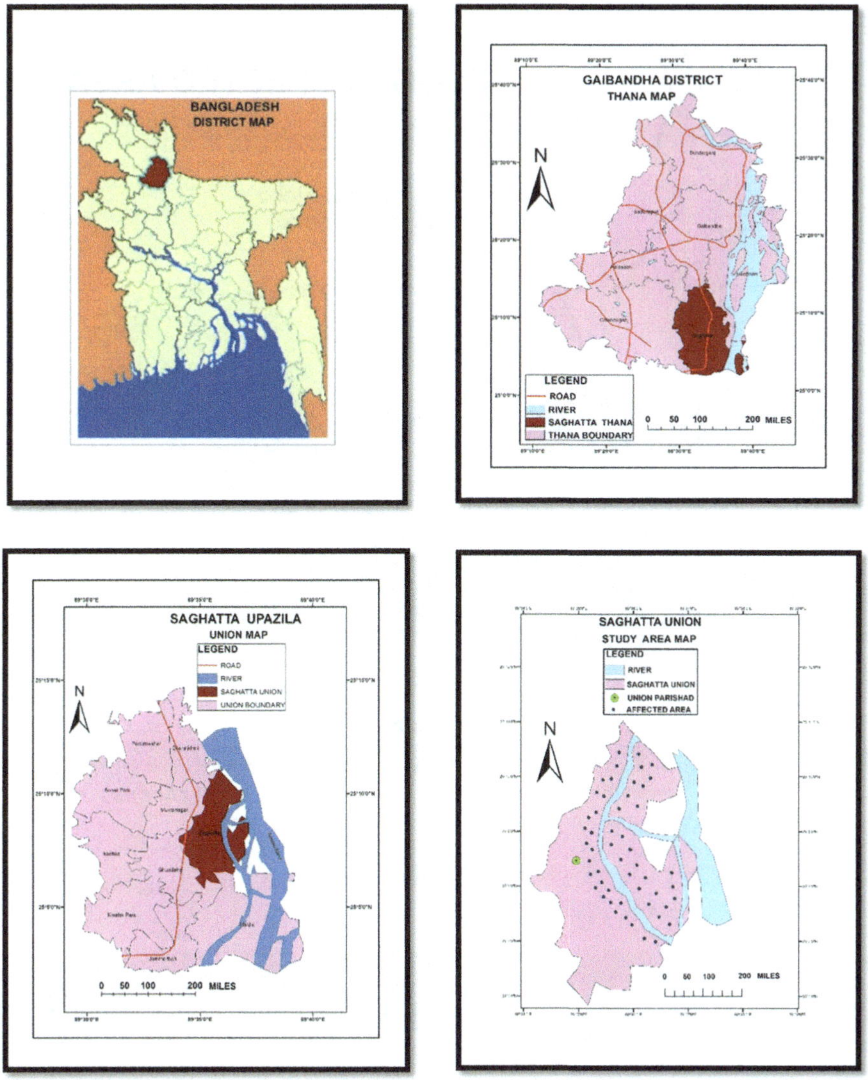

Fig. 16.3 Study area

by monsoon floods. Jamuna River is the largest braided river in Bangladesh and experiences enormous bank erosion. The floods resulting from overflow of this river are characterized as riverine/monsoon floods (Fig. 16.4). Saghata Union of Gaibandha District (the study area is 24.35 km^2) encompasses 6019 acres of land, with 5846 HHs and a population density of 925 (BBS 2013).

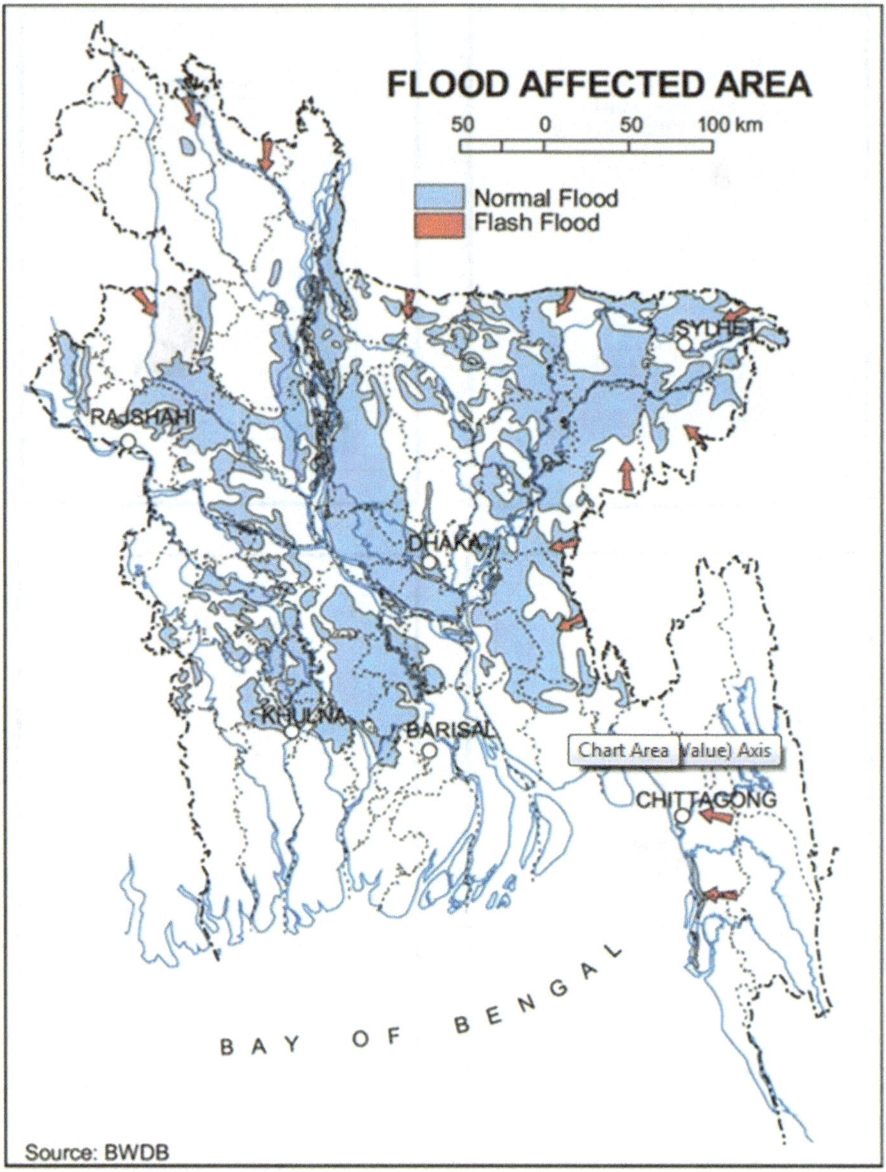

Fig. 16.4 Current flood-affected areas in Bangladesh. (Source: Bangladesh Water Development Board n.d.)

16.5 Sampling Design and Sampling Size

Sampling design is the method of collecting samples from the total population. A simple systematic sampling technique was used. In the study area, the total number of households was 5846 (BBS 2013); among them, 3125 HHs (UP 2018) were victims of the 2017 flood. A sample of 5% was considered to be the true value, and the sample size was determined by Kothari's (1999) sample size determination formula:

$$n = \frac{z2.p.q.N}{e2(N-1) + z2.p.q}$$

$$n = \frac{(2.005)2 \times 0.05 \times (1-0.05) \times 3125}{(0.05)2 \times (3125-1) + (2.005)2 \times 0.05 \times (1-0.05)}$$

$$n = \frac{4.02 \times 0.05 \times 0.95 \times 3125}{0.0025 \times 3124 + 4.02 \times 0.05 \times 0.95}$$

$$n = \frac{596.72}{8.00}$$

$$n = 74.59$$

Here,
N = Total number of households.
n = Sample size.
e = .05 (because the estimate should be within 5% of the true value)
z = 1.96 (per table of area under the normal curve for the given confidence level of 95%)
p = 0.05 (based on our experience and past data)
q = (1-e)

16.6 Data Sources and Analysis

This empirical study was based on primary and secondary data. The primary data were collected through field surveys based on a pre-designed questionnaire that consisted of a series of semi-structured questions. The sample questionnaires were taken at the household level, with 75 sample surveys conducted among the flood victim HHs following a simple systematic sampling technique (in which every tenth household was selected for the target sample). In addition, direct observations were used to conduct this study. Secondary data were collected from different sources. Both quantitative and qualitative data were gathered, mostly from primary sources and some from secondary sources. The secondary data used in the study were from the Bangladesh Bureau of Statistics (BBS), Bangladesh Water Development Board, Union Parishad, Bangladesh Red Crescent Society (BDRCS), websites, daily newspapers, peer-reviewed documents, and research papers. All collected data were processed and analyzed using Microsoft Excel 2010 and SPSS 17.0; the maps were prepared using ArcGIS 10.5.

16.7 Results and Discussion of the Study

16.7.1 Age-Sex Pyramid

The numbers of surveyed households and population were 75 and 309, respectively. The average surveyed household size was 4.12, in comparison with the average household sizes in Saghata Upazila (3.88), Gaibandha District (3.88), and Bangladesh (4.44) (BBS 2013). Household sizes in Gaibandha District were lower than the national average, likely because the recurrent flooding and riverbank erosion directly and indirectly affected the inhabitants' perceptions regarding family planning issues in Gaibandha District. The age-sex pyramid of the study area (Fig. 16.5) is progressive as is the nature of Bangladesh, where the number of births increases from year to year (Clarke 1972). Figure 16.5 illustrates the age and sex structure of the study area, with insights about political and social stability as well as economic development. Bangladesh has achieved rapid progress in many social and development indicators, although there is poverty and some low-quality public services (Rana and Islam 2016).

16.7.2 Educational and Occupational Status

Regarding education, most study participants could only sign (28.8%), whereas others were illiterate (14.24%); most of the elderly participants were illiterate. In all, 22.98% completed primary education, 18.12% completed secondary education, 10.36% completed higher secondary education, and 5.5% had graduated from national university. In contrast, the education rate of Saghata Upazila in Gaibandha District was 34.3% in 2001 and 40.6% in 2011 (BBS 2013). From this study, the

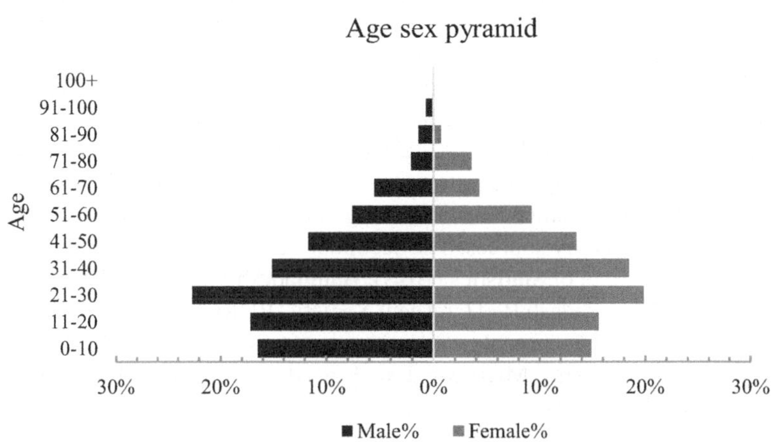

Fig. 16.5 Age-sex structure of the study area. (Source: Field Survey 2018)

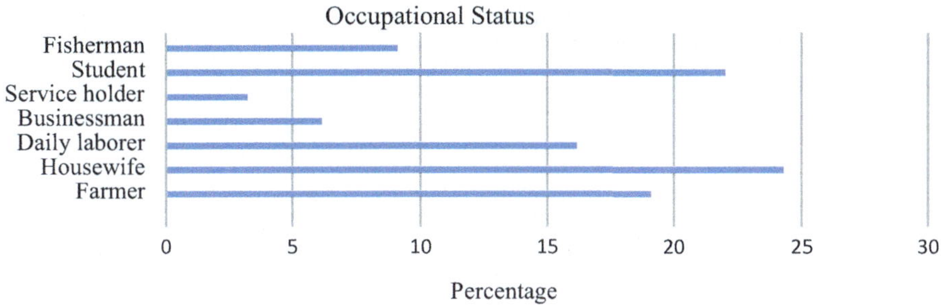

Fig. 16.6 Occupational status of the study area. (Source: Field Survey 2018)

education rate in Bangladesh seems to be increasing, likely due to public awareness and initiatives by the Government of Bangladesh to transform its huge population into human resources. Education for All is a constitutional responsibility of the government, which affirms equal rights in education. Thus, the government has emphasized education to the marginalized populations and women for gender equality to achieve this initiative (Islam 2017).

The study area belongs to Saghata Union in Gaibandha District and is typical rural area of Bangladesh. The 2017 flood victim people engaged themselves in different occupational status. The occupational pattern of the respondents of 2017 flood affected households are farmer (19.1%), housewife (24.3%), daily labor (16.18%), businessman (6.15%), student (22.01%) and service holder (3.24%) shown (Fig. 16.6). These occupations reflect the availability of opportunities, the residents' mobility, the effects of flood and river bank erosion on poverty, and underdeveloped communications. The 5.93% urbanization rate (BBS 2013) of the study area also affects the education rate and occupational patterns. Many surveyed households (41.33%) had monthly incomes that were less than Tk. 5000. Most households were living under the poverty line with daily income equivalents of less than US$1.90 (Tk. 160). Living under poverty measured whose are living equivalent of US$1.90 or less per day in 2011 purchasing price parity terms (World Bank 2017). In Bangladesh, 9% of the population lived under the poverty line in 2018 (Dhaka Tribune 2018). The household incomes of some study participants (5.33%) were above 20,000 Tk.; these households were engaged with business and services.

16.8 Housing Type

The housing types of respondents of the questionnaire survey in study area were made by concreate 2.67% semi- concrete 8%, wood and aluminium corrugated sheet 84%, and mud-wall houses 5.33%. These are the prevalent housing types due to income, erosion, and recurrent monsoon flooding. In 2011, the housing types in Saghata Upazial were reported to be concrete house (1.3%), semi- concrete house (10.1%), wood and aluminium corrugated sheet (87.6%), and bamboo and

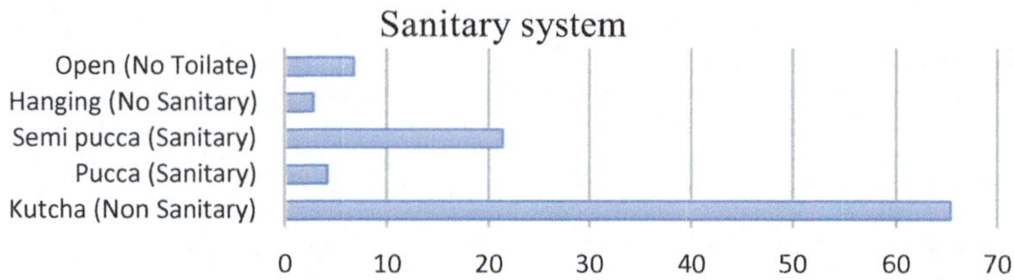

	Kutcha (Non Sanitary)	Pucca (Sanitary)	Semi pucca (Sanitary)	Hanging (No Sanitary)	Open (No Toilate)
Percentage	65.33	4	21.33	2.67	6.67

Fig. 16.7 Sanitary systems in the study area. (Source: Field Survey 2018)

straw (1.0%) (BBS 2013); thus, the conditions appear to be improving slowly based on the surveyed households.

16.9 Sanitary Systems

Bangladesh faces multiple challenges in terms of sanitation, hygiene, and pure drinking water. However, Bangladesh has made significant progress in the field of sanitation. Financial, technical, and social supports are provided by the government and nongovernmental organizations (NGOs) in the selected areas. Most of the surveyed households had non-sanitary toilets (65.33%); others had hanging toilets (2.67%) or no/open toilet facilities (6.67%) (Fig. 16.7). The census indicates that Saghata Upazila households have sanitary toilets (33.3%), non-sanitary toilets (51.1%), and no toilet facilities (15.6%) (BBS 2013). The number of households that use of open fields or have no toilets is decreasing gradually as the literacy rates of the study area and Bangladesh increase. However, the number of sanitary toilets in households remains unsatisfactory.

16.10 Drinking Water and Health Care Services

All households in the study used tube-well water for drinking. These households may use ponds and other sources for domestic purposes.

Citizens with lower incomes and who live in rural areas do not have much access to health care services. Delivery of services also varied depending on level of income. Most rural residents in Bangladesh receive health services from rural practitioners. Figure 16.8 shows the sources of health care services in the study area: rural practitioners (50%), Upazila Health Complex (40%), homeopathy service (27.14%), pharmacy (15.71%), and rural herbal practitioners (20%). Residents did not receive health care services from District Headquarters as it is very expensive.

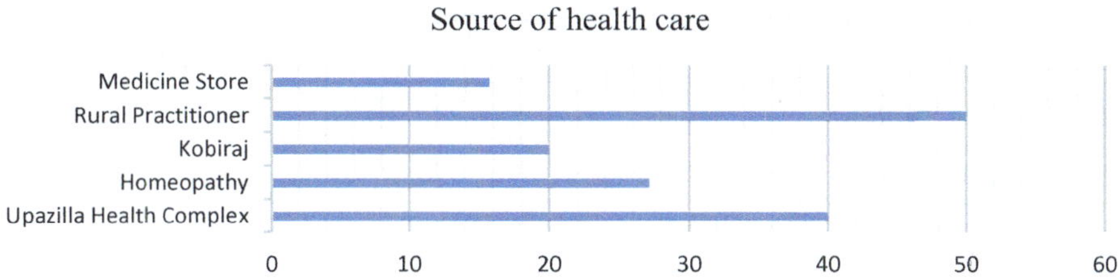

Fig. 16.8 Sources of health care in the study area. (Note: Multiple responses considered. Source: Field Survey 2018)

Table 16.1 Impacts of 2017 flood in Bangladesh

	April 2017	August 2017
Affected Districts	6	32
Affected Upazilas	60	208
Affected Unions	450	1324
Affected Paurashavas	NA	64
Affected Households	1,031,405	1,734,384
Affected People	4,667,623	8,202,025
Deaths (drowning, snake bite, etc.)	10	147

Source: MoDMR and BBS Statistical Yearbook various issues cited in CPD 2017, p.10

16.11 Nature of 2017 Flood

The 2017 flood hit Bangladesh four times in different four months, starting in April to July and August, and ending in October (Nirapad 2017c). Extreme havoc occurred in Bangladesh from 2017 flood; flash floods occurred during the pre-monsoon period and monsoon floods occurred during the monsoon period, as shown in Table 16.1.

Table 16.2 shows the impacts of monsoon floods in 2017 for Gaibandha District specifically; all Upazilas of this district were inundated and affected. The geographical location of Gaibandha District is the right bank of Jamuna River. Hatbari, Naukarchar and Hasilkandi villages of Saghata Union, Gaibandha District were characterized as having high floods in 2017.

16.12 Duration of 2017 Flood

Gaibandha is a district of Rangpur Division in northern Bangladesh that experiences monsoon floods every year. Generally, this district is affected by monsoon floods in the first week of June due to the water flow from downstream sections of the

Table 16.2 Overall impact scenario of 2017 flood in Gaibandha District

Particulars	August 2017
Affected Upazilas (n)	7
Affected Unions (n)	69
Affected Villages (n)	607
Affected People (% of total population of Gaibandha District)	21.07
Damaged Houses (n)	121,157
Affected Crop Land (hectares)	27,167
Deaths (n)	13
Affected Institutions (n)	623
Affected Roads (km)	680
Affected Embankments (km)	43
Affected Bridges (n)	23
Displacements (n)	1272
Affected Tube-wells (n)	3031

Source: DDM Report on Damage Information and Relief Distribution on Monsoon Flood, August 30, 2017 cited in Nirapad 2017d, p.4

Table 16.3 Duration of 2017 flood in study area

Duration (in days)	Respondents Number of Survey	Households affected (%)
0–5	3	4
6–10	6	8
11–15	18	24
16–20	35	46.67
>20	13	17.33
Total	75	100

Source: Field Study (2018)

Brahmaputra and the Jamuna; it is sometimes affected until August but varies from year to year. In the 2017 flood, Gaibandha District (including the study area) was affected twice, in July and August (Nirapad 2017b). The damage from floods depends on the duration, depth, and devastating nature. Table 16.3 shows the duration of the flood in the study area; most of the households (46.67%) indicated that they were affected for 16–20 days. This flood lasted for more than 20 days in the lowland areas, which had flood depths of more than 180 cm (Rasheed 2008).

16.13 Estimated Losses from 2017 Flood

South Asia was extremely affected by the 2017 flood, with more than 1000 deaths. The United Nations estimated that 41 million people were directly victimized in Bangladesh, India, and Nepal (The New York Times 2017). The northwest of

Table 16.4 Damage to crops, livestock, and fishing during 2017 flood

Items	Loss (in TK)	Affected households (%)
Crops	4,000,000	53.33
Cattle	1,50,000	20.00
Poultry	1,65,000	73.33
Fish	75,000	14.67

Note: Multiple responses considered
Source: Field Study (2018)

Bangladesh was heavily flooded during July and August (Rahman 2017). Flash floods in northeastern Bangladesh, the hoar basin, damaged 1,389,968 livestock (cow, buffalo, goat, and sheep), 597,175 ducks, 2,638,440 hens, and 213.95 metric tons of fish (Nirapad 2017a). The estimated losses of the surveyed households in Saghata Union, Gaibandha District are shown in Table 16.4.

Cattle and poultry are the main sources of protein (milk, eggs, and meat) of the surveyed households (Ghosh 2016). Most households (73.33%) lost their poultry in the 2017 flood. Rearing cattle was a very complex issue during the flood in 2017. Some flood-affected households sold their cattle at steeply discounted prices (Rahman 2017) to mitigate the effects of flood. Migration from inundated places was also very common in the study area and all over Bangladesh. Households in the study area lost their paddies (22.86%), jute (90%), and maize (47.14%).

16.14 Damage to Land

The surveyed households lost cultivable land (87.14%), uncultivable land (47.14%), homestead land (68.57%) and other land (21.43%). Small-scale and large-scale temporary flooding can cause significant soil deterioration. The 2017 flood damaged 219,840 hectares of agricultural land in hoar basin areas of six districts located in the northeastern part of Bangladesh during the pre-monsoon period (Nirapad 2017a).

16.15 Assistance to Mitigate the Effects of the 2017 Flood

Flood-affected households require shelter, assistance, and emergency response to mitigate the effects of the disaster (Ghosh and Mahbub 2017, 2018). In the study area, surveyed households received cash (42.86%), dry food (64.29%), clean drinking water (54.29%), emergency medicine (82.56%), and clothes (50%) to mitigate their losses from the 2017 flood. A total of 149 shelter centers were opened to accommodate the affected households, with 29,612 people being displaced from

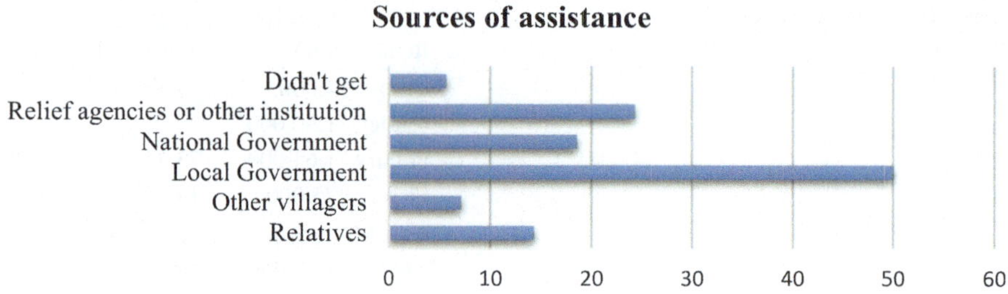

Fig. 16.9 Sources of assistance received by 2017 flood-affected households. (Note: Multiple responses Considered. Source: Field Study 2018)

homes in Gaibandha District due to the 2017 flood (Nirapad 2017c). Figure 16.9 shows the sources of assistance received by the 2017 flood-affected households in the study area. Most households (50%) received support from the local government, although 5.71% of households did not receive any support.

16.16 Initiatives to Mitigate the Effects of Flood

The households affected by the 2017 flood have taken some initiatives to prevent future flood damage. These households upgraded their home's materials (33.33%), relocated children and elderly household members (40%), rebuilt damaged embankments (13.33%), and attempted to elevate their homes (20%).

16.17 Adaptive Mechanisms During 2017 Flood

Victims of natural disasters may become unemployed, become vulnerable, or lose sources of income (Ghosh 2016). Some of the surveyed households changed their livelihood strategies and management, ad hoc basis techniques to safe livestock resources and keep them on safety places (Table 16.5). They may change their occupations in response to the disaster. Animal husbandry was the second largest income source among the surveyed households. The flood-damaged roads disrupted the transport of cattle feed and destroyed safe places to house livestock, causing them to sell cattle at discounted prices.

The historical data shows when the flood was the severe and went peak discharge stage, that time two-thirds of people in the study area loss their income sources. Which is, some flood affected peoples became involved in alternative activities to earn more money. All households in the study were concerned about their drinking water; they used water purification methods to ensure the availability of safe drinking water during the 2017 flood. Most households (82%) used antiseptics

Table 16.5 Keeping Place of Cattle and Poultry during 2017 Flood

Keeping Place of Cattle and Poultry	Percentage of respondents (%)
Road	53.6
Highland	32.14
Embankment	8.9
Other Places	5.36

Source: Field study (2018)

Table 16.6 Residence changes during the flood

Residence Change	Frequency	Percentage (%)
Yes	26	34.67
No	49	65.33
Total	75	100

Source: Field study (2018)

(Fitkiri); 18% of them boiled the water to prevent the waterborne diseases that spread rapidly during flooding.

16.18 Residence Changes During 2017 Flood

Changing residence is an adaptive mechanism to stay safe during a flood. Table 16.6 shows that only one-third of households moved their residence to a safe place. The preferences among the surveyed households for a flood-free place were roadsides, high lands, embankments, and other open places during the flood. They returned their residences to their original locations after flooding. Self-managed relocation is the first step for adaption in response to a disaster (Zaman and Khatun 2017). Their temporary locations often coincided with other villagers; generally, affected households tried to stay with their communities to feel secure (Ghosh and Mahbub 2014).

16.19 Displacement of Affected People

Participants who were affected by the 2017 flood migrated to different areas within North Bengal and outside of the region. Migration is a way to manage livelihood and income changes caused by poverty and natural disasters (Ghosh and Mahbub 2014). People relocated from the study area to escape effects of the flood, as well as to find shelter and employment opportunities. Table 16.7 illustrates the displacement of affected people from the study area to different destinations. The movement of the affected people from Saghata Union of Gaibandha District to locations within this district and outside the district are shown (Table 16.7; Fig. 16.10) as strategies to sustain themselves from the complex issues associated with the 2017 flood. These migrating people selected surrounding upazilas in the study area where employment opportunities were available, such as the nearby Bogura District of North Bengal or the large city of Dhaka, which has a large population and many jobs (Ghosh and Mahbub 2014); the cities of Chottogram and Sylhet City were also selected by the people of Gaibandha District (Fig. 16.10). Cities such as Dhaka provide many opportunities for work, such as garment worker, rickshaw puller, housekeeper, and similar jobs.

Table 16.7 Displacement of affected people

Affected village	Relocation destination (Village/Town/City)
Hatbari	Rasulpur and Edilpur Village, Fulchhari and Gaibandha Sadar Upazila, Jamalpur District, Dhaka, Chattogram, and Sylhet City.
Hasilkandi	Ghuridah, Chinirpotol, Monoharpur and Dhaperhat Village, Palashbari and Sadullapur Upazila, Bogura District, Chattogram, and Dhaka City.
Noukarchar	Jumma bari, Kachua, Sonatola, Vatgram Village, Bongaram Union, Sundarganj Upazila, Dhaka, and Sylhet City.
Munsirhat	Ramnagar, Muktanagar, Voratkhali Village, Bonarpara Union, Gobindoganj, Palashbari Upazila, Sylhet, Chattogram, and Dhaka City.

Source: Field study (2018)

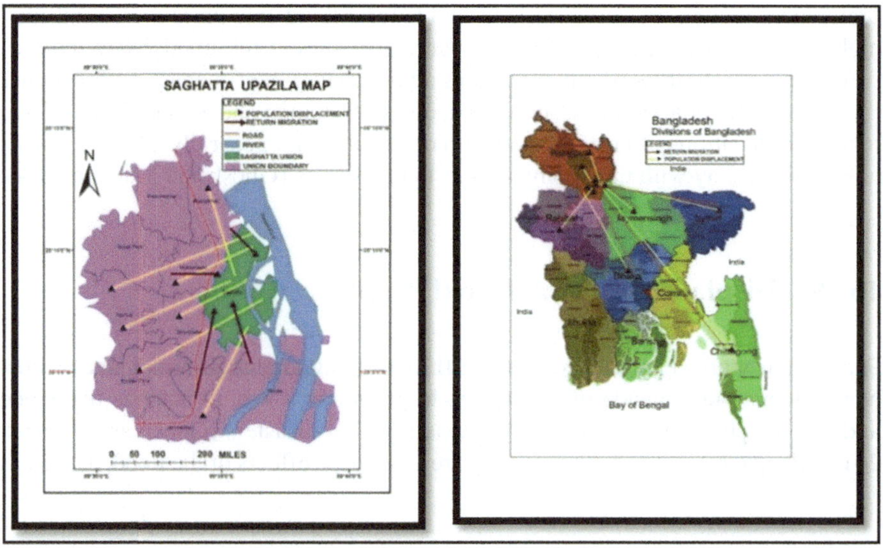

Fig. 16.10 Population displacement from study area to other regions. (Source: Field Study 2018)

People who lost their homes and property were usually employed as daily wagers (workers who are paid at the end of each day for their services) in other villages or cities. Some people moved permanently from the area due to the yearly flooding.

16.20 Strategies Taken After the 2017 Flood

Living with floods is a concept introduced as a result of climate change. Bangladesh is very vulnerable to the effects of climate change due to its geographical location and natural settings. Therefore, the people of Bangladesh are at risk of the effects of climate change, which is evidenced by the increasing number of natural disasters. Natural disasters cannot be stopped, but the effects of the disasters on lives can be mitigated. The local people of Bangladesh are adopting strategies to cope with disasters and have indigenous knowledge on how the effects can be mitigated. Local people use a variety of measures of different levels and magnitudes on short-term, long-term, or ad-hoc bases.

In the study area, the affected households have adopted strategies after the 2017 flood to protect their livelihoods. Figure 16.11 shows that the majority of households took some corrective measures to minimize flood loss. Their remaining resources were used to regain their income sources and adapt to their local environment for sustainability. Loan opportunities were limited, resulting in most of the households (64.29%) mortgaging their land for cash to create income opportunities (Fig. 16.11). Some measures after the flood were conducted by the community, local governments, and NGOs to sustain the local people. Figure 16.11 shows the individual strategies for initially recovering income and occupation; however, these are not sufficient for sustainable development at the local and community levels.

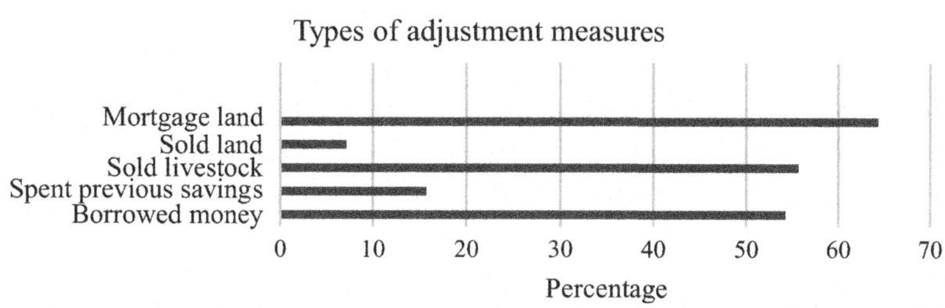

Fig. 16.11 Distribution of adaptation measures taken by respondents. (Note: Multiple Answers Considered. Source: Field Study 2018)

16.21 Steps Taken by Governments and NGOs

A number of national and international NGOs were working in the study area to support people affected by the 2017 flood, including Bangladesh Rural Advancement Committee (BRAC), ASA (Association for Social Advancement), Proshika, Grameen Bank, CARE (Cooperative for Assistance and Relief Everywhere), Bangladesh Red Crescent Society, Grameen Uddag, Ganasastha Kendra, and the Social Work Centre. Most villagers received some assistances from the various NGOs to cope with this flood, but the majority of support came from the local and national governments. Table 16.8 evaluates the level of satisfaction of the local people regarding flood management issues in the study area. The assessment table shows the minimum level of satisfaction; most respondents were dissatisfied by the assistance provided by different institutions with regard to mitigating flood-related losses and emergency responses during and after the flood. Mismanagement and insufficient assistance from different organizations occurred during and after this flood, leaving the victims without support for their sustainability.

16.22 Conclusion and Recommendations

Our study emphasizes that the 2017 flood had adverse effects on the socio-economic conditions of the affected areas. To a large extent, it also evident that periodic floods play a vital role in the livelihood patterns of the affected areas. We examined the varying underlying causes of individuals' vulnerability during the 2017 flood and other disasters. The study reveals that the 2017 flood was not managed properly in the study area; relief operations and programs for the victims' recovery were inadequate, even though flooding is a common occurrence in that area and elsewhere in other flood-prone areas of Bangladesh. A number of structural and nonstructural measures were taken, but they were largely ineffective at mitigating the effects of the 2017 flood. Policy makers must focus on non-structural and community-level

Table 16.8 Respondents' satisfaction levels with government and NGO flood disaster management

	Government Institutions		NGOs	
Satisfaction level	Frequency	Percentage	Frequency	Percentage
Highly dissatisfied	5	6.67	3	4
Dissatisfied	48	64	37	49.33
Moderately satisfied	7	9.33	20	26.67
Satisfied	15	20	13	15.71
Highly satisfied	0	0	1	1.33
Total	75	100.0	75	100.0

Source: Field Survey (2018)

flood management. Emergency preparedness can mitigate enormous flood losses; hence a proper system is needed to be effective.

Based on the results of this study, policy makers and planners have to implement modern flood management strategies in addition to traditional strategies. In addition to regional planning, activities that impact climate change should be reduced and environmentally friendly economic activities should be introduced effectively. Early flood warning systems, embankment rebuilding, and proper management of assistance from different organizations should be ensured on different levels to support vulnerable groups. A delta plan should be executed to maintain water flow and implement regional collaboration for transboundary river water management. Highly erosion-prone areas should be identified and properly zoned. Finally, local and national organizations should address the internal migration of vulnerable groups and migratory people in response to natural disasters.

References

BBS (2013) Population and housing census 2011, community report: Gaibandha. Bangladesh Bureau of Statistics. Statistics and Informatics Division. Ministry of Planning, Dhaka, Bangladesh, Dhaka

BDRCS (2017) Monsoon floods in Bangladesh – situation report-02. Bangladesh Red Crescent Society. [Online] Available at: https://reliefweb.int/report/bangladesh/monsoon-floods-bangladesh-situation-report-02-16-august-2017. Accessed 20 Apr 2018

BDRCS (2019) Baseline Study Report Integrated Flood Resilience Program. Bangladesh Red Crescent Society, Integrated Flood Resilience Program, Disaster Risk management Department. [Online] Available at: http://www.bdrcs.org/publications/baseline-study-report-integrated-flood-resilience-program. Accessed 10 Jan 2020

Brammer H (1990) Floods in Bangladesh: geographical background to the 1987 and 1988 floods. Geographical J [e-journal], 156(1):12–22. Available through: https://www.jstor.org/stable/635431. Accessed 12 Mar 2018

Clarke JI (1972) Population geography. Pergamon Press, New York/Toronto/Sydney/Paris/Frankfurt

CPD(2017) Flood 2017: assessing damage and post-flood management. [Online] Available at: https://cpd.org.bd/wp-content/uploads/2017/10/Flood-2017-Assessing-Damage-and-Post-flood-Management.pdf

Dhaka Tribune (2018) Per capita income increases by 149% in 10 years. Dhaka Tribune [Online] September 28. Available at: https://www.dhakatribune.com/business/economy/2018/09/25/per-capita-income-increases-by-149-in-10-years.Accessed 10 May 2019

Ghosh BK (2016) Riverbank erosion induced migrants' livelihood pattern changes and management strategies: a case study of Charbhadrasan Upazila, Faridpur. Oriental Geographer 60(1&2):75–86. (Printed in March 2019)

Ghosh BK, Mahbub AQM (2014) Riverbank erosion induced migration: a case study of Charbhadrasan Upazila, Faridpur. Oriental Geographer 58(1):59–71 (Printed in July 2017)

Ghosh BK, Mahbub AQM (2017, 2018) Resettlement schemes of the riverbank erosion induced displaces: a case study of Charbhadrasan Upazila, Faridpur. Dhaka University Journal of Earth and Environmental Sciences 6 & 7:25–30

Hossain A (2017) Bangladesh: living with floods. The Daily Sun Online, [Online] 21 August. Available at: https://www.dailysun.com/printversion/details/249268/2017/08/21/Bangladesh:-Livingwith-Floods. Accessed 15 Apr 2018

Islam N (2017) Sustainable development goals and the tasks for Bangladesh. In: Islam MS, Khalequzzaman M (eds) Sustainable development goals and Bangladesh environment. Dhaka, BAPA and BEN, pp 1–26

Miah MM (ed) (2004) Water management in Bangladesh: from vision to action. Bangladesh Water partnership, Dhaka

Nirapad (2017a) Hazard incidents in Bangladesh, April 2017. [Online] Available at: https://www.nirapad.org.bd/home/stroage/file/public/assets/resource/monthlyHazard/1515912671_Monthly%20Hazard%20Incidence%20Report_April%202017.pdf

Nirapad (2017b) Hazard incidents in Bangladesh, July 2017. [Online] Available at: https://www.nirapad.org.bd/home/stroage/file/public/assets/resource/monthlyHazard/1515913055_Monthly%20Hazard%20Incident%20Report_July%202017.pdf

Nirapad (2017c) Flood situation updated on August 22, 2017. [Online] Available at: https://reliefweb.int/sites/reliefweb.int/files/resources/Situation_Report%20of%20Flood%20_Updated%20on%20August_22%2C%202017.pdf

Nirapad (2017d) Hazard incidents in Bangladesh, August 2017. [Online]. Available at: https://www.nirapad.org.bd/home/stroage/file/public/assets/resource/monthlyHazard/1515913118_Monthly%20Hazard%20Incident%20Report_August%202017.pdf

Rahman MR (2017) Technological fixes for disaster management in Bangladesh. In: Khatun H, Baqeee A, Kabir H (eds) People at risk: disaster and despair. Disaster Research Training and Management Centre, University of Dhaka, Dhaka, pp 71–96

Rana S, Islam S (2016) The socio-economic condition of Bangladesh: a study on Trishal Upazila in Mymensingh. Res Human Soc Sci [e-journal] 6(12). Available at: https://www.iiste.org/Journals/index.php/RHSS/article/view/31456. Accessed 12 Mar 2018

Rasheed KBS (2008) Bangladesh: resource and environmental profile. A H Development Publishing House, Dhaka

The Daily Star (2017a) Flood hits the North. The Daily Star Online [Online] July 9. Available at: www.thedailystar.net/frontpage/bangladesh-flood-hits-the-north-1430338. Accessed 15 Apr 2018

The Daily Star (2017b) Floods wreak havoc on crops. The Daily Star Online [Online] August 21. Available at: https://www.thedailystar.net/country/floods-wreak-havoc-crops-1451692. Accessed 15 Apr 2018

The New York Times (2017) More than 1,000 died in South Asia floods this summer. The New York Times [Online] August 29. Available at: https://www.nytimes.com/2017/08/29/world/asia/floods-south-asia-india-bangladesh-nepal-houston.html. Accessed 10 May 2019

Uddin AFM, Basak JK (n.d.) Effects of riverbank erosion on livelihood. Unnayan Onneshan-The Innovators, Dhaka. Available at: http://www.unnayan.Org/…/Effect_of_Riverbank_Erosion_on_Livelihood.pdf Accessed 25 Oct 2019

UP (2018) Union Parishad assessment list 2018, Saghata Union, Saghata Upazila, Gaibandhana District, Bangladesh

World Bank (2017) Bangladesh continues to reduce poverty but at slower pace. [Online] October 24. Available at: https://www.worldbank.org/en/news/feature/2017/10/24/bangladesh-continues-to-reduce-poverty-but-at-slower-pace

Zaman M, Khatun H (2017) Thirty years of development –induced displacement in Bangladesh: an overview. In: Zaman M, Khatun H (eds) Development-induced displacement and resettlement in Bangladesh: case studies and practices. Nova Science Publishers, New York, pp xIiii–Ix

The manufacturer's authorised representative in the EU is Springer Nature Customer Service Centre GmbH, Europaplatz 3, 69115 Heidelberg, Germany. If you have any concerns regarding our products, please contact ProductSafety@springernature.com

Printed and bound by CPI Group (UK) Ltd, Croydon, CR0 4YY
28/09/2026
02224067-0005